Studies in Inorganic Chemistry 17

The Chemistry of Artificial Lighting Devices

Lamps, Phosphors and
Cathode Ray Tubes

Studies in Inorganic Chemistry

Other titles in this series

Studies in Inorganic Chemistry 17

The Chemistry of Artificial Lighting Devices

Lamps, Phosphors and Cathode Ray Tubes

R.C. Ropp

138 Mountain Avenue, Warren, NJ 07059, U.S.A.

ELSEVIER
Amsterdam — London — New York — Tokyo   1993

ELSEVIER SCIENCE PUBLISHERS B.V.
Sara Burgerhartstraat 25
P.O. Box 211, 1000 AE Amsterdam, The Netherlands

ISBN 0-444-81709-3

Transferred to digital printing 2005
Printed and bound by Antony Rowe Ltd, Eastbourne

Preface

This manuscript describes how mankind has used light and lighting to further the development of his cultural progress. I have covered the early use of artificial lighting and current methods of manufacture of both incandescent and fluorescent lamps. The protocols for manufacture of fluorescent lamp phosphors and those used in cathode-ray tubes are also presented in some detail.

Early man used torches and campfires for lighting and protection from wild animals. Twentieth-Century man uses electrically operated lamps for artificial lighting, display and amusement. The number of artificial lighting devices, both existing and obsolete, is truly amazing. This manuscript attempts to survey all of those known to date in terms of how they arose and are, or have been, used by mankind.

I have enjoyed preparing this manuscript and hope that you will find reading it both profitable and enjoyable.

Dr. R.C. Ropp
May 1993

This work is dedicated to my wife Francisca Margarita, who has constantly supported my efforts and me during these past 40 years.

Table of Contents

Cathode Ray Phosphors

Introduction

This book was written to introduce the reader to the inorganic chemistry of artificial lighting devices and the materials used to manufacture them. In Chapter 1, a brief history of the study of light is then presented, followed by a summary of the quark theory as applied to matter and its constituents. The nature of light is quantified, wherein it is shown that a photon is a force carrier between electrons and other particles. Homo Sapiens originally used light in the form of campfires, followed by use of oil lamps and candles. Finally, in the late 19th century, the first practical incandescent lamp was developed by Edison. A brief history leading to that milestone is given which includes both carbon arc and mercury discharge lamps.

In the second Chapter, the principles of design relating to both incandescent and mercury lamps is presented, followed by that of sodium vapor lamps and other gaseous discharge lamps.

In Chapter 3, the manufacture of "lamp parts" is described in depth. This includes the manufacture of tungsten metal, its formation into wire and the manufacture of tungsten coils for lamps. The fabrication of incandescent lamps is discussed in some detail, as well as the production of raw materials used in the manufacture of phosphors. Finally, the manufacture of fluorescent lamps and the protocols required to do so are elaborated.

In Chapter 4, the manufacture of lamp phosphors is addressed, including formulations, times of firing, and the processing needed to complete their manufacture before they are incorporated into fluorescent lamps. The manufacture of halophosphates is presented in entirety.

In Chapter 5, the manufacture of the various cathode ray phosphors used in the Industry is examined, including their formulations and methods of firing. These are listed in terms of "P-numbers", as devised by JEDEC, and their measured decay times. Finally, a brief discussion of the manufacture of the newest cathode ray tube phosphors is presented.

I hope that you find this manuscript interesting and rewarding to read.

Dr. R.C. Ropp - May 1993

Chapter 1

LIGHT AND THE ORIGIN OF ARTIFICIAL LIGHTING

LIGHT and light sources have always played an important role in day-to-day living for mankind. "In the beginning, there was darkness"- according to the Scriptures, and the first act of the Divine Creator was the production of light [Genesis I:1-5]. Separate creation of the sun, moon and the stars followed soon after [Genesis I:14-19]. That primitive man worshipped these natural light sources as deities is manifest. The Egyptian sun-god was named Ra and later Anton. The sun worship of the Inca, Aztec and Mayan civilizations in the Americas is well known. The Japanese sun-goddess, Amaterasu, is still worshipped today as the head of the imperial family and the founder of the state (1). Even the Japanese flag reflects this perspective, being that of the "Rising Sun".

Worship of heavenly bodies is also recorded, but was forbidden in the Hebrew Scriptures [Deuteronomy 17:2-5; Ezekiel 8:16]. However, these heavenly luminaries were considered as intelligent beings who worshipped the Creator [Psalms 148:3]. Light was also considered an attribute of GOD, as shown in many parts of the Christian Bible [Exodus 3:1-2, 14:21, 19:18; Psalms 4:6, 27:1, 119:105; Revelations 4:5; Acts 9:1-5; John 1:6-9, 8:12, 9:5].

In most modern European languages, the days of the week still remind us that each day was once sacred to an astral deity. Their number (seven) is that of the luminous heavenly bodies known to early man who had no telescopes to distinguish other possible celestial light sources. Some of the names that can be associated include: Sunday, Sonntag [Sun]; Monday, Montag, Lundi [Moon]; Mardi [Mars]; Mercredi [Mercury]; Jeudi [Jupiter]; Vendredi [Venus]; Saturday [Saturn]. Since the English language is a mixture of Teutonic as well as Mediterranean roots, other pagan gods are also represented in the week: Tuesday [Tyr], Wednesday [Wodin], Thursday [Thor], Friday [Freya or Frigga].

Numerous Scriptural references attest to the importance of light in religious ritual [Exodus 25:31-40, 27:20] and everyday affairs [Matthew 25:1-13]. The use of candles and lamps in religious observances persists in many instances today. For example: the "Eternal Light" before the sanctuaries of Catholic, Jewish and high Anglican (Episcopal) houses of worship [Exodus 27:20,

Leviticus 24:2-4], the votive candles of Catholicism, the Jewish Sabbath and Yarhzeit candles, the Hanukkah menorah, the use of lights on Christmas trees, and the halos often depicted on "Saints", all affirm the importance of light in religious functions (2). Other modern symbolic uses of light may be seen in the Eternal Flame at the Arc de Triomphe in Paris, the torch of the Statue of Liberty in New York harbor and the use of candles on birthday cakes. A common trademark, "Mazda", once used in the electric lamp industry, is the name of a Persian deity connected with light (1). Thus, light has been for mankind more than a mere means of sight and illumination.

We shall be concerned not with natural sources of light but with artificial or man-made sources. We intend to present first a history of light-sources and then will survey present-day sources and uses of light. Finally, we will summarize the manufacturing methods used in the Lighting Industry. Unfortunately, although we know the name of the first blacksmith [Tubalcain-Genesis 4:22] and that of the first viniculturist [Noah - Genesis 9:20], the name of the first inventor of artificial light is not recorded. Perhaps this is due to the fact that man used light and artificial lighting long before the art of writing was invented (which occurred about 7000 years ago), and thus written records and early accounts of the history of Man do not reflect this aspect.

1.1.- The Nature of Light

Nowadays, we know that light is comprised of "photons", which are quantized waves having some of the properties of particles. If this perception is not clear, it will be made more lucid shortly. Such was not the case originally. The concept of photons with wave properties has its roots in the study of optics and optical phenomena. Until the middle of the 17th century, light was generally thought to consist of a stream of some sort of particles or *corpuscles* emanating from light sources. Newton and many other scientists of his day supported the idea of the corpuscular theory of light. It was Newton in 1703 who showed that "ordinary" light could be dispersed into its constituent colors by a prism, but the phenomenon was not clearly grasped at that time. About the same time, the idea that light might be a *wave* phenomenon was proposed by Huygens and others. Indeed, diffraction effects that are now known to be associated with the wave nature of light were observed by Grimaldi as early as 1665. However, the significance of his observations was not understood at the time.

Early in the nineteenth century, evidence for the wave theory of light grew more persuasive. The experiments of Fresnel and Thomas Young (1815), on interference and diffraction respectively, showed conclusively that there are many optical phenomena that could be understood on the basis of a wave theory, but for which a corpuscular theory was inadequate. Young's experiments enabled him to measure the wavelength of the photons and Fresnel showed that the rectilinear propagation of light, as well as the diffraction effects observed by Grimaldi and others, could be accounted for by the behavior of waves having very short wavelength. It was Maxwell in 1873 who postulated that an oscillating electrical circuit should radiate electromagnetic waves. The speed of propagation could be calculated from purely electrical and magnetic measurements. It turned out to be equal, within the limits of experimental error of the time, to the previously measured speed of propagation of light. At that time, the evidence seemed inescapable that light consisted of extremely short wavelength waves, having an electromagnetic nature. In 1887, Heinrich Hertz, using an oscillating circuit of small dimensions, succeeded in producing electromagnetic waves which had all of the properties of light waves. Such waves could be reflected, refracted, focused by a lens, polarized etc., just as waves of light could be manipulated. Maxwell's electromagnetic theory of light and its experimental justification by Hertz constituted one of the major triumphs of science. One should note that at about this same time Edison was busy inventing the first practical incandescent light bulb.

A. Nature of a Photon

Nonetheless, the classical electromagnetic theory of light failed to account for several phenomena, including the absorption and emission of light. One example is the phenomenon of photoelectric emission. That is, the ejection of electrons from a conductor by photons incident on its surface, where the number of incident photons can be correlated to the number of electrons released. In 1905, Einstein extended an idea proposed by Planck five years earlier and postulated that the energy in a light beam was concentrated in "packets" or *photons*. The wave picture was retained, however, in that a photon was considered to have a frequency and that the energy of a photon was proportional to its frequency. Experiments by Millikan in 1908 soon confirmed Einstein's predictions.

Another striking confirmation of the photon nature of light is the Compton effect. In 1921, A.H. Compton succeeded in determining the motion of a photon and an electron both before and after a collision between them. He found that both behaved like material bodies in that both kinetic energy and momentum were conserved in the collision. The photoelectric effect and the Compton effect, then, seemed to demand a return to the corpuscular theory of light. The reconciliation of these apparently contradictory experiments has been accomplished only since about 1930 with the development of quantum electrodynamics, a comprehensive theory that includes both photon wave and particle properties. Thus, the theory of light propagation is best described by an electromagnetic wave theory while the interaction of a photon with matter is better described as a corpuscular phenomenon.

The speed of light in free space is one of the fundamental constants of nature. Its magnitude is so great (about 186,000 miles/second or 3.0 x 10^8 meters/second) that it evaded experimental measurement until 1676. Up to that time, it was believed that light traveled with an infinite speed. The first recorded attempt to measure the speed of light was a method proposed by Galileo (1605) in which two experimenters were stationed on the tops of two hills about a mile apart. Each was provided with a lantern and was to cover and uncover his lantern when the light from the other was seen. Nowadays, we know that the speed of light is just too great for this method to work satisfactorily. In 1676, Olaf Röemer, a Dutch astronomer, obtained the first evidence that light propagates with finite speed. He observed the eclipse of one of the moons of the planet Jupiter and found that the observable periodic times of eclipse were greater or lesser depending upon the positions of the Earth and Jupiter. A time difference of 22 minutes resulted in a rough calculation of about 69.9% of the refined value of the speed of light that we know today. In 1849, Fizeau made the first successful measurement using terrestrial instruments. Using a rotating toothed-wheel that chopped the beam of light, he obtained a value only slightly better than that of Röemer, namely 3.1 x 10^8 meters/second. Foucault replaced the rotating toothed-wheel by a rotating mirror and this has remained the method of choice. Michelson, an American physicist (1852-1932) used this same method to determine the best value known to date, viz-

1.1.1.-   c = 2.9979246 x 10^8 meters/second ± 2.0 meters/second

The conclusion that photons are associated with mass and matter rather than with space alone is inescapable. All bodies emit some form of electromagnetic radiation, as a result of the thermal motion of their molecules. This radiation, called thermal radiation, is a mixture of wavelengths. At a temperature of 300 °C, the most intense of these waves have a wavelength of 50,000 Å which is in the *infrared* region of the electromagnetic spectrum. When the temperature is raised to about 800 °C, such a body emits enough visible radiation to be self-luminous and appears "red-hot". However, by far the most energy is still carried by photons having wavelengths in the infra-red region of the spectrum. But at 3150 °C, which is the temperature of a tungsten filament in an incandescent light bulb, the body then appears "white-hot" and a major part of the energy is in the visible region of the spectrum.

The following diagram, given on the next page as 1.1.3., shows the emission of optical radiation as a function of "black-body" temperatures. Note that even bodies at liquid-air temperatures emit photons between 10 and 100 microns in wavelength, i.e.- 100,000 and 10^6 Å in wavelength. The earth itself at a temperature of 300 °K. has an emission between about 20,000 and 300,000 Å in wavelength, i.e.- 2000 nm. and 30,000 nm.

Another related phenomenon is that of fluorescence by "phosphors" (3). These inorganic materials are energy-converters in fluorescent lamps in that ultraviolet light is absorbed from a mercury-vapor discharge and is converted to visible light. Still another light source is that of the laser. In this case, a crystal, or metal vapor, or gaseous vapor, is made to store energy in an excited state. By suitable optical means, the energy is released by a resonance method in which the waves all cooperate to emanate at the same time. This is called "stimulated emission of coherent radiation" and is the optical basis of the laser, whether solid or gaseous. In vacuum, all electromagnetic radiation travels at the speed of light. This is given by:

1.1.2.- $c = \{1/\varepsilon_0 \mu_0\}^{1/2}$

where ε_0 is the permittivity of free space, and μ_0 is the permeability of free space. The former comes from Gauss's Law and the latter from Faraday's Law. However, the speed of light in media *other* than vacuum is always **slower** than in space.

1.1.3.-

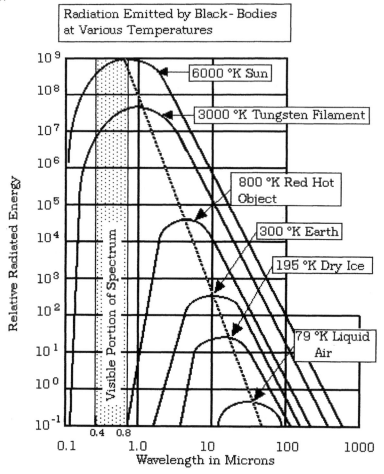

Radiation Emitted by Black-Bodies at Various Temperatures

This is believed to be due to resonance interactions between the electromagnetic fields of the electrons associated with the transparent media and that of the traveling photon. When electrons or other charged particles are accelerated to relativistic speeds, i.e.- a fraction of the speed of light, they emit photons. One example of this is the "free-electron" laser. The apparatus is built so that alternating magnetic fields of opposite polarity pervade the interior in precisely spaced intervals. As electrons are accelerated down the length of the apparatus, each separate magnetic field causes a deviation in the path of the electron by interaction of the magnetic field and that of the electromagnetic wave of the electron. This "wiggle" in the path of the accelerating electrons causes them to emit photons. The wavelength will vary according to the speed and degree of "wiggle" induced in the electrons. Thus, it should be clear that both photons and particles of mass are inexorably interwoven by the matrix of space-time. What this means is that when particles having a given "rest" mass are caused to accelerate near to a limiting speed, i.e.- the speed of light, they are prone to release that excess energy gained through the emission of photons. Einstein was the first to realize this phenomenon, which has since been proven many times over.

B. Photon Interactions - Quarks, Electrons and Photons

Because particles, i.e.- electrons, have wave properties similar to those of photons, we need to differentiate between them.. It was de Broglie in 1906 who first postulated the wave nature of particles. In 1927, Davisson and Germer first showed that electrons are reflected from the surface of a solid in the same way that x-rays are reflected. The wave hypothesis clearly required sweeping revisions of our fundamental concepts regarding the nature of matter. The best explanation to date seems to be that a particle must be regarded as an entity not entirely localized in space whereas the photon is a point-source, i.e.- is localized in space. What this means is that a particle is strongly attracted or repelled by electromagnetic fields as it moves through space whereas a photon, being a localized point source, is only weakly affected. Thus, a photon moves at a constant speed through space while an electron does not. Yet, both have electromagnetic fields thereby associated with each, which are subject to reflection, diffraction etc. The photon-packet thus interacts with the space-time continuum as it moves through space at a constant speed.

In contrast, the particle interacts with both the time-space continuum and the electromagnetic fields thereby associated with mass, and its speed is **not** constant, but subject to mass-mass (gravity) interactions as well. The wavelength of the photon (which has no mass) is a function of its internal energy as it moves through space. The particle has a mass which is determined by how much it is spread out in the space-time continuum. Consequently, it has properties we normally associate with "mass". In the 1960's, a major advance occurred when the Quark theory came to the forefront of physical theory. Actually, this was a culmination of Einstein's "Grand Unification Theory" which strove to combine both "weak" forces, gravitational forces, and "strong forces", i.e.- those present within the nucleus of atoms.

Among the "leptons" (which are particles with spin equal to 1/2) are electrons and photons (Actually, photons are force carriers between electrons, as we shall see). Photons have been classified as "bosons" while electrons are called "fermions". The former is described by Einstein-Bose statistics and the latter by Fermi-Dirac statistics. Einstein-Bose statistics are defined as "The statistical mechanics of a system of indistinguishable particles for which there is no restriction on the number of particles that may exist in the same state simultaneously", whereas Fermi-Dirac statistics are defined as "The statistics of an assembly of identical half-integer particles; such particles have wave functions antisymmetrical with respect to particle interchange and satisfy the Pauli exclusion principle".

With the advent of the Quark theory, it has been shown that sub-nuclear particles can have fractional charge, as shown in the following Table, viz-

TABLE 1-1

Fundamental Particles of Matter

1. FERMIONS (Matter constituents with spin = 1/2, 3/2, 5,2 ...)

Leptons (spin = 1/2)			Quarks (spin = 1/2)		
Particle	Mass (GeV/c^2)	Electric Charge	Flavor	Mass (GeV/c^2)	Electric Charge
Electron	5.1×10^{-4}	-1	d_{down}	0.007	-1/3
Electron-Neutrino	2×10^{-8}	0	u_{up}	0.004	2/3
			$s_{strange}$	0.20	-1/3
			c_{charm}	1.5	2/3
Muon	0.106	-1	t_{top}	>91	2/3
Muon-Neutrino	3×10^{-4}	0	b_{bottom}	4.7	-1/3
Tau	1.784	-1			
Tau Neutrino	4×10^{-2}	0			

TABLE 1-1 Continued)
Fundamental Particles of Matter

2. BOSONS (Force carriers with spin = 0, 1, 2 ...)

Electroweak force (spin = 1)			Strong force (spin = 0)		
Name	Mass (GeV/c^2)	Electric Charge	Name	Mass (GeV/c^2)	Electric Charge
Photon	0	0	Gluon	0	0
W $^-$	80.6	-1			
W $^+$	80.6	+1			
Z ^{0}	91.16	0			

3. Properties of the Interactions of Fundamental Particles

Fermionic Hadrons

Name	Quark Content	Mass (GeV/c^2)	Electric Charge	Spin
Proton	uud	0.938	+ 1	1/2
Anti-proton	uud	0.938	- 1	1/2
Neutron	udd	0.940	0	1/2
Lambda	uds	1.116	0	1/2
Omega	sss	1.672	- 1	3/2
Pion	ud	0.140	+ 1	0
Kaon	su	0.494	- 1	0
Rho	ud	0.770	+ 1	1

From the above, it should be clear that a photon is a "force-carrier" while an electron is a "matter-constituent". The notion of particles as mediators of force in nature has provided a framework for testing and developing the **Standard Model** and the associated "Big-Bang" theory of the formation of the Universe. It has also been important for the exploration in depth of many other important questions about the physical world. Complex in practice but simple in conception, this model identifies four forces in nature: 1) Electromagnetism; 2) the Strong Force; 3) the Weak Force, and 4) the Gravitational Force. Each is transmitted when a force carrier is exchanged between two elementary particles. Two of these forces function only at extremely close range (e.g.- between quarks inside the nucleus), whereas the other two are effective across long distances.

The most familiar interaction is Electromagnetism. Electromagnetic radiation in its various forms (including radiowaves, microwaves, infra-red-light, visible-light, ultraviolet-light, and x-rays) can be thought of as the exchange of massless photons between electrically charged particles, either quarks or leptons. The Strong Force, the interaction that binds quarks and hence the nucleus together, is transmitted from quark to quark by massless particles called gluons. The exchange of gluons acts on a property of quarks called "color", which is analogous to electric charge. Quarks make up the proton and the neutron, and thus the nucleus of an atom. They both have color and electric charge, and they give the proton its charge.

The Weak Force regulates the "burning' of hydrogen into helium in the interior of stars, among other processes. It is described by a similar model. The carriers of the Weak force are the electrically charged W^+ and W^- particles and the neutral Z^0, shown in the above table. In this case, these carriers do have mass, and are exchanged between quarks and leptons. The Standard Model postulates that the electromagnetic and Weak force are derived from a single, unified "Electroweak Force", a hypothesis that demands the existence of the "Top Quark". Of all of the quarks, only this one remains to be directly observed. This stems from the fact that the mass of the Top Quark is so large that accelerators do not exist which have sufficient energy in their beam of particles to reach the minimum energy required to observe this massive particle. However, accelerators are being built that are expected to be able to reach the level in energy required, namely about 90-95 Gev. i.e.- 90-95 giga-electron volts.

Orbiting the nucleus is the familiar electron, which is the only lepton that is a part of ordinary matter. Its charge is always -1, relative to the proton's charge of +1. Other leptons exist under extreme conditions: the Muon, the Tau, and the neutrino (postulated in 1931 by Wolfgang Pauli). The neutrino has no charge and nearly no mass. There are three neutrinos, since each charged lepton is believed to have an associated neutrino. Leptons feel all of the forces except the strong force. It is the electron-neutrino pair that is committed in the exchange of force between electrons exchanged within the weak force and the photon between electron-positron pairs within the electromagnetic force. This is made more cogent by the following diagram, given on the next page as 1.1.4.

1.1.4.-

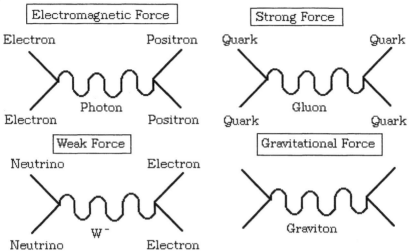

The particles mediating exchange forces are thus photons, gluons, the W-boson, and gravitons. And, as we stated previously, the electron has a mass whereas the photon does not since it is a force-carrier. The other particles, given in the above table, are observed only when atomic nuclei are bombarded at very high energies. We can summarize all of the above in that a photon is a quantum of radiation, whereas an electron is composed of matter-quanta. This aspect is made more lucid in the following Table:

TABLE 1-2

Elementary Particle Interactions

Force	Physical Phenomena	Relative Strength	Effective Range	Exchange Quanta	Matter Quanta
Strong	Nuclear fission, fusion	1.0	10^{-13} cm.	Gluons	Quarks
Electro-magnetic	Electricity, Light Magnetism,	10^{-2}	Infinite	Photon	Quarks, Charged Leptons
Weak	Radioactive Decay	10^{-5}	10^{-16} cm.	W, Z & Higgs Particles	Quarks, Leptons
Gravity	Curved Space-time	10^{-36}	Infinite	Graviton	All Particles

C. Photon Exchange

Perhaps the best distinction to be made between photons and electrons is the "Cerenkov " effect, discovered by P.A. Çerenkov in 1934. We have already pointed out that the speed of light is a fundamental physical constant, both in vacuum and in transparent materials, including air. Radioactive materials emit three kinds of emission when the nucleus transforms , namely α , β and γ particles. Alpha particles are energetic, doubly-ionized helium atoms. Beta particles are energetic electrons, while gamma particles are high-energy x-rays (or photons). When such particles are emitted from a nucleus (because that nucleus has transformed into another type), they frequently possess high energy and speed. When a charged particle such as an electron passes through a dielectric medium such as water, at a speed **greater than** the speed of light in that medium, this particle throws off a sort of "shock-wave" in the form of emitted photons. Thus, any matter-constituent particle traveling faster than the speed of light in a particular medium sheds part of its excess energy in the form of photons (which are the force-carriers).

1-2 Man's First Use of Light and Artificial Lighting

In order to address how man first began to use artificial light, we need to explain how man came into being.

A. The Origin of Homo Sapiens

Recent work (4) has shown that mankind probably originated in Eastern Africa about 160,000 - 240,000 years ago. Since it is obvious that the major attribute that separates Homo Sapiens from other animals is that of intelligence, it seems likely that the appearance of mankind was due to a massive mutation of genes which produced offspring with substantially improved intelligence and curiosity. The scientific evidence appears irrefutable in that Homo Sapiens can be traced back to a single female who was probably a Neanderthal (A debate still rages concerning whether Homo Sapiens materialized from Neanderthals or Homo Erectus - who evolved over a million years of history). However, both genetic evidence of the human family tree (4) and study of languages (5) have come to the same conclusion, namely that mankind originated in Africa and that massive migrations occurred throughout the history of man. The earliest genetic tree implies movement from East Africa into the Middle East (100,000 years ago), into the Far East, including Australia, about 60,000 years ago, and into Northern Europe and crossing into North America about 35,000 years ago. Details of the actual routes are speculative. Studies of peoples and classification of languages have reached the same conclusion, namely that early Homo Sapiens most likely evolved from a single source and that language differentiated from a single linguistic source. Prior to that time, language was extremely limited, if it existed at all.

The fact that **all** humans are mammalian mutants and suffer from a potentially fatal liver-enzyme disease may come as a shock to most people. This genetic disease is of universal occurrence in the human population. The disease is scurvy and the curative substance is ascorbic acid, or Vitamin-C. Man is one of the few mammals not biochemically equipped to make their own ascorbic acid directly in their livers. In most animals, after liver synthesis, ascorbic acid is poured directly into the bloodstream in large daily amounts. For example, goats synthesize about 13.3 grams (about 25 tablets of 500 milligram each) of Vitamin-C per day; dogs and cats = 2.8 grams daily (6 tablets), but man synthesizes 0.0 grams daily. This abnormality is due entirely to the genetic code inherited by each person.

How old is the genetic code? Genes are specific sequences of bases attached to the sugar-backbone to form the DNA molecule having the double helix or spiral staircase structure. By identifying the sequences of bases, one can relate the sequences of many organisms to one another. There are three major classifications (Kingdoms) of living things: a) Plants, b) Animals (which include mammals, birds, fishes, reptiles and insects), and c) Bacteria and related organisms. Each classification has many subspecies comprising that kingdom. In particular, there are about 1000 base-sequences known today, involving all three kingdoms. There are at least two ways of ordering this wealth of sequence data. First, one may focus on one species and produce an alignment of its DNA-transfer sequence functions. Fifteen such specie families have been thus identified, with each family consisting of 15-30 individuals. Secondly, one may identify a specific sequence and follow it through the various species. This has been done for 24 family-species in all three kingdoms. Kinship relations are revealed by alignment of sequences. The results show that individual and master sequences of DNA-transfer functions reflect kinship relations consistent with generally accepted evolutionary patterns of the appearance of life on earth. What this means is that the first life to appear consisted of one-celled creatures which evolved into the complex kingdoms of life known today. All three kingdoms have certain sequences exactly the same. Thus, the three main kingdoms appear to be "equally old". The early nodes of kingdom separation can be dated to be about 2.5 ± 0.5 billion years old, and the genetic code cannot therefore be older than 3.8 ±0.6 billion years. Because the earth is about 4.2 billion years of age, "life", as we know it, did evolve on the earth, and was not of extraterrestrial origin. Yet, this does not explain the genesis of Homo Sapiens.

In a recent study, Dr. E.L. Simons , Director of the Duke University Center for the Study of Primate Biology and History stated "New discoveries combine to indicate that all of the major steps in human evolution took place in Africa. Skeletal analysis of oldest human forebears around 3 million years ago (Homo Erectus) reveal many anatomical similarities to African Great Apes. These and biochemical resemblances indicate a common ancestry for humans and apes, perhaps only a few million years earlier. Enlarged knowledge through recent discovery of skeletons of successive stages in the line leading to modern peoples shows that many skills by which we define humanity arose much more recently in time than heretofore believed" (6). It has been generally agreed that the chain of succession includes : early man ape (Homo Australopithecus- the first hominid)- age = 2-3 million years; Java apeman (Homo Erectus)- age = 1.5 million years; Neanderthal man- age = 600,000 years, and finally, Homo Sapiens-age = 140-200,000 years. If this is true, then we are no more than 8000 generations of age (using 25 years as one generation). I, myself, will see at least five generations.

Undoubtedly, the remnants of an Ice Age had a marked effect upon the dates and paths of human migration. During an Ice Age, migration was restricted because of its effects upon the Earth's climate and the necessity for providing some kind of covering from the cold if Man was to exist and prosper. Over the past 900,000 years, four glacial periods have occurred. The following Table shows the periods of the Ice Ages comprising the Pleistocene Epoch. It is likely that Neanderthals also migrated into Europe and Eastern Asia during the Third Interglacial Period (about 175,000 years ago). Neanderthals made tools and weapons of flint (Old Stone Age) for hunting and domestic use. They knew the use of fire. They had many of the attributes that we consider, nowadays, to be essential to civilization.

TABLE 1-3

Recent Glacial Ages

	Began (Years Ago)	Lasted	Climate
1st Ice Age	600,000	64,000 years	cold
1st Interim	536,000	60,000	warm
2nd Ice Age	476,000	156,000	cold
2nd Interim	320,000	90,000	warm
3rd Ice Age	230,000	55,000	cold
3rd Interim	175,000	60,000	warm
4th Ice Age	115,000	75,000	cold
4th Interim	40,000	-----	warm

They had family groups. They buried their dead. They occupied caves as the weather became colder. The matter of speech is one of conjecture. By the time the Fourth Glacial Period arrived, Neanderthals were scattered over Africa, Europe and most of Asia. Nevertheless, there appears to be a change-over about 40,000 years ago from Neanderthals (particularly in Europe where the records are the most clear) to Homo Sapiens. This date coincides with the onset of the 4th Interim Period. It seems likely that Homo Sapiens migrated into Europe, and elsewhere, while the Neanderthals were declining and died out.

B. History of Lighting

Quite obviously, early man used fire. It is likely that the first fires originated from natural causes. Lightning strikes are known to cause forest fires. Early man undoubtedly used fires in caves, as has been determined from remnants of charcoal left therein. From there, he learned how to make torches that would burn a long time so that they could be carried from place to place. Thus, combustion in some form remained the sole practical source of light until the late nineteenth century.

1. Combustion and Flames

It is interesting to speculate how man learned to control combustion. In the beginning, cave-fires were undoubtedly started from fire-brands or torches carried from natural fires found in the environment. Later, Man learned how to start fires by either rubbing or twirling a hard piece of wood against a softer one, or by striking two chips of rock together. The latter may have come about as Man learned how to flake stone such as flint to form arrow-heads. At any rate, striking stones have been dated from before 7000 B.C. and the flintlock was used until about 1600 A.D. to ignite gunpowder to discharge fire-arms. The friction match was only discovered about 1827 when a mixture of sulfur and phosphorous sesquisulfide was found to be ignitable by friction. The final step in our progression of "fire-starters" was the invention of the mechanical lighter wherein a steel wheel causes a mischmetal rod to emit a series of sparks.

From cave-fires, man graduated to torches and similar "firebrands" such as oil lamps. The early fuels for lamps were vegetable in origin (olive oil or oil from nuts). At a later stage, animal oils (especially from the whale) were used. Although Pliny notes the use of mineral oil (actually fossil vegetable) in 50 A.D., kerosene was not introduced as an illuminant until after 1853, when the first oil well was drilled in both Pennsylvania and Ontario, Canada. Leonardo da Vinci is said to be the first to employ a glass chimney to protect oil flames from drafts in 1490. Street lighting in Paris and London first used oil lamps in 1736.

The candle is much younger than the oil lamp. Although the Greeks and Romans used threads coated with pitch and wax, it apparently took its modern form in 400 A.D. in Phoenicia. For centuries, such candles were made from natural materials such as tallow, beeswax, spermaceti (from the heads of whales) and/or vegetable waxes such as wax-myrtle, bayberries, etc. The fruit of the shrub *Aleurites Mouccana*, or candle nut, can be burned entirely as a candle. Synthetic paraffin, derived from petroleum, was introduced in the late 1850's. The original photometric "standard candle" was based on a spermatici candle 7/8 inches in diameter consuming 120 grains per hour. The candle is now defined in terms of the incandescence of a "blackbody" at the freezing point of platinum (2043 °K). The luminance is 60 lumens/cm^2, so that one candle emits one lumen/steradian for a total of 4π lumens.

It is surprising that natural gas was used by the Chinese for illumination many centuries before the Christian era in Europe, i.e. - around 900 A.D. Bamboo pipes were used to transport the gas to the point of illumination. The first use of gas for illumination in "modern" times was by Jean Pierre Mincklers at the University of Louvain in 1784. This same use of gas in the U.S. was made by David Melville of Newport, R.I. in 1812. Gas manufacture by greatly improved methods was accomplished by William Murdock in England and Phillippe LeBon in France in the early nineteenth century. Gas street lighting was introduced in Baltimore in 1821 and in New York City in 1827. However, methods had not improved much over the Chinese methods of antiquity since hollow logs were used as pipes. A great improvement in gas flames was made by Carl von Auerbach in London in 1887 with the introduction of the thoria mantle. In this case, the thoria mantle afforded greatly improved brightness since the mantle has an apparent temperature of 1700-1900 °C. when operated in a gas flame of 700-900 °C. We shall discuss this in more detail later. Acetylene for illumination was introduced in 1892 by Henri Morisson in France and Thomas L. Willson in the U.S. It is difficult to realize that in 1916 more than 50% of the gas consumed in the U.S. was used for illumination rather than for heating. As late as 1921, ten million American homes were lighted with gas. The "gas-works" were as much a fixture in each community as was the railroad. Only in recent years have huge pipe lines brought in natural gas (consisting mostly of methane) to replace locally manufactured (mostly hydrogen and carbon monoxide) gas. Nonetheless, the days of gas illumination are over.

The structure and properties of flames is quite complex (7). Combustion is defined as any chemical reaction which evolves light and heat. Usually, oxygen is one of the reactants, although other oxidizing sources, such as the reaction between hydrogen and fluorine gases, may also be used. A flame produced by combustion of carbonaceous material produces both CO, CO_2 and water as by-products. The spectral structure and properties of **dry** CO *flames* consist of a continuum extending from the visible far into the ultraviolet (2500-5500 Å). Its exact origin is still open to speculation. There are, in addition, bands due to CO_2. If water vapor is introduced, hydroxyl bands also appear whose heads are at 3064, 2811, 3428, 3122, 2875, and 2608 Å (given in order of intensity), but shaded towards the red. In hydrocarbon flames, bands due to C_2, CH, and OH also appear. The C_2 (Swan) bands are favored in gas-rich mixtures during burning and are predominantly in the visible portion of the spectrum (4737, 5165, and 5635 Å) which makes them appear green to the eye. These bands are shaded toward the violet. The CH spectrum has two main bands with heads at 4315 Å and 3872 Å, and appears violet-blue to the eye. HCO bands are also observed in the spectra of acetylene, methane and higher hydrocarbons. If the oxygen content is low enough in a flame, the generally blue flame becomes yellow and strongly luminous due to the presence of hot solid incandescent carbon particles formed by reduction in the flame during combustion. These may be collected as "soot" if a cold object is introduced into the flame. Since the yellow flame color due to carbon-incandescence is the best type of flame for lighting, it was necessary for "illuminators" derived from petroleum to be introduced into manufactured gas for maximum derived illumination from gas flames.

According to Lax and Pirani (8), the luminous efficiency of an oil or illuminating gas flame is 1.26 lumens per watt, while that for acetylene is around 2.5 lumens per watt. However, Weitz (9) gives much lower values for these luminous sources as:

1.2.1.-	Source of Light	Luminous Efficiency (in Lumens per watt)
	Candle	0.10
	Kerosene oil lamp	0.30
	Acetylene flame	0.70
	Incandescent tungsten filament	245.0

The Encyclopedia Britannica (10) gives the following figures for the cost of one million lumen-hours from various illuminating sources now known:

1.2.2.-	Source of Light	Cost
	Candles	$ 400.00
	Kerosene lamp	$ 13.50
	Tungsten incandescent lamp	$ 2.55

The color temperature of a candle or kerosene flame is about 1900 °K, whereas that of a tungsten incandescent lamp is closer to 3150 °K. The great strides made in both illuminating efficiency and cost should be evident from the above. We shall discuss the methods and inroads made on this subject somewhat later.

2. Incandescence and Blackbodies

Incandescence is radiation emitted by a body because it is hot. As we said before, charged particles (fermions) such as electrons obey Fermi-Dirac statistics whereas uncharged particles (bosons) such as photons obey Einstein-Bose statistics. Application of the latter theory to a "gas" of photons in equilibrium with a body at a temperature, T, leads directly to the Planck Radiation Law for the volume density of radiant energy:

1.2.3.- $\qquad d E_\upsilon = 8\pi h\upsilon^3 \, d\upsilon \, /c^3 (e^{h\upsilon/kT} - 1)$

where υ is the wavelength, h is Planck's constant and k is Boltzmann's constant. For the radiation emitted from a surface, the equation is:

1.2.4.- $\qquad d E_\lambda = c_1 \lambda^{-5} \, d\lambda \, / \, e^{c^2/\lambda T} - 1$

where $c_1 = 2\pi h c^2 = 3.7413 \times 10^{-12}$ watt-cm^2 and $c^2 = hc/k = 14{,}388$ micron-degree K.

Here E_λ is the power radiated per unit area in a wavelength range of $d\lambda$. Integration of 1.2.6. over all wavelengths gives the Stefan-Boltzman Law:

1.2.5.- $\qquad E = \sigma T^4$

where $\sigma = 2\pi^5 k^4 /15 \, h^3 \, c^2 = 5.6686 \times 10^{-12}$ watt/cm^2-deg^4. Differentiation of the distribution function given in 1.2.5. to find the wavelength of maximum emission results in:

1.2.6- $\qquad \lambda_m T = c_2/ \, 4.965 = 2898$ micron-degree

Substitution of λ_m into 1.2.6. gives for the maximum radiation:

1.2.7.- $\qquad E_{\lambda m} (T) = 1.285 \times 10^{-15} T^5$ (watts/cm2 per micron range in λ).

20

Equations 1.2.6. and 1.2.7. taken together constitute the Wien Displacement Law by which it is possible to represent the Planck distribution at any temperature by the "universal curve", in terms of λ/λ_m and $E_\lambda/E_{\lambda m}$, shown as by the following diagram:

1.2.8.-

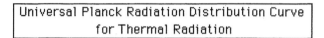

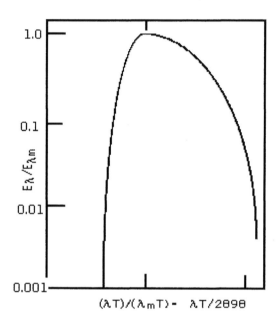

$(\lambda T)/(\lambda_m T) = \lambda T/2898$

Note that the curve falls off from the peak much more rapidly on the short wavelength side than on the long wavelength side. The integrated curve shows that about 25% of the total energy is at wavelengths less than λ_m.

Extensive tables of values calculated from the Planck Law are also available. From the point of view of light sources, we are interested not only in the total amount of energy emitted from a hot body, but also the effect of this radiation on the human eye. That is, we must weight the radiation function by the eye-sensitivity curve.

The quotient of these two quantities gives the luminous efficiency in lumens/watt, as shown in the following diagram:

Figure 1.2.9.-

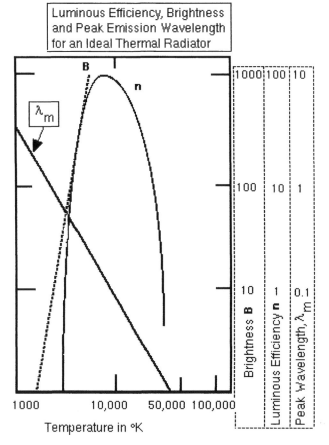

Here, **B** is the brightness, λ_m is the total radiated power and **n** is the luminous efficiency. The maximum luminous efficiency (95.05 lumens/watt) is reached at a temperature of 6625 °K. Although both the brightness and total radiated power both increase with increasing temperature, the latter decreases faster than the former. Hence, efficiency decreases faster above the maximum luminous temperature. At the point of maximum efficiency, the peak emission wavelength is 4370 Å., i.e.- a "blue-white". Further increases in temperature causes the peak to shift into the near-ultraviolet, and the efficiency drops dramatically. The low efficiency at low temperatures is correspondingly due to the predominance of infra-red radiation.

The constants given in 1.2.9. are for a calculated hypothetical "black-body" which has a maximum theoretical radiating power. We have already given the actual emission curves of real black-body emitters in 1.1.2. In general, a free surface of a real material will radiate a smaller quantity of radiation than the hypothetical one, and the ratio of the emission intensity at a given wavelength to that of the blackbody having the same wavelength and temperature is the "spectral emissivity" of the material. In practice, a blackbody can be realized by providing a small opening in a spherical enclosure whose walls are all at the same temperature. In this case, multiple reflections of the radiation at the wall make the emission independent of the material used if the geometrical considerations are proper (11,12). The thermal emissivity is related to other electrical and optical properties of the material, including the volume concentration of electrons and the electrical resistivity (11,13). Data for various materials have been compiled in the literature (11).

In practice, the theoretical limit of 95 lumens per watt for incandescence cannot be reached because there is no known solid material which can be operated at the temperature of 6625 °K because of melting point and vapor pressure considerations.

The search for better materials has therefore been one of striving for higher operating temperatures. Fortunately, the emissivity of tungsten is higher in the visible region of the spectrum than the infra-red region. A comparison is shown in the following Table:

TABLE 1-4

Characteristics of a Blackbody and a Tungsten Radiator

Blackbody				Tungsten Filament				
Temp	Bright	Power	Effic.	Temp(°K)		Emissivity		Effic.
(°K)	(candles/ cm^2)	(watts/ cm^2-ster)	(lumens per watt)	Brightness	Color	Visual	Total	lumens per watt
1800	11.78	18.99	0.620	1684	1825	.454	.236	1.19
1900	24.24	23.57	1.028	1771	1929	.453	.248	1.94
2000	46.49	28.94	1.607	1857	2033	.452	.259	2.84
2100	83.96	35.17	2.387	1943	2137	.450	.269	4.08
2300	235.9	50.61	4.655	2109	2347	.448	.286	7.24
2400	370.8	60.01	6.179	2192	2452	.447	.294	9.39
2500	563.1	70.65	7.971	2274	2557	.446	.301	11.72

TABLE 1-4 (CONTINUED)

Characteristics of a Blackbody and a Tungsten Radiator

Blackbody				Tungsten Filament				
Temp	Bright	Power	Effic.	Temp(°K)		Emissivity		Effic.
(°K)	(candles/ cm^2)	(watts/ cm^2-ster)	(lumens per watt)	Brightness	Color	Visual	Total	lumens per watt
2600	828.1	82.65	10.03	2356	2663	.444	.308	14.34
2700	1186	96.12	12.34	2457	2770	.443	.315	17.60
2800	1656	111.2	14.89	2516	2878	.442	.321	20.53
2854	1964	120.2	16.34	2595	2986	.441	.328	23.64
2900	2260	127.9	17.67	2673	3094	.440	.334	27.25
3000	3023	146.5	20.64	2750	3202	.438	.337	30.95
3100	3970	167.0	23.77	2827	3311	.437	.341	34.70
3200	5127	189.6	27.04	2903	3422	.436	.344	38.90
3300	6522	214.5	30.40	2978	3533	.435	.348	43.20
3400	8182	241.7	33.85	3053	3646	.434	.351	47.15
3500	10140	271.4	37.36					50.70
3600	12410	303.8	40.85	3165	3817	.433	.354	53.10
3655	13700	322.8	42.44					
4000	25,300	463.0	54.56					
6500	3 x10^5	3228	94.98					
8000	6.6x10^5	7408	89.52					
10000	1.3x10^6	1.8x10^4	73.14					
15000	3.5x10^6	9.1x10^4	38.91					
20000	6.2x10^6	2.9x10^5	21.54					
40000	1.8x10^7	4.6x10^6	3.95					

At a temperature of 2800 °K, the luminous emissivity of tungsten is 0.442 while the total emissivity is only 0.321. The spectral distribution from a tungsten filament at a temperature of 2800 °K. is most like that of a blackbody at 2878 °K. This is usually expressed by stating that its "color-temperature" is 2878 °K. However, the luminous efficiency of tungsten at its melting point (3410 °C or 3683 °K) is 53.1 lumens per watt.

The fact that the color temperature of tungsten is always greater than its true (thermodynamic) blackbody temperature is an additional reason for its superiority as a filament material in incandescent lamps, over, for example, carbon. Note also that the efficiency of the blackbody drops dramatically as the temperature exceeds about 10,000 °K. The emissivity of many non-metals, such as oxides, often show a strong dependence on wavelength (11). Some special cases of such "selective-radiators" will be discussed below. The thermal radiation of hot gases (including our sun) is also of importance and becomes rather complicated (14). Thus, incandescence is not only of interest in flames and photo-flash lamps, but in normal incandescent lamps as well.

Because photo-flash lamps have been obsoleted by flashlamps, and because photo-flash was used in photography long before incandescent lamps became practical, we will discuss the subject of photo-flash lamps first, and that of incandescent lamp development later on. And in the next Chapter, we will discuss the development of flashlamps in considerable detail (flashlamps could also be called spark-lamps).

C. Pyrotechnics and Photoflash Lamps as Related to Incandescence

Flames normally require an external supply of oxygen gas. However, there are some reactions between hydrogen and fluorine gases which provide quite lively combustion with considerable heat as a by-product. Solid oxidants such as potassium nitrate or chlorate may also be used to support very intense combustion of various fuels such as sulfur, charcoal and shellac. The color of the light produced can be controlled by additions of Sr for red, Ba for green, Na for yellow, Cu for blue, etc.

Light sources for photography have usually employed intense illumination sources having short periods of duration. In 1857, Moule employed a mixture of sulfur, potassium nitrate (saltpeter) and a sulfide (which comes close to being gunpowder) for photographic purposes. Crookes in 1859 used magnesium metal ribbon for the same purpose. Later, "flashlight powder" consisting of magnesium metal powder mixed with potassium chlorate as an oxidant became the standard. The disadvantages of such sources were many: smoke, smell, danger of fire and the fact that the duration was too long. Indeed, early photographs always showed persons with their eyes closed because they usually had time to react before the flashlight powder ceased burning. Similar sources were also used for aerial photography and illuminating flares at sea.

The first enclosed photoflash bulbs, consisting of magnesium ribbon or aluminum foil in an oxygen atmosphere were introduced by Keisling in 1898, and Smith in 1899 (15). Such lamps were improved to the stage of commercial practicality by Ostermeyer in 1928. The metal was ignited by an explosive "primer" coated on a tungsten filament. Passage of a small current through the filament set off the primer and this in turn ignited the metal. The primer was usually zirconium metal powder plus potassium chlorate. In 1930, the Philips organization introduced a closed all-gaseous photoflash lamp consisting of a mixture of carbon disulfide vapor and nitrogen monoxide, but the output proved to be too low for practical purposes. Van Leimpt and de Vriend (16) studied the combustion of a variety of metals and alloys in an atmosphere of pure oxygen. Results are summarized in the following Table, viz-

TABLE 1-5

Metal	Light Yield (lum-sec/mg)	Heat of Combustion (cal/gm)	Efficiency (lumen/watt)	Density (gm/cc^3)	Light Yield (lum-sec/mm^3)
Mg	700	5.958 cal	28	1.745	1220
Al	750	6.93	26		2030
Zr	441	2.86	36		2820
W	39	1.056	9.0		750
Mo	68	1.88	8.7		695
Ta	185	1.373	32		3070
Ce	65	1.661	9.3		450
Th	138	1.26	26		1550
Ti	500	4.59	25.5		2250
C	63	7.833	1.9		142
Al + 8% Mg	850		29.5		
Al+4%Cd	890		30.8		
Al+2.8%Zr	810		26.5		
91Al+7Mg+2Zr	1100		32.8		
Al+2.5%Ti	933		24.5		
Al+1%Ca	770		24.5		
Al+1%Li	990		25.0		
Al+16%Zn	860		29.3		

In this Table, the light-yield in terms of both the weight of material and the volume are given. Al has the highest light yield per unit of weight. In the second part of the Table, it can be seen that the yield can be increased by the addition of several alloying agents. Al + 8% Mg has been used by some manufacturers. With a knowledge of the heats of combustion, these light yields may be converted into efficiencies. Mg, Al, Th, and Ti all give values of 26-28 lumens per watt. Of the materials listed, Ta and Zr have the highest efficiency, but a lower yield per unit of weight. Efficiency was not, however, the major factor in photoflash bulb applications. Only Al and Zr have been used commercially. The real reason for use of zirconium seems not to be efficiency, but the ability to pack a lot of foil or wire into a given bulb volume.

The following are comparative figures for two commercial flash lamps which were otherwise comparable. Note that under equal conditions, Al burns five times faster than Zr. The spectrum of a photoflash lamp (aluminum) consists mainly of a broad continuous spectrum resulting from the incandescence of hot Al_2O_3 (boiling point = 3500 °C.).

1.2.10.- Lamp		Total Output	Efficiency (per mg)
Metal	Type		
Al	M-2	7500 lumen-sec.	430 lumen-sec.
Zr	M-25	17,000	374

In addition, many other lines and bands are superimposed on the continuum. These have been identified as being due to: AlO (4400-5400 Å), Al metal ((3844 and 3961 Å), MgO (if Mg is present), Fe, Na and K (as impurities). Some of these lines have been observed to show reversal during the latter part of the flash. This point has been interpreted as evidence that the emitter is in the vapor state. The color-temperature has been given variously as 3500-4000 K., with that of Zr being about 150 °K higher than that of Al. It may be noted that an increase of oxygen pressure to ten atmospheres increases the value for Al from 3800 °K to 4400 °K.

Rautenberg and Johnson (17) have also studied the light production of the Al-O_2 reaction in an effort to improve the light production and/or color. However, their efforts only produced a small change of 3% and they state:

"We conclude that in Al-oxygen filled photoflash lamps having no filtering, the color-temperature can be deliberately altered only in a minor way. The apparent limitation on reaction temperature in lamps of conventional design (to the melting point of Al_2O_3) puts a limit on the peak radiative output of a given size lamp. It is apparent that higher color temperature and radiative output can probably be obtained only in other systems in which a higher temperature can be reached".

Helwig (18) studied the effect of oxygen pressure, bulb volume and foil-weight on output of aluminum photoflash lamps. If w is the weight in mg. of the Al-foil, then the optimum oxygen pressure is:

1.2.11.- p(mm. Hg) = 18,000/w

The amount of oxygen required to combine chemically with the foil is given by:

1.2.12.- p(mm. Hg) $\cdot$ V (cm^3)   = 520 w

Combination of these two equations gives for the optimum volume:

1.2.13.- $V_{Opt.}$ (c m^3) = w^2/35

The pressure developed in the bulb during combustion, as a result of the high temperature, is:

1.2.14.- $P_{Atm.}$ = 19 w/ V

and can reach very high numbers indeed. It is for this reason that most of the commercial flash lamps were coated with an external expandable plastic, similar to that used in automobile windshields. The usual oxygen filling pressures were in the range of 1.5 atmospheres.

The properties of commercial flash lamps depended upon many factors such as bulb volume, weight of foil, physical state of foil and packing density, oxygen pressure, etc. Lamps were rated according to the time required to reach peak output.

Such lamps included:

"Special" - 13-15 millisec. (M2, M25, AG1)

"Medium" - 20 millisec. (No. 2, 5, 8, 11, 22, M5)

"Slow" - 30 millisec. (No. 3, 50)

"Focal-plane" - Peak output reached in 20 millisec. and then maintained for 20 millisec.(No. 6), or 50 millisec. (No. 31), or 17 millisec.

M5, M25, M3 and AG1 were made with Zr, the others with Al. Output peaks as high as 100,000 lumen-sec. and 5.2 million peak lumens were achieved in large bulbs. No. 50 is 2.62 inches in diameter and 5.38 inches overall length. The midget M5 and M25 lamps produced 16,000 lumen-seconds and 1.2 million peak lumens.

We have included this section on photoflash lamps to illustrate the diverse usages to which combustive reactions have been applied. At the present time, photoflash lamps are little used, having been replaced by electronic **flashlamps** which operate by excitation of xenon gas within a flashbulb. Such bulbs have the advantage of producing repetitive flashes, while photoflash lamps give only one and must be replaced. For this and other reasons, it is currently difficult to purchase or otherwise obtain photoflash lamps, since most photography stores have little demand for such lamps.

Flashlamps are capable of providing much higher peak intensities than are available with other sources other than the best photoflash lamps. A plasma is produced by causing a spark streamer to form between the electrodes. As the discharge channel grows, the electrical resistance drops sharply. The electrons in the plasma equilibrate in a high temperature distribution very quickly, and ionize and heat the plasma through collision.

Radiation from a flashlamp is made up of both line and continuum components. The line radiation corresponds to discrete transitions between the bound energy states of the gas atoms and ions (bound-bound transitions). The continuum is made up primarily of recombination radiation from gas ions capturing electrons into bound states (free-bound transitions) and of bremsstralung radiation from electrons accelerated during collisions with ions (free-free transitions). The spectral distribution of the emitted light depends in complex ways on electron and ion densities and temperatures. A typical discharge circuit is shown in the following diagram:

1.2.15.-

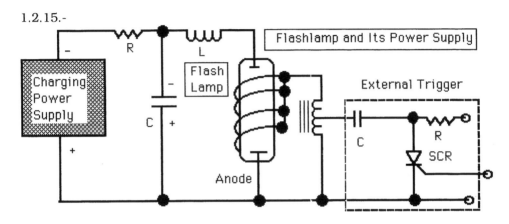

The blackbody temperature rivals that of photoflash lamps, i.e.- 4000 - 5000 °K. Xenon gas is generally chosen as the gas fill since it yields a higher input/output efficiency than any of the other gases. Typical efficiencies range from 25 to 60%, depending upon lamp type and operating conditions. For short-pulse, high current density applications, higher peak brightness can also be achieved by using the lower atomic weight noble gasses such as kryton, argon, neon and helium. All of these lamps require kilovolts for operation. Both peak output and efficiency increases with increasing power density. If we assume local thermodynamic equilibrium in the plasma, the blackbody characteristic of the plasma becomes that of the electron temperature in the plasma (modified only by the emissivity of the hot components, which in this case can be nearly equal to 1.0). In normal operation, the electron temperature will range from 9000 to 12,000 °K. For low power density flashlamps, the emissivity of the continuum is low but increases rapidly as the power density is increased. However, since the "explosion" energy time is a few milliseconds, the overall power required is low.

It is for this reason, and that of repetition, that flashlamps have replaced photoflash lamps in photography. Such flashlamps can now be made very small to fit directly upon the camera (We will discuss their design more specifically in the next Chapter, see page 215 for details). We now turn our attention to another aspect related to combustion, that of incandescent filaments and incandescent lamps.

D. Early Incandescent Lamps

Although there may be a general impression that Thomas A. Edison "invented the incandescent lamp in 1879", it is more correct to state that Edison invented the first *practical* incandescent lamp in that year. Such lamps have had a much longer history (10,19,20). In 1809, Sir Humphrey Davy made a lamp utilizing an incandescent platinum filament in air; in 1809, de La Rue used a partial vacuum. The first carbon-rod lamp was made by Jobard in Belgium in 1838. In 1840, Sir William Grove lighted the auditorium of the Royal English Institution with platinum lamps employing a water-seal to avoid entry of outside air (21). The first patent for an incandescent lamp was issued in 1841 to Frederick de Moleyns of Cheltenham, England. The emitter was powdered charcoal. A British patent was also issued in 1845 to J.W. Starr of Cincinatti, Ohio for an incandescent lamp with metallic filament. Starr also made a lamp in the same year consisting of a carbon rod operating in the vacuum above mercury in a barometer-type device. Other early lamps are shown in the following table:

TABLE 1-6

Early Incandescent Lamps and Their Inventors

Inventor	Year	Filament Material
W.E. Staite	1848	Platinum, Iridium
W. Petrie	1849	Iridium
E.C. Shepard	1850	Carbon cylinder
M.J. Roberts	1852	Graphite rod
Gardiner & Blossom	1858	Platinum
M.G. Farmer	1859	Platinum
J.W. Swan	1860	Carbonized paper & cardboard
Isaac Adams	1865	Carbon strip
James Gordon	1879	Platinum-iridium alloy
Lodyguine	1872	Graphite rod + nitrogen gas

One can see from the above that there were a large number of people working on incandescent lamps before Edison became active in 1877. Lodyguine, a Russian, was apparently the first to use nitrogen gas filling and the Admiralty Dockyard in St. Petersberg was lighted by 200 lamps of his design in 1872.

A major factor in this work was the development of the vacuum pump. It is not generally realized that **the** major problem to be overcome in developing a practical lamp, as conceived by Edison, was that of oxidation of the filaments. This was solved by the invention of a new type of mercury pump by Sprengel in 1865, i.e.- the "mercury diffusion" pump. The use of this pump was perfected by Sir William Crookes in 1875 during work on his famous "radiometer" which required a "good" vacuum of "one millionth of an atmosphere", or about 10^{-3} mm. Hg. The necessity of driving out all occluded gases, particularly oxygen, in the filament and glass bulb was realized in 1879 by both Edison and Joseph W. Swan independently, and was in part responsible for their success.

At least six contemporaries of Edison were also active in developing the incandescent lamp (22):

1. Moses G. Farmer (U.S.) - In 1879, he was using carbon rods in an exhausted or nitrogen-filled bulb, but had difficulty with seals.

2. Hiram S. Maxim (U.S.) - In 1878, he was using carbon rods in hydrocarbon vapor. Maxim is better known for his later work on machine-guns. He later became an English citizen and was knighted.

3. St. George Lane-Fox (England) - In 1878, he was employing a high-resistance Pt-Ir alloy wire in nitrogen gas or air, within a stoppered glass bulb, and also used carbon-impregnated asbestos as a filament in a nitrogen atmosphere.

4. William E. Sawyer and Albon Mann (U.S.) - This team of inventors used a variety of incandescent emitters, including carbonized paper cut into a horseshoe shape and live willow-twigs. Their best lamp of 1879 employed a long carbon pencil, in a

nitrogen atmosphere, which was automatically replaced as it was consumed.

5. <u>Joseph W. Swan (England)</u> - Swan seems to have been closest to Edison in arriving at a practical lamp. He employed improved vacuum techniques and also realized the need for high resistance in the filament. His 1878 lamp employed a slender carbon rod in an evacuated sealed glass bulb. The patent situation in England was such that in 1883 it proved expedient to form the Ediswan Company (Edison and Swan United Electric Light Co., Ltd.) which was the operating company in Britain.

Yet, the fact remains that none of the lamps described above were practical. They were plagued by short life due to filament emitter breakage, and by excessive blackening of the interior of the glass bulb.

E. The First Practical Incandescent Lamp (1879-1905)

Thomas A. Edison, "The Wizard of Menlo Park, N.J." first tackled the incandescent lamp problem in 1877. His prior work had included many years of labor on devices such as: 1. the Telegraph; 2. the Telephone; 3. the Phonograph; 4. the Mimeograph. Work on motion pictures and the Edison battery came later. It may be truthfully said that Edison was better at improving on the ideas of others and actually making them work, than in developing startlingly new ideas of his own. His observation of thermionic electron emission in his carbon-filament lamp in 1883 was patented (U.S.P. 307,031) but was not followed up. This effect was later studied and applied by Sir Ambrose Fleming as the thermionic diode (1904). One must remember that even at this time Ohm's Law was not universally accepted, and it was not until 1884 that the Stefan-Boltzmann law was discovered. Furthermore, it was not until 1900 that the Planck distribution law was derived. Thus, the subject of incandescence was still a mystery for nearly all of the early lamp workers.

However, Edison did observe sparks induced by currents in other electrical circuits and concluded that he had found a "new etheric force" which could be used for wireless signaling at a distance. The resulting patent (U.S.P. 465,971) was later purchased by the Marconi Wireless Co. in 1903. Nonetheless in 1877, Edison became convinced that a practical electric lamp could be developed and that it would operate by incandescence, in contrast to the several forms of arc-lamps that were being developed at the same time. He, as well as others, realized that, although arc-lamps might be applicable to street lighting and other outdoor usage, those for home use had to be made much smaller and more manageable. He also realized that electric lamps had to operate on parallel circuits rather than on series circuits to be practical. Because of this, he took the position that only "high-voltage" lamps would be justifiable in order to avoid excessive losses in the power lines due to otherwise high currents. Thus, Edison stressed the use of long thin emitters with high resistance, in contrast to the short carbon rods used by many of his contemporaries. The word "filament" was indeed introduced by Edison and his selection of 110 volts as the ideal voltage is still with us today. The use of long filaments also leads to a reduction in "end-losses" due to conduction at the supports, although it may not have been understood at that time. Edison's application of improved vacuum techniques was another important factor leading to his success, where others had failed.

Not all of Edison's work was with carbon filaments. Indeed, on November 25, 1878, an interview with Edison printed in the <u>NEW YORK SUN</u> stated that "carbon won't do". Edison tried various materials including: B, Si, Ru, Pt, Zr,Ir. Cr, Rh, Os,Ti, ZrO_2 and TiO_2 without success. In the spring of 1879, he used thin platinum wire coated with ZrO_2. But the oxide, which was there to retard evaporation, become conductive and shorted out the Pt coil. Work on carbon began in August 1879, beginning with a mixture of lamp black and tar. This lamp had sufficient promise so that Edison changed his mind and became convinced that a practical carbon filament could be developed. Thus, a search for a practical source of carbon began and included many sources as shown in the following, given as 1.2.16. on the next page.

These materials were carbonized by slow heating in a furnace from which air was excluded. Edison's first success came in October 1879 when a filament made from cotton sewing thread burned for about 40 hours. Even better success was achieved later that year with horse-shoe elements cut from cardboard. These lamps surpassed the 100 hour mark and some reached as high as the 170th hour before failure.(21). However, Edison was not pleased entirely with these lamps since they operated only at 70-80 volts and did not meet his goal of at least 100 volts.

1.2.16.- Threads of:

Absorbent cotton	Grass fibers	Cork	Celluloid
Absorbent cotton	Hard woods	Hardwoods	Fishline
Chinese hemp	Parchment	Twine	Wool
Cotton sewing	Linen twine	Parchment	Rattan
Jute silks & fiber	Teak	Horsehair	Tar paper
New Zealand flax	Tar paper	Manila	Boxwood
Raw Flax	Tissue paper	Corn silks	Jute fiber
Spruce	Celluloid	Grass fibers	Rattan
Vulcanized rubber	Linen twine	Cords	Cardboard
Wrapping paper	Italian Silk	Cardboard	Hollywood
Linen fibers	Tissue paper	Cords	Celluloid

Word of his developments reached the press and Edison therefore reluctantly held a public demonstration on December 31, 1879. The steamship, Columbia, was lighted with such card-board filament lamps in May 1880. Finally, in July of 1880, Edison reached his goal of a "high-voltage" lamp when he used filaments of Japanese Bamboo (which was patented December 27, 1881). It is this material which was used commercially by the Edison Lamp company until 1894. Bright (20) summarized the situation as-

"Thomas A. Edison was the first inventor to discover how a substance could be satisfactorily used in a commercial lamp.... His illuminant consisted of a high-resistance carbon conductor in a filamentary form. The distinction between a carbon rod and a carbon filament was a real one and provided the basis for Edison's patent victory as well as his commercial success. Of the principle features of the successful Edison lamp of 1880, only the form of the conductor had not appeared in previous lamps, except for the questionable Góbel lamp of 1854. The vacuum-sealed glass globe, the material of the illuminant, the platinum lead-in wires and the major other characteristics were well known....the first Edison lamps sold commercially were made from carbonized paper, a material which Swan had utilized in his experiments by 1860. Sawyer and Mann had also used carbonized paper before Edison began his experiments .... Sealed glass globes were likewise no novelty, although the stoppered type were more widely employed."

Many of Edison's competitors continued their work, following his "triumph". The majority made lamps for operation at 40-70 volts for several years, primarily because they were easier to make (22):

"All of the illuminants used by Sawyer and Mann were comparatively stubby compared to Edison's long slender illuminant. Although Mann was sympathetic to the idea of trying longer and thinner carbons of high resistance, Sawyer's insistence that the resistance be kept as low as possible confined their attention to short, thick carbons for the most part. Even after Edison's disclosure, Sawyer expressed the belief that only a low-resistance illuminant was feasible. Several years passed before he admitted the superiority of the Edison-type lamp. An outstanding invention of Sawyer and Mann seems to have been their first discovery of preparing carbons by "flashing" them in an atmosphere of hydrocarbon gas."

The latter process was patented by Sawyer and Mann on January 7, 1879. It depended upon the fact that pyrolysis of the hydrocarbon (gasoline) vapor and deposition of carbon on the filament occurs fastest just where the filament is thinnest and hence hottest. Hence, the filament could be made uniform over all of its length. Both Maxim and Lane-Fox also used a very similar process, and Maxim's patent was issued in October 1880. In other respects, Maxim's commercial lamps were much the same as Edison's cardboard type, on which they were probably based. The second commercial installation of incandescent lamps was made by Maxim and the U.S. Electric Lighting Co. in 1880 in New York City. Lane-Fox carbonized grass fibers by a similar hydrocarbon technique, which was patented in England on March 10, 1879. As for Swan, he also shifted from his slender rods to a carbon filament similar to that of Edison some time after Edison's announcement of success. Swan also developed a process of "parchmentizing" cotton thread in dilute sulfuric acid in 1880. This resulted in a structureless, more efficient, filament than Edison's bamboo one. At any rate, after a patent interference contest, the carbonized paper or cardboard patent was finally awarded to Sawyer and Mann on May 12, 1885.

George Westinghouse took no part in the early development of the incandescent lamp, but in 1883 began manufacturing such lamps at his Union Switch and Signal Co. In 1888, he acquired the patent rights of Maxim (through the U.S. Electric Lighting Company) and those of Farmer, including the Sawyer-Mann Electric Company (through the Thompson-Houston Electric Company), and established his electric lamp manufacturing Company (although it was still called the Sawyer-Mann Electric Company). The General Electric Company (GE) was formed in 1892 by the merger of the Edison General Electric Co. and the Thomson-Houston Electric Co. The "Edison Lamp Works" in Harrison, NJ later became the receiving tube manufacturing factory of Radio Corporation of America (RCA) when RCA acquired the radio-receiving tube manufacturing facilities from G.E. and Westinghouse in 1929. RCA had been formed in 1919 solely as a sales agency for the tubes manufactured by G.E. and Westinghouse.

In 1907, the Sawyer-Mann Electric Co. was renamed the Westinghouse Lamp Co. and moved from New York City to Bloomfield, NJ. In 1936, it was renamed the Lamp Division of the Westinghouse Electric Corporation. Westinghouse lost control of his company in the financial panic of 1907-1908.

According to Bright (20)-

> "Westinghouse was a prolific inventor, rivaling Edison himself in his breadth of interest.... Edison opposed the use of alternating current, saying that it was too dangerous compared to direct current.... The same mistake was not made by George Westinghouse, who became the principal sponsor of this method of power distribution in the United States".

Thus, although Edison clearly saw the necessity for efficient power generation and distribution to promote the general use of electricity (as evidenced by the many Edison electric utility companies still scattered around the U.S.) he did not grasp the advantages of voltage conversion by transformers made possible with alternating current. Although Westinghouse used 89,000 of the famous "glass-stoppered" lamps to light the Chicago World's Fair in 1893, this was done for the sake of avoiding the Edison patents and such lamps were never sold commercially due to the "uncertain quality". The really important thing about this installation was the success of the alternating current system.

The next stage in the development of carbon filaments was the use of non-fibrous materials (19,20). Most of this work had been done in England. Swan's "parchmentized" cotton has already been mentioned. In 1882, Desmond G. Fitzgerald soaked paper in zinc chloride solution to homogenize it, followed by washing in dilute HCl, and drying. In 1883, Swan developed a process of squirting a viscous solution of nitrocellulose through a die into an alcohol coagulating bath. In the U.S., Edward Weston developed a similar process in 1884 involving sheets of guncotton. Such "tamadine" filaments were used in the Westinghouse stopper-lamps of 1883. In 1886, Alexander Bernstein developed a method of electrolytically depositing carbon upon a fine metallic wire immersed in a liquid hydrocarbon. In 1888, Leigh S. Powell combined the Fitzgerald and Swan procedures by dissolving cotton in hot $ZnCl_2$ solution and squirting filaments into an alcohol solution. In all cases, the filaments were carbonized before being put into lamps. The status of carbon-filament development then remained essentially stationary until about 1904. The preparation and application of such filaments has been described at length in a book by Ram (23).

The luminous efficiency of Edison's first lamps was quite low, about 1.7 lumens per watt. Because the efficiency changed quite markedly during life because of bulb blackening, this number is the initial efficiency and the average was about 1.1 lumens per watt throughout life. The use of "squirted" filaments and the hydrocarbon "flashing" technique gave efficiencies of 3.4 lumens per watt which were typical for many years. Such filaments operated at a temperature of about 1600 °C. The ordinary carbonizing temperature was 2700 °C. Considerable information on the development of carbon-filament incandescent lamps may be found in books by Barham (24) and Soloman (25).

In 1893, J.W. Howell (26) discovered that heating ordinary "flashed" carbon filaments to a high temperature led to a positive temperature coefficient of resistance rather than to a negative one. In 1904, Willis R. Whitney, Director of the G.E. Research Laboratories used a new furnace of his own design to heat untreated (unflashed) filaments to 3500 °C. They were then flashed as usual at 2700 °C, and finally reheated to 3500 °C, before usage. This treatment drove off all of the impurities from the carbon, but the most important effect was the physical change in the graphite "shell" surrounding the core of the filament, which was more pure, dense, tough, and flexible than before the high temperature treatment. This led to the introduction of the "GEM" filament (General Electric Metallized) in 1905 (19b,20). However, there was no metal in these filaments. The name only meant that the temperature coefficient was positive, like a metal, and that hence such filaments could withstand voltages variations far better than ordinary carbon filaments. They could be operated at 1900 °C (color temperature ≈ 2150 K.) with an efficiency of 4.0 - 4.2 lumens per watt. Blackening of the inside of the lamp was also reduced significantly. The GEM lamp also had advantages over fragile osmium filaments and the complicated Nernst "glowers" which were being introduced at the same time. However, the death-knell for the carbon filament was sounded in 1907 by the introduction of the tungsten filament and carbon lamp sales declined dramatically after this date. The GEM lamp was removed from the market in 1918. Nevertheless, its introduction served to establish the reputation of G.E. during this period.

F. Metal Filaments and Incandescent Oxides (1883 - 1907)

Although carbon became the standard for the incandescent lamp, and was used for nearly thirty years, it still possessed undesirable features. For example, although its sublimation point is quite high, its vapor pressure is also high at the operating temperature so that bulb blackening was always a severe problem in carbon filament lamps. The search for better materials therefore continued, particularly in Europe, for improved filament materials which could be operated at higher temperatures. The major parameter to be considered was the melting point of the material used.

The following Table compares some of the thermal properties of potential filament materials:

TABLE 1-7

Thermal Properties of Metals and Compounds Potentially Useful for Filaments

Material	Melting Point (°C.)	Temperature for Vapor Pressure =1x10^{-6} mm Hg (°C.)	Melting Point (°C.)			
			Carbide	Nitride	Boride	Silicide
U	1130		2475			
Pd	1552					
V	1700		2830	2050	2100 d	1750
Pt	1769	1759				
Ti	1800		3250	2930	2980	1540
Th	1850		2655			
Zr	1960		3175	2980	3040	1520

TABLE 1-7 (CONTINUED)

Thermal Properties of Metals and Compounds Potentially Useful for Filaments

Material	Melting Point (°C.)	Temperature for Vapor Pressure $=1 \times 10^{-6}$ mm Hg (°C.)	Melting Point (°C.) Carbide	Nitride	Boride	Silicide
Rh	1960					
Hf	2207		**3890**	**3310**	**3259**	
B	2300			3000		
Nb	2415	2326	3500	2050	2900 d	1950
I r	2454					
Ru	2500					
M o	2610	2048	2690		2180	1870 d
Os	2700					
Ta	2996	2513	3880	3090	3000 d	**2400**
Re	3180					
W	**3410**	**2860**	2700		2860	2150
C	3670	2261			2450	2600
ZrO$_2$	2715					
ThO$_2$	3050					

Here, we have arranged the metals in order of increasing melting points and have included "intermetallics" such as carbides, nitrides, borides and silicides of a few of these metals as well. In this Table, "d" indicates decomposition in air. Those numbers given in bold are those with the highest value for each column. It is clear that the metal with highest melting point is tungsten and that the temperature at which its vapor pressure reaches 1×10^{-6} mm. of mercury is also the highest of any of the metals.

Note also that of the "intermetallic" compounds given, those of hafnium have the highest melting point. If oxides are to be considered, only those which become conducting at elevated temperatures will be useful. The problem with oxides and the intermetallic compounds is one of formability and even today such compounds are not easily formed into long thin filaments. One material of specific interest today is that of the ceramic superconductors, as exemplified by: $Y_{1.0}Ba_{1.8}Cu_{3.0}O_{7-\partial}$, where ∂ is a small decimal. The problem of forming "wires" of such materials to use in superconducting applications has remained one of the major obstacles to its success. The best solution to date has been the use of silver metal, as a tube, combined with superconducting particles.

At the time that new filament materials for improved incandescent lamps were being investigated, the problems of forming filaments successfully were equally as great and most of the materials of interest could not be formed into fine wires by the methods then known. This situation was aggregated by the relatively low resistance of metals, compared to carbon, and brittleness of metal filaments, due in many cases to impurities present.

Many attempts were made to improve upon carbon filaments (20). As early as 1883, F.G. Ansell in England tried electrodeposition and subsequent oxidation of Ca, Al and Mg on carbon filaments. In 1889, T.D. Bottome of the U.S. made composite filaments of carbon and W or Mo, but these did not permit operation at higher temperatures compared to carbon alone. The Russian inventor, Lodyguine, was hired by Westinghouse and tried cladding C or Pt with W, Mo, Os and Cr in an effort to avoid the Edison patent. None of the lamps were successful. Rudolph Langhans, a German, made composite filaments of carbon with Si or SiC during 1890-94, which were sold for a short time by the Premier Electric Lamp Syndicate in England. In 1907, the "Helion" lamp, consisting of a carbon filament coated with SiC, was developed by H.C. Walker and W.G. Clark in the U.S. This lamp, produced commercially (20), was said to produce 1.0 candlepower per watt, i.e.- 12 lumens per watt, but this remains difficult to believe because of the temperature that such a lamp would have had to operate.

The only non-metallic filament lamp which seemed for a time to offer a serious challenge to the carbon filament lamp was the "glower" invented in 1897-99 by Walter Nernst in Germany (20). Many early workers, including Lane-Fox, Edison, Ansell, Welbach and Langhans, had unsuccessfully tried to use oxides as emitters. Nernst succeeded by giving up the idea of making a simple filament. He used a small rod of oxides which was heated externally until it became sufficiently conductive to be self-sustaining. Originally, a flame was used, but this was later replaced by an external coil (usually of Pt) with an automatic cutout. Because of the extreme negative temperature coefficient of resistance of Pt, an external "ballast", consisting of an iron wire operating in a hydrogen atmosphere, was necessary (27). Many different oxides could be used. One early mixture consisted of 15% Y_2O_3 and 85% ZrO_2. Later, a mixture of oxides of Th, Zr, Y, and Ce was employed. The operating temperature was 2350 °C. and the efficiency was about 5.0 lumens per watt, including ballast losses. The life of the rods was about 800 hours on alternating current and only 300 hours on direct current (due to electrolysis of the oxides during operation). The lamp could be restarted by replacing the rods. A vacuum was not necessary, and, in A.C. operation, air was used. These lamps were produced by AEG in Germany and the Nernst Lamp Co. (owned by Westinghouse) from 1902 to 1912. Solomon (25) devotes an entire chapter to these lamps.

Although incandescent oxides have long since been replaced by metal filament-type lamps for visible light sources, they still find application as infra-red radiators. In this region of the spectrum, most oxides are better thermal radiators than are metals. In infra-red spectrophotometers, the source has remained a "Nernst glower". Recipes for preparing such lamps are therefore still to be found in relatively current scientific papers where infra-red emitting sources are important (28). The spectral energy distributions from them and SiC sources have been studied by Friedel and Sharkey (29).

The first practical metallic filament used in lamps was osmium (20,25,26). This metal has the highest melting point of all of the Pt-group metals. In 1898, when patented by the Austrian, Carl Auer von Welsbach, it was thought to be the highest of all of the metals. However, no known metallurgical technique was then known for making wires from this brittle metal. After trying a cladding of Pt and and other sheathes, Welbach devised an extruding or "squirting" process which made the Os-lamp possible. Production was begun in Austria and Germany in 1902, but Os-lamps were not made in the U.S. Because of the higher operating temperature, compared to carbon, an efficiency of 5.5 lumens per watt was realized. Absence of "blackening", i.e.- maintenance of output, was also much better than carbon lamps. However, these lamps were much more expensive, i.e. $1.25 - 2.00 as compared to $0.17 for carbon filament lamps. The Os filament was therefore reclaimed because of the scarcity of Os metal. Thus, Os-lamps were generally rented, not sold. The filaments were also very brittle when cold and subject to easy breakage. The low resistance of such lamps made long thin filaments necessary so that lamps operating at 110 volts could not be produced at first. The lamps could also be burned only in a vertical position because of filament sagging during operation. Finally, in 1906, the Os filament was replaced by "Osram", an alloy of Os and W. In 1901, Wilhelm Sander of Berlin developed a squirted filament of Zr, but these never were commercially successful. His patent covered U and Th as well.

In 1902, W. von Bolten and O. Feuerlein at the Siemans & Halske Co. succeeded in developing techniques for purifying Ta and making it ductile (20,24,25). The melting point of tantalum is about 300 °C. higher than that of Os, but the price in 1905, when lamps were placed on the market, was about $5000 per pound of Ta-metal. Siemans and Halske gained control of all possible world sources and retained manufacture of Ta filaments during the entire period that lamps were manufactured and marketed. The U.S. licensees for assembling these filaments into lamps were G.E. and its affiliate, the National Electric Light Co. Lamps were produced from 1906 to 1913. Operating temperature was 1900 °C and the initial efficiency was about 5.0 lumens per watt, which declined to about 4.25 lumens per watt during its useful life of 600-800 hours. But, on alternating current, Ta recrystallized rapidly and became brittle so that its useful life was shortened to 200-300 hours. The reason for this behavior still remains somewhat obscure. However, it has been shown (30) that for direct-current operation of W and other metallic filaments, a surface structure develops due to surface migration of atoms. These are mostly oxide atoms which tend to migrate to the cold end of the filament where they can do no harm. But for alternating current, where there is no temperature gradient during operation, these oxide atoms tend to cluster at grain boundaries and make the filament brittle.

The melting point of W exceeds that of all known metals. According to Bright (20):

> "Although W was known for over a 100 years, it did not become available in relatively large quantities at moderate prices until after about 1890. Despite its apparent attractiveness for incandescent lamp filaments, it extreme brittleness and difficulties in forming it into filaments were not overcome until around 1904, when a number of alternative methods were devised". (We will address these methods below, see p. 45).

In 1902-04, Alexander Just and Franz Hanaman at the Techniche Hochschule in Vienna developed two techniques for preparing W-filaments. One was an extrusion-sintering process, using an organic binder similar to that used by

Welbach for osmium. The other involved deposition of W on a carbon filament from an atmosphere of tungsten oxychloride, and subsequent removal of the carbon, leaving a hollow tube of tungsten. Even though these filaments were very fragile, they yielded lamps with initial efficiencies of 7.85 lumens per watt and a life of about 800 hours. Thus, W as a filament material proved early on to be superior to the carbon filament lamps being marketed at the time.

Hans Kruzel of Vienna also described in 1906 another extrusion method in which the organic binder was omitted. A "paste" was used, composed only of colloidal particles of W and water, formed by striking an electric arc under water between tungsten electrodes. Fritz Blau and Herman Remané of the Austrian Welsbach company also developed in 1905 sintering procedures for W, using either pressed metal powder or a paste formed by treating WO_3 with NH_4OH. At first, Welsbach used W to form "Osram", an alloy with Os, but the better performance and lower cost of W-filaments eventually led to complete elimination of Os in lamp filaments. A similar process occurred with the so-called "Z-lamps" of Hollefreund and Zernig in Germany. These consisted of Zr-coated carbon or ZrC-coatings which were improved by the addition of W. Finally, only W was used alone. Von Bolton of Siemans and Halske Co. applied in 1904 for a patent covering lamps with filaments consisting of purified *ductile* wires of a number of metals, including tungsten, but failed to state how this was to be accomplished. The Thomson-Houston Co. in England and W.D. Coolidge of G.E. used a method employing an amalgam to hold the particles together temporarily. At Siemans and Halske, a method was developed in 1903 involving the use of a ductile W-Ni alloy, from which the Ni was later removed.

In 1906, Westinghouse bought the Austrian firm, Welsbach & Co., and then conducted a profitable European business for several years. However the patent situation in the U.S. was dominated by G.E. who controlled the patents of Just and Hanaman, Kuzel, and the German Welsbach Co. Their first W-filament lamps were introduced in 1907 and the 60 watt lamp sold for $1.75, while the 40 watt lamp was priced at $1.25. These prices were substantially above those in Europe. Westinghouse produced tungsten-filament lamps in the U.S. as a G.E. licensee.

Despite its efficiency, tungsten-filament lamps in 1907-10 were finding a hard time replacing the carbon lamp. Several factors played a role:

1. The higher initial cost

2. The lower cost of electricity compared to Europe

3. The practice of electric utility companies in supplying lamps to their customers.

Bright stated (20) -

"For some years, they were reluctant to encourage use of the new lamps, and customers accustomed to free carbon lamps were unwilling to buy W-lamps. Eventually, the electric companies discovered that tungsten lamps.... had no real downward effect on lighting load, for consumers took the opportunity to improve their lighting rather than increase their bills".

Additionally, it was not too clear from a technical standpoint that tungsten at that time was the answer to lighting mastery.

Solomon in his book (25) published in 1912 said:

"The metallic-filament lamp is beyond question destined to have a very great effect on the electric lighting industry (p. 200)....Whether or not the tungsten filament lamp represents the last word in metal-filament lamps is on the knees of the gods (p. 306)....That the carbon-filament lamp will be entirely killed seems improbable....It seems, at any rate, feasible that the resources of the carbon-filament lamp are not yet exhausted and that we shall ultimately find the best filament for an incandescent lamp in the only element which we have not yet succeeded in melting (p. 307)".

One other point ought to be mentioned. Just as the long length of filaments of low resistivity metals was sometimes a problem for 110-120 voltage lamps, it was an advantage for operation at low voltages. For example, after the introduction of W-filament lamps, 6 volt lamps to replace the oil and acetylene lamps used on automobiles became possible. Portable "flashlights" ("electric-torch" in England) became practical as well. These applications were not practical with high-resistance carbon filaments. Fragility was still a problem until ductile tungsten became available.

G. The Market Conquest of Tungsten Filament Lamps (1910-1958)

The "giant-step" step ahead for tungsten-filaments, which sounded the demise of gas lighting, carbon lamps, and extruded or pressed W-filaments, was the development in 1910 by W.D. Coolidge of the G.E. Research Labs. of the long awaited procedure for making ductile W-filaments. We will address the metallurgy of W more thoroughly in a later chapter. Von Bolton of Siemans and Halske Co. believed that W was like Ta, and would be ductile if pure enough. This has been shown to be true for extremely pure (99.990-99.997%) single-crystal W, but not for commercial (99.95-99.98%) polycrystalline W, which is **not** ductile at room temperature. The answer, found by Coolidge, lies in working the metal while it is hot, but the working conditions are critical due to the somewhat unusual properties of tungsten metal. Most metals, such as copper for example, are ductile at room temperature and become "work-hardened", caused by a change in grain structure when worked cold. Upon reheating above an "annealing point", the original structure and ductility are then restored. Ordinary tungsten, if heated to the proper temperature, can be worked and becomes fibrous in structure. If overworked, it tends to fall apart. This then must be remedied by reheating, which causes recrystallization and hardening rather than the "annealed" state in the usual metallurgical sense. The necessary temperature was found to be a function of the amount of work the tungsten had already received. Upon heating in lamps, the filament wire then reverts to a uniaxial structure due to recrystallization.

The procedure worked out by Coolidge involved several steps of reduction, pressing, sintering, swaging and drawing. Most of these are still used today. It is to be noted that if W is heated in air, it will oxidize to WO_3 . This necessitates processing tungsten bars and wires in pure hydrogen gas. Lamp filaments are very peculiar and the purest tungsten does not work as well for this purpose. High temperatures causes grain growth within the wire. If these extend across the diameter of the filament, slippage along the grain boundaries occur, sag of the wire occurs, and current is restricted, causing "hot" spots with subsequent burnout of the filament. The nature of the recrystallization is also important in determining the amount of "sag" which may occur during operation of the filament. Coolidge was the first to add "doping" agents to control crystal growth and sag in the filament. Various additives may be used, including ThO_2, SiO_2, and Al_2O_3 (33). Although thoriated W-wire can withstand severe vibration better than untreated wire, it cannot be used at the highest operating temperatures because the thoria becomes reduced. ThO_2 was introduced in 1916 by Jeffreys in England, while the $KAlSiO_4$ used in present-day "non-sag" W-wire was the contribution of Aladar Pacz of G.E. in 1917.

Lamps containing the ductile W-wire were placed on the market in 1911. The efficiency of 8.4 lumens per watt, at 1000 thousand hours of life, was an important gain over competing light sources. In addition, the new filaments were cheaper, less fragile than carbon filaments, and allowed much better control over lamp voltage. Prior to this time, each lamp had to be individually rated for voltage. It now became possible to standardize on a few voltages, i.e.- 110, 115, and 120 volts, and to standardize on wattage as well.

The next improvement in the incandescent lamp followed soon after. This was the use of gas rather than of vacuum within the glass bulb. Actually, we have already mentioned the use of nitrogen gas. The Siemans Company in Germany introduced hydrogen gas in its carbon filament lamps in 1886. Other concerns used gases to avoid the Edison patents. Thus, in 1884, the Star Electric Co. used a heavy hydrocarbon gas in its "New Sunbeam" lamp, while the Waring Electric Co. used bromine gas in its "Novak" lamp. In the last case, bulb discoloration was reduced by chemical action, but the bromine became depleted and the lamp became a vacuum lamp despite its name, and was so ruled by the courts. In 1894, Professor William A. Anthony of Cooper Union studied the problem of bulb blackening in carbon filament lamps and concluded that use of gas of a high molecular weight would retard the blackening process by causing some of the vaporized carbon atoms to return to the hot filament (31). However, the use of gas filling also results in increased heat dissipation from the filament and hence lower efficiency of the lamp. At the operating temperature of carbon filaments, the one effect offsets the other, and all attempts to make the gas-filled carbon filament lamp better than the vacuum lamp failed. However, carbon monoxide was used by AEG in 1901, nitrogen gas in France in 1908, and Hg-vapor was employed by R. Hopfelt in Germany in 1908 (20) in carbon-filament lamps.

Irving Langmuir, working in the G.E. Research Labs., attacked the problem of bulb blackening in tungsten lamps in 1909. He first studied the effect of various gases on W-filaments and convinced himself that the blackening observed in properly exhausted lamps was not due to chemical attack of the filaments, but to simple evaporation of W metal. However, if water vapor is present, the situation is quite different. At the temperature of the hot filament, the following reaction takes place:

1.2.17.- $W + 3H_2O \Rightarrow WO_3 + 6H\cdot$

where H· is atomic hydrogen. The WO_3 , which is white or yellow in appearance, is deposited on the cool glass wall where it reacts with the H· to form a black deposit of tungsten metal plus water, so that the vicious "water-cycle" can be repeated. Langmuir also demonstrated that several gases could be used to retard the evaporation of the metal. The important question then was whether conditions could be found so that the additional heat loss would not be great enough to prevent realization of the greater efficiency that should accompany higher temperature operation. Langmuir thus began an intensive study of the heat losses from heated wires in gaseous atmospheres (32). He found that these "convective" heat losses were only slightly affected by the wire diameter and concluded from this that the gas near to the filament was stagnant (later called the "Langmuir layer"). Heat from the wire was transferred by conduction through this gas layer and then is dissipated at its outer surface by convection. The thickness of the layer, and hence the effective area for heat transfer is only a slowly increasing function of the wire diameter. The heat loss per unit of radiating area is thus greater for thin wires than for thick wires. The next question was how to use a filament with large diameter and still have the desired high electrical resistance. The solution to this problem, and to that of an efficient gas-filled lamp, was to make the filament with a large "effective" diameter (as far as the gas was concerned) by coiling up the wire into a coiled filament (33). Coiled filaments have an advantage over straight ones even in vacuum lamps because of the partial blackbody effect.

According to Langmuir, the thickness of the stagnant gas layer, D, is related to the wire, or coil, diameter, d, by the relation:

1.2.18.- $\ln (D/d) = 2B/D$

where B is a constant which for air at atmospheric pressure is about 0.43.

Some illustrative examples are given in the following:

1.2.19.-

D(cm)	d (cm)	s= (D-d)/2 (in cm.)
1.0	0.425	0.288
0.5	0.0795	0.210
0.3	0.0171	0.141
0.2	0.00272	0.0986

The relatively small dependence of D (or s) on d is easily seen, The thickness of the Langmuir layer is a few millimeters for all practical cases. The heat loss from the film is then:

1.2.20.- $W \text{ (watts)} = \{2\pi\lambda / \ln (D\text{-}d)\} (\phi_2 - \phi_1) = 7.31 \lambda D (\phi_2 - \phi_1)$

where ϕ is the filament length, and the subscripts refer to the temperature of the bulb and filament, respectively.

Geiss (34) gives for the case of heat-loss in gas-filled lamps the empirical equation:

1.2.21.- $W = C \phi d^{0.3} T^{1.6}$

where T is now the filament temperature. It may be noted that at the temperatures of interest, the light radiation increases as the 4.8th power of T, while the rate of evaporation increases as the 34.3th power. Langmuir's constant, B, is directly proportional to the gas viscosity (which in turn depends on the molecular weight and the pressure). For the quantity ($\phi_2 - \phi_1$) in 1.2.20., Geiss gives:

1.2.22.- $(\phi_T - \phi_{300}) = \alpha (T / 2700)^{1.65}$

Values of α and of B for the three gases having most interest in gas-filled incandescent lamps are given in the following:

1.2.23.-	Gas	α (watts/cm)	B (cm)
			(p= 570 mm).
	Nitrogen	1.70	0.57
	Argon	1.22	0.51
	Krypton	0.73	0.27

Fonda (35) has also shown that the mass transfer through the stagnant gas layer, i.e.- the amount of tungsten evaporation, obeys the same law as the heat transfer, and that therefore, at a given filament temperature, the evaporation rate, v, is:

1.2.24.- $$v = C / \{ [M\ p\ d][\ln(D/d]\ \}$$

where M is the molecular weight and p is the pressure of the gas. Thus, it can be seen that the use of a heavy gas such as Krypton decreases both the heat loss and the evaporation rate of the filament.

When gas-filled W-lamps were first introduced in the U.S. in 1913, argon was not available in commercial quantities, and nitrogen was used instead. At first, only the larger sizes of lamps, i.e.- 200 watts and above, were gas-filled. In 1917, argon, which has lower conductivity than N_2 as shown above, essentially replaced nitrogen gas. However, a small amount of N_2 must be present to prevent arcing due to the lower breakdown voltage of argon, compared to nitrogen gas. The first gas-filled 100-watt lamps had an efficiency of 12.6 lumens per watt, as compared to 10.0 lumens per watt for corresponding vacuum lamps.The color temperature was about 2850 °K., compared to about 2500 °K. for the vacuum lamp.

As stated above, the efficacy of the gas decreases as the size of the filament decreases. Hence, gas-filling has not been used for 110 volt lamps smaller than about 40 watts. However, gas filling is sometimes used in small lamps for special purposes such as flashing lamps where it is desired that the filament cools rapidly. In that case, hydrogen gas may be used efficiently. Vacuum is also used even in large lamps if it is desired for some reason that the glass envelope should operate at a lower temperature than is normal. In the following Table, data on gas losses, end-losses and other efficiency factors are presented, viz-

Wattage[*]	Lum. Eff. in Visible (lumens/ watt)	% Output in Visible (%)	Gas Loss (%)	End Loss (%)	Base+ Bulb Loss (%)	Filament Heat Content in Joules
500 C	19.8	12.08.8		1.8	7.1	182.0
300 C	18.8	11.1	11.6	1.8	6.8	80.0
200 C	18.5	10.2	13.7	1.7	7.2	39.7
100 CC	16.3	10.0	11.5	1.3	5.2	14.2
60 CC	13.9	7.5	13.5	1.2	4.5	5.5
40 C	11.6	7.4	20.0	1.6	7.1	2.5
25 V	10.6	8.7	----	1.5	4.5	2.8
10 V	7.9	7.1	----	1.5	5.0	0.62

TABLE 1-8

[*]- C = coil; CC = coiled coil; V = vacuum.

The smaller gas losses for larger filament size also results in the fact that, for a given voltage, a low-wattage lamp will have higher efficiency than a high voltage one. thus, 120 volt lamps are more efficient than those for 230 volts. Those designed for operation at 30 or 60 volts are even more efficient, i.e.- for train lighting. However, if the voltage is too low, end losses (conduction) become too high and thus an optimum voltage of about 12-15 volts does exist (36). Schilling (37) also shows an optimum of 10 volts for the brightness of a 50 watt projection lamp. According to Weitz (36), the efficiency of a 15 volt 100-watt lamp should be 34% higher than that of a similar 120 volt lamp.

However, the high currents in such a low voltage system would obviously result in higher power line losses and wiring costs. Presumably, the 120 volt system used in the U.S. is the best compromise between these and other factors. On the other hand, in Europe, where power is more expensive, the use of 230 volts is more usual. According to Weitz (36), the efficiency of a 100 watt lamp designed for 230 volts will be about 15% less than one designed for 120 volts.

In principle, even higher efficiencies (or longer life with the same efficiency) could be achieved with krypton rather than argon gas. An even greater advantage would accrue with xenon gas. Smaller bulb sizes could be used because of the lower thermal conductivity, and would be desirable because of the relatively higher cost and scarcity of the xenon gas. However, Geiss (34) only found an increase of 3-4% for krypton gas, even though other workers have given values of 5-15% or even higher (38,39). Nevertheless, krypton gas is not used in the U.S. because of the higher cost of such lamps. It has found limited usage in Europe where such lamps give about 12 lumens per watt in the 25 watt size, 16 lumens per watt for 60 watts, and 20 lumens per watt for the 300 watt size. In these lamps, a mixture of krypton with 10-20% xenon is used. The color-temperature is somewhat higher than for the argon-filled lamp made in the U.S. and the light is somewhat "whiter". However, the cost is just about double that of lamps manufactured in the United States.

A comparison of the efficiencies of various gas-filled lamps can be seen in the following diagram:

1.2.25.-

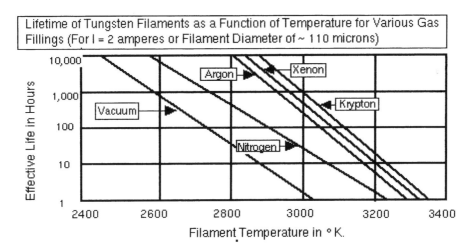

It should be clear that the use of gas-filled incandescent lamps does allow higher efficiencies and can, by adjustment of the lamp operational parameters, markedly improve effective life of the lamp as well.

Another step in improvement of the tungsten filament lamp came in 1936 when the coiled filament was coiled once again to give the coiled-coil filament, which is now standard in the wattage sizes most commonly employed today, i.e.- 50 watt to 200 watt gas-filled lamps.

This resulted in an improved "stagnant layer" as first described by Langmuir, and an increase in efficiency from 15.3 to 16.0 lumens per watt (~ 5%) for the 100 watt size, and from 12.5 to 13.8 Lumens per watt (~ 10%) for the 60 watt size. Non-sag wire is essential in coiled-coil filaments to prevent shorting of adjacent turns in the coil. The final step to date for General Service lamps has been the introduction of the "vertical or axial" filament in 1956 - 1958. The filament in this case is parallel to the rotational axis of symmetry of the pear-shaped bulb now being manufactured. Since most lamps are burned either base-up or base-down, it also results in the gas convection currents within the lamp being parallel to the axis of the filament, and permits a slightly higher filament operating temperature. This type of mounting results in less loss of light by the base or socket, and the result is about a 6% increase in efficiency.

Thus, the initial increases in incandescent lamp design were major, but after the introduction of tungsten-filament lamps, the increases in efficiency became much smaller. A summary of the general trend of improvement in incandescent lamps is summarized in the following Table as presented on the next page.

The design of straight tungsten filaments is a fairly straight-forward business, if one uses the Tables of Jones and Langmuir (40). Later measurements by Forsythe and Watson (41) differ slightly. However, the design of coiled or coiled-coil filaments is entirely another matter (42). The effects of supports in conducting heat away, i.e.- the "end-losses" has been studied by many workers (42,44,45). The heat capacity of the filament is also of interest in connection with the flicker of light, caused by alternating current operation (~ 5% for a 100 watt lamp at 60 cycles per second). This factor is much more complicated in coiled-coil filaments where some of the emitted radiation can strike other parts of the filament. Also of interest is the inrush of current when voltage is applied to a cold filament and the rise and decay times of the emission. We shall address these details in a later chapter.

TABLE 1-9

Luminous Efficiencies of Incandescent Lamps

Filament	Date	Lumens per Watt[*]
Carbonized bamboo	1881	1.7
Squirted cellulose	1885	3.4
"GEM" lamp	1905	4.2
Nernst glower	1902	5.0
Osmium	1902	5.5
Tantalum	1906	5.0
Extruded tungsten	1907	7.8
Drawn tungsten-vacuum	1911	8.4
Drawn tungsten-vacuum	1917	10.0
Tungsten - Gas-filled	1917	12.6
Tungsten - Gas-filled	1936	15.3
Tungsten: Coiled-Coil	1936	16.0

* Figures for tungsten lamps are for 100 watt lamps.

1-3 Electric Discharges in Gases

Although incandescent lamps are the more common variety of "electric lamps" to be found today, such was not always the case. According to the Encyclopedia Britannica (10), the first electric lamp was of the gas discharge type-

"In 1650, Otto von Guericke discovered that light could be produced by electricity or electrical excitation. He demonstrated that when a globe of sulfur was rotated, the friction produced by the pressure of his hand produced a luminous glow. About 1700, Francis Hawksbee used this discovery, together with a glass globe, partially exhausted by means of a vacuum pump (also invented by von Guericke) to produce the first "electric" light. When a globe was rotated at high speed and rubbed by hand, it emitted a faint glow of light".

Throughout the 19th century, the development of arc lamps and of incandescent lamps proceeded side-by-side (10,16,21,45) Gas discharge lamps of various types, including fluorescent lamps, are used today for a wide variety of applications (47,48,49,50). The literature is very voluminous concerning electrical conduction in gases and vapors (51,52,53,54,55). Most efficient gas discharges involve "arcs" of high current density and relatively low voltage drop, in contrast to corona, streamer discharges and glow-discharges which involve low currents and relatively high voltages.

As distinguished from incandescence, the emission spectrum of a gas discharge is usually typical of the electron-shell structure of the gaseous atoms carrying the current. Most gas discharges have a negative voltage-current characteristic and therefore require a "ballast" to stabilize the discharge. A case does exist where the electrodes themselves become so hot due to electron bombardment so that incandescence from them dominates the radiation emitted. Such an example would be that of the carbon arc.

A. Carbon Arcs

Carbon arcs operate by ionizing air molecules (which carry the current between the electrodes) to incandescence. The incandescence of carbon particles within the arc also serves to promote radiance. The first electric arc was discovered by Humphrey Davy in 1802. The power source was the voltaic cell, or battery, discovered only two years earlier by Volta. Davy's electrodes consisted of soft wood charcoal and were consumed very rapidly. The first patent on an arc lamp was granted to Thomas Wright in England in 1845. Bunsen, in 1840, provided improved carbon electrodes for such lamps in Germany, while two workers, Leon Foucault in1840 and F.P.E. Carré in1876, did the same in France. The earliest worker in the U.S. seems to have been E.C. Shephard in 1858. A major problem of automatically adjusting the electrodes as they were consumed was solved by a design of V.E.M. Serrin in 1857. The application of carbon arc lamps, of course, had to await the development of electromagnetic generators. Nonetheless, a Serrin lamp was installed in 1862 in the Dungeness lighthouse in England.

The first permanent installation of carbon arc electric lighting was in the Gramme workshop in Paris in 1873. The Paris Exposition of 1878 was lighted by the famous "electric-candle" invented in 1876 by the Russian, Paul Jablockoff. This lamp consisted of two parallel carbon rods insulated from one another by a clay coating. The clay vaporized as the carbons were consumed. Alternating current was required. In the U.S., carbon arc lamps were developed by William Wallace working together with Moses G. Farmer in 1875. Other workers included Edward Weston of Newark, N.J., Elihu Thomson and Edwin J. Houston (who were founders of the Thomson-Houston Co. as mentioned above in connection with carbon filament incandescent lamps). However, the most practical carbon arc lamp was invented by Charles F. Brush in 1877. His lamps were used to light the Public Square in Cleveland in 1879. As early as 1846, W.E. Straite in England found that enclosing the arc by means of a glass globe to restrict the amount of oxygen resulted in longer carbon electrode life. But, such enclosed lamps were not practical at first because of blackening of the globe internally by carbon and other particles during operation of the lamp. Successful enclosed carbon arc lamps were later developed by William Jandus in Cleveland, Ohio during the time of 1886 to 1893. In 1894, Louis B. Marks of New York City increased the electrode life to about 150 hours. The operating voltage was 80 volts and the efficiency was about 4.25 lumens per watt for alternating current operation and 8.25 lumens per watt for direct current operation. A common size consumed about 510-550 watts each. This large consumption of power and high output made such lamps unsuitable for home usage. The performance of carbon arc lamps of this era is discussed in Solomon's book (25). It is well to note that the lighting companies of the era also produced carbon arc lamps in addition to carbon filament lamps.

As early as 1844, Bunsen studied the effect of introducing various chemicals into the carbon arc. The color of the light from the Jablockoff candle was also influenced by the vaporized clay insulation. Early attempts to successfully introduce additives into the carbon arc failed because of the formation of a "crust" on the carbon rods during operation. It was Hugo Bremer of Germany who solved this problem in 1898-1900 by addition of a "flux" of boron or fluoride to eliminate crust formation. Various elements could now be used to obtain various colors, including: Ba or Mg for white, Sr for red, Ca or Fe for blue, and Ni for ultraviolet (56). Efficiencies of 15-18 lumens per watt for white and 30-35 lumens per watt for yellow made these lamps the most efficient light sources of the day. At first, these "flaming arcs" had very short electrode life, with noxious fumes. Only after about 1908 was it determined how to enclose them and obtain an electrode life of 100 hours. Indeed today, such carbon arc light sources persist as the "search-lights" commonly seen and used for advertising display, particularly at automobile dealer displays.

The properties and spectra of carbon arcs have been studied by a number of workers (19,39,57,58,59,60) over many years. There are three general types that were used:

1. Low-Intensity Arcs

Typical operating conditions are 20-254 amperes at 5-55 volts. The light source is the incandescent **positive** electrode, which is normally larger than the cathode. The anode is heated to the sublimation point so that the color temperature is about 3800 °K. The temperature of the gas in the main part of the arc stream is about 6000 °C, and is very conductive (normal voltage gradient is about 20 volts per centimeter). The anode exerts a cooling effect on the arc so that the temperature and conductivity in its vicinity is lower. This accounts for the anode voltage drop of about 35 volts. The anode is heated by electrons having fallen through this potential, and by Joule heating of about 40 amperes per cm^2. This current is supported by thermionic emission. Thermal conduction from the arc is the other means of heating of the anode. The overall brightness is determined by the anode temperature and is therefore normally about 160-180 candles per mm^2. An increase in current increase the area of the anode crater, and the total output of light, but not the brightness. The overall efficiency is about 16 lumens per watt, not including ballast losses. The positive carbon may contain a mixture of soft carbon and a potassium compound. These do not contribute to the overall light output, but increase the stability of the arc by anchoring it on the soft carbon core, and by lowering the effective ionization potential (due to the presence of potassium). The positive electrode consumption is typically about 2.0 inches per hour.

2. Flame Arcs

If materials are introduced into the positive core, they vaporize into the arc stream. Because they usually possess a low ionization potential, the anode voltage drop is reduced to about 15 volts, with a total arc voltage of 37.5 volts. Thus, the anode temperature is lower and its contribution to the total light output is relatively unimportant. Instead, the light results from ionization and recombination of the additive in the arc stream, and the emission color is therefore determined by the nature of the additive used in the carbon electrode. The entire arc is luminous and the efficiency can be very high, up to 80 lumens per watt. But, the brightness is low despite the high output. The standard type of carbon (such as National Carbon's 8 mm-"Motion Picture Studio" carbon rods) gives a total output of 100,000 lumens for 1400 watts input, at a color temperature of 4650 °K. (39, 56).

3. High Intensity Arcs

These are sometimes known as "Beck" arcs. The current is normally 40-150 amperes and the voltages used about 50-75 volts. This corresponds to about 2 to 12 kilowatts input. Current density at the anode is about 100 amperes per cm^2. The current is so high that the entire anode area is covered by the molten anode spot and the vaporization of the flame material, usually CeF_3 or CeO_2 , is very rapid. This action results in a very deep positive crater. The deep crater and a tongue of yellow-white flame issuing from it are characteristic of this type of arc. The flame is actually 32% of the total light output, but it contributes little to the useful output of light since, with the usual optical projection system used, only the crater emission and the hot gas contained in it is utilized. The temperature in the crater is much higher than the anode temperature of the low intensity arc, so that the color temperature is normally 5900 - 6400 °K. The crater brightness of the normal arcs of this type of discharge is typically 400-1200 candles per mm^2, and may be as high as 2000 candles per mm^2 (which is higher than that of the sun at 1650 candles per mm^2). The efficiency is usually in the range of 28-34 Lumens per watt. However, Finkelnburg (61) has claimed efficiencies as high as 90 lumens per watt, crater temperatures of 8000 °K., and arc stream temperatures of 12,000 °K., using 200 kilowatt arcs.

The flame-arc may be operated on alternating current, but the other two require direct current operation. Because of its characteristics, the flame arc is normally used for illumination in motion picture and television studios. The other two, which are more nearly point sources, are used in optical projection systems like movie projectors and searchlights. Mechanical devices must be provided in the latter for striking the arc by touching the electrodes together and then separating them and adjusting the electrode position as it is consumed. In many cases, the anode is rotated to assure uniform burning and stability of the arc. Control of the arc stream by magnetic means may also be used. A carbon arc, cooled by water, with automatically loaded anodes, has also been described (62).

The spectrum of the flame arc is greatly influenced by the additives employed in the anode. They generally consist of the corresponding line and band spectra of the species, similar to the case of NaCl in a Bunsen burner flame. That of the other two types is primarily a broad band, typically that of the incandescence of the appropriate color temperature. However, a strong peak at 3800-3900 Å, due to cyanogen, i.e.- CN, and a small peak at about 2500 Å is usually present in spectrum of these arcs. In the case of the low intensity arc, the incandescent emitter is clearly the solid carbon anode. But, in the case of the high intensity arc, most of the emission appears to arise from the luminous excited gas in the crater. The origin of the continuous spectrum and the high color temperature has been open to some speculation. Bassett in 1921 (57) first proposed that that CeC_2 was formed at the anode. This material has a higher volatilization temperature than carbon itself, and might explain the the higher color temperature. A similar situation existed in a case described by Riehl (63) when an arc was struck between a carbon cathode and a thorium metal anode. The latter was converted to ThO_2 and the color temperature of the emission corresponded to the melting point of the oxide, i.e.- 5000 °K. However, it was also suggested that incandescent particles might exist in the crater of the high intensity arc. It is now known, however, that a wide variety of core materials can be used with nearly identical results. Yet, not all of these form carbides with high melting points. Both Finkelnburg and Maecker (60,64) suggested that the continuous spectrum resulted from energy states of excited electrons, i.e.- "bremstralung". If this were the case, as pointed out by MacPherson (59), then the easily excited alkali metals should be the best core-materials. Such is not the case, however, and it has been determined that elements with a large number of strong emission lines are best. Then, the continuous spectrum is strongest in the wavelength region where the lines are most numerous. MacPherson thus suggested that the emission is connected with these lines, as broadened in the dense, hot gas. The crater of the high intensity arc thus serves as the necessary blackbody cavity, or "hohlraum".

The carbon arc plasma was studied further by Finkelnburg (61) who showed that although a core material is helpful to overall output, it is not required at all. The important effect in the high intensity arc, in his opinion, is the vapor stream from the rapidly consumed positively charged carbon particles. The rapid velocity of this jetstream, which may exceed 10^3 cm/sec., greatly exceeds that of the positive ions in a normal discharge by a factor of 100 or more. This alters conditions at the anode dramatically and causes an increase in anode voltage drop. Such jets arise from compressive forces exerted on the arc by its own magnetic field and can arise at any restriction. Finkelnburg has also shown that the brightness may be increases by a factor of 3.3 if the pressure of the surrounding gas is increased to 10 atmospheres. It seems clear then that the luminoscity of the arc stream arises from the excited gases within the crater, as well as to any core material that may be present in the anode.

B. Mercury Arcs

As early as 1821, Sir Humphrey Davy produced light by maintaining a discharge in air between an wire and a mercury pool (65). Note that he also was responsible for the first incandescent lamp (Pt-filament in 1802) , the first carbon arc in 1802 and the first arc employing a gasous discharge, mercury. Thus, he may be said to be truly the real *"Father of the Electric Light"*. J.T. Way in 1856-57 patented in England an arc lamp which used a stream of mercury drops wherein the arcs were formed between drops of mercury. Mercury vapor was used for the first time by Rapieff in 1879. A mercury arc lamp was also designed by Arons in 1892. However, the first commercially successful mercury vapor lamp was patented in the United States by Peter Cooper-Hewitt (U.S. Patent 682,696). His early experiments were financed by George Westinghouse, and the Cooper-Hewitt Electric Co. was organized in 1902 by these two as principal stockholders. After the death of Westinghouse in 1919, the Cooper-Hewitt Electric Co. was acquired by General Electric, who operated it at Hoboken, N.J. for about 20 years as the General Electric Vapor Lamp Co. Note, however, that this lamp was not the familiar "fluorescent lamp" which now employs phosphors for visible light generation, but was a forerunner of such lamps, which were not invented until about 1939. (See p. 66 for a design of the Cooper-Hewitt lamp).

The original Cooper-Hewitt lamp consisted of a sealed evacuated glass tube about 4 feet long, with slightly enlarged ends (20). The tube was normally inclined from the horizontal about 15°. The lower end contained a pool of liquid mercury which served as the cathode while the other end had a metal anode. The lamp was started by tipping it so that the mercury ran down to bridge the two electrodes to start a current flowing. the heat thus generated vaporized some of the mercury and when the lamp resumed its normal position, the arc first struck across the small gap separating the broken stream and then transferred to an end-to-end vapor discharge between the electrodes. Later lamps used various devices to avoid hand-tilting, including electromagnetic tilting devices, inductive "kick" devices to give a high voltage pulse, and an external starting band at the cathode (66). The original lamps consumed 385 watts (3.5 amperes @ 110 volts DC), had an efficiency of 12.5 lumens per watt, and a life of 2000 hours. The quality of the light produced was poor because of red emission, characteristic of the emitted mercury spectrum. Lamps were made in sizes varying from 1/8 inches in diameter and 3 inches long to 3 inches in diameter and 12 feet long. An alternating current version was also manufactured, having two anodes (both located at the same end) connected as a full-wave rectufier with an inductance filter. An improved U-shaped lamp was developed by Bastian and Salisbury in England in 1904 (20). Additionally, Westinghouse mercury vapor lamps were used for street light in Paris in 1910 (67).

Several versions of the Cooper-Hewitt lamp and their characteristics have been described by Buttolph (66). An important feature of the original design was the enlarged bulb at the end of the lamp. This was used as a condensing chamber at the cathode to control the internal mercury vapor pressure. These lamps, which used glass as an envelope, operated at relatively low pressures of 0.25 mm Hg. This corresponds to a temperature of about 100 °C. in the internal gaseous stream. The voltage gradient was about 1.3 volts per inch, the brightness was 15 candles per in^2 , and the efficiency in 1920 was 14.2 lumens per watt. As of 1939, when the new fluorescent lamps and improved high pressure mercury vapor lamps came into the marketplace, the Cooper-Hewitt lamp had been improved to the point where the arc efficiency was now 25 lumens per watt and an overall efficiency (including ballast losses) of 19.4 lumens per watt for a 350-watt, 50 inch lamp (66).

Lamps similar to those of Cooper-Hewitt but made of quartz (with mercury pools at each end) were designed and made in Germany by Küch (68) in 1906. These lamps operated at a higher temperature of 400 °C, and an internal pressure of about 1.0 atmosphere (760 mm. of Hg) because of the superior temperature quality and resistance of quartz compared to that of glass. The luminous efficiency of these "high-pressure" lamps was 26 lumens per watt and they had an operating life of about 1000 hours. Lamps of this type were manufactured by Heraeus in Germany and also by G.E. in the U.S. The use of quartz permitted operation of a voltage gradient of 25-30 volts per inch, the brightness was 1500 candles per in^2 (2.3 candles per mm^2), and the mercury emission-line spectrum was accompanied by a continuous background. Because of the use of quartz as an envelope, large quantities of ultraviolet radiation was also produced. By 1930, such lamps made by G.E. were called "Uvilarcs". Yet, these were not the same type as the high-pressure mercury lamps being manufactured today, since they were also started by tilting, as were the original Cooper-Hewitt lamps.

The Cooper-Hewitt lamps (low-pressure) and the Küch lamps (high-pressure) were characterized by a cathode consisting of a liquid mercury "pool" whose temperature determined the internal vapor pressure. Such a construction has one advantage in that material sputtered or evaporated away is replenished by condensation, and hence the cathode life approaches infinity. However, use of a liquid cathode inhibits the lamp orientation during operation. In addition, the density of liquid mercury is very high and its presence may cause breakage during transportation of such lamps. Such pool cathodes have also been used in high-power industrial rectifiers for alternating to direct current operation. Although it is generally agreed that the mechanism of the mercury cathode is one of field electron emission, rather than thermionic emission, some details of its operation are still open to question (69). Somerville (70) has stated- "In two recent publications, the pressure at the cathode was estimated at 200 atmospheres at the one end, and at 0.001 mm. Hg at the other. This amounts to a difference of 10^9".

The arc in tubes with a mercury pool have been "started" by a number of means:

1. Mechanically, i.e.-by tipping or by withdrawing a wire dipping into the mercury, and connected to the anode by a resistor. This method was used by Weintraub prior to 1911 (71,72) and by N.V. Philips Gloeilampenfabrieken in the Netherlands as late as 1936 (73).

2. A stationary electrode near the cathode to provide a source of ionization. This method was used by Weintraub and Kruh (71) prior to 1911. Kruh used a Pt-wire coated with alkaline earth oxides.
3. External heaters, either gas flame or electrical (72.74).

4. A non-condensible gas-fill such as argon to carry the current until sufficient merciry has been vaporized to sustain the discharge (75).

5. A high-voltage high-frequency discharge, such as a "Tesla-coil" applied externally, or a high-voltage direct current spark produced between the electrodes (76).

6. Magnetic initiation of the arc may also be used under some conditions (77). Tubes, called "Permatrons" also exist in which the discharge is controlled, but not initiated, by a magnetic field (R.F. Earhart and C.B. Green *Phil. Mag.* **7** 106 (1929); W.P. Overbeck, *Trans. AIEE* **58** 224 (1939).

7. An internal "grid" electrode can be used to which a voltage pulse is applied (78).

8. A "starting-band" outside the envelope to which a high voltage "kick" is applied. This is generally applied as an inductive pulse. Such "capacitative starters" were first used by Cooper-Hewitt in 1903, and later by many other workers (79). Particles of SiC or similar materials are cemented to the inside of the envelope at the mercury level to amke starting easier. Electronic tubes working on this pribnciple have been called "capacitrons" or kathetrons".

9. Germeshausen (80) improved the external capacitive starter in 1939 by coating a wire with a thin insulating glass sheath and immersing it in the mercury pool. Such "dielectric ignitors" are more sensitive than the external type because of their higher capacitance. however, formation of mercury oxide can interfere with their operation.

10. Slepian in 1933 invented the Westinghouse "Ignitron" rectifier, using a "resistance-ignitor" consisting of arod of high-resistance

material such as SiC or BN which dipped into the mercury pool (81). such rectifiers have since found wide industrial usage.

For alternating current operation, the arc must be re-ignited for each cycle of operation.

The specific action of the dielectric and resistive ignitors is still open to question, although the main function is to create a high field strength at the surface of the mercury pool so as to create a mercury vapor discharge. Tonks and Frenkel (82) have shown that, for a field strength higher than about 53 kilovolt per cm. on the mercury surface, the surface becomes unstable. this undoubtedly plays a role in arc initiation. The normal cathode "spot" moves continually and erratically over the surface of the mercury at a velocity of about 10 meters per second. The spot temperature is about 300-400 °C. and the current density is about 10^6 amperes/cm^2 (70). Cobine (57) states that the minimum stable current is 3.0 amperes, unless some sort of "spot-fixer" is employed to localize the spot. It is clear that the mercury pool cathode is inherently unstable and reëstablishes itself many times a second (83). It is for this reason that a large inductance is normally used in series with such devices. The inductive voltage produced when the arc momentarily stops causes it to reignite. The spot is actually composed of many small spots, which emit a continuous spectrum , in addition to the usual Hg line spectrum. The rate of evaporation from the spot has been estimated to be about 2.5×10^{-4} gram/coulomb, or one atom for every eight electrons (if all the current is carried by electrons). The cathode spot is thus far from a state of equilibrium. Kenty (49) has also shown that the flow of Hg-ions to the walls exerts a strong pumping action in these low-pressure arcs and that sudden changes in current can produce large changes in internal pressure, and even arc extinction.

The Cooper-Hewitt low-pressure mercury-pool lamp enjoyed considerable popularity after its introduction to the marketplace in the early 1900's because of its high efficiency compared to the then-available incandescent lamps. However, mercury vapor lamp usage declined after the introduction of ductile tungsten filament lamps, except for usage as ultraviolet sources. They were revived, however, when new, more efficient designs were introduced in the 1930's. These new lamps differed from the Cooper-Hewitt lamps (low-pressure) and the Küch lamps (high-pressure) in several important aspects:

1. Use of a rare gas (usually Ar) as a "fill" gas for starting and supplying heat for vaporization of the mercury.

2. Elimination of the liquid cathodes. This required development of solid electrodes with adequate life and absence of sputtering. Such cathodes were made to function both during the starting (warm-up) period when the arc voltage drop is high, and during normal operation which is charactristic of a rare gas discharge.

3. In high-pressure mercury-vapor (HPMV) lamps, as opposed to those operating at low-pressure, the operational internal mercury pressure is not controlled by the temperature but by limiting the amount of mercury in the lamp so that all of it is vaporized. The vapor in such lamps is therefore unsaturated and the arc stream is "superheated".

4. HPMV lamps were designed to use much higher pressures and operational "loadings", i.e.- higher operating temperatures , which mandated the use of quartz (actually vitreous silica) as an outer envelope, and the development of reliable and cheap techniques for metal-quartz seals capable of withstanding the required temperatures and currents.

The luminous efficiency of Hg-lamps was studied by Krefft (84) who showed the effect of mercury vapor pressure on the efficiency and voltage gradient of such lamps. This is shown in the following diagram, given as 1.3.1. on the next page.

The Cooper-Hewitt "low-pressure" lamps operated near to the peak "B" shown there, while the Küch "high-pressure" lamps operated at nearly one atmosphere and hence had higher efficiency. Note that a minimum occurs near to "C", and that the efficiency increases rapidly beyond "D". Also shown is the voltage gradient which occurs in the discharge, as a function of temperature. Modern high-pressure-mercury-vapor (HPMV) lamps operate within the region "E" as shown below. We shall discuss the principles relating to design of mercury vapor lamps in more detail in the next chapter.

1.3.1.-

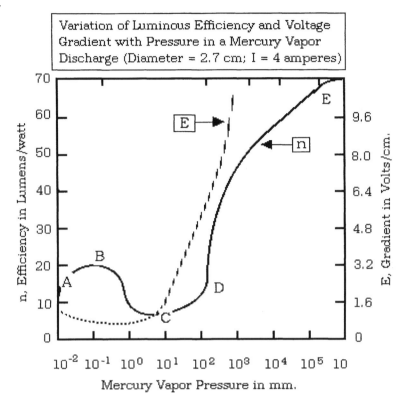

Variation of Luminous Efficiency and Voltage
Gradient with Pressure in a Mercury Vapor
Discharge (Diameter = 2.7 cm; I = 4 amperes)

Mercury Vapor Pressure in mm.

We can summarize the various methods of starting mercury vapor lamps in the
following diagram, given as 1.3.2. on the next page. Here, we have
summarized most of the prior methods used to reduce the starting voltage of
the various arc-lamps.

It should be clear, then, that artificial lighting did not make much head-way
until after the beginning of the nineteenth century. Prior to that time, most
dwellings had no internal sources of light except that of candles and that of
the fireplace. Most people woke at dawn and slept shortly after sunset. Even
the "rich" or Royalty used indoor torches infrequently, and usually on state-
occasions. It was solely during the latter part of the 19th century that the use
of gas-lighting became prevalent. Only with the advent of electrical power
generation did the widespread night-time lighting of cities occur. Street-
lighting became commonplace and the facades of buildings were flooded with

1.3.2.-

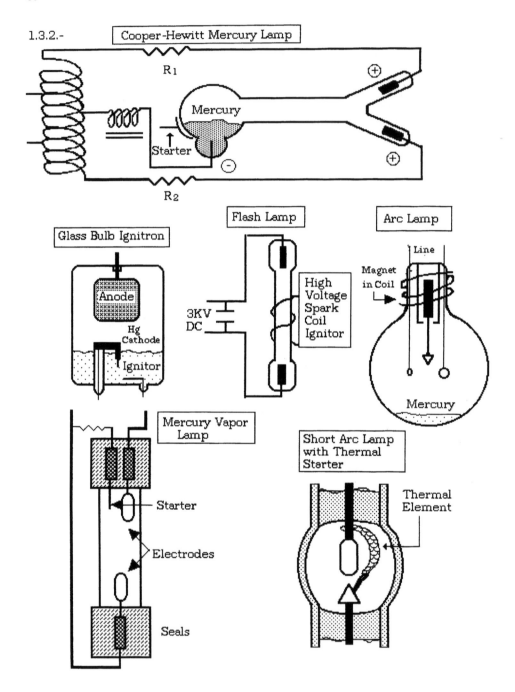

Cooper-Hewitt Mercury Lamp

R_1

Mercury

Starter

R_2

Glass Bulb Ignitron

Anode

Hg Cathode

Ignitor

Flash Lamp

3KV DC

High Voltage Spark Coil Ignitor

Arc Lamp

Line

Magnet in Coil

Mercury

Mercury Vapor Lamp

Starter

Electrodes

Seals

Short Arc Lamp with Thermal Starter

Thermal Element

light. Today, we have Las Vegas as an example where external lighting has been carried to an extreme. Nowadays, most cities "never sleep" due to the advent of luminous sources and the development of outdoor lighting devices.

REFERENCES CITED

1. V. Fermi, *"Encyclopedia of Religion"*- Littlefield, Adams and Co., Paterson, NJ (1959); see also- *"Larousse Encyclopedia of Mythology"*- Prometheus Press, New York (1959).

2. R. Patei, *Man and Temple in Ancient Jewish Myth and Ritual*,Thomas Nelson & sons, London (1947)

3. See for example: R.C. Ropp, *Luminescence and the Solid State*, Elsevier Science Publ., Amsterdam & New York (1991).

4. See for example: L. L. Cavalli-Sforza, *Sci. American*, **265** 104 (Nov. 1991); "Drift, Admixture and Selection in Human Evolution", A.M. Bowcock et al, *Proceedings Nat. Acad. Sci.*,**88** 839-843 (1991).

5. Merritt Ruhlen, *A Guide to the World's Languages*, Acad. Press Inc., New York (1987); See also: "Tracking the Mother of 5,000 Tongues", *Insight Mag.*, **Feb. 5**, p. 54-55 (1990).

6. "Human Origins"- E.L. Simons, *Science*, **241** 1343 (1989).

7. A.G. Gaydon and H.G. Wolfhard- *"Spectroscopy and Combustion Theory"*, Chapman & Hall, London (1948); See also: B. Lewis and G. von Elbe, *"Flames, Their Structure, Radiation and Temperature*, Chapman & Hall, London (1953).

8. E. Lax and M. Pirani in *Handbuch der Physik*,- Edited by Geiger & Scheel, Vol. 19, Springer Inc., Berlin (1928).

9. C.E. Weitz in *General Electric Co. Application Engineering Bulletin* **LD-1** (Jan. 1956).

10. *"Lighting and Artificial Lighting"*, Encyclopedia Britannica, 1952 Edition.

11. G.A.W. Rodgers, *Handbuch der Physik* **26** 129-170, Springer Publ. Berlin, (1958).

12. C.S. Williams, *J.Opt. Soc. Am.* **51** 564 (1961).

13. S. Roberts, *Phys. Rev.* **114** 104 (1959).

14. G.N. Plass, *Proc. IRE,***47** 1442 (1959)

15. G.D. Rieck and L. H. Verbeek *Artificial Light and Photography,* Philips Tech. Library, Elsevier Press, New York (1952).

16. J.A.M. Van Liempt and J.A. de Vriend, *Rec. Trav. Chim. Pays-Bas* **53** 839, 895 (1934); loc cit **54** 239 (1935), loc cit **55** 239 (1936); loc cit **56** 126 (1937); loc cit **58** 423 (1939).

17. T.H. Rautenberg and P.D. Johnson, *J. Opt. Soc. Am.* **50** 602 (1960).

18. H.J. Helwig, *Tech.-Wiss. Abhandl. Osram Ges.* **6** 93 (1953); see also: H kocker and G Kocher, ibid, **7** 297 (1958).

19. H. Schroeder, *History of Electric Light,* Smithsonian Institution Publ. #2717 (1923); 18b- see also J.W. Howell and H. Schroeder, *History of the Incandescent Lamp,* Maqua Co., Schenectady, N.Y. (1927).

20. A.A. Bright, *The Electric Lamp Industry,* MacMillan, NY (1949).

21. William Grove, *Phil. Mag* **27** 442 (1845).

22. F. Jehl, *Menlo Park Reminiscences-* publ. by the Edison Institute, Dearborn, Mich (1936).

23. G.S. Ram, *The Incandescent Lamp and Its Manufacture,* D. van Nostrand, NY (1894).

24. G.B. Barham, *The Development of the Incandescent Electric Lamp,*Scott, Greenwood & Son, London (1912).

25. M. Solomon, *Electric Lamps*, D. van Nostrand, NY (1912)

26. J.W. Howell, *Electricity,* **12** 117 (1897).

27. R.O. Jenkins, *Brit. J. Appl. Phys.* **9** 391 (1958).

28. H.P. Griffith, *Phil. Mag.* **50** 262 (1925); see also C. Tingwaldt, *Phys. Zeit.* **36** 627 (1935) & C.W. Munday, *J. Sci. Inst.* **25** 418 (1948).

29. R.A. Friedel and G.A. Sharky, *Rev. Sci. Inst.* **18** 928 (1947).

30. R.P. Johnson, *Phys. Rev.* **54** 459 (1938); R.W. Schmidt, *Z. Phyk.* **120** 69 (1942).

31. William A. Anthony, *Electricity* **6** 139 (1894)

32. I. Langmuir, *Phys. Rev.* **34** 401 (1912); *Trans. Electrochem. Soc.,* **20** 225 (1911); *loc. cit.* **23** 299 (1913); *Trans. AIEE* **31** 1229 (1912).

33. H.A. Jones *G.E. Rev.* **28** 650 (1925); See also: R.O. Jenkins, *Brit. J. Appl. Phys.* **9** 391 (1958).

34. W. Geiss, *Philips Tech. Rev.,* **1** 97,316 (1936); *loc. cit.* **6** 334 (1941).

35. G.R. Fonda, *Phys. Rev.* **21** 343 (1923); *loc. cit.* **31** 260 (1941).

36. C.E. Weitz, *G.E. Appl. Eng. Bull.* **LD-1** (Jan. 1956).

37. W. Schilling, *Lichttechnik* **12** 610 (1960).

38. G. Claude, *Comtes Rend.* **200** 1585 (1935).

39. I.E.S. Lighting Handbook, **3rd Ed.** (1959).

40. H.A. Jones and I. Langmuir, *G.E. Rev.* **30** 310, 354, & 408 (1927); See also: G.L. Davis, *Nature* **196** 565 (1962).

41. W.E. Forsythe and E.M. Watson, *J. Opt. soc. Amer.* **24** 114 (1934); W.E. Forsythe and E. Q. Adams, *loc. cit.* **35** 108 (1945).

42. W.E. Anderson, *Illum. Eng.* **48** 402 (1953)

43. I, Langmiur, S. MacLane and K. Blodgett, *Phys. Rev.***35** 478 (1930); J.W. Clark and R.E. Neuber, *J.Appl. Phys.* **21** 1084 (1950).

44. A.G. Worthing, *J. Franklin Inst.* **194** 597 (1922); G. Stead, *J. IEE (London)* **58** 107 (1920); V. Bush and K.E. Gould, *Phys. Rev.***29** 337 (1927); G. Ribaud and S. Nikitine, *Ann. de Phys.* **7** 5 (1927); I. Langmuir and J.B. Taylor, *J. Opt. Soc. Am.* **25** 321 (1935); *Phys.Rev.* **50** 68 (1936); H.G. Baerwald, *Phil Mag.***21** 621 (1952).

45. J. Fisher, *Z. Tech. Phys.* **19** 25. 57, 105 (1939); *Arch. Elektrotech.* **33** 48 (1939); **40** 141, 262 (1951); *Z. Angew. Phys.* **4** 90 (1952).

46. H. Schroeder, *"History of Electric Light"*, Smithsonian Inst. Publ. No. 2717 (1923).

47. W. Uyterhoeven, *"Elecktrische Gasentladungslampen"*, Springer, Berlin (1938).

48. H. Cotton, *"Electric Discharge Lamps"* , Wiley, New York (1946).

49. W.E. Forsythe and E. Q. Adams, *"Fluorescent and Other Gaseous Discharge Lamps"* , Murray Hill Books, New York (1948).

50. J. Funke and P.J. Oranje, *"Gas Discharge Lamps"*, Philips Tech. Library, Elsevier Press, New York (1951).

51. L.B. Loeb, *"Fundamental Processes of Electrical Discharge in Gases"*, John Wiley, New York (1939).

52. J.D. Cobine, *"Gaseous Conductors"*, McGraw-Hill, New York (1941).

53. J.M. Meek and J.D. Craggs, *"Electrical Breakdown in Gases"*, Oxford Univ. Press (1953).

54. S.C. Brown, *"Basic Data of Plasma Physics"*, Wiley, New York (1959).

55. W. Elenbaas, *"The High pressure Mercury Discharge"*, Interscience, New York (1951).

56. W.C. Moore, *Trans. Electrochem. Soc.* **27** 435 (1915); See also: W.R. Mott, *Trans. Electrochem. Soc.***31** 365 (1917); D.B. Joy, F.T. Bowditch and A.C. Downes, *J. Motion Picture Engns.* **22** 58 (1934).

57. J.D. Cobine, *Gaseous Conductors*, McGraw-Hill, NY (1941).

58. W.W. Coblentz, M.J. Dorcas, and C.W., Hughes, *U.S. Bur. Stds. Sci. Paper #* *539*(1926); See also: N.K. Chaney, V.C. Hamister and S.W. Glass, *Trans. Electrochem. Soc.* **67** 017 (1935); C.G. suits, *Physics* **6** 315 (1935); W.C. Kalb, *Trans. AIEE*, **53** 1173 (1934); ibid, **56** 319 (1937); W.E. Forsythe, *Trans. Illum. Eng. Soc.* **35** 127 (1940); H.G. MacPherson, *J. Opt. Soc. Am.*, **30** 189 (1940); J.T. MacGregor-Morris, *J.AIE (London)* **91** 183 (1944); M.R. Null and W.W. Lozier, *J. Opt. soc. Am.* **52** 1156 (1962).

59. H.G. MacPherson, *J.Appl. Phys.* **13** 97 (1942)

60. W. Finkelnburg, *Zeit. Phys.***112** 305 (1939); ibid, **113** 562 (1939);ibid, ibid, **114** 734 (1940);**116** 214 (1941); ibid, **117** 344 (1942); ibid, **119** 206 (1943); ibid, **122** 36 (1944); ibid, **122** 714 (1944).

61. W. Finkelnburg, *J. Appl. Phys.,***20** 468 (1949); W. Finkelnburg and S.M. Segal, *Phys. Rev.* **80** 258 (1949); ibid, **83** 582 (1951).

62. W. Finkelnburg and J.T. Latil, *J. Opt. Soc. Am.* **44** 1(1954).

63. N. Riehl, *Z. Angew. Phys.* **7** 582 (1955).

64. H. Maecker , *Zeit Phys.***114** 500 (1939).

65. H.K. Bourne, *Discharge Lamps for Photography and Projection*, Chapman & Hall, London (1948).

66. L.J. Buttolph, *G.E. Review* **23** 741 (1920); ibid, **23** 858 (1920); ibid, **23** 909 (1920); ibid, **42** 160 (1939); *Rev. Sci. Instr.***1** 487 (1930); *Illum. Engng.* **33** 161 (1939); R.C. Kelting and L.J. Buttolph, *Illum. Engn.* **33** 161 (1938).

67. *La Lumiére Electrique*, **11** 410 (1910);   **25** 306 (1914); M. LeBlanc, *Bull. Soc. France Elect.***1** 89 (1921).

68. R. Küch and T. Retschinsky, *Ann. d, Phys.* **20** 563 (1906), ibid, **22** 595 (1907).

69. L. Tonks, *Physics*, **6** 294 (1942); *Phys. Rev.* **50** 226 (1936); See also: C.G. Smith, *Phys. Rev.* **62** 48 (1942); J.R. Haynes, *Phys. Rev.***73** 891 (1948); K.D. Frome, *Proc. Phys. Soc.(London)* **62b** 805 (1949); C.J. Gallagher, *J. Appl. Phys.* **21** 768 (1950); R.M. St. John and J.G. Winans, *Phys. Rev.* **94** 1097 (1954); R.V. Bertele, *Brit. J. Appl. Phys.* **3** 127 (1952); D. Zei and J.G. Winans, *J. Appl. Phys.* **30** 1813 (1959); J.H. Rich, *J. Appl. Phys.* **32** 1023 (1961).

70. J.M. Somerville, *The Electric Arc*, MacMillan, New York (1959).

71. O. Kruh, *Electrotech. u. Maschinenbau* **29** 615 (1911).

72. W. Gurski, *Lichttechnik* **12** 663 (1938).

73. J.G.W. Mulder, *Philips Tech. Rev.* **1** 65 (1936).

74. E.L. Harrington, *J. Opt. Soc. Am.* **7** 689 (1923); See also: J.H. Vincent and G.D. Biggs, *J. Sci. Instr.* **1** 242 (1924).

75. H. Geroges, *Comtes Rend.* **170** 458 (1920).

76. E.Edels, *Brit. j. Appl. Phys.***2** 171 (1951).

77. R.J. Strutt, *Proc. Roy. Soc.,***89A** 68 (1913); See also: R.E.B. Makinson, J.M. Somerville, K.R. Makinson and P. Thonemann, *J. Appl. Phys.* **17** 567 ((1946).

78. B.O. Baker and K.G. Cook *Brit. J. Appl. Phys.* **13** 603 (1962).

79. H.E. Edgerton and K.J. Germeshausen *Rev. Sci. Instr.* **3** 535 (1932); See also: T.S. Gray and W.B. Nottingham, *Rev. Sci. Instr.* **9** 105 (1938); M.A. Townsend, *J. Appl. Phys.* **12** 209 (1941);W.B. Nottingham, *Rev. Sci. Instr.* **14** 161 (1947).

80. K.J. Germeshausan, *Phys. Rev.* **55** 228 (1939); See also: K.J. Germeshausan, U.S.P. 2,325, 603 (1943); U.S.P. 2, 398, 422 (1946); N. Warmholtz, *Philips Tech, Rev.* **8** 346 (1946); loc. cit. **9** 105 (1947/48); ibid *Philips Res. Rpt.* **2** 426 (1947).

81. J. Slepian and L.R. Ludwig, *Trans. AIEE* **52** 693 (1933); See also: J.M. Cage, *G.E. Rev.* **38** 464 (1935); A.H. Toepfer *Trans. AIEE* **56** 810 (1937); G. Mierdal, *Wiss. Veroff Siemans Werke* **15** 35 (1936); E.G.F. Arnott, *J. Appl. Phys.* **12** 660 (1941); N. Warmoltz, *Philips Res. Repts.***6** 388 (1951); W.W. Rigoud, *J. Appl. Phys.* **22** 787 (1951).

82. L. Tonks, Phys. Rev. 48 562 (1935); ibid, *J. Franklin Inst.* **221** 613 (1936); J. Frenkel, *Phys. Zeits Sowjetunion* **8** 675 (1935).

83. P.L. Copeland and W. H. Sparing, *J. Appl. Phys.* **16** 302 (1945); See also: C.W. Lufcy and P.L. Copeland, *J. Appl. Phys.* **16** 740 (1945); P.L. Copeland, *Rev. Sci. Instr.* **16** 154 (1945); J.D. Cobine and G.A. Farrell, *J. Appl. Phys.* **31** 2296 (1960); G.A. Farrell and G.H. Reiling, *J. Appl. Phys.* **32** 1528 (1961).

84. H. Krefft, *Z. Techn. Phys.***15** 554 (1934); See also: ibid, *Z. Techn. Phys.***19** 345 (1938); ibid, *Tech Wiss.Abhandl. Osram Konzern.***4** 33 (1936.)

CHAPTER 2

DESIGN PRINCIPLES RELATING TO LAMPS AND ARTIFICIAL LIGHTING

In this chapter, we will examine principles relating to optimal design of lamps capable of providing illumination through conversion of electrical energy to visible light. Such lamps include the following classes:

1. Incandescent lamps
2. Mercury discharge lamps
3. Sodium discharge lamps
4. Other gaseous discharge lamps

2.1.- General Principles Relating to Design of Incandescent Lamps

The lifetime of an incandescent lamp is the most important factor relating to its usage. How long it can be used is determined by the elapsed time required for filament "burnout" or failure to occur. This in turn is limited by the rate and degree of evaporation of material from the hot filament. Many studies of filament life have been undertaken (1). Generally, it has been found that burnout occurs when the filament diameter has been reduced by a certain critical percentage, which for tungsten filament lamps is about -10% for vacuum lamps and -3% for gaseous-filled lamps. A "hot-spot" in the tungsten wire then develops and the filament is destroyed by localized melting. Since the amount of metal in the filament varies as the square of the diameter, while the evaporation rate is directly proportional to the surface area present, i.e.- the first power of the diameter, it is clear that the larger the filament, the longer will be the effective life of the lamp. Because of this factor, higher wattage lamps will always have an advantage over smaller ones. Additionally, the fact that smaller gas losses occur with larger filaments contributes to their longer life. Obviously, the tungsten wire diameter, which is controlled during the wire-drawing process, is the major consideration here. Thus, larger sizes of wire are used to make larger filaments, and hence larger wattage lamps. We will discuss the wire-drawing process in a later chapter.

The steps involved in the manufacture of incandescent lamps are shown in the following diagram:

2.1.1.-

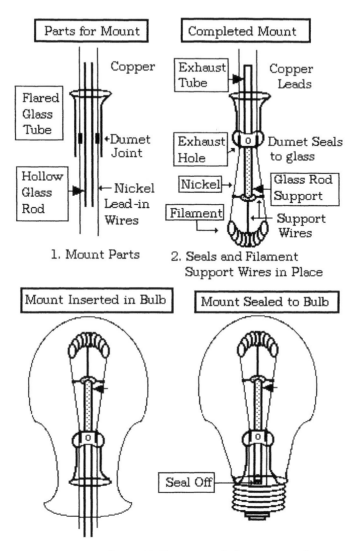

1. Mount Parts

2. Seals and Filament
Support Wires in Place

3. Mount inserted into Bulb;
Sealed to Flare; Exhausted of
air; Refilled with inert gas;
Sealed off; Ready for Basing

4. Lamp complete
with Base

Note that the filament mount parts are first assembled, the lead-in wires are
sealed-in via Dumet metal to the glass flare, and the filament support wires
(tungsten) are attached. Then the filament is attached to the support wires

and welded to the nickel lead-in wires. The completed mount is next placed within the glass bulb, and the bulb is sealed, evacuated, refilled with inert gas and then sealed off. Attachment of the base then completes the lamp.

According to Elenbaas (1), the rate of evaporation of an incandescent tungsten filament in a gaseous atomsphere is related to the heat dissipation by conduction and also convection in the gas, since both processes occur in the Langmuir layer surrounding the filament. Furthermore, the rate of evaporation, multiplied by the gaseous pressure is proportional to the dissipated heat, as confirmed by experiment. It was found, however, that as the pressure fell, the heat dissipation reached a constant. This was due to the fact that at this pressure the Langmuir layer had extended itself almost to the walls of the bulb. When further reduction in pressure occurred, the thickness of the layer remained the same. The result was that both the heat dissipation and the product of rate of evaporation and pressure remained the same. It is for this reason that many of the incandescent lamp filaments have been modified in form and position in the lamp, as shown in the following diagram:

2.1.2.-

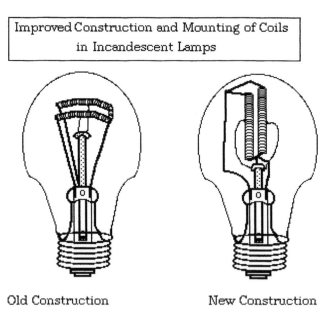

Improved Construction and Mounting of Coils
in Incandescent Lamps

Old Construction   New Construction

In this "new" construction, the vertical coils have improved heat dissipation, and hence longer lamp life. This also resulted in an increase in lumen output of 1650 lumens for a 100-watt lamp, compared to 1580 lumens for the old construction.

Some of the other factors relating to incandescent lamp operation are shown in the following Table, (which is Table 1-7 of the last chapter):

TABLE 2-1

Wattage* & Filament	Luminous Efficiency - lumens/watt	% Output in Visible	Gas Loss in %	End Loss in %	Base + Bulb Loss in %	Filament Heat Loss in Joules
500 C	19.8	12.0	8.8	1.8	7.1	182.0
300 C	18.8	11.1	11.6	1.8	6.8	80.0
200 C	18.5	10.2	13.7	1.7	7.2	39.7
100 CC	16.3	10.0	11.5	1.3	5.2	14.2
60 CC	13.9	7.5	13.5	1.2	4.5	5.5
40 C	11.6	7.4	20.0	1.6	7.1	2.5
25 V	10.6	8.7	----	1.5	4.5	2.8
10 V	7.9	7.1	----	1.5	5.0	0.62

* C = coil; CC = coiled coil; V = vacuum.

It is clear that the luminous efficiency is considerably higher for the higher wattage lamps than the smaller ones. However, so is the filament heat loss which is related to the amount of material that may be evaporated from the filament. Since the luminous efficiency and the evaporation rate of the filament both increase with increasing temperature, it is relatively easy to trade off efficiency for long life, and vice-versa. We will address the methods of producing tungsten "coiled-coils" somewhat later on.

A. Design of Lamp Life

In "GENERAL SERVICE" lamps, a life of 750-1000 hours is normally considered acceptable as the best compromise between efficiency and lamp life, power costs, lamp costs and other factors (2). Some lamps are rated at 1350 hours, while "extended service" lamps can have a life of 2500 hours, or longer. Note that we are speaking of the "average" life of such lamps. In reality, the life of a specific lamp is determined by many factors, and the individual lives of an assemblage of lamps falls under a Gaussian curve where some lamps have much shorter lives than the mean, and others much longer life than the mean life of the lamps being measured. The trick is to get the individual lives of the lamps to fall under a narrow curve. It is no trick, however, to make lamps with essentially infinite life but very low efficiency by simply lowering the filament temperature. As shown in the following diagram, one can design a lamp for a given set of operating conditions, viz-

2.1.3.-

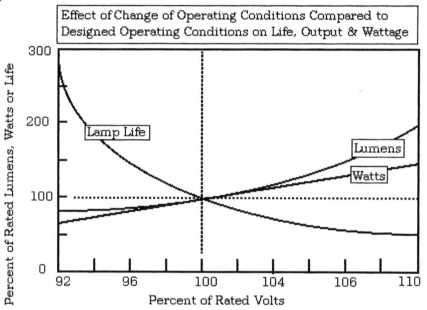

If one then operates the lamp either above, or below, its rated voltage, one obtains changes in both wattage of operation and lumen output, as well as a change in life, compared to the values expected for its designed operation. Note that it is the life of the lamp which is affected most seriously by a relatively small change in operating voltage. On the other hand, the efficiency of such lamps may be easily doubled if a life of only a few hours is satisfactory.

Some data for typical commercial lamps are given in the following Table, as shown on the next page, viz-

TABLE 2-2

Characteristics of Some Commercial Incandescent Lamps (115 Volts AC)

Type of Bulb	Wattage	Efficiency in lumens/watt	Designed Life in hours	Color Temp. in °K.
PHOTOFLOOD				
No. 4 (PS 35)	1000	32.2	10	3400
No. 2 (PS 25)	500	34.0	6	3400
No. 1 (A 21)	250	34.4	3	3400
MOTION PICTURE STUDIO				
G96	10,000	33.0	65	3350
G64	5,000	33.0	75	3350
G48	2,000	28.0	100	3200
G40 (PS25)	1,000	28.0	50	3200
A23	500	26.0	60	3200
GENERAL LIGHTING SERVICE				
PS52C	1,000	21.5	1000	2994
PS40C	500	19.8	1000	2994
A25CC	200	18.5	750	2894
A21CC	100	16.3	750	2849
A19CC	60	13.9	1000	2772
A19C	40	11.6	1000	2750
A19C(VAC)	25	10.6	1000	2583
S14C(VAC)	10	7.9	1500	2422

In the above Table (as before), C= coil, CC= coiled coil and VAC = vacuum lamps. Note also that some photo-flood lamps are rated for only a few hours, i.e. 3 to 10 hours; those for the motion-picture studio are still only rated for up to 100 hours. These are examples of high efficiency-high intensity lamps which are used and then replaced when the filament fails. However, for General Service lamps, the mean-life is usually designed at 1,000 hours.

The following diagram shows the luminous efficiency of tungsten filament lamps of various constructions as a function of design life and filament current. Since the filament diameter i.e.- the resistivity of the wire, is the

2.1.4.-

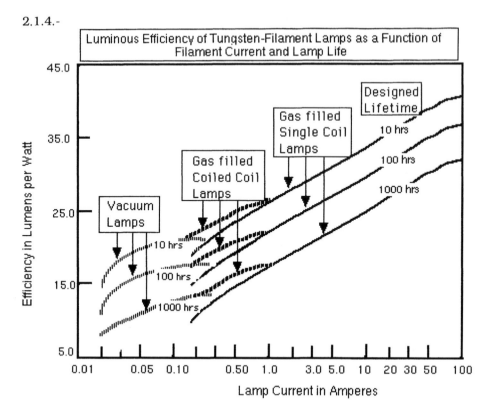

most important factor in determining the losses in the lamp, the required heating current is of more interest than the lamp voltage or wattage. It is easy to see in the diagram that vacuum lamps fall into the 0.02 to 0.15 ampere range, while gas filled coiled-coil lamps fall into the 0.10 to 1.00 ampere range. For high brightness (efficiency) lamps, one uses the 1.0 to 100 ampere range with a single coil winding for the filament. Obviously, this diagram is a gross oversimplification of the design parameters needed to produce a given tungsten filament lamp, but it does aptly illustrate the point.

It should be clear that coiled-coil lamps used for general service are only slightly better over a small range of filament currents than vacuum lamps with single coils . For small deviations of less than 10% of the normal voltage, the life of a gas-filled tungsten filament varies as the - 13.1th power of the voltage while the efficiency varies as the 1.8th power. The total light output varies as the 3.38th power of the voltage (3,4). At the end of rated life (at constant voltage operation), the **efficiency** of such a lamp is about 86% of its initial value while the **total light output** is only about 82% (3). This loss is partly due to bulb blackening, but also in part due to the higher filament temperature because of the increase due to evaporation of tungsten from the filament.

Although the largest commercial incandescent lamp normally constructed has been about 10 kilowatts, experimental lamps as large as 75 kW. have been made. Such lamps had an envelope about 20 inches in diameter and were 42 inches high, with a filament weighing 2.7 lbs. They weighed about 50 lbs. The output was 2.4×10^6 lumens at an efficiency of 32 lumens/watt. In contrast, some of the smallest commercial lamps are the so-called "grain-of-wheat" lamps which are 2 mm in diameter and 8.7 mm long. Additionally, lamps as small as 0.4 mm in diameter and 1.7 mm long have been manufactured (called "pinlites") which weigh 2.5 mg. with an output of 60 milli-lumens for a wattage of 22.5 milli-watts (or an efficiency of 2.7 lumens/watt with a filament temperature of 2173 °K) . Even smaller lamps have been made to fit into digital watches (with liquid-crystal displays) so that the current time can be read under darkened conditions.

B. Lamp Life and Internal Blackening

Blackening in vacuum lamps is uniform over all of the bulb area, but, in gas-filled lamps, it is localized at the top of the bulb as a result of the internal gas convection currents. If gas-filled lamps are operated base-up, the deposited metal collects mainly in the neck and base area and is not as troublesome as for horizontal or base-down operation. Some very large incandescent lamps contain a quantity of loose tungsten powder which can be used periodically to scrub the deposit off the glass walls by mechanical action, so as reduce the light absorption.

Projection and other high power lamps sometimes contain a wire-mesh screen connected to one side of the filament. The use of such screens is very old, three patents having been issued to Edison on their usage between 1881 and 1883 (5). In 1936, they were rediscovered by Wright (6) who applied them to reduce blackening because it was believed (7) that they collected evaporated charged tungsten atoms, thus preventing them from reaching the bulb wall. It is true that positive ions are emitted from the hot tungsten filament (8), but not in the quantity previously asserted. The following thermodynamic equation is applicable to this case:

2.1.5.- $\qquad \Phi_+ + \Phi_- = \Phi_0 + V_i$

where Φ_+ = "work function" for positive ion emission, Φ_- = normal electron work function, Φ_0 = atomic work function and V_i = ionization energy. For tungsten:

$$\Phi_- = 4.52 \text{ e.v.}$$
$$\Phi_0 = 8.3 \text{ e.v.}$$
$$V_i = 8.1 \text{ e.v.}$$

This gives a result for Φ_+ as 11.9 e.v. which is a very high value. Reimann (9) gives the following results:

2.1.6.- $\qquad \log V \text{ (gm./cm}^2 \text{ sec.)} = 7.814 - (4.1 \times 10^4)/ T$

and: $\qquad \log i_+ \text{ (amp./cm}^2) = 6.95 \times 10^4) /T$

One can then obtain the ratio of tungsten ions (N_+) to atoms (N_a) as:

2.1.7.- $\qquad \log(N_+ / N_a) = -3.5838 - 3120 / T$

Thus, at normal filament temperatures, only one out of about 40,000 evaporated atoms is charged. It is therefore difficult to explain the action of the screen on this basis. The actual action of these screens appears to be the collection of neutral atoms by means of the convection currents always present in the gas sheath. Therefore, in practice, the screen is always mounted **above** the filament (3,5). As additional proof, Leighton and Makulec (5) found no difference if the screen were isolated electrically from the filament or connected to the positive or negative end (for the case of D.C. operation).

C. Seals and Getters

A very important factor in lowering the price of incandescent lamps was the development of materials to eliminate the use of platinum seals as used in early lamps (10,11). Although many different materials were tried, "Dumet" wire, as developed in 1912 by Colin G. Fink of G.E., was the best and is still used today in all soft-glass lamp seals. This material consists of a nickel-iron core, with a sheath of copper. The leads are copper while the filament support leads are nickel (see 2.1.1. given above).

Still another important development, from the viewpoint of cost through reduced exhaust time and of quality (as exhibited by the reduction of bulb blackening), was the use of "getters" to react with residual (and potentially harmful) gases in the lamp. As early as 1882, Fitzgerald used a strip of Mg-metal connected to a separate third terminal which was heated to combine with residual oxygen. In 1894, Arturo Malignani used red phosphorous in Italy to improve the vacuum in normal lamps. The patent rights were later bought by G.E. in 1896. Use of this technique resulted in lamps which could be exhausted in one minute with better results than had been formerly obtained in 30 minutes. Originally, the phosphorous was placed in the exhaust tube and driven into the lamp by heating. In 1909, John Marshall of G.E. coated the filament itself with a phosphorous and water mixture. After the lamp had been sealed off and the coating was vaporized, the residual pressure dropped from 25μ to less than 1μ. The materials, SiO_2, Al, MnO, As, S and I_2, also function as getters, but not in the same manner. The most probable action of these materials is to cause adsorption at the glass wall. The presence of a glow-discharge when the getter is flashed apparently also plays a role (12). This area needs more study.

A different type of getter was invented in 1910 by the Austrian, Franz Skaupy (11). He used K_3TlCl_6 in a hollow of the glass filament support. Other halide compounds such tungsten oxychloride may also be used. The use of gaseous bromine in early carbon-filament lamps was mentioned in the previous chapter. In the case of tungsten-filament lamps, the halogen combines chemically with the evaporated metal and produces a solid deposit which is lighter in color than the metal, thereby obviating "blackening". The compound, KI, was also used by Colin G. Fink in 1912 for the same purpose, In the same year, Harry Needham of G.E. used cryolite, Na_3AlF_6, for the same function. $Ba(ClO_3)_2$ was used by Ernst Freidrich in Germany in 1913. F.W. Gill of G.E. employed NaCl in 1915. Additionally, barium azide (i.e.- BaN_6), CaF_2, $Ca(BF_4)_2$ and KBF_4 have all been used as getters. However, all of these getter-compounds were used in conjunction with phosphorous. The action of the fluoride or cryolite getter apparently is to prevent agglomerization of the tungsten deposit on the glass wall as a "metallic" deposit.

The more-or-less standard lamp getters used today are red phosphorous plus cryolite. However, some manufacturers use Na_3FeF_6 and K_3FeF_6. Other improvements in incandescent lamps have been:

> 1. Exhaust through the base rather than through the bulb (which leaves a tip at the top) results in a better appearing lamp. This enhancement was applied in the early 1930's.

> 2. The use of etching, i.e.- "inside-frost", or application of a film of finely-divided silica on the internal glass wall of the bulb to diffuse the light resulted in better appearing and more useful lamps. Moreover, the efficiency of the lamp is not affected.

> 3. The emission color of an incandescent lamp may be obviously modified by use of a colored glass envelope, an external pigmented coating, or addition of a pigment to the internal silica coating. All of these are in use today. Such lamps are used for decorative purposes or to improve color-rendition, ex.- "Daylight" lamps with blue envelopes. However, the measured efficiency of such lamps is obviously affected, although the filament is still operated under design conditions.

Indeed, the "Soft-White™" lamps being marketed today by G.E. incorporate a number of these "improvements" (even though they are obviously used to generate sales rather than to improve quality).

Present day incandescent lamps are still far from the theoretical maximum efficiency of 95 lumens/watt, largely because of filament temperature limitations. The following Table, given on the next page, illustrates this predicament. Note that tungsten has one of the highest melting points of all of the elements, and that only the carbides of Ta and Hf exceed this temperature. Although efficiencies as high as 32-34 lumens/watt may be obtained for tungsten filament lamps, they are accompanied by short life.

TABLE 2-3

Thermal Properties of Metals and Compounds Potentially Useful For Filaments in Incandescent Lamps

Material	Melting Point (°C.)	Temperature for Vapor Pressure = 1 x 10⁻⁶ mm Hg	Melting Point , °C			
			Carbide	Nitride	Boride	Silicide
U	1130		2475			
Pd	1552					
V	1700		2830	2050	2100 d	1750
Pt	1769	1759 °C.				
Ti	1800		3250	2930	2980	1540
Th	1850		2655			
Zr	1960		3175	2980	3040	1520
Rh	1960					
Hf	2207		**3890**	**3310**	**3259**	
B	2300			3000		
Nb	2415	2326	3500	2050	2900 d	1950
Ir	2454					
Ru	2500					
Mo	2610	2048	2690		2180	1870 d
Os	2700					
Ta	2996	2513	3880	3090	3000 d	**2400**
Re	3180					
W	**3410**	**2860**	2700		2860	2150
C	3670	2261			2450	2600
ZrO₂	2715					
ThO₂	3050					

At a "reasonable" life of 1000 hours, one only obtains about 20-22 lumens/watt for very large lamps and 16 lumens/watt for the usual 100 watt size. In the latter case, the emission peak lies almost exactly at 1μ, i.e.- 10,000 Å, while only 10% of the total radiated energy lies in the visible region of the spectrum. For this reason, the search to find better materials for incandescent lamps has continued throughout the years, although it has been clear from the start that serious material limitations do exist.

D. Iodine Cycle Lamps

A more recent improvement in incandescent lamp design has been the introduction of iodine vapor into some types of commercially produced lamps. This resulted from prior studies of the effects of bromine and chlorine to reduce blackening, as discussed previously. At GEC in England, iodine was introduced earlier in lamps because of its effect in reducing arcing. However, considerable manufacturing difficulties were encountered. In 1959, Zubler and Mosby of G.E. (13) found that if conditions were adjusted properly, the maintenance of lamp output would be almost 100% throughout life when I_2 is added to the lamp. A regenerative cycle is set up within the lamp where the evaporated tungsten metal reacts with the I_2 to form WI_4 , which in turn reacts with the hot filament to deposit tungsten again. Such action prevents or slows "hot-spots" from forming during the lifetime of the lamp. The cycle is represented by:

2.1.8.- $W + I_2 \leftrightarrow WI_4$

If the temperature of the bulb wall is high enough, this reaction proceeds in the forward direction and so prevents formation of tungsten deposit. The higher wall temperature also serves to keep the iodine vaporized. It has been found that the bulb wall temperature must be at least 250 °C., with about 600 °C. being optimum. This means that the bulb or envelope must be comparatively small and constructed either of quartz, Vycor™ or Pyrex™ glass. Lamps of this type are therefore usually cylindrical with axial coiled-coil filaments. Since the bulb diameter is only slightly larger than the Langmuir sheath, gas convection does not occur. Such lamps are usually operated in a horizontal position to prevent separation of the iodine vapor from the fill gas by thermal diffusion. The high operating temperature of the bulb also presents problems in electrode seal design since it has been found that cold metal surfaces must be avoided. The cycle discussed above can occur with any of the halides.

The temperatures at which the reaction occurs rapidly in both directions are as follows, viz-

2.1.9.-   Halide   Required Bulb Temperature

F_2	3500 °C
Cl_2	2000 °C
Br_2	1600 °C
I_2	850 °C

With the exception of fluorine, the normal filament temperature is clearly higher than the required bulb wall temperature. Nevertheless, it is the effect of the cold-ends of the filament that makes the use of iodine practical. With chlorine, for example, metal would be transferred from any part of the filament cooler than 2000 °C. to the hot part. This would result in failure of the filament at the cooler ends. With iodine, this end effect is alleviated. However, temperature variation over the length of the filament must still be minimized, and as a result such filaments are designed to be rather short in order to slow transfer of tungsten from hotter portions to those parts at lower temperatures. This overall transfer from hot to cooler regions occurs because evaporation is naturally greater at the hotter regions while redeposition is more-or-less uniform over all of the filament length regardless of temperature. According to van Tijen (14), the life of an iodine-lamp is proportional to:

2.1.10.-   Life = $f([T_m]^{-32} / \Delta T)$

where T_m is the maximum temperature of the filament and ΔT is the temperature spread from the hottest to the coldest part. The importance of ΔT is obvious. The normal iodine concentration employed is about 0.25 $\mu moles/cm^3$ of internal volume. The presence of other metals such as Hg, Ni, Fe, Ta and Mo interfere with the desired cycle and hence must be avoided as impurities or as filament supports, etc. in these lamps. Getters also cannot be employed. Most iodine lamps contain a rare-gas at a cold-fill pressure near atmospheric pressure, i.e. 600 mm Hg. Because of the higher operating temperature as compared to normal lamps, the internal operating pressure in this case is about 1800 mm. Hg, or 2.4 atmospheres. A 500-watt, 120-volt lamp is only about 95 mm. long and 10 mm. in diameter. The efficiency for a 1000 hours of design life is 21 lumens/watt, compared to 19.8 lumens/watt for a standard 500 watt lamp which is much larger. In the case of a 1500 watt, 277 volt lamp (10 inches long and 0.50 inches in diameter), the efficiency is 22.0 lumens/watt, an increase of 25% over that of a comparable non-iodine lamp. Maintenance of output is also much improved, being closer to 95-98%% instead of 80-82%. Efficiency varies as the fill pressure varies to the 0.12th power, in contrast to the 1.8th power for normal lamps. Internal pressures as high as 10.0 atmospheres have been employed, yielding a further gain of 30% in efficiency. For a designed lifetime of 5 hours (photo-flood operation) an efficiency of 45 lumens/watt has been obtained. However, it is well to note here that the cost of iodine-lamps is greater than that of standard lamps. Iodine lamps have, of late, found their greatest usage as automobile headlights where both efficiency and maintenance of output is essential. In 1992, iodine-lamps are beginning to be used for general lighting where the fill-gas is krypton and 10,000 hours life are quoted. Such lamps, manufactured by Philips, have integral ballasts built in.

E. Thermal Reflectors for Incandescent Lamps

The radiation emitted by an incandescent source includes both near-ultraviolet, visible, and near-infra-red radiation. Most of the radiation is in the near-infra-red part of the spectrum (only about 10% is in the visible). Both ultraviolet and infra-red, obviously, add nothing to the visible output, but are major dissipators of power input. Hence, generation of radiation in these parts of the spectrum lowers the overall luminous efficiency of the lamp. Even the radiation near the extremes of the visible region, i.e.- 4100 and 7800 Å, adds little to the luminosity of the filament.

If a blackbody were to be enclosed in a perfectly reflecting shell which reflected and focused all non-visible radiation back to the emitter where it was absorbed, while transmitting all of the visible radiation, **then** the luminous efficiency would be increased. (Note that the absorptivity in this case, according to Kirchoff's Law, is equal to the emissivity, and is therefore unity). This would result in an increase in the temperature of the emitter, if the power input were maintained in the same manner as in the absence of the reflector (This arises because the emitter would be required to dissipate the excess energy into the restricted visible wavelength range, and not the infra-red or ultraviolet). Basically, this would be a way of converting both ultraviolet and infra-red radiation into visible light by thermal conversion. Alternatively, if the emitter temperature was to remain the same when the reflector was installed, then the power input would have to be reduced.

Hisdal (15) made calculations for such perfect reflectors in 1962. If the transmitted range is made to be 3800-7600Å, the expected maximum efficiency is about 203 lumens/watt and occurs at an emitter temperature of about 4800 °K. This is 2.14 times that of the same lamp with no reflector, operating under the same conditions. (Note that such lamps have a maximum efficiency of 95 lumens/watt at 6600°K). For a narrower range of 4100-7100 Å, the corresponding values are 245 lumens/watt and 4000 °K. If a constant temperature of 3000 °K is considered (as is found in the normal tungsten lamp), the numbers are as shown in 2.1.10. A large increase in efficiency is thus predicted.

2.1.11.-	Case	Range	Efficiency	
	No reflector	3500-12,000 Å	20.7	Lumens/watt
	Reflector	3800-7600	176.0	
	Reflector	4000-7100	232.0	

Since the emission in the ultraviolet from a 3000 °K. filament is quite small, it was also of interest to consider not only a "bandpass" system but one in which a "cutoff" point of reflection has occurred. In this case, the maximum efficiency found by Hisdal was 410 lumens/watt for a cutoff wavelength of 6000 Å . If the hypothetical blackbody is replaced by a **real** radiator like tungsten, the result is essentially the same, i.e.- 407 lumens/watt, as compared to 28.6 lumens/watt when no reflector is present.

These predicted gains are impressive, but difficult to realize in practice. Real reflectors do not have 100% reflectivity beyond the cutoff wavelength, 100% transmission above it, nor a sharp transition between the two regions. Furthermore, deviations from 100% reflectivity become very important when the emitter itself has an emissivity and absorptivity of only 30-40% (as is true for real metals). Thus. for a blackbody emitter in a reflector where r= 90% for long wavelengths and 10% for short wavelengths, the maximum efficiency would be 125 lumens/watt instead of the 410 lumens/watt found for the case of the 100% cutoff frequency, i.e.- when the cutoff wavelength changes more slowly in the wavelength of interest, the maximum efficiency achievable is corresponding smaller. And for tungsten, this value falls to 84 lumens/watt. Additionally, the reflected radiation should ideally be completely focused back on the emitter. But for the usual bulb shapes, this presents problems, even for the case of the iodine-lamp which is cylindrical.

Studer and Cusano (16) performed experiments on the use of such thermal reflectors several years before Hisdal's calculations were published. They used films of TiO_2 deposited on the glass envelope. Films of optimum thickness had a reflectivity of 8% at 5500 Å and 30% at 11,000 Å. Double films, i.e.- one on the outer and one on the inner surface, can reflect 40% of the infra-red radiation from a tungsten filament while reflecting only 8% at 5500 Å. This resulted in increases of up to 20% in output for filaments operating in vacuum, or a reduction of 10% in power input for the same light output. However, they stated:

> "Because this result was obtained only when non-frosted spherical bulbs with accurately positioned filaments were used, it is questionable whether this process will have any practical application in connection with incandescent lamps".

Nonetheless, they did not anticipate certain advancements in how such films would be deposited in the future. It is this achievement that accounts for the success of this approach. The most recent announcement in this area was made by G.E. in 1991 (17), via a commercial announcement, viz-

> "G.E. scientists have pioneered the first truly practical way to coat a lamp with a selective light filter that lets visible light through while blocking infra-red and reflecting it back onto the filament. The reflected IR radiation sharply lowers the amount of electrical current required to keep the filament hot enough to produce the light, resulting in energy cost savings of up to 60% over conventional incandescent lamps. The technology has found its first commercial application in a new family of incandescent lamps manufactured by G.E. Lighting - Halogen IR PAR 38 lamps. the new lamp, available in spot and flood versions, has been in production since January 1990".

"The lamp consists of an inner iodine-lamp, made from quartz, nested in an outer reflecting bulb. G.E.'s IR-coating is placed on the outer surface of the quartz bulb, which is precisely shaped to provide focusing onto the hot filament. The result is that the new IR PAR 60 watt lamp can provide the same amount of useful light as a standard 90 watt iodine-lamp, or a 150 watt PAR standard (parabolic aluminum reflector) floodlight. Furthermore, the new lamp offers 33% energy cost savings over the standard iodine-lamp, and 60% energy cost savings over conventional 150 watt tungsten-filament floodlamps. The reflective IR-coating technique uses chemical vapor deposition (CVD) to deposit alternating layers of high- and low-refractive index material on the outer surface of the quartz bulb".

It should be clear that the technology described above has been ongoing for many years. But, it was only the advent of CVD techniques, originally used to manufacture integrated circuits by deposition of both passive and active layers of materials on the silicon substrate (and its attendant knowhow concerning mechanisms of chemical interactions occurring in the formation of thin films) that made possible the use of such films. Essentially, a reflecting interference layer is formed in place, where the composition and thickness of each alternate layer is important.

F. Other Filament Materials

Rhenium (Re) is a refractory metal intermediate to W and Os in the periodic table. It also has the highest melting point, next to tungsten (W). It was only discovered in 1925 and hence was not available to early lamp workers or it certainly would have been tried by them. In 1991, the cost was still 350 times that of tungsten. The vapor pressure of Re is higher than that of W but it is more resistant to the "water-cycle", that is- it does not react with water when it is at high temperatures to the degree that W does. Its electrical resistivity is higher than W but its thermal emissivity is slightly lower. Re suffers work-hardening to a greater extent than any other metal, which makes it much more difficult to draw into wire. Although experimental incandescent filaments of Re have been made (18), the lamps seemed to offer little advantage over W-filament lamps, if any. On the other hand, its resistance to the water-cycle has made it useful for electron-emitting filaments in ionization-type vacuum gauges and in mass spectrometers. G.E. has claimed that W-Re alloys were superior as ignitors in photoflash lamps because of this property. However, this has not been confirmed by production of any commercial lamps.

It has seemed obvious that if a better incandescent emitter than tungsten is to be found, it is the refractory compounds that will perform better, rather than the metals. Various oxides, nitrides, borides and carbides have been considered for this purpose. Melting points for some of these compounds were given in Table 2-3. The two most interesting are TaC and HfC, both of which have melting points about 500 °C. above that of tungsten. However, melting points (or vapor pressure) are not the only criteria for incandescent emitters. Fok (19) has commented on the superiority of semiconductors as incandescent emitters because of lack of absorption in the infra-red (and hence emission) if the forbidden energy gap is great enough, i.e. - > 1.8 e.v. Infra-red absorption due to lattice vibrations could be minimized, **if** the right substrate were to be found. Also, it is necessary that non-fragile filaments of appropriate resistivity can be prepared reproducibly. This last requirement rules out oxides and ceramics to a great extent. (The extensive work being carried out for high-T_c ceramic superconductors may aid in this area). Additionally, brittleness of the carbides still is great enough to present a major stumbling block in the preparation of a lamp. To avoid the problems of making TaC filaments, Peek (20) described a lamp in which a massive piece of TaC (in argon atmosphere at two atmospheres pressure) was heated by radio-frequency induction. Such a lamp produced about 43 candles/mm^2 , which is equivalent to about 78 lumens/watt, with a lifetime of about 100 hours (A corresponding tungsten filament at a color temperature of 2850 °K. emits about 9.0 candles/mm^2). Young (21) described experimental lamps in which plates of W or TaC were heated by electron bombardment in a vacuum. These lamps required an auxiliary filament for starting, but once the plates became hot, they were able to supply the necessary thermionic emission and were self-sustaining. Unfortunately, TaC is not stable in vacuum above about 2500 °K. Young, as well as some Russian workers (22), described similar lamps where the plates were heated by a low pressure discharge. Again, there were problems of decomposition, evaporation and/or sputtering.

Because of their high temperature properties, TaC and related carbides have been extensively studied (23,24). Although TaC filaments may be made by heating Ta wire in a hydrocarbon atmosphere, their brittleness and stability have always presented considerable problems. TaC lamps were made by Siemans-Halske as early as 1902, but successful commercialization is still uncertain. Becker and Ewest made TaC lamps at Osram in Germany as early as 1930. Work was done on both HfC and TaC lamps at G.E. during the 1930's (25). In 1937, Russian workers were able to operate TaC filaments at 3300-3400 °K for 100 hours (26). In 1952, Cooper (27) improved the filament stability greatly by using a hydrocarbon **plus** hydrogen- gas mixture as the lamp atmosphere. The purpose of the carbon was, of course, to keep the Ta carburized, rather than reverting back to the lower melting point metal. Likewise, the purpose of the hydrogen was to prevent deposition of the carbon on the glass envelope and the filament supports (the hydrogen is atomic at the filament temperature). A life of 25 hours at 3600-3800 °K. was claimed, or 50 hours at 3500 °K. Unfortunately hydrogen increases heat dissipation considerably and such lamps did not prove successful. Additions of ZrC, HfC, etc. (made by using alloy wire) were claimed to improve performance. An atmosphere containing CN radicals (such as H_2 + HCN) was said to be beneficial, but the toxicity problems of manufacturing and public use of such lamps is not feasible.

Another development in TaC lamps has been the use of halogens. Cooper, Bird and Brewer (28,29) found that such additions prevented carbon deposition at a filament temperature of 3600 K. or higher (The limit for hydrogen-hydrocarbon mixtures has been found to be no more than 3600 °K beyond which catastrophic degradation occurred). Use of more than one halogen was found to be advantageous. "Routine" operation of TaC lamps at 3500-3700 °K produced "satisfactory" efficiencies and "useful lives" and "considerably higher intrinsic" brightness over tungsten was claimed, even though actual numbers given did not make these points clear. The complexity of these lamps, even if they were to become commercial, probably means that they would be used for special applications such as photography and studio applications. However, the fact that TaC is the first material to ever surpass tungsten since its introduction in 1906 is, in itself, rather remarkable.

Before we leave the subject of incandescent lamps, we should point out that the largest usage of such lamps, other than General Service lamps used in buildings and homes for lighting purposes, are those used in automobiles. Some eighty lamps are used as "tail and brake" lights, "instrument panel" lights and "headlights". A recent innovation (30) involves the use of a so-called "Light Engine" which involves the use of a **single** light-source coupled via fiber optics to the various points of illumination required in the lighting system, as shown in the following diagram:

2.1.12.-

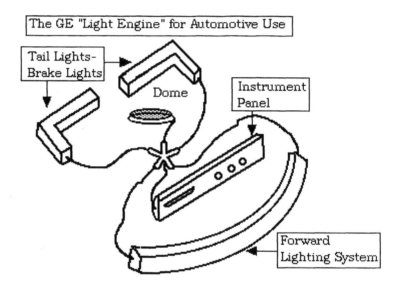

Obviously, the light source must be of very high brightness in order to be used both for forward lighting, i.e.- headlamps and taillights. However, the internal lamps need to be of much lower brightness, which is presumably obtained by attenuation of the fiber-optic cable supplying the light. Nonetheless, although such a system has been proposed, it has not yet made its way into the marketplace in 1992, even though such high-brightness lamps are available, as is the fiber-optic system.

Additionally, the advent of electronic recording on magnetic tape, including such devices as hand-held cam-corders (Camera-Recorders) has greatly obviated the need for high intensity lamps for film projection and similar uses. Indeed, the Television Industry uses such devices for on-the-scene recording of events and subsequent playback without the use of incandescent lighting. Only in the Studio itself is the very high intensity lighting required. The same may be said for the Movie Industry, except for movie screen projectors. Even these will become outmoded as time passes in favor of the very large cathode-ray tube projection screen, i.e.- Sony's "Jumbo-Tron", or similar devices.

2.2.- General Principles Relating to Design of Mercury Discharge Lamps

The element, mercury, has a density of 13.53 gram/cc^3, a melting point of 234.6 °K. (- 34.6 °C.), a boiling point of 630 °K. (357 °C) and an ionization potential of 10.44 volts. Normally, Hg is a liquid metal and sublimes at room temperature. Hg exhibits a definite vapor pressure, even at normal atmospheric conditions, and is classified as a toxic heavy metal whose presence in the environment cannot be tolerated where people are concerned. Although part of the Hg that may be absorbed by a given person is excreted in the urine, a significant part may remain and be stored within the internal organs. Thus, its health effects are long-term and those exposed to Hg fumes must be monitored on a continuous basis. Hg also has a substantial vapor pressure at elevated temperatures, as shown in the following diagram, given as 2.2.1. on the next page.

Note that at 1000 °C., the pressure may exceed 100 atmospheres. Even at relatively low temperatures, the vapor pressure of mercury is substantial. It is this property which has made possible both the low-pressure and high-pressure lamps referred to previously in Chapter 1. This is also the basis for present-day HPMV lamps, for xenon-mercury lamps and for current fluorescent lamps as well.

As we showed in the last Chapter, the luminous efficiency of Hg-lamps had been studied by Krefft (31) who determined the effect of mercury vapor pressure on the efficiency and voltage gradient of such lamps. These data are shown in the following diagram, as presented on the next page as 2.2.2., and is a repeat of that already presented as 1.3.1. in the first Chapter. The Cooper-Hewitt "low-pressure" lamps of Chapter 1 operated near to the peak "B" shown in 2.2.2, while the Küch "high-pressure" lamps operated at nearly one atmosphere and hence had higher efficiency.

2.2.1.-

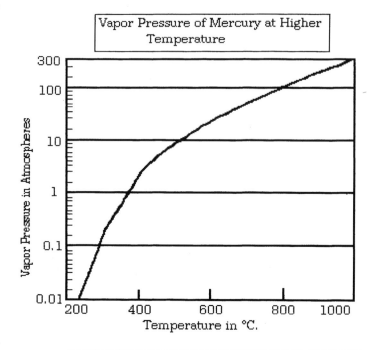

Vapor Pressure of Mercury at Higher Temperature

2.2.2.-

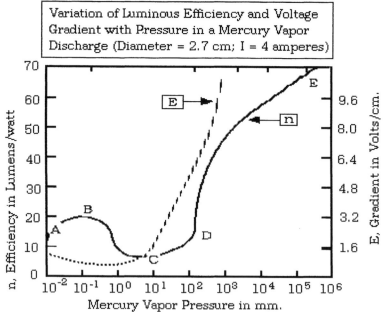

Variation of Luminous Efficiency and Voltage Gradient with Pressure in a Mercury Vapor Discharge (Diameter = 2.7 cm; I = 4 amperes)

Note that a minimum occurs near to "C", and that the efficiency increases rapidly beyond "D". Also shown is the voltage gradient which occurs in the discharge, as a function of temperature. Modern high-pressure-mercury-vapor (HPMV) lamps operate within the region "E" as shown above. The at-first rather peculiar shape of the curve shown in 2.2.2. may be explained on the basis of the fundamental processes occurring in the discharge.

In the following diagram a simplified energy level diagram of the mercury atom is shown:

2.2.3.-

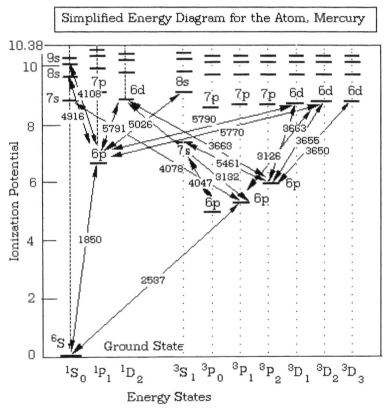

Simplified Energy Diagram for the Atom, Mercury

Not all of the transitions are depicted, just the main ones. It can be seen that the two "resonance" lines, i.e.- those that terminate on the ground state, are in the ultraviolet portion of the spectrum, and thus do not contribute to the luminous output of the discharge. The lines which contribute most to the visible output, i.e.- 4047 Å, 4358 Å, 5461 Å, and 5791 Å, all have the first excited level of 6^3P_0, 6^3P_1, 6^3P_2, or 6^1P_1 as their terminal state. Another group of strong lines in the near ultraviolet near to 3650 Å also have the 6^3P_2 level as their final state. It should also be noted that both 6^3P_0 and 6^3P_2 are metastable, i.e.- there are no allowed transitions to the ground state.

At very low pressures, the electron mean-free path in a discharge is very long. In a typical fluorescent-lamp discharge where the temperature is about 40 °C. and the internal pressure is about 6.0 microns, the mean-free-path is about 5.0 cm. Thus, if only mercury vapor were present, most of the energy would be lost to the walls. The presence of the rare-gas fill, at about 1-3 mm. of pressure, limits the mean-free-path to about 0.01 cm. Most of the electron-atom collisions are elastic, i.e.- only changes in kinetic energy occur, with no changes in energy level involved.

Thus, the maximum energy that can be lost by an electron is the fraction of its initial energy given by:

2.2.4.- $4 \, nM/(M + m)^2 \approx 4m/M$

where m and M are the mass of the electron and the atom, respectively. (If the atom is also in motion, there is an additional factor of $[1-(T_g - T_e)]$, where T_g is the temperature of the gas and T_e is the effective temperature of the electrons in the discharge). Since m/M is very small, i.e.- about 10^{-4} to 10^{-5}, the fraction of energy lost per collision is also very small. Because the ionization potential of Hg is much less than most of the rare gases, the infrequent electronic excitations and ionizations which do occur are those with Hg-atoms rather than rare gas atoms. Under these conditions, the gas temperature is close to room temperature whereas the electron temperature is very high, i.e.- 10,000 to 30,000 °K.

This can be seen in the following diagram, as given on the next page as 2.2.5., Note that both the electron-temperature and the gas-temperature are shown. Since 1.0 ev is equivalent to 11,600 °K., the average electron energy is therefore a few volts, while some will have energies many times this value.

2.2.5-

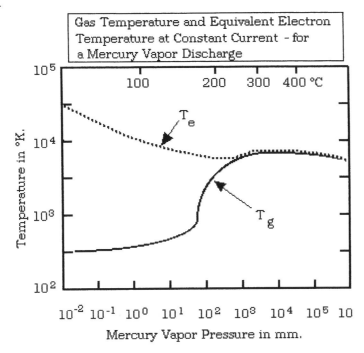

The meaning of such an electron temperature is that the average velocity of the electrons, due to energy extracted from the electric field, is the same as that in a field-free electron gas at this equivalent temperature, T, viz-

2.2.6.- $v = (2kT/m)^{1/2} = 5.50 \times 10^5 T^{1/2}$ cm/sec.

The voltage gradient in the discharge is of the order of one volt per cm. Since self-absorption will be low at these pressures, most of the radiation will be in the resonance lines at 1850 and 2537 Å, and the luminous efficiency will also be low, around 5-6 lumens per watt.

This is the situation shown at "A" in 2.2.2. As the vapor pressure is increased, reabsorption of the resonance lines becomes more important and the population of the higher energy levels is increased (see 2.2.3.) with the result that the efficiency increases. Above a vapor pressure of about 0.1 mm. (Point B in 2.2.2.- corresponding to the Cooper-Hewitt lamp of Chapter 1), the effect of elastic collisions becomes so great that the efficiency is reduced by energy loss to the glass bulb walls (Point C in 2.2.2.). The gas temperature (see 2.2.6.)

simultaneously begins to increase, but the electron temperature decreases in this region as a result of the number of increased collisions. When the pressure exceeds that of Point C in 2.2.2., the gas temperature has become so high, i.e.- ~ 1000 °K., that thermal excitation of the Hg atoms has become important and the luminous efficiency of the discharge again increases. Further increase in vapor pressure causes further increase in gas temperature until the electrons and atoms finally come into essential equilibrium at a temperature of about 6000 °K. (Actually, since the electrons transfer energy gained from the electric field to the atoms, the electron temperature is always slightly in excess of the gas temperature). It is this temperature which is mainly determined by the energy level structure and hence varies only slightly with vapor pressure. However, it tends to saturate above 10 atmospheres (see 2,2.2. & 2.2.3.). It should be clear, then that both "low-pressure" and "high-pressure" discharges differ considerably from each other in their fundamental mechanism in that the former consists mostly of resonant transitions within the energy levels of the atom while the latter has energy transitions between populated upper energy levels. Thus, the emitted spectrum of the former consists mostly of ultraviolet radiation, whereas the latter has a considerable portion of the radiation in the visible and the near-infra-red. Below Point C in 2.2.2., the electron temperature is always much higher than the gas temperature and excitation of mercury atoms is almost entirely by electron impact. For the higher pressures, temperature equilibrium is approached and excitation becomes thermal in nature.Thus, a better distinction between "low-pressure" and "high-pressure" arcs might be "non-thermal" and "thermal". The radiative properties of a very dense gas or vapor discharge approach those of a solid (26). High-pressure gas discharges thus afford a means of achieving quasi-thermal radiation whose "color-temperature" is much higher than can be supported by that of a real incandescent solid. The special properties of Hg- gas discharges, as exemplified in 2.2.5., is a main reason why they have found such wide-spread usage as lighting sources. The continuous component of emission of a high-pressure discharge results in an improved color rendition.

The emission spectra of a "low" pressure discharge may be seen in the following diagram, given as 2.2.7. on the next page, while that for the "high" pressure discharge is given as 2.2.8. immediately following.

2.2.7.-

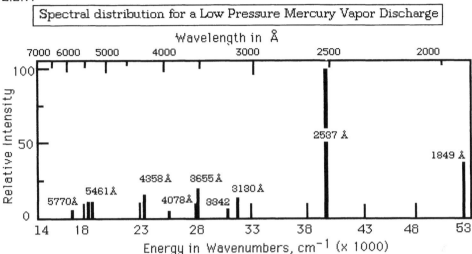

2.2.8.-

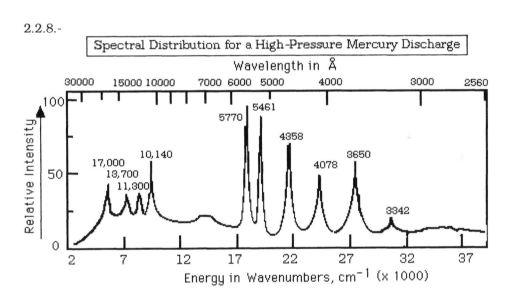

Note the "continuum" present in the high pressure mercury discharge, which is not present in the low pressure mercury discharge. This factor will be examined in more detail later. Also, the "lines" are broadened in relation to those of the low-pressure discharge. Calculation of the properties of a low-pressure discharge involves detailed knowledge concerning the excitation and recombination processes involved. These include collision cross-sections, state-lifetimes, etc. Another distinction between the two regimes is the fact that, for pressures below point-C in 2.2.2., the discharge fills the entire volume. But, for higher pressures, the arc stream becomes self-constricted and can be much smaller than the container (Such containers are usually a quartz tube whose length will vary depending upon the exact application).

Constriction of the arc is a very fortunate occurrence in the high pressure mercury discharge since otherwise it would be very difficult to find a practical container for the hot plasma. (Constriction of the arc plasma can be obtained by other means for low pressure arc streams as well). As will become apparent, the ability of the container to withstand the large amount of heat to be dissipated nevertheless still sets an important limit of the performance of the high-pressure Hg-arc light source. However, as Elenbaas (32) has shown, most of the properties of a high-pressure discharge can be obtained by applying thermodynamic principles to a simple model.

A. Processes in the High Pressure Mercury Vapor Discharge

Since the processes in the high-pressure arc discharge are essentially thermal in nature, there are two results from statistical thermodynamics which may be applied. Firstly, the ratio of the concentration of atoms, n_a , excited to a level, V_a , above the ground state is related to the concentration of n_0 atoms in the ground state by a Boltzmann relation:

2.2.9.- $\qquad n_a/n_0 = (g_a/g_0) \exp (-eV_a/kT$

where g_a and g_0 represent the statistical weights of the two levels, as related to $(2J + 1)$ where J is the quantum number, k is Boltzmann's constant, and T is the effective temperature. Secondly, the concentration of ions, n_i , and the concentration of electrons, n_e , is related to n_a (obviously $n_i = n_e$) by the Saha equation, as modified by Fowler (33), viz-

2.2.10.- $\qquad n_i n_e/n_a = \{2\pi\, mkT/h^2\}^{3/2} [2U' (T)/U(T)] \exp (-eV_i /kT)$

where V_i is the ionization potential, m is the mass of the atom involved, k is the Boltzmann constant and h is the Plánck constant. Both U and U' have the value:

2.2.11.- $U(T) = \Sigma_k \, g_k \exp(-eV_k / kT)$

where $U(T)$ is for the atom and $U'(T)$ is the expression for the ion. Except at very high temperatures, the factor: $[2U'(T)/U(T)]$ may be replaced by the expression: $G = (g_e \, g_i \, /g_a)$, where g_i and g_a refer to the ground states of the ion and atom, respectively and g_e is, of course, 2.0. For Hg, $G = 4$ and $V_i = 10.38$ e.v. In terms of the pressure, p (in atmospheres) and the degree of ionization, $\alpha = n_i /n_a = n_e/n_a$, and

2.2.12.- $\log \{\alpha^2 p/1 - \alpha^2\} = 5/2 \log T - 52{,}340/T - 5.850$

For $T = 6000$ °K, the right hand of the equation equals - 5.127, so that α is only 2.7×10^{-3} for a pressure of 1.0 atmosphere, and 10^{-4} for a pressure of 100 atmospheres.

The most important principle in the design of HPMV lamps is the loading, or wattage per centimeter length of the discharge. It is this which determines the luminous efficiency attained. The tube diameter, or the vapor pressure (provided that the latter is sufficiently high, i.e.- above a half-atmosphere) have only secondary influences. Such behavior is a result of the self-constriction of the discharge. Because of this constriction, the major sources of heat loss from the arc column are due to radiation and to conduction through the annular region of cooler vapor surrounding the column. Of the total power radiated by the arc, only a fraction is useful. The remainder is absorbed by the cool vapor, the container wall, and the outside air (The latter is particularly true for the very short ultraviolet wavelengths). In practice, the conduction loss is found to be essentially independent of loading, pressure and tube diameter (32,34). The radiated power, R, is related to the power input, P, by the relation:

2.2.13.- $R = 0.72 \, (P-10)$

where both are in terms of watts per cm. of arc length. From this, it is obvious that the radiant efficiency will increase as the loading increases. Thus, for a given wattage input, it is advantageous to have the arc as short as possible (We will consider electrode losses later). Elenbaas (34) gives a curve for the luminous efficiency for the range 15<P<50 watts which can be fitted by the equation:

2.2.14.- $L/W = 35 \log_{10} (P-10)$

where L is the discharge length and W is the wattage loading. The equation given by Marden, Meister and Beese (35) for the same situation is:

2.2.15.- $L/W = 42.4 \log_{10} (P/4.2)$

Thus, the efficiencies given by the last equation are smaller than those of Elenbaas. These same workers stated that the arc temperature, T(°K), is roughly equal to $1000\ P^{1/2}$, while the brightness of the arc is given by (36):

2.2.16.- $B(candles/cm^2) = 2.5(watts/cm^3)^{1.03}$

B. The Elenbaas Equations and the High-Pressure Mercury Discharge

Elenbaas (32) developed an extensive theory of the high pressure mercury discharge based on a very simplified model in which the rather complicated energy level diagram of the Hg-atom is replaced by a system consisting of only three levels. The excited level is 7.80 e.v. above the ground state and radiates to another level 5.55 e.v. above the ground state, which corresponds to emission at 5550 Å. The position of this excited level and of other parameters was adjusted empirically on the basis of a large set of experimental data. For the temperature of the discharge, Elenbaas found:

2.2.17.- $T = 5310/\{1.0 - 0.135 \log[P-10/m]$

where P is the loading in watts/cm and m is the amount of mercury (assumed to be completely vaporized) in mg./cm. In this case, the pressure, p, is directly proportional to m/d^2, where d is the arc-tube diameter. The proportionality constant is a function of the temperature distribution across the arc tube. For this, Elenbaas gives:

2.2.18.- $p\ (atmospheres) = 1.25\ m/d^2$

where d is in centimeters. According to Kenty (37), a more accurate expression is:

2.2.19.- $p = [0.75(m/d^2) P^{1/4}]/ md^{0.1}$

if the discharge is not cooled. For water-cooled arcs, the expression:

2.2.20.- $p = [E-100]/3$

is more accurate. In terms of the current, I, and the voltage gradient, E (where $P = EI$), Elenbaas found for the gradient:

2.2.21.- $E = \{6 P^{1/2}/(P-10)^{1/3}\}\{m^{7/12}/d^{3/2}\}$

where the terms remain as defined above. E then has a maximum of 12.1 $m^{7/12}/d^{3/2}$ volt/cm when $P = 30$ watt/cm. For higher loadings, the volt-ampere characteristic remains positive. Equation 2.2.20. was stated to be accurate above $P = 15$ watt/cm.

For very high values of P, i.e.- if $E \approx P^{1/6}$ (or $I^{1/5}$), Equation 2.2.17. states that for a constant current the temperature will be a maximum when $(P-10)/m$ is a maximum (see Figure 2.2.5.). If 2.2.20. is multiplied by I to obtain P, and then is solved for m, it is found that T will be maximum when $(P-10)^{3/7}/ P^{6/7}$ is a maximum, or when $P = 20$. Under these conditions:

2.2.22.- $p \approx 3(d^{1/3}/I^{12/7}$ atmospheres and $m \approx 2.4(d^{3/2}/I)^{12/7}$ mg./cm.

Elenbaas (32) also considered the effect of the rare gas used for starting the discharge. Since the ionization potential of Hg is below that of any of the rare gases, as may be seen in the following Table as given on the next page, ionization of any of these gases is negligible, even in the warm-up period.

Note that any or all of these vapors could be used as the basis of a lamp. However, only Na and Li have emission in a part of the spectrum where the human eye is efficient, and only Na will produce a lamp with "brightness", that is- emission where the human eye responds sufficiently, i.e.- "eye-brightness".

Of the other vapors, only Hg has a vapor pressure sufficient to support a gaseous discharge at ambient temperatures.

Table 2-4

Properties of Some Gases and Vapors of Interest in Gaseous Discharges

Element	Ionization Potential	Potential of Metastable Level (Volts)	Wavelength of Resonance Lines (Å)
Li	5.37 volts		6708
Na	5.12		5890, 5896
K	4.32		7645, 7699
Rb	4.16		7800, 7948
Zn	9.36	3.99, 4.06	2139, 3076
Cd	8.96	3.71, 3.93	2288, 3261
Hg	10.38	4.64, 5.44	1849, 2537
Tl	6.07	0.96	2768, 3776
He	24.47	19.77, 20.55	584, 592
Ne	21.47	16.53, 16.62	736, 744
Ar	15.69	11.49, 11.66	1048, 1067
Kr	13.94	9.86, 10.51	1165, 1236
Xe	12.08	8.28, 9.40	1295, 1469

Zn and Cd, for example, must be heated to about 400 °C. before the gaseous discharge becomes stable. The rare gas discharges fall into the far-ultraviolet and therefore can be used as fill-gases. Their presence in the Hg-discharge does have two effects however. First, the electron mobility is reduced (although this is a small effect), and second, the heat loss by conduction is increased. Therefore, the latter has an effect on the potential gradient which increases slightly, thereby decreasing the radiation emitted. However, the total effect is not serious enough to affect overall performance.

Elenbaas also considered the effect of convection in the gas surrounding the arc column. For a vertical discharge, the additional dissipation due to cooling at the **bottom end** is given by:

2.2.23.- $P'_{conv.}$ (watts) $= 1.5 \times 10^{-2} \, m^2 (P-10)^{3/4}$

In addition, there is a loss due to local circulation along the arc column. According to Elenbaas, it increases as $(P-10)^{3/4}$ and is also an increasing function of m^2/d for both horizontal and vertical arcs. Other workers have also given data on the effect of orientation of commercial HPMV lamps (38). Of even more importance than the small influence of convection on the power dissipation (and hence the efficiency of the lamp) is the effect of arc stability. The effect of turbulence in the surrounding gas sets an upper limit on the usable pressure which forms the arc, and is (32):

2.2.24.- $m = 10 \, d^{1/2}$

Under this condition, the pressure is:

2.2.25.- $p = 12.5/d^{3/2}$.

Also, in this case, the heat loss per unit of arc length is:

2.2.26.- $p_{conv.} = K(P-10)^{3/4}$

where $K = 0.39$ for vertical discharges and $K = 0.63$ for horizontal discharges. In the vertical case, the end-loss of $1.5 \, d(P-10)^{3/4}$ must be added (see 2.2.23,) As the mercury vapor pressure is increased, not only are the normal emission lines of Hg shifted and broadened, but a continuum also appears and partially "fills-in" between the lines. This is shown in the following diagram, given as 2.2.27. on the next page.

The exact origin of the continuum is still not completely clear. One obvious suggestion has been that it arises from recombination of free electrons (having random energies) with Hg ions (39). However, not only is the calculated energy less than that observed, but the intensity of such radiation should increase proportionally with the pressure like that of the emission lines (it does not). Rössler (40) therefore concluded that the continuous emission was molecular radiation due to impact of an excited atom with a normal atom, ions not being involved. The intensity of such emission should vary as the square of the pressure and hence should become pronounced at

2.2.27.-

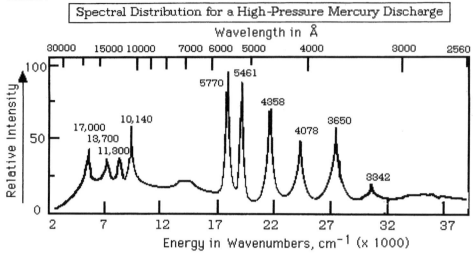

| Spectral Distribution for a High-Pressure Mercury Discharge |

Wavelength in Å

high pressures. Band emission due to some kind of molecular process (peaks at 3350 Å and 4500-5000 Å) has indeed been observed at lower pressures (66,67), but the exact nature of such processes still remains open to debate. The HPMV continuum probably arises from more than one process and needs further study (39,40,41).

According to Elenbaas (32), the energy in the continuum and the emission lines is given by:

2.2.28.- $P_{Cont.}$ = $(P-10)\{1+ m/d^2/4 + m/d^2\}$

P_{Lines} = $(P-10)\{3/4 + m/d^2\}$

$P_{Cont.}/P_{Lines}$ = $(1+ m/d^2)/3$

Since there are no strong Hg lines in the red portion of the visible spectrum, the intensity in this region may be used as a measure of the continuum output. The "red-ratio" is usually expressed as the reduction in luminous output produced by interposition of selected filters.

According to Marden, Beese and Meister (36),

2.2.29.-   % red = $0.8 + 2.1 \times 10^{-3} P$

The radiation appearing in the continuum does have an effect on the discharge temperature and the energy gradient. According to Elenbaas (32), m in 2.2.17. should be multiplied by the factor $(1+m/4d^2)$, resulting in a reduction in temperature, while the same factor, raised to the 1/3 power, should be placed on the right side of 2.2.20.

In addition, the gradient is modified by a term, exp (-e ΔV_i /4kT), where ΔV_i is the reduction in ionization potential at high pressures. This gives (according to Elenbaas) ΔV_i = 0.035, 0.17, 0.45 or 0.75 e.v. at m/d^2 = 5, 10, 20 or 40, respectively. Additionally, the yellow lines of Hg disappear at m/d^2 = 250, corresponding to ΔV_i = 1.58 e.v. As a result, the gradient never becomes more than 1.5 times that value without the continuum correction.

The theory of the arc column is complicated because of variations of temperature and electron concentration in the radial direction. Francis, Isaacs and Nelson (42) concluded that the luminous efficiency of the HPMV arc column approached a limit of 84 or 85 lumens/watt asymptotically as the loading was increased to very high values such as 400 watts/cm. They noted that this is the efficiency of a blackbody at a temperature of 6500 °K., which is roughly the electron temperature of the discharge. It is true that, at very high loadings, the behavior of the arc column approaches that of a blackbody. And, the possibility exists that the efficiency could surpass the 85 lumen/watt limit because of the presence of line radiation. However, this does not occur because of saturation and line broadening at high loadings. It may be noted that an arc efficiency of 73 lumens/watt was achieved by Francis, Isaacs and Nelson at a loading of 50-100 volts/cm.

Let us now consider specific design of HPMV lamps. It is obvious from the above discussion that high efficiency is favored by high loading (refer to 2.2.14., 2.2.15. and 2.2.18.). What this means is that, for a given wattage lamp, the arc length should be made as small as possible, i.e. "Short-Arc" lamps. Also, it is found experimentally (32) that the luminous efficiency, at constant power- P and diameter- d, increases as m increases (Actually, the voltage gradient, E. increases as m, the mass of Hg, increases). It is also found that, at constant P and E, the efficiency increases with increasing arc-tube diameter. These results are shown in the following diagram:

2.2.30.-

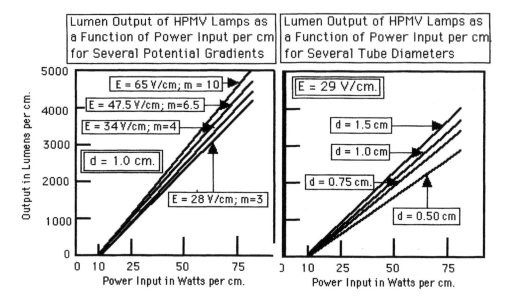

From this diagram, it is clear that the amount of Hg used should therefore be made as high as possible, consistent with good arc stability, since it increases the voltage gradient in the lamp. Furthermore, the "short-arc" or "compact-arc" design should be best, since a large diameter, coupled with a short-length, promotes high efficiency.

However, such lamps are very expensive to make and are not used as much as other designs. Additionally, as the length of the arc is decreased, the voltage drop at the electrodes becomes the over-riding consideration in reducing the overall efficiency of the lamp (this factor was not included in the above discussion). Maintenance of output during life of the lamp is negatively affected because of increase in metal sputtered from the electrodes to the wall. Thus, it is clear that some sort of optimum arc length exists.

Normally, the problem is to design a lamp of wattage, P, to operate at a voltage, V. Generally, V is taken as about 60% of the available open-circuit voltage, in order to have good stability (43,44). The remainder appears across the inductive ballast where the losses are usually small compared to dissipation of power in the lamp. Voltage drop at the two electrodes is about 15 volts. The lamp current is thus given by: $I = P/V$, where the power factor of the arc is neglected [In general, the power factor of the arc increases as the length decreases and the diameter increases (44)]. Assuming that the maximum amount of mercury is to be used (as given by 2.2.23.), we then neglect losses due to convection. Calculation shows that the factor of Equation 2.2.21. , i.e.-

2.2.31.- $\qquad P^{1/2} / (P-10)^{1/3}$

for the voltage gradient varies only from 2.265 at P= 15 to 2.018 at P = 30, and then to 2.359 at P = 150. i.e.- it is close to 2.00 for all cases. This gives the result for 2.2.22. as:

2.2.32.- $\qquad E = 46 / d^{29/24}$ and $E \cdot L = V - 15$

where L is the arc length. Thus:

2.2.33.- $\qquad 46 \, L = (V-15) \, d^{29/24}$

This fixes the shape, but not the size of the arc tube. From 2.2.13., the total radiated power is: $0.72 \, (E \bullet I - 10) \, L$. The radiant efficiency of the entire lamp (not just the arc column) is:

2.2.34.- $\qquad \xi = 0.72 \, \{(1- 15/V - 10L/P\}$

It is this result that shows that the efficiency increases either as the lamp voltage or the lamp wattage increases. But, it also increases as L is decreased. Therefore, there is **no optimum arc length** for the column.

However, we have not accounted for the fact that any power which is not radiated must be dissipated through the lamp envelope by conduction. By defining the maximum loading on the envelope as C_d , corresponding to a constant wattage per unit area, we are able to show:

2.2.35.- $C_d = (P-15\ I)/\ L$ and: $L_{min} = W/C_d\{(V-15)/V\}$.

Thus, 2.2.34. becomes:

2.2.36.- $\xi = 0.72\ \{(V-15/V)\{\ 1\ -10/C_d\}$

One concludes that the maximum attainable efficiency in practice increases as V increases, because electrode losses are minimized. It also increases as the permissible arc-tube loading, i.e.- P as determined by C_d increases. A material such as quartz with a large value of C_d permits higher efficiency than one with a smaller value such as glass. If the value of $10/C_d \ll 1.0$, little effect on lamp performance will be seen (at least according to this approximation). Equations 2.2.33. and 2.2.34. may be solved simultaneously to give:

2.2.37.- $d_{min} = (46\ W/\ C_d\ V)^{24/53} = 46\ I/C_d)^{0.45}$

and: $L_{min} = (V-15)(I/24\ C_d)^{0.55}$

If the emission is assumed to be entirely thermal, the luminous efficiency may be calculated from the radiant efficiency. We may also neglect variation of temperature since the efficiency of a blackbody only varies from 88.9 to 95.0 lumens per watt as the temperature varies from 5500 to 6500 °K. Equation 2.2.36. thus becomes:

2.2.38.- $\eta = 64\ \{(V-15/V)\{\ P\ -10/P\}$

At a power loading of 200 watts/cm (where the last term is 0.95), then the maximum efficiency of lamps designed for operation will be that shown in the following Table.

TABLE 2-5

Expected Operation and Efficiency of Designed Lamps

Open Circuit Voltage	Arc-tube Voltage	Efficiency in Lumens per Watt
115 volts AC	65 volts AC	46.8
230	135	54.0
460	265	57.4

Note that these results were obtained by using Elenbaas' equations, as given above. These values are reasonably consistent with actual results obtained for quartz lamps of high wattage, but not for small lamps. However, it will be shown that convection appreciably affects the lamp efficiency. This is a factor that we chose to ignore in the beginning. Additionally, the effects of the continuum radiation and the influence of the loading on the voltage gradient were ignored. We will now make a more complete use of Elenbaas' empirical results.

If the value of d is chosen, then the values of m and p follow from equations 2.2.24. and 2.2.25. The temperature, T, and voltage gradient, E, may also be calculated in terms of the loading, P, from equations 2.2.17. and 2.2.21. After the continuum corrections have been applied. the current, I, may then be obtained from: I = P / E.   In the following diagram, given as 2.2.39. on the next page, some of these results have been plotted.

The arc tube loading is plotted against the arc tube diameter, as a matter of convenience. Actually, this diagram is a combination of several, where d, p, m, and E are plotted against P, the arc loading which results. Since each of these cases are interdependent, the result will depend upon how each point is chosen. There are several factors to be noted, viz-

> 1. For a given arc tube diameter, the optimum mass, m, of Hg to be added is given as a function of the arc tube diameter.

> 2. The approximate gradient which results is shown at the top of the diagram, along with the internal pressure which results in the arc tube.

> 3. At the right is shown the approximate efficiency which results as a function of the arc-stream temperature.

> 4. The maximum loading, C_d, on the envelope is also shown. For glass, , C_d = 8; for quartz, , C_d = 30. These are design limits that cannot be exceeded.

5. The current which results can be chosen as a function of the gradient, E, and the arc tube diameter.

2.2.39.-

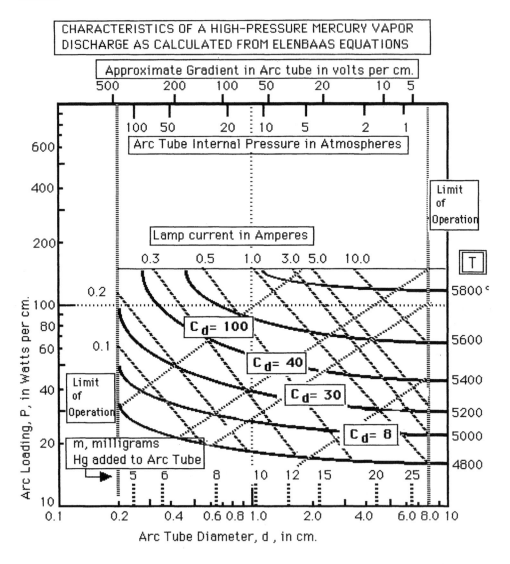

Let us choose certain parameters to show how this diagram can be used. Suppose we start with a quartz tube of 2.0 cm. in diameter. We intend to design a 200 watt lamp. Thus, the tube needs to be 4.4 cm. in length, since the arc-loading will be 45 watts per cm. The design parameters are thus expected to be the following:

2.2.40- For a 2.0 cm. diameter tube of 4.4 cm. length:

      a. Add 14 mg. of Hg for optimum performance

      b. Current is expected to be about 0.75 ampere

      c. The gradient will be about 35 volts/cm and the operating internal pressure will be about 5 atmospheres.

      d. The color temperature (efficiency) will be about 5300 °K.

It should be clear that it is possible to design HPMV lamps of specific operating characteristics, but one needs to point out that the above diagram is not completely accurate. If one wished to use such a diagram, it would need to be plotted much more carefully. The actual equations to be used are given in the following Table:

TABLE 2-6
Summary of Formulas for Design of High-Pressure Mercury Vapor Lamps
Given: W (wattage) - V (voltage) - C_d (permissible loading of arc tube in watts/cm times π)

1. Current: I (amperes) $\approx$ W/ 0.9 V
2. Arc Loading: P (watts/cm) = 5.5 $C_d^{0.60}$ $(W/V)^{0.45}$

3. Arc Tube Diameter: d (cm) = P/C_d
4. Arc Tube Length: L (cm.) = W/P {V-15/V} - {1.5[$(P-10)^{3/4}$]/C_d}

5. Voltage Gradient: E (volts/cm) = P•V/W
6. Operating Pressure: p (atmospheres) = $12.5/d^{3/2}$

7. Amount of Hg per cm. of Arc: m (mg./cm) = $10D^{1/2}$ = $10(P/C_d)^{3/2}$

8. Luminous Efficiency: L/W = 35{ [(V-15)/V] - [$1.5P(P-10)^{3/4}/C_d$ W] }
                       • log {(P-10)(1- [$0.39/(P-10)^{1/4}$] }

In 2.2.39., it may be noted that the decrease in temperature with decreasing diameter over most of the region considered is a result of the continuum correction, i.e.- T is a maximum at d = $5^{2/3}$ = 2.92 cm. It may also be noted that the variation in T is small, i.e.- only 1000 °K for a variation of 2 decades in current. From examination of data for actual lamps, as given by Bourne (43), it appears that permissible loading on glass arc tubes is about 2.4-2.5 watts/cm^2 of surface. For quartz, it is about 14 watts/cm^2. Noel (38) has stated that 14 watts/cm^2 was originally used in commercial quartz lamps, but this led to short life, particularly for horizontal operation. The figure was then accordingly reduced to 9-10 watts/cm^2. This number, corresponding to a value of C_d = 30 is typical of present-day quartz lamps. Elenbaas (32) indicates a value of C_d = 10 for glass and about 30 for large quartz tubes.

The following Table presents **calculated** characteristics of HPMV lamps similar to those being currently manufactured:

TABLE 2-7

Electrical and Optical Characteristics of HPMV Lamps

Operating			Diameter			Loading					
Watts	Volts	Amps	Arc Length (cm)	Tube (cm)	Arc (cm)	I in amp/ cm^2	Grad. volts /cm	Tube- watts /cm^2	Arc- Watts /cm	Eff.- Lum/ watt	Bright ness- C/mm^2

HIGH PRESSURE LAMPS IN GLASS @ 1.0 ATMOSPHERE PRESSURE

100	135	0.9	6.5	2.0	0.4	7.2	21	2.4	15	30	1.2
400	135	3.2	15.9	3.2	1.1		8.5	2.5	25	38.8	0.85
500	135	3.7	18	3.5	0.7	9.6	7.5	2.5	28	47	1.5
2000	135	15.5	40	6.5	1.5	8.8	3.5	2.4	50	55	2.0

HIGH PRESSURE LAMPS IN QUARTZ @ 5-10 ATMOSPHERES PRESSURE

100	135	0.9	2.5	0.9	0.2	28.5	54	14.1	40	40	10
500	390	1.5	10	1.1	0.3	21.2	39	14.5	50	50	10
1000	2200	1.5	44	1.1	0.3	21.2	50	14.5	45	60	10

Thus, it should be clear that the limiting factor for HPMV operation is that of C_d, i.e.- the tube-material characteristics. Thouret(60) has considered the pressure and thermal stresses in cylindrical and spherical envelopes and concluded that a safety factor of about 10 over the "ultimate strength" of the material should be employed.

We can now calculate the radiant efficiency, taking into account both electrode and convection losses in the lamp. For a vertical discharge, we may write:

2.2.41.- $\xi = 0.72 \{(P-10)- P_{Conv.}\} L/ 15 I + P_{Conv.} + P L$

where the numerator is the total wattage input, i.e.- W and $P = EI = EW/V$, while the convection (Conv.) terms are given by Equations 2.2.23., 2.2.24. and 2.2.26. Therefore:

2.2.42.- $\xi = \{0.72 L(P-10)(1 - 0.39 (P-10)^{1/4})\}/[15 I + 1.5 d (P-10)^{3/4} + P L]$

d may be eliminated by setting it equal to P/C_d , while in terms of the applied voltage:

2.2.43.- $L = W_{arc} /P = I (V-15)/P - 1.5 (P-10)^{3/4} /C_d$

The final result can therefore be written as:

2.2.44- $\xi = 0.72 \{[(V-15)(P-10)]/P V\} - [1.5(P-10)^{7/4}/CIV]\{1-[0.39/(P-10)^{1/4}]\}$

The first bracket includes the effect of electrode losses and heat conduction from the arc column (first term) and convection losses at the bottom of the arc (second term), while the second bracket accounts for convection losses from the arc-column. For a horizontal discharge, the second term in the first bracket disappears while the constant in the second bracket becomes 0.63. The parameters are: P, I, and V (but P and I are related to each other according to the lines of constant C_d given in 2.2.39).

Instead of obtaining the relationship between P and I from 2.2.39., careful analysis shows, except for glass tubes with I < 1.0 ampere, that the following equation gives good results:

2.2.45.- $P = 5.5 C_d^{0.60} \cdot I^{0.45}$

From 2.2.40. and $P = C_d \bullet d$, one can obtain a similar result except that C has the exponent = 0.55. The modification due to electrode and convection losses in the lamp now becomes:

2.2.46.- $\xi = 0.72$ {[(V-15)(P-10)]/P V] - [1.5(P-10)$^{7/4}$/C_dIV]}{1-[0.39/(P-10)$^{1/4}$]}

$\qquad$ Term-a $\qquad\qquad$ Term-b $\qquad\qquad\qquad$ Term-c

where the terms are enclosed in brackets, i.e.- [ ]. As the voltage is increased, Term- a increases while Term-b decreases, both of which make the efficiency larger. As the loading is increased, both Terms-a and -c become larger, which causes the efficiency, ξ, to increase as well. Term- b is a maximum for P = 530/11 = 48.2 watt/cm. For higher values of P, Term- b also indicates higher efficiency as the loading is increased. The opposite effect for P < 48 will, in all cases, be offset by Terms- a and c. Term- b also shows that the efficiency is slightly reduced by an increase in C_d , but this will be offset by the higher values of P permissible with higher values of C_d . One thus concludes that high radiant efficiency is favored by high voltage, so as to minimize electrodes losses, and by high arc loadings. Since high loadings accompany high currents (see 2.2.39. and Equation 2.2.45.), the radiant efficiency should increase as the current increases.

We are actually more interested in the luminous efficiency than the radiant efficiency. But, it is difficult to know how to calculate this, since the emission spectrum departs so greatly from that of a black-body. In general, a decrease in arc-tube diameter (and therefore the loading) corresponds to a lower effective temperature but with a higher pressure (with a larger contribution from the continuum). Therefore, the only recourse seems to be to revert to Elenbaas' empirical result, as given in 2.2.14. Making use of 2.2.45. and this result, we introduce a correction for electrode losses. This results in the following equation:

2.2.47.- $\quad$ L/W = {35(V-15)/V} $\bullet$ {log {(5.5 $C_d$$^{0.60}$ $\bullet$ I$^{0.45}$ - 10)}

Although the effects of convection are included in 2.2.46., they are not included in 2.2.47. If this is done, the result is:

2.2.48.- L/W=35{[(V-15)/V]-[1.5P(P-10)$^{3/4}$/CW]}$\bullet$log{(P-10)$\bullet$[1-(0.3/(P-10)$^{3/4}$)]}

for the vertical case, with P still given by 2.2.45. Efficiencies as calculated from Equations 2.2.47. and 2.2.48. are presented in the following diagram, given as 2.2.49. on the next page.

Here, we have plotted the case for a glass envelope both for uncorrected convection and corrected convection for three types of lamps, namely: 65 Volt, 135 Volt and 265 Volt lamps. Note that the uncorrected version of the curve nearly mirrors that of the corrected version, except that it is higher. For this reason, the other two parts of the diagram do not show this feature (although it is indicated to a certain extent). Note also that specific wattage commercial-lamps are shown for comparison on two parts of the diagram, i.e.- that for glass and for quartz envelopes. It should be clear that the equations developed by Elenbaas do result in HPMV lamps that have outputs and characteristics very close to those which are manufactured today by the Lamp Industry.

Also shown are calculations and expected luminous outputs for lamps employing an envelope with a C_d of 100. Note that at least an increase of 150% would be expected **if** such a material could be found. At the present time, no such material exists. Perhaps, the use of alumina or a similar oxide material might be effective. However, at the present time, there is no incentive, since the cost undoubtedly will be higher as well.

From the results shown in 2.2.49., the following conclusions can be drawn:

1. The luminous efficiency is always increased by an increase in loading, P, or the current. For a given voltage, the efficiency is therefore greater as the wattage is increased.

2. For a given wattage, the loading (and therefore the current) will be higher for low-voltage lamps. But because of electrode losses, the efficiency is also generally lower for low-voltage lamps. As a result of these conflicting tendencies, the highest efficiency is usually achieved at intermediate (135 volts) voltages.

2.2.49.-

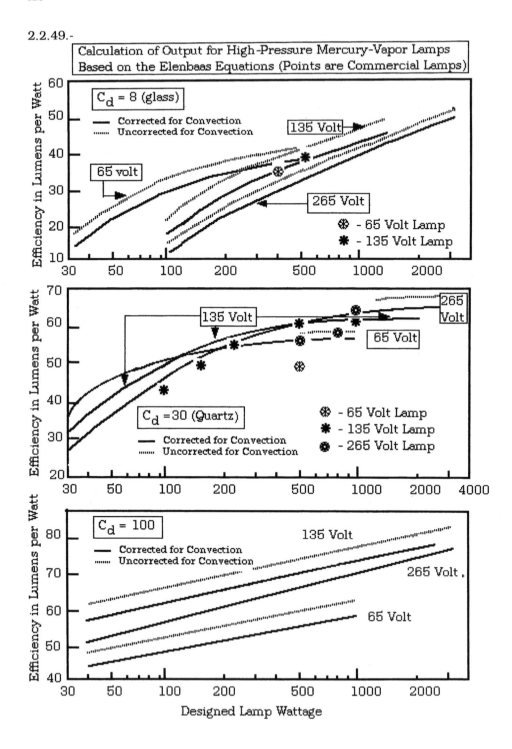

Calculation of Output for High-Pressure Mercury-Vapor Lamps Based on the Elenbaas Equations (Points are Commercial Lamps)

3. Use of arc-tubes that can withstand high loadings obviously leads to high efficiency. However, it is found that the effect is fairly small and becomes increasingly so as wattage or C_d is increased. Going from glass to quartz only increases the calculated efficiencies of a 1000 watt lamp from 45 to 58 lumens/ watt, while increasing C_d to 100 (a dissipation of 32 watts/cm^2 of surface) gives only an increase to 69 lumens/watt.

4. It is clear that correction for convection losses is important. It will be noted that in general the difference between horizontal and vertical operation is expected to be small. Except for low-voltage and high-C_d lamps, the efficiency is normally greater for vertical operation. In practice, however, the difference is considerably greater than is calculated. Frazer et al. (38) state that if a 400-watt quartz lamp, operated in a vertical position, is changed to a horizontal position (keeping the ballast the same), the wattage changes to 392 watts and the output to 93%. By changing the wattage to 410 watts, the output is restored to 100% that of vertical operation.

The values calculated in this way give good agreement with those of experimental lamps (as shown in 2.2.49.). Table 2-8 details the calculated results for 400 watt lamps having arc tubes of these three materials (the last being hypothetical).

TABLE 2-8

Calculated Characteristics for 400 Watt HPMV Lamps Operated Vertically

Material = :	Glass (C_d = 8)		Quartz(C_d=30)		"X" (C_d=100)		Units
Operating Voltage:	135	265	135	265	135	265	Volts
Arc Tube Length	9.94	15.28	4.27	6.75	1.96	3.34	cm.
Arc Tube Diameter	3.79	2.85	2.23	1.67	1.40	1.00	cm.
Hg Weight	194	258	64	87	23	33	mg.
Operating Pressure	1.7	2.6	3.8	5.8	7.5	12.5	Atm.
Current*	2.96	1.51	2.96	1.51	2.96	1.51	Amp.
Voltage Gradient	10.2	15.1	22.6	33.1	47.3	66.2	Volt/cm
Arc Loading	30.2	22.8	67.0	50.0	140.0	100.0	Watt/cm
Electrode Loss	44.4	22.6	44.4	22.6	44.4	2.96	Watts
End Convection Loss	54.4	28.9	69.5	39.9	80.8	43.8	Watts
Arc Convection Loss	37.1	40.3	34.5	41.8	29.4	38.1	Watts

TABLE 2-8 (CONTINUED)

Calculated Characteristics for 400 Watt HPMV Lamps Operated Vertically

Material = :	Glass (C_d = 8)		Quartz(C_d=30)		"X" (Cd=100)		Units
Conduction Loss	99.4	152.6	42.7	67.5	19.6	33.4	Watts
Radiation	164.7	155.4	208.9	228.2	225.8	262.1	Watts
Total Power	400.0	400.0	400.0	400.0	400.0	400.0	Watts
Operating Voltage:	135	265	135	265	135	265	Volts
Radiation Efficiency	0.296	0.280	0.376	0.411	0.406	0.508	
Luminous Efficiency							
Calculated:	37.3	32.9	51.8	50.0	62.7	62.0	Lumens/ Watt
Experimental:	38.8		53.8	50.6			Lumens/ Watt

Additionally, the properties of 9 types of commercial lamps are shown in Table 2-9 for comparison:

TABLE 2-9

Electrical and Optical Characteristics of Selected HPMV Commercial Lamps

Design Name	Watts	Oper. Volts	I in Amps	Arc Length (cm)	Tube Diam (cm)	Grad. volts /cm	Tube- watts /cm²	Arc- Watts /cm	Eff.- Lum/ watt	Maint. in % *
H24	415	70	6.6	6.7	2.2	10.4	9.0	62	48.2	86
H38-A	100	130	0.9	2.5	0.8	52	15.9	40	36.5	81
H37-22	175	135	1.45	5.1	1.1	27	9.6	34	44.6	86
H37-5	250	135	2.1	5.4	1.5	25	9.6	46	48.0	86
H33-1	400	135	3.2	7.0	1.8	19	10.1	57	53.8	88
H34-12	1000	135	8.2	12.7	2.2	10.6	11.4	79	55.0	81
H40-17	425	265	1.7	8.9	1.5	30	10.2	48	50.6	88
H35-18	700	265	2.8	12.7	1.8	21	9.7	55	52.9	86
H36-15	1000	265	4.0	15.2	2.2	17.4	9.6	66	57.0	81

* Measured from "rated initial lumens" - i.e.- 100 hours burning in vertical position) to 16,000 hours life.

Note that at least 5 of the commercial lamps show differences of 1.0 lumen/watt or less to the calculated values given in Table 2-8. It should be clear that the empirical method of Elenbaas does produce designs of HPMV lamps that are in general agreement with actual practice. Note also that an marked increase, according to these calculations, does not result in marked increase in luminous output.

It is clear that the agreement would not have been as good if the corrections for convection losses had not been included. Furthermore, the results for glass-enclosed arcs indicate that conduction from the arc-stream is the major loss while that for quartz (and other materials) is not so-severe. Better arc-tube materials, i.e. higher values of C_d , would permit higher efficiencies, but the benefit to be gained would be fairly small. According to Thouret (45), the permissible loading for sapphire (α- alumina) is essentially the same as quartz. Thus, although its melting point is higher, i.e.- 2050 °C vs: 1470 °C for quartz, the cost would be several times higher for sapphire. It should be noted that "Lucalox™" and "Coram™" are tradenames for polycrystalline translucent α-alumina currently being used for manufacture of high-pressure sodium-vapor lamps. However, regardless of cost, the permissible loading factor is either not known or has not been determined.

In conclusion, Francis, Isaacs and Nelson (42) experimentally achieved 84 lumens/watt in quartz at P = 400 watts/cm. with arc tubes of 2-3 cm. in diameter. This makes C_d about 15 to 22.5 times the loading used in commercial lamps. Although no commercial air-cooled HPMV lamp has an efficiency in excess of 60 lumens/watt (see Tables 2-7 and 2-8), it would seem that higher efficiency could be achieved at the expense of life for some special purposes.

C. Current Mercury Lamps Being Manufactured

Mercury-vapor lamps currently being manufactured can be divided into several different categories. These include:

1. Low Pressure Lamps

2. Photochemical Lamps

3. High-pressure Arcs in Glass

4. High-Pressure Arcs in Quartz

5. High Pressure Arcs with Forced Cooling

6. Short-Arc Lamps

1. Low Pressure Lamps

As discussed before, most of the energy of low pressure lamps is radiated in the resonance lines of 1849 and 2537 Å (46). The luminous efficiency is therefore very low, being about 5-6 lumens/watt. Metastable mercury atoms play an important role in this type of discharge. The ultraviolet output from such discharges may be used to excite luminescent inorganic materials which emit in the visible region of the spectrum. Such lamps, i.e.-"fluorescent lamps", will be discussed in more detail below. In the next Chapter, we intend to discuss the manufacture of the luminescent inorganic materials (phosphors) in greater detail as well.

Maximum 2537 Å output is obtained at an operating temperature of 40 °C where the vapor pressure is about 6μ and the gas fill is either Ar, Ne or a mixture of both. The ultraviolet output from such discharges can be used directly for germicidal and other purposes (47,48) and have been called "Sterilamps™" (49), especially when the outer glass tube has been selected to transmit the ultraviolet output. Similar uses include generation of ozone (50) and atmospheric ions (51), the latter of which have physiological effects on the human body. Because of the low voltage gradient in this low-pressure mercury vapor discharge, such lamps generally have an extended form, in comparison to the diameter of the tubing. A common length is either 15, 30 inches or four feet. The electrodes can either operate hot or cold, the discharge being maintained by ion bombardment from the discharge. However, a starter circuit of some sort is required. Radiation of 1849 Å is required for ozone generation, while 2537 Å is near to the peak of germicidal action.

A common electrode design is shown in the following diagram:

2.2.50.-

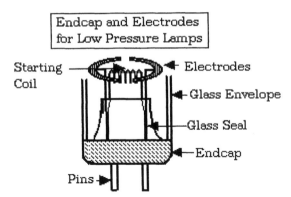

Note that both a "hot" cathode, i.e.- starting coil, and cold electrodes are included in this design. When the lamp is "started", a current flows through the cathode and thermionic emission serves to start the Hg-discharge. Such cathodes are generally coated with a mixture of alkaline earth oxides to improve the thermionic emission of electrons. The current is then switched to the discharge through the electrodes to maintain the discharge. This type of configuration is used for nearly all low-pressure discharge lamps, whether fluorescent, germicidal or other usages.

In contrast, the erythemal peak (skin-tanning) is closer to 2950-3000 Å so that a phosphor (52) is required to convert 2537 Å to this region of the spectrum. However, since most such lamps represent dangerous sources of potential damage to the human eye, the majority of manufacturers today use glass that cuts-off radiation below about 3400 Å in order to avoid a potential source of danger to the general public, particularly in so-called "Tanning Salons". Therefore, a "Blacklight" phosphor (53) having emission in the 3700-3900 Å region is generally used. The same tanning effect on the human body is seen, except that the erythemal action is much slower.

Other configurations for these lamps have been suggested. Rather than have a long length of tubing to form the lamp, Laporte (54) suggested a compact lamp consisting of an externally-heated cathode with a grid for accelerating electrons to a sufficient energy to excite the Hg-atoms present. These lamps, however, proved to be inefficient because of losses in cathode-heating power and also due to the fact that low voltages (about 11.5 volts) were needed for operation. This voltage is actually the resonance potential of the argon fill gas, and energy is transferred during operation to the much less numerous Hg-atoms present. Overall luminous efficiency of these 43-watt lamps, even with a phosphor, proved to be only 5.8 lumens/watt. Marden and Nicholson (55) in 1931 also described a low-voltage Hg-lamp (20 volts) in which the discharge occurred between two indirectly-heated oxide-coated cathodes operating on AC with a ballast. Because at least half of the current was required to heat the cathodes, efficiency was low and manufacture of such lamps was discontinued. These design problems have now been solved and, in 1992, a number of such lamps have appeared due primarily to their greater efficiency and l;onger life, compared to incandescent bulbs. The reasons why such lamps are expected to replace incandescent bulbs is shown in the following diagram:

2.2.51.-

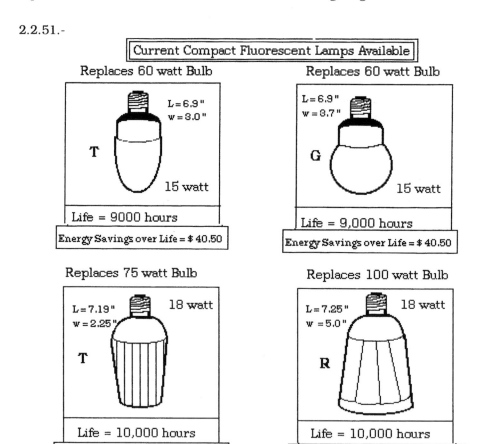

While an incandescent bulb costs about $0.39, comparable so-called "compact fluorescents" range from $ 12.95 to $ 18.95. However, as can be seen, in spite of the higher cost, these compact fluorescent lamps save from $41 - $ 57 in power costs over that of the incandescent bulbs. Furthermore, they do not have to be changed every 6 months because of "burn-out". It is likely that as these lamps become more widely available and as the the costs come down, they will entirely replace the tungsten filament bulbs used world-wide and incandescent bulbs will become a thing of the past.

2. Photochemical Lamps

Many chemical reactions are aided by the use of ultraviolet radiation (56). Energy at 3650 Å corresponds to 3.4 e.v. or 78 Kcal./mol. Many types of sensitized paper used in the photocopying trade have their peak sensitivity in the 3500-4000 Å range so that special lamps were developed for such applications (57). Data for some typical lamps are given in the following Table, viz-

TABLE 2-10

Typical High-Pressure Mercury Vapor Lamps for Photochemical Applications

Manufacturer	Type	Arc Tube Material	Watts	Volts	Amperes	Length (cm.)	Loading (Watts/cm)
Philips	HOG	Glass	2000	550	4.2	122	16.4
Philips	HOK	Quartz	3000	1100	3.2	130	23.0
U.S.	H9F	Glass	3000	535	6.1	122	24.6

These lamps were typically one inch in diameter and several feet long. The pressure is about 0.50 atmospheres and therefore are roughly comparable to the old Küch lamps, but without the liquid cathodes. Although not designed as a visible source, the luminous efficiency was about 44 lumens/watt. Such lamps were actually used for a lighting installation in 1947 (58) and produced about 60 times as much light as a 40-watt low-pressure fluorescent lamp of essentially the same size.

3. High Pressure Arcs in Glass

The first commercial HPMV lamps of this type (59,60) were made with hard glass envelopes, special glass being developed for the purpose (61). Jenkins and others have presented data (62,63) on the ultraviolet transmission of glasses. This type of lamp was first introduced in England by G.E.C. in 1932, while Osram in Germany and Philips in Holland were not far behind (64). The English lamp consumed 400 watts, with 250 watt and 150 watt sizes being introduced later (65). The initial efficiencies were 45, 36 and 32 lumens/watt, respectively with an approximate life of 1500 hours. This was later increased to 3000 hours (59). Experimental lamps of 1000 watts and 2500 watts were also made (59). However, such lamps were not introduced into the U.S. until 1934 (66). The 400 watt lamp, of Westinghouse manufacture (67), was designated the H1- lamp and the 250 watt lamp as H2. Life was stated to be 2000 hours with efficiencies of 40 and 30 lumens/watt, respectively. Maintenance after 1400 hours of operation was said to be 80%. These lamps were appreciably higher in efficiency than the old Cooper-Hewitt and Küch lamps.

In lamps of this type, the arc tube is mounted in another larger glass envelope which serves several purposes. Firstly, it prevents injury or danger of fires from contact with the hot arc-tube. Since these lamps operate at about 1.0 atmosphere internal pressure, the coolest part of the arc tube is about 360 °C., with the hottest part being about 550 °C. Secondly, use of a protective atmosphere in the outer envelope protects the hot current lead-in's from oxidation, especially where molybdenum metal ribbon-seals are used for the arc tubes. Thirdly, variation of the gas pressure in the outer envelope permits control of heat dissipation from the arc (68). Increasing the pressure increases heat dissipation and hence lengthens the "startup-time" of the lamp for it to warmup to full internal operating pressure. Additionally, the "restrike-time", i.e.- that time required for the internal pressure to fall low enough for the ballast voltage to restart the lamp, is also lowered considerably. Increased pressure in the outer envelope also has an effect on the arc-over voltage between the leads in the outer envelope.

In general, glass mercury arcs must be operated in a vertical position. Because of the fairly long arc length, excessive "bowing" of the arc stream will occur for horizontal operation which causes melting of the glass arc-tube unless some sort of external magnetic field control of the arc is used. However, if the internal operating pressure is lowered (at some sacrifice to efficiency), a fairly short arc may be operated in a horizontal position. Since the permissible loading is only about 2.5 watts/cm^2, the efficiency is correspondingly limited. Operating pressure is about one atmosphere because higher pressure would result in higher arc resistance and therefore higher heat dissipation than could be supported by the glass.

As stated previously, these glass HPMV lamps used solid rather than liquid cathodes as in the Cooper-Hewlitt lamps. A major problem has been that of sputtering of the metal by ion bombardment. Wehmelt (69) in Germany solved this problem in 1903 by the development of the "oxide-coated" cathode. A mixture of alkaline-earth oxides was coated on the tungsten-metal cathodes to improve thermionic emission of electrons. The electronic work-function is about 1.0 e.v. as compared to 4.5 e.v. for tungsten. At a given temperature, the emission is therefore greater than that of tungsten by the factor: exp 40,600/T. However, such electrodes are sensitive to contamination by ion bombardment. Early on, glass HPMV lamps had electrodes consisting of rods of sintered alkaline earth oxides inserted into tungsten coils (64, 65). The oxide served as the emitter during the cathode half-cycle while the tungsten served as the anode on alternate half-cycles. Although these electrodes were a great improvement over the old liquid cathodes, lamp life was still limited by blackening as a result of electrode sputtering. This was "cured" by the introduction of a small "starting" electrode, connected by a resistor to the main electrode at the opposite end. At starting, the full applied voltage thus appears between this electrode and the adjacent electrode. The small discharge thus formed supplies initial ionization for starting the main discharge. As soon as the current starts to flow to the starting electrode, its voltage drops due to the resistor and the discharge transfers to the main electrodes.

Nowadays, nearly all HPMV lamps are made with quartz arc-tubes, glass arc tubes having been discarded universally. This has come about due to the lower efficiencies experienced with glass arc-tubes and improvement in both cost and quality of quartz tubing.

4. High Pressure Arcs in Quartz

Lamps of this type are similar in general design to the older glass arc-tube lamps except that the quartz permits higher loading and efficiencies. Both glass and quartz lamps use an outer glass envelope, and an auxiliary starting electrode. Bourne (59) has stated that the safe operating pressure lies between 5 and 10 atmospheres. In practice, the value is somewhat lower, around 3-5 atmospheres. This may be due in part to the fact that about 70% of these lamps are burned horizontally for street lighting. The operating temperature inside the quartz arc-tube may be as high as 850 °C. Although the superiority of quartz over glass was obvious for a long time, practical lamps had to await the development of satisfactory seals for the lead-ins which were capable of operation at the required high temperatures (34,59,70). It was D. Gabor in 1931 who invented the molybdenum seals used almost universally today.

Because of the large reduction in efficiency with reduced wattage, glass HPMV lamps were not made smaller than about 250 watts. With quartz, however, smaller lamps could be made. The first lamp of this type, i.e.- 75 watts- 230 volts, was announced by Philips in The Netherlands in 1936 (71). This lamp had an efficiency of 40 lumens/watt. Lamps of the same general type, i.e.- 85 watts with 35 lumens/watt luminous efficiency, were soon introduced by Westinghouse. The life at first was only 500 hours. In these early lamps, a quartz diaphragm was placed in front of the electrodes to prevent material sputtered from them being deposited on the main part of the quartz arc-tube. The 85 watt lamp was soon superseded by a 100 watt-130 volt lamp in 1937, followed in 1938 by a 250 watt and a 400 watt lamp (100), all rated at 1000 hours of life.

The oxide-coated electrodes used with glass arc-tubes gave very poor performance in the newer quartz HPMV lamps, where the shorter length made blackening much more serious. A new electrode, consisting of a sliver of thorium held to a tungsten rod and protected from the discharge by a tungsten coil, was introduced by Gustin and Freeman of Westinghouse (72) in 1939 and soon became standard in the U.S. Actually, the use of thermionic electron emitters consisting of Thorium on W in vacuum tubes was pioneered by Langmuir (73). Such emitters are very susceptible to ion bombardment and hence a large reserve of Th must be provided for use in operation of HPMV lamps. Following World War II, HPMV lamps were redesigned in the US for lower loadings in order to increase the life to 2000-3000 hours (74). At about the same time, the glass used for the outer envelope was changed and the bulb was made larger. This was done to control more easily the diffusion of water vapor from the outer glass envelope to the hot quartz tube where hydrogen from breakdown of the water at the hot surface could diffuse through the quartz and impair lamp performance. In Europe, efforts continued to improve "oxide" electrodes for HPMV lamps (75). Considerable success was achieved at Osram in East Germany, and the new "Life-Guard" electrode was introduced in the U.S. by both Westinghouse and G.E. who were licensees of Osram. In the new electrode, the emission material was protected from the discharge by a tightly-wound overlying tungsten coil, much as in the prior thorium electrode. The alkaline earth material was also made more refractory by the addition of ThO_2 and SiO_2. According to Martt, Gottschalk and Green (76), the operating temperature of the new oxide electrode was reduced to about 1670 °C., compared to 2200 °C. for the thorium electrode. Also, with this new electrode, low temperature starting (which is important in street lighting applications) was greatly improved, blackening was much reduced, and luminous maintenance was much improved. With the old thorium electrodes, rated life had been increased to 6000 hours @ 60-70% maintenance (77), while the new alkaline earth electrodes gave a maintenance of 85% at 9000 hours (75). The best HPMV lamps of today have a useful life of 16,000 hours with a maintenance of about 81-88%. Table 2-9 gave some of the characteristics of commercial lamps of this type. The actual life of a commercial HPMV lamp is, of course, a function of the number of starts to which it is subjected. Electrode damage during starting is much more severe than during normal operation of the lamp, but is also a function of ballast design (78). If a HPMV lamp is installed and left to burn undisturbed, the life will be essentially unlimited, but the light output will, of course, decline with time.

In Table 2-9, the atmosphere in the outer envelope of the 100 watt lamp is vacuum (to conserve heat). In the 175 and 250 watt sizes, an outer fill gas pressure of nitrogen at about 100 mm. of Hg is used. For the larger wattage sizes, the pressure used is about 500 mm. of nitrogen gas. In the arc-tube, the fill gas used is argon at a pressure of 20-25 mm. A higher argon pressure results in improved maintenance of luminous output by reducing electrode sputtering during starting and warm-up. Conversely, increased argon pressure also results in increased starting voltage. In practice then, the actual pressure used is a compromise (75).

As compared to tungsten lamps, the efficiency of HPMV lamps is very high and the operating cost low. On the other hand, mercury lamps are expensive and also require a ballast, so that initial cost is high. Till and Unglert (75) have shown that, for street lighting or industrial applications, the overall cost of mercury lighting is cheaper than either incandescent or fluorescent lighting (provided that the color is not objectionable - we will have more to say about this subject below). The relative cost of incandescent lighting is 178-186% that of mercury lighting. However, the cost of the ballast needed and its effect and its effect on lamp operation are significant factors. The following Table, lists some of the characteristics of typical HPMV ballasts.

TABLE 2-11

Characteristics of Some Typical HPMV Lamp Ballasts *

Lamp Type	Lamp Wattage	Ballast Watts	Ballast Eff. in %	Luminous Efficiency in Lumens per Watt	
				Lamp	Total
H38-4	100	125	80.0	36.5	29.2
H39-22	175	205	85.4	44.5	38.0
H37-5	250	285	87.7	48.0	42.1
H33-1	400	450	89.0	53.7	47.8
H35-18	700	780	89.7	52.9	47.5
H34-12	1000	1100	90.9	55.0	50.0
H36-15	1000	1100	90.9	57.0	51.8

*Data from Sola Electric Co. for constant wattage - 60 cps. ballasts for single lamps.

However, the ballast cost is recouped because of the lower operating costs of HPMV lamps. Mercury lamps cannot be used for applications where instantaneous output is required when voltage is applied. Furthermore, since the discharge is essentially extinguished twice each cycle, the stroboscopic effect is greater than for tungsten filament lamps (79). However, this is also true (to a lesser extent) for fluorescent lighting as well. Despite these limitations, HPMV lamps excel as compact high-efficiency long-lived light sources.

> ## 5. High Pressure Arcs with Forced Cooling

If outside cooling such as water or compressed air is employed, the loading on quartz arc-tubes may be increased to very high values. This is shown in the following Table:

TABLE 2-12
Electrical and Optical Characteristics of Specially Loaded HPMV Lamps

Operating			Diameter			Loading					
Watts	Volts	Amps	Arc Length (cm)	Tube (cm)	Arc (cm)	I in amp/cm²	Grad. volts/cm	Tube-watts/cm²	Arc-Watts/cm	Eff.- Lum/watt	Bright-ness: c*/mm²
"SUPER HIGH PRESSURE" WATER-COOLED LAMPS@50-100 ATMOSPHERES											
500	400	1.4	1.25	0.6	0.09	220	320	213	400	60	300
1000	840	1.4	2.5	0.6			346	213	400	65	300
2000	1600	1.4	5.0	0.6	0.09	220	320	213	400	65	300
"SHORT-ARC" HIGH PRESSURE LAMPS @ 25-35 ATMOSPHERES											
100	60	1.6	0.3	1.2	0.12	142	200	22	330	45	120
500	60	8	0.5	3.5	0.2	255	120	13	1000	50	200
1000	60	32	0.5	6.0	0.3	455	120	18	4000	50	560

Note that the loading is above 200 watts/cm² with pressures of the order of 50-100 atmospheres. The arc-tube is usually 2.0 mm. ID and 6 mm. OD, so that such lamps may be called "Capillary Lamps". The first such lamps were made at Philips in 1935 (80). The most popular sizes in England and The Netherlands seem to be 500 and 1000 watts, but in the U.S. only the 1000 watt (AH6 or BH6) lamp made by G.E. has been generally available. Experimental lamps consuming up to 10,000 watts were studied by Marden, Beese and Meister (81), but high wattage was only achievable by making the discharge longer. The loading was constant at about 400 watts and the gradient was about 330 volts/ cm. Noel (82) has also given data for arcs in tubes of 1.0 mm. ID and 3 mm. OD. Other water-cooled lamps have been made in the U.S. by the Huggins Co. (Menlo Park, Cal.) and were available in sizes of 1.00, 1.25 and 1.50 mm. ID and gave a brightness correspondingly higher than that of the AH6 lamp (83).

In lamps of this type, not all of the mercury is evaporated during operation. The electrodes are generally tungsten rods projecting above the liquid Hg at the cooler ends of the lamp. These lamps are generally started horizontally and then rotated to a vertical position (43). Because of the high energy dissipation per unit volume, and the external cooling required, both the warm-up and restrike times are very short, less than five seconds. The life is fairly short, about 75-100 hours, and usually results from the devitrification and failure of the quartz. (The operating temperature at the inner wall is close to 1000 °C.). If an explosion should occur, the damage is usually small because of the small lamp volume and the cushioning effect of the surrounding cooling-water.

The efficiency of water-cooled HPMV lamps is around 60-65 lumens/watt and the brightness is about 300 candles (C)/mm^2 (or 30,000 stilb.). However, in experimental lamps, efficiencies as high as 78 lumens/watt and a brightness as great as 1800 candles/mm^2 , compared to 1650 of the Sun, have been achieved (84). Because of the very high pressures, the contribution of the continuum radiation is much greater than in other types of HPMV lamps (see 2.2.27. for example). According to Noel (82), the water-cooled arc at very high loadings in a small tube of 1.0 mm ID has a color temperature of 12,500 °K. The spectral distribution, however, still departs appreciably from a true black-body distribution (see also (84)). The efficiency of water-cooled HPMV lamps is but slightly greater than the normal quartz HPMV lamps. Although the brightness is high, equal values can be obtained with the type of lamps to be described in the next section. In view of this, the short life, and the complication of maintaining efficiency water-cooling, such lamps have found only limited applications in the Lighting Industry.

6. Short-Arc Lamps with Mercury

In this type of lamp, the loading on the arc is increased by decreasing the electrode separation, keeping the voltage constant at about 60-75 volts. The diameter of the bulb is increased at the same time to withstand the increased heat dissipation. Because of the small electrode spacing (usually a few mm.), the bulb is generally spherical in shape and the arc approximates a point source. The operating pressure is usually about 30 atmospheres and the bulb wall must therefore be relatively thick. The brightness of these "Short-Arc" or "Compact-Source" lamps is very great, but the efficiency is not as high as in other quartz lamps (see Table 2-12). This is due mainly to electrode losses. The exact amount of electrode voltage drop in these arcs is not well known, but it must be appreciable compared to the total voltage across the lamp. The geometry is such that processes at the wall of the tube are not very important compared to those at the electrodes. Thus, these discharges are "electrode-stabilized" rather than "wall-stabilized" as is the case for other mercury lamps. According to Elenbaas (34), this approach to HPMV lamp design was first tried with glass envelopes, but with no success. The first successful lamps, using quartz, were made in Germany in 1936 by Rompe and Thouret (85), but the first commercial development occurred in England (86,87). Development in the U.S. did not come until after World War II and was pioneered by Westinghouse (88). Lamps of 250 and 500 watts usually have an outer glass envelope, as do "long-arc" HPMV lamps. But for the largest sizes, there are no outer envelopes. The electrodes are normally inserted into the bulb from diametrically opposite lead-ins, but single-ended lamps have been discussed by Bourne (43) and by Francis (87). The following Table gives some of the characteristics of Short-Arc lamps, viz-

TABLE 2-13
Characteristics of Some Short-Arc Mercury Lamps

Watts (watts)	Volts (Amps.)	Current (cm)	Bulb Diam (mm)	Wall (mm)	Arc Length (cm)	Press. (Atmos)	Life in Hrs	Eff. in (Lumens / watt)	Brightness (Candles / mm^2)
75	51	1.7	1.4	1.8	3.0	60	400	33	44
100*	20	5.0	1.2	1.8	0.3	60	100	10	1000
250	37	8.0	1.6	2.3	2.5	30	200	40	175
500	73	7.0	3.2	3.2	4.5	35	200	43	200
800	70	12.0	4.4	3.2	8.5	10	600	50	125
1000	65	18.0	5.4	3.5	6.5	25	300	50	250
2500	65	45.0	7.3	3.8	10.0	15	400	50	225
2500	50	50.0	7.3	3.8	4.5	30	400	48	550
5000	55	100.0	8.6	4.5	5.0	21	400	46	780

The loading per unit area of bulb in commercial lamps is usually about 15 watts/ cm^2. Envelope design has been discussed by Bourne (43) and by Thouret (45), and right-angle lamps by Aldington (86).

Lamp failure is generally caused by devitrification of the quartz, followed by an explosion. Blackening of the bulb by electrode sputtering during starting and operation increases the heat absorption and accelerates the deterioration of the lamp. Although forced air cooling is possible, lamps are usually designed for convective cooling. The fixture used is designed so that the bulb temperature does not exceed 850 °C. These lamps are normally designed for vertical operation. Although they can be operated in any position, if they are operated more than 10° from the vertical, convection currents in the bulb then cause hot gases to strike the bulb wall directly, rather than the electrode and lead-in area. Because of the violence of lamp failure, it has been recommended that they be removed from service after the output has reached a predetermined level, which depends upon the particular application (89). In addition to the potential for explosion, the powerful ultraviolet radiation and the high bulb temperature create other hazards.

As might be suspected, electrode design is a major factor in short-arc mercury lamps (43). Conditions are quite different at the cathode and at the anode, where the heat dissipation is about seven times that of the cathode (90). For lamps designed to operate on DC, each electrode must be optimized for the function it is to perform. Additionally, it is necessary to specify which will be the upper electrode. For AC operation however, each electrode must alternately perform both functions and hence cannot be optimized for either. In early low-wattage lamps, oxide-activated electrodes were used. Today, thoriated tungsten is normally employed. The electrode tip operates at a temperature close to the melting point of the metal and the growth of small "pips" on them can be a problem. In back of the electrode tip, which may be ball or chisel-shaped, there is normally provided a tight-fitting tungsten coil which may contain a sliver of thorium (88). The larger area thus provided not only aids in dissipating the heat, but also helps in starting the lamp. Since this coil can heat quicker than the massive tip, the discharge normally starts on it and then transfers to the tip (43). The following is one design used for these types of lamps. Lead-ins and seals are also a major problem in short-arc design, viz-

2.2.52.-

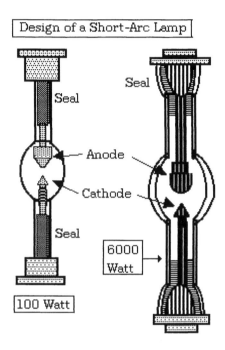

Design of a Short-Arc Lamp

Normally, large xenon lamps are designed to operate on DC because of higher arc brightness, better arc stability and better lumen maintenance. Better performance is obtained with the arc axis in a vertical position. Arc stability is not affected but with the anode uppermost, the convection flow of hot gas from the arc is spread out by the large area of the anode. If the lamp is tilted, hot gas can concentrate heat on the walls of the lamp with catastrophic failure. Actually, seal-design is an art in itself (43,88). The high currents in these discharges must be carried by the molybdenum ribbons necessary for the seals, which are usually only 0.0006 inches thick. Thus, the current density may be as high as 80,000 amperes/in^2. Despite this factor, it has been stated (87) that seals capable of carrying 500 amperes have been made. Both Bourne (43) and Freeman (88) show photographs of 10 Kilowatt lamps while Nelson (86) speaks of 25 Kilowatt lamps "in development" and 50 Kilowatt lamps as "experimental". Despite this, the largest lamps commercially available are about 2500 or 5000 watt lamps.

Early low-wattage short-arc mercury lamps sometimes had starting electrodes similar to those of conventional HPMV lamps, but these are not now used. Aldington (86) also described the use of a heated-coil to start the arc. Mercury short-arc lamps, with a fill gas of argon at a pressure of 5-6 cm. of Hg, normally take about 6 minutes to fully warm up after initial starting. an equal time is required for cool-down. The problem is aggravated by the fact that while the starting voltage is low when the arc is first struck, the current cannot be allowed to exceed a certain value because of electrode sputtering during starting. This problem could be alleviated by use of an external heating coil (and sometimes is solved by this method).

Another solution is to introduce xenon gas at a pressure of a few atmospheres, i.e.- around 5 atmospheres pressure. Such a lamp starts as a xenon short-arc lamp but the mercury pressure builds up rapidly within 2 minutes. Since the ionization potential of Hg is less than that of Xe (see Table 2-4 for example), the normal spectrum and other operating conditions are influenced only to a slight degree by the presence of xenon gas (Retzer and Gerung (88) have discussed this aspect thoroughly). Additionally, xenon decreases the cool-down time, due to higher thermal conductivity, and leads to better maintenance of output by retarding electrode sputtering. This probably results from better starting characteristics, not operational ones. Although the warm-up time is reduced in this way, the lamp can no longer be started with low voltage. High-frequency pulses of 30-50 kilovolts are required for either a hot or cold lamp. But, in this way, restarting of a hot lamp can be almost instantaneous, i.e.- 3 seconds or less. However, such high voltages can only be applied to double-ended lamps. Although mercury lamps are subject to dangerous explosion when hot, mercury-xenon lamps are worse and are always potential bombs. Hence, they are always shipped in individual protective cases which should only be removed when the lamp is to be installed in its operating housing.

The characteristics of some short-arc mercury lamps were presented in Table 2-13. Since, in general, the wall thickness cannot be increased proportionally as the bulb diameter is increased, the permissible operating pressure for high-wattage lamps is usually lower than for small lamps. It should be emphasized that short-arc lamps are not intended for applications where high efficiency or long life is essential. They are especially made to approximate point sources of high brightness and are used in optical systems where this is important such as searchlight and other projection systems. The cost of a 2500 watt lamp can be as high as $500 so that the cost of lamp operation may be up to $1/hour. According to Aldington (86) the maximum brightness that can be produced is given approximately by:

2.2.53.- $B(candles/cm^2) = 3 E^{1.5} I^{1.7}$

or since V is essentially constant for these lamps:

2.2.54.- $B \approx W^{0.7} / L^{1.5}$

High brightness is thus favored by high wattage and small electrode spacing. The spacing, however, also has a strong influence on arc stability and lamp life. In Table 2-13 is shown one lamp (marked with a "*") where the inter-electrode spacing is much smaller than the others. This produces an average brightness of 1000 candles/mm^2, which is roughly a factor of 4 times that of the other lamps shown.

It should be noted that short-arc mercury lamps containing xenon as a fill gas are generally preferred to those containing only xenon, i.e.- the so-called "xenon-lamps". The reason for this is that the mercury lamps are easier to start and maintain their efficiency better during long term operation. We will discuss the xenon short arc lamps in more detail below.

D. The Color of Mercury Lamps and Its Improvement

We have already mentioned the deficiency of the mercury discharge as concerns radiation in the red region of the spectrum. It is very difficult to find data for lamps of various types which may be directly compared. For low-pressure lamps, the energy in each of the emission lines is usually given. For high pressure mercury discharges, where line broadening is important and where the continuum is appreciable, arbitrary wavelength bands are usually employed. Often, data for the ultraviolet region of the spectrum are not included, for it is difficult to separate out the effect of envelope transmission even on the visible emission spectrum. In the following Table, some data for various types of lamps is presented, gathered under such difficulties. Note that, for the low-pressure mercury vapor discharge lamps, 80-95% of the output energy lies the 2537Å line. In contrast, the photochemical lamp, which is a moderately high-pressure discharge lamp, has little of this line energy and its spectrum is spread more evenly across the spectrum. The same may be said for the other high-pressure discharge lamps, except that their arc-tubes and envelopes tend to absorb a major part of the ultraviolet lines that are emitted. Nevertheless, little of the energy lies in the red region of the spectrum, for any of these lamps.

TABLE 2- 14

Spectral Distribution of Emission Energy for Mercury Vapor Lamps
in Terms of Percent of Input Power

| | Low-Pressure | | Photochem. | High Pressure Discharge | |
| | | | | Glass Arc Tube | Quartz Arc Tube |
Wavelength	40 Watt	80 Watt	(2000 Watt)	(450 Watt)	(400 Watt)
1849 Å		8.75			
2483					
2537	55.0	50.0	1.75		
2652			1.45		
2699			0.30		
2753			0.30		
2804			0.95		
2894			0.45		
2967	0.09	0.06	1.05		0.06
3022	0.09	0.06	1.90		0.23
3130	0.09	0.06	4.25		0.29
3342	0.09	0.06	0.45	0.15	0.50
3655	0.32	0.30	5.00	1.56	4.01
4047-4048	0.45	0.62	1.50	1.29	1.52
4358	1.05	1.13	2.65	2,18	3.06
5461	0.60	0.75	3.20	3.42	3.47
5770-5791	0.12	0.25	3.50	3.78	3.90
Continuum	-----	------	-----	------	2.38
Total Watts in All Lines	57.91	63.02	28.70	12.23	19.42

A better comparison of the high-pressure discharge may be seen in the following Table, viz-

TABLE 2-15

Spectral Distribution of Emission Energy for High-Pressure Mercury Vapor Lamps in Terms of Percent of Input Power

Wavelength	400 Watt Glass	400 Watt Quartz	1000 Watt	Water-Cooled
2200-2300				0.06
2300-2400				0.24
2400-2500				0.86
2500-2600				0.30
2600-2700				0.73
2700-2800				0.71
2800-2900		0.01	0.01	1.43
2900-3000		0.06	0.03	1.76
3000-3100		0.23	0.13	1.55
3100-3200		0.73	0.59	2.02
3200-3300		0.21	0.11	0.94
3300-3400	0.01	0.50	0.28	1.69
3400-3500		0.17	0.10	0.60
3500-3600	0.04	0.20	0.14	1.10
3600-3700	0.80	4.01	4.38	3.49
3700-3800	0.06	0.18	0.16	1.46
3800-4000	0.06	0.30	0.24	1.65
4000-4100	0.85	1.52	1.35	2.55
4100-4300	0.09	0.27	0.19	1.75
4300-4400	2.08	3.06	2.85	4.08
4400-5400	0.32	0.76	0.61	4.75
5400-5500	2.83	3.47	3.22	3.12
5500-5700	0.12	0.27	0.23	1.69
5700-5800	2.80	3.90	4.46	2.25
5800-7000	0.63	1.07	1.12	4.40
TOTAL:	10.69	21.27	20.20	45.18

The above data have been presented to show that the emission present from the HPMV lamps lies chiefly in the visible region of the spectrum. Because of this, these lamps are used primarily for lighting purposes. However, the lack of red wavelengths severely restricts their general usage because of the color-rendition of objects by reflection. Faces appear "greenish" at first sight and color reflection deviates considerably from its true chromaticity. The same is not true for Short-Arc mercury lamps where the contribution of the continuum is appreciable. This fact is shown in the following Table:

TABLE 2-16
Spectral Energy Distribution for Mercury Short-Arc Lamps in Terms of Percent of Input Power

Wavelength Range,Å	250 Watt	2500 Watt	Wavelength Range,Å	250 Watt	2500 Watt	Wavelength Range,Å	2500 Watt
2000-2200		0.024	7000-7200	0.60	0.596	12000-12500	1.48
2200-2400		0.068	7200-7400	0.60	0.580	12500-13000	1.24
2400-2600		0.116	7400-7600	0.60	0.584	13000-13500	1.22
2600-2800		0.100	7600-7800	0.60	0.600	13500-14000	1.23
2800-3000		0.288	7800-8000	0.60	0.612	14000-14500	1.11
3000-3200	0.08	0.920	8000-8200	0.68	0.660	14500-15000	1.12
3200-3400	0.20	0.604	8200-8400	0.68	0.668	15000-15500	1.14
3400-3600	0.28	0.608	8400-8600	0.68	0.696	15500-16000	1.00
3600-3800	1.56	2.064	8600-8800	1.32	1.300	16000-16500	1.00
3800-4000	0.40	0.456	8800-9000	1.24	1.224	16500-17000	1.04
4000-4200	1.52	1.746	9000-9200	1.24	1.220	17000-17500	1.16
4200-4400	2.32	2.312	9200-9400	1.24	1.208	17500-18000	0.78
4400-4600	0.56	0.552	9400-9600	1.20	1.236	18000-18500	0.63
4600-4800	0.40	0.404	9600-9800	1.24	1.556	18500-19000	0.50
4800-5000	0.44	0.440	9800-10000	1.56	2.464	19000-19500	0.50
5000-5200	0.40	0.404	10000-10200	2.48	0.908	19500-20000	0.75
5200-5400	0.76	0.740	10200-10400	0.92	0.812		
5400-5600	2.28	2.260	10400-10600	0.80	0.778		
5600-5800	2.40	2.384	10600-10800	0.80	0.828		
5800-6000	0.96	0.964	10800-11000	0.84	0.988		
6000-6200	0.60	0.580	11000-11200	1.00	0.688		
6200-6400	0.56	0.572	11200-11400	0.68	0.672		
6400-6600	0.64	0.686	11400-11600	0.68	0.548		
6600-6800	0.64	0.640	11600-11800	0.64	0.508		
6800-7000	0.64	0.628	11800-12000	0.52			

Note that the emission from these lamps extends from the far-ultraviolet into the blue, green and red regions of the visible portion of the spectrum, and then into the near and middle portions of the infra-red regions of the spectrum. Thus, the short-arc mercury vapor lamp may be said to be a truly wide-range emitting source.

If one is concerned with color of the source, rather than color rendition, then the simplest way to specify such is the C.I.E. trichromatic coefficient system. As a matter of fact, it has only been recently that a color-rendition system has been specified in terms of "% of true color" (53). The C.I.E. system, however, does give an unambiguous method of specifying color as perceived by the human eye.

The following diagram, given on the next page as 2.2.55., is a CIE diagram, showing the locus of colors for mercury discharge lamps. The locus of Plánckian (blackbody) emitters is shown from T = 1000 °K to T = ∞. Also shown are the C.I.E. Standard Illuminants: "A"- T= 2854 °K, typical of a gas-filled tungsten-coil incandescent lamp; "B"- T = 5000 °K., typical of noon-sunlight; "C"- T = 6500 °K., typical of "Daylight" (Sunlight plus skylight); and "E"- typical of equal energy at all wavelengths. Additionally, the locus of colors corresponding to mercury- lamps whose internal Hg-pressure is the temperature, in °C., given on the curve. These are for vapor pressures corresponding to -20 °C to 600 °C, as determined by Jerome (91). At low pressures, the color of a normal mercury lamp is very blue. shifting to a greenish color at higher pressures of the Hg-discharge. However, the color never approaches that of a black-body. Noel (82) has shown that the color depends not only on the internal pressure of the lamp, but also on the voltage gradient present in the lamp. This factor is related, of course, to the amount of continuum radiation contributing to the overall output.

In contrast, the short-arc mercury lamps have a color temperature much closer to the black-body line, as shown by the point, SA, in 2.2.55. Thus, these lamps are much closer to natural light sources than any of the other mercury discharge lamps. This is due, of course, to the much higher degree of the continuum output to the overall spectrum of the lamp, as well as the infra-red portion of the spectrum. Nevertheless, as general sources for indoor lighting purposes, high-pressure mercury discharge lamps have not been found wholly

2.2.55.-

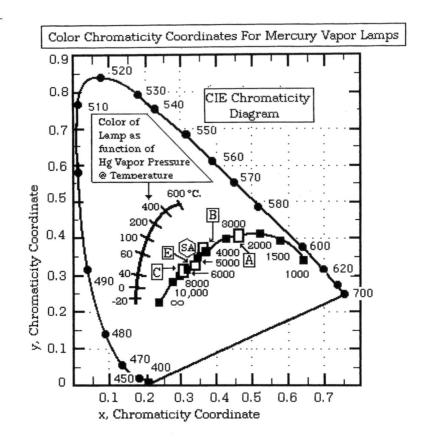

Color Chromaticity Coordinates For Mercury Vapor Lamps

suitable. Even for street lighting at night, such lamps have been deemed to be "off-color" and methods of color-correction have been sought.

There are, in general, three methods of color correction of the high-pressure mercury discharge. These are:

2.2.56.-   1. Use of Phosphors and Filters

2. Combination with Incandescence

3. Additives to the Discharge

1. Use of Phosphors and Filters

Since HPMV lamps emit a considerable amount of ultraviolet (UV) radiation which does not contribute to the visible output, any conversion of UV can only add to the total visible radiation. Furthermore, if the phosphor radiation lies in the red region of the visible spectrum, a color correction may be obtained as well. Thus, both increase and color-correction would be expected to result if a red phosphor were to be used in HPMV lamps. Actually, this was tried with Cooper-Hewitt lamps in 1902, using Rhodamine on an external reflector. Attempts were also made in the early 1930's to use phosphors coated on the internal glass wall of the outer envelope of HPMV lamps. However, it was soon determined that the then available phosphors did not perform well because of the high temperature present in these lamps.

As shown in the following Table, at least 45% of the total output power lies in the UV region of the spectrum.

TABLE 2-17

Relative Peak Heights, and Relative Energy Present in the Output of a 400 Watt Quartz HPMV Arc-Tube

Wavelength Range in Å.	Relative Energy (%)	Percent of Input Power	Wavelength in Å	Energy Relative to 4358Å Line
2450-2500	1.3	0.42		
2500-2620	13.0	4.20		
2620-2670	3.0	0.97	3341	1.0
2670-2725	0.8	0.26	3650	51.6
2725-2775	0.6	0.19	3655	29.5
2775-2830	1.9	0.61	3663	23.5
2860-2925	0.9	0.29	3906	0.3
2935-2990	2.5	0.81	4047	61.5
2990-3045	4.4	1.42	4947	13.4
3075-3175	10.0	3.23	4078	3.2
3290-3390	1.4	4,52	4339	8.2
3600-3700	16.0	5.17	4358	100.0
4000-4100	7.3	2.36	4916	1.4

TABLE 2-17 (CONTINUED)

Relative Peak Heights, and Relative Energy Present in the Output of a 400 Watt Quartz HPMV Arc-Tube

Wavelength Range in Å.	Relative Energy (%)	Percent of Input Power	Wavelength in Å	Energy Relative to 4358Å Line
4300-4400	12.0	3.88	5461	84.3
5440-5480	13.0	4.20	5770	37.7
5740-5810	12.0	3.99	5791	36.4
TOTAL =	100.1	36.41	-----	-----

Thus, the use of phosphors to convert ultraviolet light to visible light appeared to be feasible. Since the major deterrent to use of phosphors was the ability to emit while being held at rather high temperatures within the outer envelope, it became necessary to determine the "Temperature Dependence of Emission" of a number of phosphors. However, none of the then currently available phosphors were found to be entirely suitable. The first practical phosphor developed for this purpose had to wait until about 1950 when the "Fluorogermanate" phosphor , i.e.- $Mg_3 F_2 GeO_4:Mn$, was developed at Westinghouse. At about the same time, a magnesium arsenate phosphor, i.e.- $Mg_2AsO_4:Mn$, was developed at Sylvania. These phosphors are activated by Mn^{4+} and have five emission bands at:

2.2.57.-Emission of Mn^{4+} Activated Phosphors =   6230 Å

6300 Å

6400 Å

6450 Å

6550 Å

These emission lines fill in the gap in the red region of the HPMV lamp emission spectrum and are still in use today. The emission intensity of these phosphors is maintained until a temperature of about 300 °C is reached where it begins to drop off beyond that point. Ideally, since visible light is being added to the total output, the efficiency should also be improved. However, these phosphors absorb some of the blue light generated by the discharge, and the particulate coating on the outer envelope has the tendency to scatter some of the light back into the discharge, where it is absorbed. Thus, the effect on efficiency was measureable. However, the color temperature was improved and the color-coordinates of these "Fluorescent HPMV lamps" tended to cluster close to the points "C" and "SA" shown in 2.2.55.

However, the fluorescent HPMV lamps made from these materials are still not on the blackbody locus, being localized around: x= 0.30-0.35 and y = 0.35-0.40. For a better color, "Deluxe" HPMV lamps utilizing a purple filter on the outer envelope were introduced in 1958 (93). Color improvement by the use of an absorbing filter will obviously effect a loss in total light output, and hence is not very satisfactory from a design viewpoint. Some lamps with a yellow filter have also been made for some special applications (91).

The most recently developed phosphor in use today has a phosphate-vanadate composition, i.e.- $YPO_4 \bullet YVO_4 : Eu^{3+}$ (92). Its emission spectrum consists of but one major peak at 6125 Å. However, its temperature dependence of emission is actually superior to the arsenate and germanate phosphors described above. Furthermore, it does not absorb in the blue region of the spectrum, is less expensive, and has a much broader absorption band. The latter property allows it to utilize the shorter UV wavelengths as well. The phosphate-vanadate phosphor also has a higher quantum efficiency than either the fluorogermanate or the arsenate phosphors. We will, in a later Chapter, discuss the manufacture of these materials and compare them to one another. A comparison of the amount of red emission of the earlier commercial HPMV lamps (as a percent of total light output) is shown in the following Table:

TABLE 2-18
Comparison of Red Emission in some 400-Watt HPMV Lamps

Type of Lamp	Output in Lumens	% Red (Approximate)
Clear	21,500	2.0
High-Output White	24,000	5.0
Standard White	21,000	8.0-10.0
Deluxe White	15,000	12.0-15.0

As shown here, an increase in "red-ratio" above about 5% resulted in a decrease in luminous output of some 28.6%. This was due in part to the optical properties of the fluorogermanate then in use. This situation has been improved markedly by the use of the phosphate-vanadate phosphor.

2. Combination with Incandescence

The question of the most natural or pleasing color for an indoor artificial light source is not an easy one. In general, most people prefer a "warm", i.e.- low color temperature, source to be of low intensity, while a "cool" one, i.e.- high color temperature, should have high intensity. Since HPMV lamps have too much blue radiation and are deficient in red emission, while incandescent lamps have too much red radiation and are deficient in the blue portion of the spectrum, it is natural to try to combine them to obtain a more pleasing combination than either alone. It is true that HPMV lamps have very poor color rendition and should be improved by such a combination. This potential combination may be effected in one of three ways:

1. Normal HPMV and tungsten lamps may be placed in the same, or adjacent, fixtures, but with no electrical connection between them except a common power source.

2. Normal tungsten lamps may be used to partially or completely replace the inductive ballast (94) of the HPMV lamp.

3. Rather than use separate outer glass envelopes as in "2", the tungsten filament lamp may be placed in the same outer envelope, as the quartz arc-tube of the HPMV lamp, to serve as both a resistive ballast and an incandescent source.

Lamps of the last type have been called "mixed", "blended", "dual", or "self-ballasted" lamps. They were apparently first invented in England in 1936 (95), but were produced in larger quantities in Holland and Germany (96). The outer envelope was filled with an argon-nitrogen mixture usual in incandescent lamps, rather than nitrogen as is usual in HPMV lamps. In addition to requiring external ballast, such HPMV lamps emitted light instantaneously when voltage was applied. The additional heat from the filament lowered the warm-up time for the arc to reach equilibrium from about 5 to 2 minutes. However, if power was interrupted, the lamp had to cool before restarting was possible.

Nevertheless, such self-ballasted lamps do have problems. Since the filament and the arc are in electrical series, and no step-up transformer is available, they are usually suitable only for 220 volt or higher operation. This probably explains the lack of interest in the U.S. and England where 115 volts is normal. With an inductive ballast, a reverse voltage across the arc already exists when the current falls to zero so that restarting is facilitated. But with a resistive ballast, both current and voltage go through zero together so that a "dead-time" exists before the voltage increases enough in the reverse direction to allow the arc to strike. Furthermore, the lamp efficiency is intermediate to that of normal HPMV lamps because neither element can be designed to its normal optimum. This arises because an operating compromise must be sought for each element, due to its dual function. Since, on starting, most of the applied voltage appears on the filament, it must be operated at a lower temperature, and hence lower efficiency, while operation of the high-pressure arc with a resistive ballast is not as good as that with a normal inductive ballast. As one can see from 2.2.55., if one wished to obtain a color temperature similar to that of "SA", i.e.- the short-arc lamp, one must either operate the incandescent filament at 2854 °K, and the temperature of the Hg-arc at about 60 ° C, or raise the temperature of the Hg-arc and lower the temperature of the tungsten filament. But, the efficiency of an ideal tungsten filament at 2000 °K is only 2.5 lumens/watt, compared to 19.0 at 2854 °K. Furthermore, there is a problem with lifetime of the lamp. Tungsten filament lamps are cheap and have a normal life of 1000 hours. HPMV lamps are relatively expensive and have a useful life of over 16,000 hours. If they are combined into a single bulb, one is likely to end up discarding the expensive Hg-arc because of the failure of the cheap filament part of the lamp, which is not a very economical situation.

The first commercial self-ballasted lamp introduced by Philips and Osram were 250 watt and 160 watt sizes @ 225 volts. They had efficiencies of 20.0 and 18.8 lumens/watt and a useful life of 2000 hours. (Note that about the same efficiency has been achieved by present-day tungsten filament lamps). Osram later introduced a 500 watt lamp with an efficiency of 22 lumens/watt. In these lamps, the filament and arc-tube contributed equally to the luminous output. About 35% of the applied voltage (or 78 volts) was across the arc-tube in normal operation while 28% of the power was dissipated in the arc-stream. In view of the high cost, the relatively low efficiency, and the necessity for high voltage operation, it is not surprising that such lamps are no longer on the market.

Strauss and Thoret (97) have pointed out that some of the problems normally associated with design of self-ballasted lamps can be relieved if a red phosphor is used to supply red radiation in addition to the tungsten filament. [The red-ratio for a tungsten filament is 29-31% over a wide range of temperatures (97)]. In this case, the filament only has to supply about 17% of the total luminous output and the overall luminous efficiency is therefore higher. They describe lamps designed for 10,000 hours life (5 hours operating time per start) with a "red-ratio" of 13% (the same as Illuminant C). Efficiencies of 29, 30, and 33.5 lumens/watt were obtained for lamps of 450, 750 and 1750 watts, respectively. However, they also point out that evaporation from the overloaded filament during one start is roughly the same as that for 35 hours of normal burning. The filament for a 10,000 hour lamp must therefore be designed for 80,000 hours of normal life. Osram made such lamps in 160, 250, 500, and 1000 watt sizes with efficiencies of 16.9, 19.2, 24.0, and 28.0 lumens/watt, respectively, with a rated life of 5000 hours (3 hours/start).

Another suggestion has come from Nelson (98) who advocated the use of tungsten-iodine lamps for improved self-ballasted lamps. Such a lamp consisted of two quartz tubes, one containing mercury and the other iodine and a tungsten filament. These were placed side by side for maximum heat transfer. An experimental 450 watt lamp (250 watt Hg-arc plus 200 watt tungsten filament) was described which had 11% red-ratio, an efficiency of 26 lumens/watt and a life of 5000 hours. Nevertheless, this lamp has not been commercially successful. Nor have any self-ballasted lamps been manufactured and sold commercially in the United States.

3. Additives to the Discharge

Since the arc-column of the HPMV discharge has a high temperature, i.e. > 6000 °K., it has been suggested that introduction of small particles of a refractory material would benefit the overall output by adding their incandescence to the overall visible emission. It was hoped that if the particle sizes were small enough, they could be maintained in the arc-stream by the convection currents already present. Dusts of: diamond, graphite, MgO, BeO, Al_2O_3, Er_2O_3, ZrO_2, ThO_2, and the Welsbach mixture- i.e.- (ThO_2 and CeO_2) were tried experimentally by Thorington (99) without any real success. Nelson (98) also tried similar experiments with tungsten powder, using mechanical vibration to keep the particles suspended. He reported that "the light form the discharge is strongly affected by the radiation from the hot particles, but the degree of blackening is excessive".

Much older is the suggestion for the addition of materials that will vaporize and/or ionize in the discharge so as to contribute their characteristic emission to the output. Several requirements must be met by such additives, viz-

1. The emitted radiation must lie in the desired red wavelength region of the spectrum for color improvement, including that between the 4358Å and 5467 Å Hg-lines.

2. The desired emission must be efficiently excited by the conditions prevailing in the Hg-discharge.

3. The vapor pressure at the operating temperature of the discharge must be such that the number of additive atoms in the discharge is proper relative to the number of Hg atoms present. Furthermore, a compound composed of the additive must be capable of exhibiting a significant vapor pressure at the operating temperature of the Hg-arc.

4. The additive must not react with mercury, the electrodes, or the arc-tube so as to cause devitrification, excessive blackening, or disappearance of the additive.

These requirements impose severe limitations on choice of the additive. The choice of a useful additive is not easy and has been fraught with severe experimental difficulties. About the only compounds found to be stable enough in the harsh conditions of the mercury arc-stream has been the iodides of various elements. Perhaps this is due to the fact that the electrodes are composed mostly of tungsten, and it is known that the iodine-cycle prevails in such lamps. However, many, if not all, iodides are hygroscopic and are subject to degradation in the presence of oxygen as well. Thus, a successful iodide compound needs to be kept completely dry before it is added to the quartz arc-tube. Additionally, the quartz arc-tube must be free of any water vapor present, since formation of oxides by most additives other than mercury (which is the primary source of loss of the additive from the arc-stream) is a major problem.

As early as 1912, Steinmetz (100) proposed the addition of LiI and other alkali iodides to mercury lamps. In the same year, Wolke (101) used a Hg-Cd amalgam and obtained a lamp with 56 lumens/wattt output. In 1915, Nernst (102) suggested the use of a complicated mixture of 70% $ZnCl_2$, 15% $CaCl_2$, 5% $TlCl_2$, and 5% $CsCl_2$. Schröter (103) in 1920 used alkalis, Cd, and metallic oxides to obtain about 68 lumens/watt from a mercury lamp. Perhaps, the greatest interest has centered on Cd and Zn as additives. Their energy levels are similar to that of Hg (see Table 2-4 for a comparison) in that they do possess metastable levels so that the resonance lines in the near-ultraviolet do not appear at high pressures in the arc-stream. The wavelengths of the main visible lines are shown in the following:

2.2.58.- Visible Lines Present in Discharge (in Angstroms)

Hg-	4047	4378					5467	5770	5791	
Cd-			4678		4800	5078				6438
Zn-			4680	4722	4811					6362

It can be seen that the lines of Cd and Zn compliment those of mercury quite well. However, Winch and Palmer (104) found in 1934 that the addition of sufficient Zn or Cd to give a red ratio of 4.5% caused a reduction in efficiency of about 40%. Spanner (105) also described a 400 watt lamp with hard glass arc-tube which had improved color rendition when Cd was added, but it had only 75% of the efficiency of a corresponding mercury lamp. The life of this lamp was also reduced as compared to mercury lamps of the same genre. Elenbaas (32) has also given data on such discharges and showed that as the amount of Cd was increased, the arc-temperature dropped and the voltage-gradient rose. In another paper, Elenbaas (106) gave spectral data for 1000 watt water-cooled lamps with: 20% Cd-25% Zn, or 13% Cd-17% Zn at 70 hours of life. At one time, there was interest in Cd additions to short-arc lamps (86,88). In this case, the %-red may be increased from 5.5 to 11.5% with only a 4-5% reduction in efficiency. However, none of the above lamps ever became commercial, due perhaps to the fact that Zn and Cd additions appear to be responsible for accelerated the devitrification of the quartz arc-tube.

Thallium has a strong emission line at 5350 Å. Although no color improvement might be expected, it was determined (107,108) in 1940 that the overall luminous efficiency was increased, apparently because excited Hg-atoms can transfer energy efficiently to Tl-atoms, i.e.- sensitized luminescence. Schnetzler (109) obtained 70 lumens/watt in a 250 watt lamp, but with an increase in loading of three times normal to keep the metal vaporized. Larson et al (110) have shown that the use of thallium iodide obviates this problem even in normal sized arc-tubes, although the outer bulb must be evacuated to conserve heat. An efficiency of 78 lumens/watt for a 400 watt lamp was obtained in this manner. In addition to the green line, a weak line from thallium appears at 6550 Å and contributed 5% red to the overall output. The maintenance was 75% after 4000 hours, with a useful life of 12,000 hours.

Sodium has also been added to the mercury arc stream. Nelson (98) obtained efficiencies greater than 100 lumens/watt in which the 5893 Å line of sodium was a major contributor. However, he also stated that the Na-emission was mainly due to thermally excited Na-atoms surrounding the arc-stream and that the discharge was mostly due to Hg-atoms in the arc-stream. Unfortunately, the life of these lamps was very short due to the extremely rapid degree of attack by sodium to the quartz body. One possible method of circumventing this would be to employ alumina (Al_2O_3) tubes as bodies for the arc-stream. However, with the advent of high-pressure sodium arcs (see next section), the incentive for doing so has largely vanished.

Another class of additives that have been thoroughly investigated are the rare earth metals. The rare earths are characterized by the electronic configuration:

2.2.59.- $|Xenon\ Core|\ 6s^2 = |........d^{10}4f^n5s^25p^6|\ 6s^2$

where (| |) indicates inner shells involved with the Xenon core of electrons. The mechanism of ionization involves the $6s^2$ orbitals so that divalent states of these lanthanides are known. However, the most prevalent state is the trivalent one in which one of the $4f^n$ electron orbitals is ionized. Because these orbitals are shielded by the $5s^2 5p^6$ states, line emission spectra of the rare earths is the norm.

The metals, i.e.- non-ionized states, have very complicated energy levels and thus appear, at first sight, to be uniquely suited for usage in lamps. However, the metals are extremely reactive and form oxides (which are very refractory) through reaction with water vapor normally present in the air and embedded in the quartz employed for arc-tubes. Furthermore, the vapor pressure of the metals is not sufficient, at the temperatures of the arc-stream, to get enough metal vaporized within the mercury arc.

Forming the normal trivalent state of the lanthanides by removal of 6s electrons drastically reduces the number of allowed energy levels. The following diagram, given on the next page as 2.2.61, is a plot of the Free-Ion States of the lanthanides.

Note that there are still numerous levels present in these trivalent states. However, it is found that not all work well in the Hg-arc stream. For example, neither Ce^{3+}, Gd^{3+} nor Yb^{3+} would be expected to produce emission in the visible part of the spectrum. Because the emission from mercury comes from the meta-stable levels of the ion, Hg^{2+}, the same might be expected for these ions as well. Experimentally, this is found to be true. Thus, the ions Nd^{3+}, Eu^{3+}, and Tb^{3+} do not work well in the discharge while the ions Pr^{3+}, Sm^{3+}, Dy^{3+}, Ho^{3+}, Er^{3+}, and Tm^{3+} produce spectra which are additive to the usual mercury spectrum. A great deal of work was accomplished in the 1960's to optimize both the output and color rendition of these lamps. As a result, such lamps are now made with the following additives, usually as the iodides (with the exception of Hg):

2.2.60.- Hg^{2+}, Tl^+, In^{3+}, Dy^{3+}, Sc^{3+}, and Na^+

Note that only one rare earth is represented. The purpose of indium, i.e.- In^{3+} is to provide a strong line at 4511 Å while Sc^{3+} provides a red emission line at

155

2.2.61.-

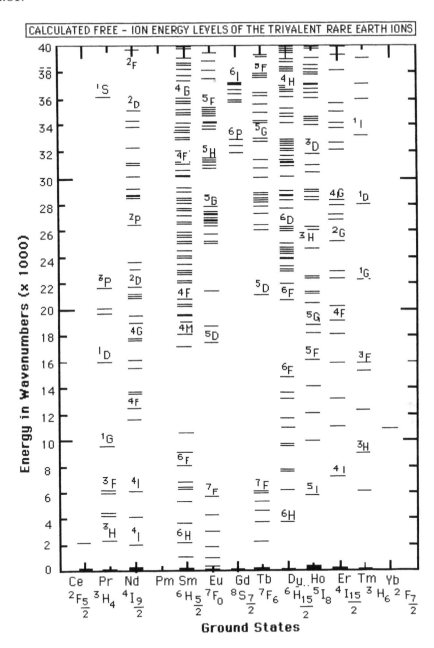

CALCULATED FREE – ION ENERGY LEVELS OF THE TRIVALENT RARE EARTH IONS

5950 Å. Dy^{3+} produces a yellow line at 5450 Å while the purpose of Tl^+ is to produce line emission over most of the visible spectrum to enhance that of Hg^{2+}.

As shown in the following diagram, the emission spectrum of such a "Metalarc™" lamp has a good color-rendition index due to emission lines clear across the visible portion of the spectrum, viz-

2.2.62.-

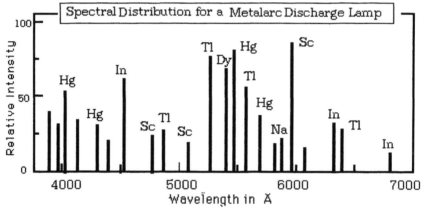

The differences from that emission expected according to the energy levels shown in 2.2.61. and that actually used in 2.2.62. are due to the fact that emission from meta-stable levels is involved. Note that Na^+ is also added, but in limited quantities. The addition of Na^+ and Tl^+ to a mercury arc discharge increases the output from 50-70 lumens/watt in the mercury discharge to 120-150 lumens/watt for the Na-Hg-Tl lamp. However, such a lamp remains "greenish" with a poor color rendition. Through the use of the other metal additives, one can obtain about 90 lumens/watt with **good** color rendition.

The loading of the arc-tube, however, is increased up to three times that of normal mercury discharge lamps and requires a certain lamp construction as shown in the following diagram, given as 2.2.63. on the next page. This may be contrasted to that of the Cooper-Hewitt lamp, also shown there as 2.2.64..

Note that the outer glass bulb of the HPMV lamp is evacuated to conserve heat, and the ends of the quartz arc-tube are coated with a heat-retentive

2.2.63.-

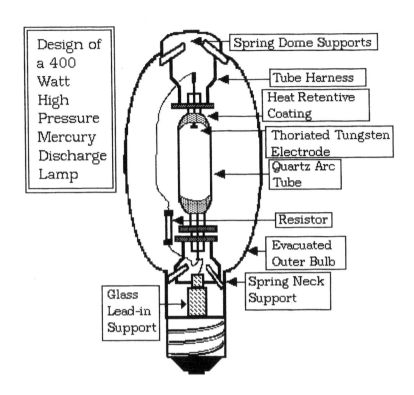

Design of a 400 Watt High Pressure Mercury Discharge Lamp

Spring Dome Supports

Tube Harness

Heat Retentive Coating

Thoriated Tungsten Electrode

Quartz Arc Tube

Resistor

Evacuated Outer Bulb

Spring Neck Support

Glass Lead-in Support

2.2.64.-

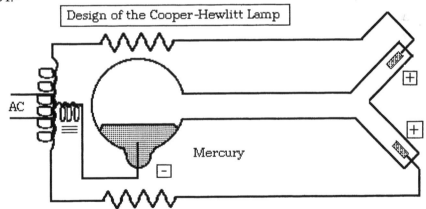

Design of the Cooper-Hewlitt Lamp

AC

Mercury

coating. With the additives shown in 2.2.62., such a lamp has good color rendition, i.e.- about 80%, a life in excess of 16,000 hours and an output above

85 lumens/watt. The Hewitt-Cooper lamp, on the other hand, had an output of 25 lumens/watt with a life of about 1000 hours, for a 350 watt lamp.

From these diagrams, it should be clear that the quality of present-day high-pressure mercury vapor lamps has progressed enormously since the days of their original conception in 1902.

2.3- Sodium Arcs and Sodium Vapor Lamps

As we have already pointed out in Table 2-4, the potential of sodium vapor as a gaseous discharge has been known for some time, having been first described by A. H. Compton and C. C. Voorhis in 1919 (111). These workers investigated some 24 gases or vapors and concluded that only Na, Hg and Ne offered any promise for development into commercial lamps having luminous efficiencies better than incandescent lamps. By using a separate oven to supply external heat, they obtained:

2.3.1.-   Na Vapor = 214 lumens/watt

                  Hg Vapor = 126 lumens/watt

        ·   Ne Vapor = 23 lumens/watt

Later, Druyvesteyn (112), who used similar external heating and who circumvented the electrode loss factor by measuring the gradient in the positive column, found that as much as 87.4% of the total electrical energy input appeared in the yellow sodium lines (the so-called "D"-lines). These lines, a doublet at 5890 Å and 5896 Å, are in an eye-sensitive region of the spectrum, and the corresponding theoretical luminous efficiency approaches 455 lumens/watt. Although such a high value cannot be expected from a practical lamps, it nevertheless demonstrates that good efficiency should be easily attainable in practice.

As shown in the following diagram, the sodium D-lines are resonance lines. That is- the energy level diagram is such that emission from the excited state causes the ion to revert to its spectroscopic ground state. This is just the opposite of that of the mercury atom, where the resonance lines produce

2.3.2.-

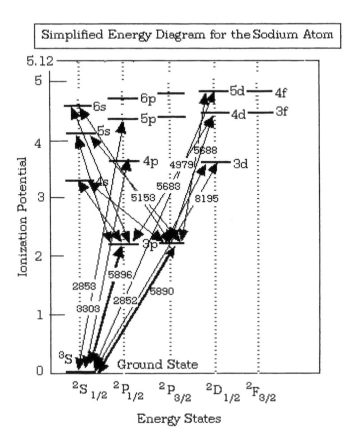

Simplified Energy Diagram for the Sodium Atom

Ionization Potential

Ground State

$^2S_{1/2}$ $^2P_{1/2}$ $^2P_{3/2}$ $^2D_{1/2}$ $^2F_{3/2}$

Energy States

ultraviolet radiation. Sodium is the only element with reasonable vapor pressure whose resonance lines are in the visible portion of the spectrum. This fact is responsible for the high efficiency, but also introduces other complications. In mercury discharges, the resonance lines must be suppressed if good luminous efficiency is to be achieved. In sodium discharges, the opposite is true. This means that the sodium vapor pressure should be kept as low as possible to prevent self-absorption due to unexcited atoms. The current density should also be kept low so that the concentration of excited atoms is low (which promotes cumulative excitation rather than radiation to the ground state). These conditions are just the opposite to those for mercury discharges (112), a fact recognized long ago by Compton and Voorhis (111). The experimentally determined vapor pressure for optimum operation is around 4.0 μ, corresponding to an arc-stream temperature of 270 °C. This may be compared to that of mercury, which is about 40 °C.

160

Heller (113) gives the following comparison of a sodium-lamp and a water-cooled mercury-lamp:

2.3.3.- Comparison of Electrical Properties of Sodium and Mercury Lamps

	Sodium	Mercury
Power Input (watts):	100	1400
Arc-Stream Temperature, °C.:	280	8600
Arc-Tube Pressure (Atmospheres):	10^{-5}	200
Current Density (Amp./cm^2):	0.4	280
Brightness (Candles/cm^2):	10-20	180,000
Efficiency (Lumens/watt):	68	78

For the reasons given above, all sodium lamps manufactured up until about the past 20 years were mainly low-pressure discharges. Then, high-pressure sodium discharge lamps began to replace these. However, the technical problems of development of a suitable high-pressure lamp were severe and several years were spent before such lamps became practical (see description given below).

A. Low Pressure Sodium Vapor Discharge Lamps

The major advantage of low-pressure sodium lamps is that of luminous efficiency, due to the chromaticity of the yellow emission lines. But, this also leads to extremely poor color rendition. If the color rendition of mercury lamps is "not good", then sodium lamps can only be described as "lousy". In the sodium light, objects can only appear yellow, brown or black. For this reason, low pressure sodium vapor discharge lamps were only used for highway lighting where color was of no concern. Even then, they were used extensively only in England and Holland, but not in the U. S.

To attain good efficiency and lamp-life in practical lamps, rather sophisticated design is required. One of the major requirements for high efficiency is the conservation of heat so as to maintain the vapor pressure of sodium in the arc-stream. Since hot sodium vapor is very reactive, a major problem in early lamp development was the glass used to make the arc-tube, so as to have sufficiently long life of the finished lamp. In general, silicate glasses cannot be used and borate or alumino-borate glasses have been used instead. Crystalline quartz is not attacked by hot sodium vapor, but fused silica rapidly blackens and disintegrates. Since the best sodium-resistant glasses are not water resistant, lamp envelopes which were exposed externally to the atmosphere have usually consisted of a coating or glaze of special sodium-resistant glass on the **inside** of a more conventional glass envelope. Obviously, this contributed significantly to the overall cost of the lamps.

The efficiency of sodium-vapor lamps is a function not only of the current density within the lamp, but also is a function of the vapor pressure of the sodium within the operating lamp. The following diagram, given as 2.3.4 on the next page as 2.3.4., shows the dependency of arc efficiency on both of these parameters.

The dashed lines correspond to constant values of the equation: $E \cdot d = 4W/J$ where E is the gradient in volts/cm, and d is the tube diameter in cm. Maintaining internal sodium vapor pressure within the lamp is the most critical of all of the operating parameters. It is for this reason that the use of external envelopes, whether evacuated (Dewar flasks) or not is mandatory to conserve heat.

The early development of low-pressure sodium lamps has been reviewed by Uyterhovnen (114) and by van der Werfhorst (115). Compton and Voorhis (111) made progress toward development of practical sodium-resistant glasses. M. Pirani and E. Lax in Germany also did early work on sodium lamps in 1922. Mackay and Charlton of G.E. studied "low-voltage" discharges in sodium vapor, using an externally heated cathode. If the heater power is neglected, they obtained an efficiency of 138 lumens/watt for the arc-stream. The first practical glass for sodium lamps was developed by Pirani in 1931, and sodium vapor lamps were first used for street lighting by Philips in

2.3.4.-

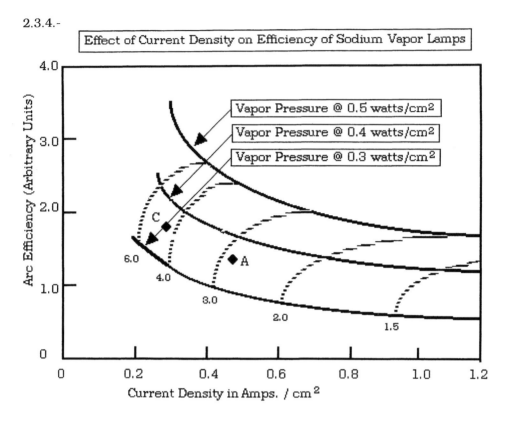

Holland in 1932 (115). These were so-called "low-voltage" lamps. The more efficient "positive-column" lamps were introduced in 1933 (116). In 1934, Druyvesteyn and Warmholtz (117) studied the sodium discharge experimentally. The first commercial sodium lamps developed in the U.S. used an external vacuum or Dewar cylinder to conserve heat (118). They were of the "low-voltage DC" type and most of the light originated from the "negative-glow" region. These were also known as "cathodic-lamps" as there was no positive column in the discharge. They were soon superseded by more efficient types, including those whose emission originated from the positive column. The only other sodium lamps developed in the U.S. used (119) two externally heated oxide-coated cathodes which used 10 amperes of heating current. Each cathode had a surrounding molybdenum anode and operated on AC current. Because they were designed for a very high operating current of 6.6 amperes for street lighting applications, this necessitated a large diameter to length ratio. Thus, these lamps were a combination of cathodic and anodic column emissions, with a luminous efficiency of about 68 lumens/watt.

In Holland, the Philips design (116,120,121) was quite different with a current of only 0.6-0.9 amperes but a higher voltage. The arc-tube was long and narrow, in the form of a "U"-tube. The oxide-coated cathodes were self-heated and the efficiency obtained was 78 lumens/watt. The following Table gives some of the electrical characteristics of these early sodium lamps.

TABLE 2-19

Electrical Characteristics of Some Sodium Lamps

Type	Wattage		Electrical		Dimensions		
Low Voltage- DC	Arc	Lamp	Lamp Volts	Current	Length (cm)	Diameter (cm)	Amps/ cm^2
GE-"4000 lumen"	65	82	13	5.0	5.0	6.3	0.160
Philips		100	10-30				
High Current Externally Heated Cathodes							
GE- "6000 lumen"	105	147	23	5.0		6.3	0.160
GE-10,000 lumen	140	186	25	6.6	23.0	7.6	0.145
"U"-Shaped with External Dewar							
Philips SO45		45	80	0.6	30	1.4	0.39
Philips SO60		60	110	0.6	42	1.4	0.39
Philips SO85		85	165	0.6	65	1.4	0.39
Philips SO140		140	165	0.9	84	1.7	0.40
GEC-Osram		45		0.6			0.32
GEC-Osram		60		0.60			0.32
GEC-Osram		85		0.60			0.32
Philips SOI85		85	162	0.57	52		
GEC-Osram		140		0.9			0.32
Philips SOI140		140	160	0.88	68		
Philips SOI300		300	215	1.6	150	2.0	

164

TABLE 2-19 (CONTINUED)
Electrical Characteristics of Some Sodium Lamps

Type Linear "Power-Groove" Type	Wattage		Electrical Lamp Volts	Current	Dimensions Length (cm)	Diameter (cm)	Amps/cm²
	Arc	Lamp					
AEI		100	70	1.6	32.2		0.80
AEI		150	102	1.6	53.5		0.80
AEI		200	136	1.6	75.0		0.80
AEI		250	170	1.6	96.5		0.80
Osram Na220W		220		1.5			

And in the following table is presented the luminous output and operating properties of these same lamps. As can be seen, a great variety of lamps were made commercially for the street-lighting market. These lamps used an external capacitative starting electrode.

TABLE 2-20
Operating Characteristics of Some Sodium Lamps

Type Low Voltage- DC	Efficiency (Lumens/watt)			Lamp Operating Properties	
	Arc	Lamp	Ballast	Life in Hours	Ave. Maint. @ 4000 hours
GE-"4000 lumen"	63	50	13	1000	
Philips		41.5			
High Current Externally Heated Cathodes					
GE- "6000 lumen"	69	49		1500	
GE-10,000 lumen	90	68		2250	
U"-Shaped with External Dewar					
Philips SO45		60	41	2500	86
Philips SO60		72	53	4000	85
Philips SO85		80	64		89
Philips SO140		78	66		89

TABLE 2-20 (CONTINUED)
Operating Characteristics of Some Sodium Lamps

Type	Efficiency (Lumens/watt)			Lamp Operating Properties	
"U"-shaped-Integral Type	Arc	Lamp	Ballast	Life in Hours	Ave. Maint. @ 4000 hours
GEC-Osram		73	50		87
GEC-Osram		82	61	4000	86
GEC-Osram		94	75		88
Philips SOI85		83			
GEC-Osram		93	80		88
Philips SOI140		79			
Osram Na200W		107	90		
GEC		90			
Philips SOI300		113			89
Linear "Power-Groove" Type					
AEI					
AEI		80			
AEI		92			
AEI		100		4000	92
AEI		102			
Osram Na220W		118	106		

A fill-gas was necessary in sodium lamps not only for starting but also to decrease the mean free path sufficiently so that the energy of the electrons is spent in excitation and ionization of the sodium vapor rather than being dissipated at the glass walls. Neon was usually employed because it leads to more efficient lamps (115). A small amount of argon was also introduced to make starting easier. Because of the relatively low wattage, the warm-up time was rather long, as much as 15-20 minutes being required to reach 80% of the full output. Since a change of 1.0 °C. could cause as much as 4% change in output, the outer Dewar flask was quite important to circumvent sensitivity to air temperature and winds.

Immediately after starting a low-pressure sodium lamp, the emission is essentially due to neon emission. Even at equilibrium, the Ne-lines at 5870-7000 Å constitute about 3-5% of the total luminous flux, in addition to those of sodium at 5890 and 5896 Å. Of the sodium emission, 85% consists of these resonance lines, with 3.5% at 6154-6164 Å, 5.4% at 5683-5688 Å, 1.4% at 5149-5154 Å, 2.4% at 4994-4983 Å, 0.4% at 4748-4752 Å, 1.0% at 4665-4669 Å and 0.6% at 4494-4500 Å (all the lines are doublets). It is clear that the interplay is very complicated in these lamps (116). It is the radial electric field that drives the positive sodium ions to the wall.

Thus, Uyterhoeven (122) states:

> "In good sodium lamps, the presence of sodium ions is limited by a suitable combination of current density, wall temperature and tube diameter to a thin layer near the wall. The light emission therefore takes place in this thin layer".

The core of the discharge is largely neon (122). In this way, the effects of self-absorption and cumulative excitation are reduced. This rather complex situation is also clarified by oscillographs of the voltage and light output during a single cycle (89, 107,123). The neon emission occurs later in the cycle than the sodium emission. Additionally, high-frequency electron and ion oscillations in sodium lamps generate radio noise in the power supply lines which must be filtered out (124).

In 1935, Fonda (125) tried various additives in an attempt to improve the emission color of sodium vapor lamps while still maintaining high efficiency. Mg, Zn and Tl were not successful, principally because of low vapor pressure. Cs and K were also unsuccessful because of their low excitation potential. However, Cd and Hg showed promise. An alloy of 48% Hg, 45% Cd and 7% Na gave pure white light with an efficiency of 36.8 lumens/watt, compared to the yellow light of pure sodium vapor at 69.3 lumens/watt, both of these being the same type of lamp (these numbers do not include the cathode heating power used). For the alloy lamp, a life of 1000 hours was claimed, but these lamps were never made commercially.

The design of sodium lamps remained essentially static for the 20 year period of 1936-1956. There were small improvements made in performance as shown by the following data for the "U"-shaped Dewar lamps:

	1936	1956
Efficiency:	71.5 Lumens/watt (0 Hours)	73.0 lumens/watt (100 hr.)
Maintenance:	57 Lumens/watt (1250 hr.)	65 lumens/watt (2000 hr.)
Average Life:	2500 hours	4000 hours

In 1956, Nelson and Rigden (126) of GEC in England showed that the efficiency could be improved by discarding the detachable Dewar then in use and using instead a single-walled highly-evacuated (and gettered) envelope. This is shown in the following diagram:

2.3.5.-

Influence of the Number of Sleeves Used, Under Optimum Watts on Lamp on Luminous Efficiency of Sodium Vapor Lamps						
Cross Section	# of Sleeves	Opt. Watts	Lamp		Discharge Tube	
			Flux in Lumens	Lumens per Watt	Flux in Lumens	Lumens per Watt
(cross section: 0 sleeves)	0	127	11,300	89	11,600	91.5
(cross section: 1 sleeve)	1	96	9,400	98	9,900	103
(cross section: 2 sleeves)	2	80	8,150	102	8,800	110
(cross section: 3 sleeves)	3	70	7,200	103	8,000	114

In this design, a glass sleeve was placed over both of the arms of the "U"-shaped arc tube. Since the absorption of glass is high for infrared radiation, and since at least half the heat absorbed is reradiated inward, these sleeves served as "heat-shields". The appearance of the lamps was also improved since the detachable Dewar often became dirty during lamp life due to "breathing" when the lamp was turned off and on.

It was also possible to increase the arc-tube diameter and hence the current density, without making the assembly larger than the old Dewar assembly. As a result, the efficiency was increased by about 10% without changing the lamp fixtures at all. This is also shown in 2.3.4. as a change from point "A" to point "C". Additionally, a 280 watt lamp consisting of two 140 watt arc tubes in a single envelope was also made (62), which gave an efficiency of 90 lumens/watt .

About 1960, Philips began to manufacture these so-called "Integral Lamps", which used glass sleeves as heat shields (127). Osram also made such lamps.. The following diagram shows some data obtained on this type of lamp:

2.3.6.-

	With Dewar (1956)	Integral Type (1958)	Integral Type with Sleeve (1960)
Efficiency@100 hrs. in lumens/watt	79	82	100
% Maint.-4000 hrs.	66	83	88

In this design, the evacuated sleeve is the first one which surrounds both arms of the "U"-shaped lamp. A singular improvement (128) is obtained by the addition of just **one** outer sleeve, and the addition of the second and third sleeve makes correspondingly smaller differences in luminous output. The reason for this behavior lies in the fact that as the heat loss is reduced the power input required for optimum vapor pressure is also reduced. This results in a decreased output from the lamp. Thus, the total lumens emitted drops, even though the efficiency of the lamp increases. If absorption by these sleeves increases according to the number of reflecting surfaces present, then heat retention (and arc-temperature) should be increased. However, it is found that after one sleeve is present, an increase in surfaces also decreases the total light output severely through internal reflection even though the luminous efficiency of the arc-stream increases. Therefore in practice, Philips used just one sleeve to contain both legs of the arc-tube. The following data were given by van der Weijer (128) for 140 watt lamps:

	With Dewar-1956	Integral Type - 1958	Integral type with Sleeve - 1960
Efficiency-100 hr	79	82	100
% Maintenance-4000 hrs.	66	83	88

A considerable improvement in both efficiency and maintenance of output was seen during this period of time. Note that an efficiency of 100 lumens/watt is shown. A 300-watt lamp with an efficiency of 113 lumens/watt was also claimed by de Boer and van de Weijer (128).

These authors also showed that the use of a thin gold film on the envelope, which reflected about 92% of the long wavelength infra-red radiation while transmitting 50% of the sodium light, increased the efficiency of a 140-watt lamp from 103 lumens/watt to 125 lumens/watt. However, this was accompanied by a drop in total light output from 14,000 to 4,400 lumens and a drop in wattage from 136 to 35 watts. If correction was made for the sodium light absorption by the gold film, the actual efficiency of light generation came out to be 225 lumens/watt. Nonetheless, these lamps are not attractive commercially.

The fill gas pressure in low-pressure sodium lamps is usually about 9-10 mm. As mentioned earlier, about 1% of argon gas is normally added to the neon to lower the breakdown voltage for starting. Verwey and van de Weijer (127) give a plot of breakdown voltage as a function of gas composition and pressure. The latter author also showed (127) that as the pressure in an optimally loaded lamp was reduced from 15 to 1 mm. Hg-pressure, the output increases from 12,900 to 13,800 lumens, while the power required drops from 137 to 121 watts. This corresponded to an increase in efficiency from 94 to 114 lumens/watt. He also found that the operating voltage is a minimum at 4 mm. However, this increase in efficiency cannot be utilized commercially. It has been found that the glass used in these arc-tubes preferentially absorbs argon during life (128) at a rate depending upon the voltage gradient. Thus, there is a "gas-life" of such lamps as well as a "cathode-life" to be considered. At low pressures, this loss drastically affects the lamp life as well as the maintenance of output.

Since the pressure of neon is 10^3 - 10^4 times that of the sodium vapor pressure, the mean free path of sodium atoms is very small. Therefore, it is necessary to keep the sodium distributed over the complete length of the lamp-tube. Migration of sodium vapor out of the arc-discharge is a major problem and it is for this reason that most low-pressure sodium vapor lamps are operated in a horizontal position. On the other hand, Philips (127) and AEI (129) provide dimples in the glass spaced periodically along its length to provide reservoirs for sodium. The interplay between sodium and neon vapors is contradictory. If the excitation of sodium is to be large and the efficiency therefore high, then the voltage-gradient (watts/cm) in the discharge containing sodium needs to be less than that of neon. On the other hand, if this is the case, the power input to a section of the discharge deficient in sodium would be expected to be higher than that in the rest of the column and hence sodium would be driven away from this region, thereby aggregating the situation. Therefore, the most efficient sodium positive-column will be therefore unstable and some compromise must be achieved. Van de Weijer (127) studied this situation in detail and found that only for pressures between 2.5 and 9 mm. pressure is the input to the positive column higher in the presence of sodium than in its absence. He also gave 0.83 for the power factor of the discharge in rare gas alone and 0.93 for the factor when sodium is present. Only in this region is the discharge stable as far as diffusion of sodium out of the arc is concerned. It is because of this (and that of gas clean-up) that the very high efficiency obtained at low pressures cannot be utilized in practice. Van de Weijer considered 6 mm.-Hg pressure as a good compromise and showed the following improvement in efficiency and maintenance (using neon gas with 0.5% argon):
2.3.7.-

			Efficiency in Lumens/Watt			
Hours:	100	1000	2000	3000	4000	5000
Pressure						
9 mm.:	97.7	94.6	91.9	88.2	81.7	81.4
6 mm.:	102.7	97.9	96.0	94.6	95.8	93.9

It was pointed out above that for efficient excitation of the resonance sodium emission lines, a low current density is required to prevent cumulative excitation of the upper levels. By reference to 2.3.4., if one draws a line between "A" and "C", one can see that an arc-loading of about 0.36 watts/cm^2 is used in practice. The efficiency is also higher at low current density for another reason, i.e.- the increase in electron temperature.

However, the electron temperature, and therefore the efficiency, may increased in another way. That is, one can change the wall-losses of the discharge by increasing the ratio of wall-surface to volume of the lamp. This can be done by establishing a non-circular discharge, and was first accomplished by G.E. in their "Power-Groove" low pressure mercury vapor fluorescent lamps in 1957 (as will be described in detail somewhat later). The same principle was applied by Weston(129) of AEI in England to low pressure sodium lamps in 1959. These lamps were made "linear" (double-ended) rather than "U"-shaped and thus resemble fluorescent lamps, except for the outer vacuum jacket. Weston stated that the discharge becomes self-constricted, i.e.- circular, and the advantages lost, if the ratio of current to inner tube diameter is greater than about 0.2 amperes/mm. (at a temperature of 230 °C). The "grooves" are formed in such a way as to also provide a cool section for sodium vapor condensation so as to maintain a constant supply over the entire length of the discharge. A series of lamps were designed, as shown in Tables 2-16 & 2-17, with efficiencies up to 118 lumens/watt. This high efficiency was achieved despite the fact that that the current density is higher than usual at 0.8 amps./cm^2 so that the loading is 0.215 watts/cm^2. Thus, during the period of 1956 to 1966, the efficiency of low pressure sodium lamps was increased by some 20%. Still, one of the major problems has remained that of glass quality. Better glasses might result in further improvement. One such type that may be applicable to this problem is the polymerized phosphate glasses developed by Ropp (130), which do not have internal diffusion mechanisms because of their polymeric nature. They do not absorb Na$^+$ from solution, but would have to be tested for resistance to attack by hot sodium vapors.

The latest improvements of efficiency from Philips has been about 120-140 lumens/watt, but the advent of high pressure sodium vapor lamps has curtailed further introduction of improved low pressure sodium vapor lamps into the marketplace. Low pressure sodium vapor lamps are undoubtedly the most efficient light sources known to date, with the possible exception of low pressure fluorescent lamps with green phosphors and equally bad color. Compared to high pressure mercury vapor lamps, also used in street lighting, they have a two to one ratio in efficiency, but a shorter life and a longer warm-up time.

B. High Pressure Sodium Vapor Discharge Lamps

Work on this type of lamp originated only about 45 years ago (131). The major problem in using a high pressure discharge for sodium vapor is the container. Most studies have involved wall-stabilized discharges, using a xenon gas fill. To generate the necessary internal pressure, the lamp needs to operate at about 750-850 °C. Internal discharge pressures up to 920 mm-Hg, which corresponds to about 900 °C. in the discharge, have been generated and current densities up to 30 amperes/cm^2 have been used. The only material capable of withstanding this sort of operating conditions has been a translucent polycrystalline alumina, i.e.- Al_2O_3, called "Lucalox™" by G.E. As shown below, it is manufactured by isostatic pressing of oxide powder in the form of a long thin tube.

Studies by Schmidt (131), Nelson (132), and Rigden (133) discovered that the luminous discharge by sodium vapor at high pressures becomes strongly broadened and is a self-reversed resonant doublet with overall wavelengths as great as several hundred angstroms. In an operating lamp, xenon gas is used along with mercury vapor to lower the starting voltage. The operating pressure is about 100-200 torr and is wall stabilized with a relatively high loading factor of 20-30 amp/cm^2. The following diagram, given as 2.3.8. on the next page, shows the spectrum obtained from a high pressure discharge.

Note that the normal resonance doublet of sodium at 5890 and 5896 Å. is missing. At this pressure of the sodium vapor, continuum emission is present and self-absorption of the "D-lines" exists so that a big "hole" is missing from this part of the spectrum. In the red and green parts of the spectrum, lines exist which are due to mercury vapor as well. Thus, the emission appears to be fairly "white". It is this factor which accounts for the fact that these lamps have replaced both high-pressure mercury vapor and low-pressure sodium

2.3.8.-

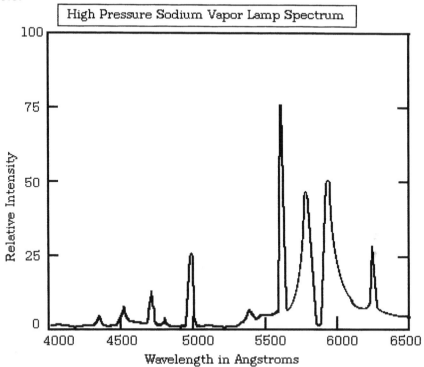

vapor lamps as "street-lamps" today. An efficiency of over 140 lumens/watt has become common (500 watt discharge lamps). Indeed, it has become difficult to find any other lamps used for street lighting purposes, except where color-rendition is more important than light intensity. Then Metalarc™ lamps are used. The translucent alumina tubes are manufactured by a specific process which includes:

2.3.9.- <u>Process for Preparation of Lucalox™ Tubes</u>

     1. Slurry Preparation
     2. Isostatic Pressing to form a "green" tubing
     3. Cut to Length
     4. Sinter in Hydrogen Gas
     5. Grind and Polish

A typical slurry preparation uses alumina plus a spinel compound, i.e.-
$MgAl_2O_4$, in a water-based suspension. It is necessary to spray-dry the
alumina, as received, in order to obtain good flow properties. A typical slurry
preparation may use:

2.3.10.- <u>Typical Slurry Preparation</u>

1. Linde-A (0.3 µ) Alumina powder = 400.0 gram
2. Spinel ($MgAl_2O_4$)   = 2.0 gram
3. Polyvinyl alcohol   = 4.0 gram
4. Lomar™ PWA   = 8.0 gram
5. Polyethylene glycol   = 1.0 gram
6. Distilled water   = 234 ml.

Spinel is added to facilitate the sintering properties of the alumina; Polyvinyl
alcohol is used as the binding agent to obtain good "green-strength" and thus
facilitate handling prior to sintering; Lomar™ PWA is a quaternary ammonium
organic-amine having no metal ions present and is used as a dispersing agent;
Polyethylene glycol is used as a "softening agent", i.e.- lubricant (Tradename:
Carbowax - 200). The polyethylene glycol and Lomar™ PWA are best dissolved
in water before use. About half of the alumina is added to all of the spinel in
100 ml. of distilled water. Then 40 ml. of a 10% solution of Lomar™ PWA plus
20 ml. of a 10% polyvinyl alcohol are added with stirring. This slurry mixture
is then blended in a stainless-steel Waring Blender for about 30 seconds. The
remaining alumina is then added in equal increments with about 15 ml of
water added each time, while continuing the Waring Blender treatment. The
resulting slurry is then stored prior to spray drying.

Spray-drying is critical to the overall process to obtain a uniform spherical
agglomerated powder with good flow properties. Such a powder will then
completely fill the constricted cavity needed to form the tube, sans void
formation. Typical operating conditions include: Atomizer rotation = 30-
40,000 rpm. at about 295 °C inlet temperature and 150 °C. exhaust chamber
temperature. Slurry feed rate should be about 60 ml./min. Spray dried
powders made under these conditions were found to be composed of spheres
of agglomerated powder about 20 to 50µ in diameter. Since the temporary
binder tends to produce brittle agglomerates upon drying, the lubricant is

added to obviate this condition. A special mold is needed for the process of forming the green-tubes. One such is shown in the following diagram:

2.3.11.-

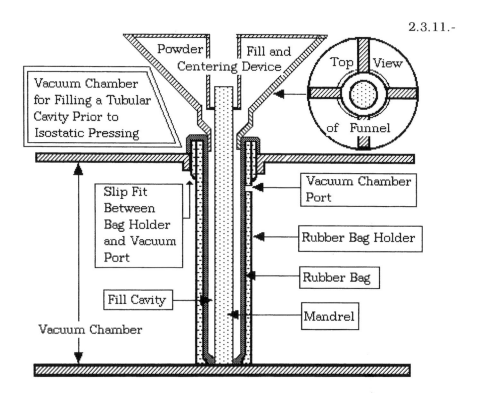

The mold consists of a polished rod (mandrel) of circular cross-section placed in a tubular rubber bag of about 1/16 inch wall thickness with one end in a closed nipple configuration to receive one end of the mandrel. The rubber bag is inserted into the mandrel and then is put into place in the holder, via the centering funnel. Vacuum compaction is used to completely fill the cavity with powder, while circumventing "voids" in the finished piece. The rubber bag assembly is then removed and placed in a suitable press and pressed isostatically at about 30,000 psi. for 20 seconds at 150 °C. The assembly is then removed from the press and the rubber bag is stripped from the green-piece.

It is cut to length, if necessary, and then pre-sintered for 2 hours in air at 1300 °C. to completely remove the organic binders. A final sinter of 2 hours in hydrogen gas at 1600 °C. brings the final density of the tubes close to theoretical density of alumina. Process yields of greater than 95% are to be expected.

The above binding and dispersion agents were chosen to maximize the "green" strength of the pressed pieces to facilitate handling and to minimize residual organic residues after sintering. Typical dimensions of the finished tubes were:

2.3.12.-   Dimensions Required for 400 Watt Lamps

    Length   = 6.00 inches
    Outside diameter = 0.483 ± 0.003 inches
    Wall thickness   = 0.044 ± 0.001 inches

The density of the pressed tubes was found to be about 2.00 grams/cc which, when corrected for a binder loss of 3.5%, gives a linear and volume shrinkage of 21% and 50.3%, respectively. Final density of the fired tubes turns out to be very close to that of pure alumina at 3.98 grams/cc. Obviously, the overall tube-fabrication process needs to take these shrinkage factors into consideration by adjustment of process variables, i.e.- sizes, so that the final size of the alumina tubes remains close to the values given above.

The resulting tubes are then end-polished with a fine abrasive. This is essential to prepare end-seals capable of withstanding the operating temperatures of the lamp throughout its entire lifetime. Most high pressure sodium vapor lamp failures can be traced to faulty seals, or seals that degrade over the life of the lamp. Niobium is the only metal that has been found to form suitable seals to alumina and only certain materials have been found to be capable of making seals that withstand operating parameters of these lamps. The whole process may be summarized as given in 2.3.13. on the next page.

Various designs of a high pressure sodium vapor discharge lamp have been studied (134). These included the following designs, given as 2.3.14. on the next page. The metal parts, made from tantalum and niobium are first fabricated and then cleaned by degreasing in trichloroethylene. They are then acid-cleaned in 60% HNO_3 plus 15% HF. Rinsing in distilled water follows and drying in alcohol is the final step. Tungsten electrodes are spot-welded

177

2.3.13.-

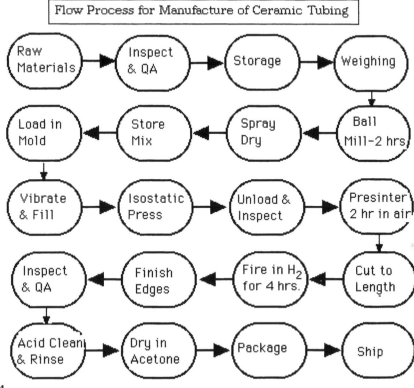

Flow Process for Manufacture of Ceramic Tubing

2.3.14.-

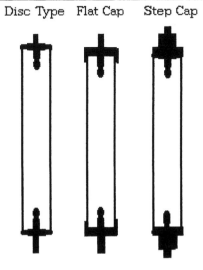

Three Types of Arc Tube Designs

Disc Type   Flat Cap   Step Cap

into place and then are coated with yttria suspended in a suitable solvent such as acetone. The alumina tubes are first cleaned by boiling in 60% HNO_3, then are rinsed in distilled water and dried with alcohol.

Seals are made using an alumina-yttria eutectic of 65% Al_2O_3 - 35% Y_2O_3 , which melts at about 1750 °C. A paste is made by grinding together the two oxides, and adding amyl acetate with further mixing. The paste is applied to surfaces of both the alumina tube and the niobium metal seal cap and allowed to dry. The two are placed in a fixture to hold them in close proximity during firing. The next step involves loading them into a vacuum furnace which is then evacuated. In this way, any residual organic matter is completely volatilized. The furnace temperature is then raised at a rate of about 50 °C./min to a maximum temperature of 1760 °C., held for 2 minutes, and the temperature is quickly lowered to 1680 °C. and held for about one minute to allow the seal to form and freeze in place. At the upper temperature, the eutectic quickly wets both surfaces and then forms the seal at the lower temperature. The temperature is then lowered at a rate of 25 °C. to cool down. After 2-3 hours when the temperature is about 200 °C., the vacuum furnace is back-filled with about 2 torr of argon or xenon gas. At about 100 °C., the gas pressure is increased to one atmosphere. Teflon plugs are inserted in the metal tubulation before the tubes are removed. The arc-tubes are then leak-tested. It has been found possible to make vacuum-tight ceramic to metal seals with the following binary eutectics, viz-

2.3.15.-   Binary Eutectic Seal Mixtures

Eutectic Mixture (Alumina +)	Phases Present (α- Al_2O_3 +)	% Alumina	Maximum Temp.
Y_2O_3	Garnet	80	1860 °C
Nd_2O_3	Perovskite	77	1730-1785
Sm_2O_3	Perovskite	77	1760
Gd_2O_3	Perovskite	77	1760
Eu_2O_3	Perovskite	77	1660-1760
Pr_2O_3	Perovskite	77	1500-1850
Ho_2O_3	Garnet	80	1500-1850
Dy_2O_3	Garnet	80	1660-1760

However, yttria is normally used because of cost and availability. Other sealants have included an alkali-resistant glass composed of calcium aluminate plus silica. Once the arc-tube has been fabricated, the lamp can be made. The tube is loaded with about 60.0 mg of a 5% mercury - sodium metal amalgam alloy, after it has been evacuated to about $3\text{-}5 \times 10^{-6}$ torr. It is then back-filled with about 15 torr of xenon or argon, which serves as the starting gas (xenon produces more efficient lamps). The metal tubulation is then externally sealed by brazing. It is then mounted in a stainless-steel frame which is set in place in an outer glass envelope, and the end-cap is mounted. The outer envelope is then evacuated and the final seals of the lamp put into place. The final step is to activate the barium-aluminum metal getter.

The following diagram, given as 2.3.18. on the next page, shows the arc-tube mounted to form the finished lamp. Note that the outer glass envelope is evacuated to conserve heat. There are several parameters which can be varied to affect the operation of the lamp. These include:

1. Length of the Electrode Assembly (which affects both the cold-spot temperature and the length of the luminous arc.

2. Design of the end-caps and seals used

3. Discharge arc loading

Following the methods used by Rössler (135) and Burns and Nelson (136), the total energy radiated can be obtained by integrating the spectral energy distribution (SED) of a typical lamp operating at a given wattage. From the SED of the lamp, the integrated energy can be obtained from the following equation:

2.3.16.- $\qquad E_T = \sum (iE)\, \Delta \lambda \quad \text{and} \quad Y = \sum (iE)\, \Delta \lambda \cdot y$

where Y and y are the usual color-coordinate constants, E_T is the total integrated energy output, and $\Delta \lambda$ is the wavelength range of the doublet. Thus, the visible radiated energy, E_r , can be calculated from the equation:

2.3.17.- $\qquad E_r = (L \times E_T) / Y \times f$

2.3.18.-

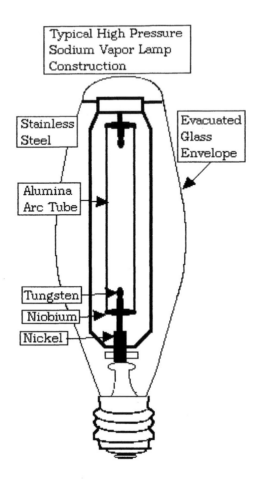

Typical High Pressure
Sodium Vapor Lamp
Construction

Stainless
Steel

Evacuated
Glass
Envelope

Alumina
Arc Tube

Tungsten

Niobium

Nickel

where f is the mechanical equivalent of light equal to 680 lumens/watt. If ñ is equal to Y/E_T , then the intrinsic efficiency can be obtained form:

2.3.19.- $E_1 = f \times ñ$

The transmission loss has been found to be about 7%, of which 5% is loss through the arc-tube and 2% is loss through the outer glass envelope.

The following Table shows typical operating characteristics for the three types of arc-tube constructions shown in 2.3.14. as given on page 177 , vis-

TABLE 2-21

Operating Characteristics of 400 Watt High Pressure Sodium Vapor Lamps

End-Members:	Flat-Cap	Disc	Step-Cap
Seal:	ceramic	ceramic	ceramic
Loading: Watts/cm^2	19.2	19.4	19.4
Lamp Wattage	400	400	400
Voltage across Arc-Tube	100	97.5	103
Current (amperes)	5.2	5.0	4.8
Cold Spot Temp., °C	685	590	615
Length of Electrode	35 mm.	45 mm.	40 mm.
Operating Temperature of Seal, °C.	735	730	820
Color Temp., °K.	2151	2026	2140
Color Rendering Index	35	22	33
Width of Reversal Line, Å	140	80	120
Chromaticity- X =	0.512	0.519	0.511
Y =	0.412	0.407	0.415
Z =	).071	0.075	0.74
Hg Lines	Moderate	Strong	Strong
Calculated Operating Pressure in Plasma (Torr)	148	78	118
Total Efficiency (Lumens/ Watt)	410	398	392
Luminous Efficiency	118	111	104

Note that the temperature of the seal is specified, as well as the Cold-Spot temperature. The latter is a function of the length of the tungsten electrodes as placed within the arc-tube. In our case, this was deliberately varied in each type of tube construction.

If we assume that the electrode loss is about 10% of the operating voltage, or 40 watts, the energy balance can be calculated as shown in the following Table, as given in the table on the next page.

TABLE 2-22

Energy Balance for High Pressure Sodium Vapor Lamps

End-Members:	Flat-Cap	Disc	Step-Cap
Light Output, Lumens	47,000	44,500	41,500
Visible Radiated Energy in Watts	115	112	103

Infra-red and Non-Visible Energy Loss in Watts	285	288	297
Electrode Loss, Watts	40	40	40
Transmission Loss, Watts	8.0	7.8	7.2
Conduction Loss, Watts	237	241	251

Radiated Efficiency in Visible Region in %	28.4	27.9	25.8

Luminous Efficiency in Lumens/watt	117.5	111.3	103.8

The values given for calculated luminous efficiency in Table 2-21 may be compared to those measured in Table 2-22. The agreement is quite remarkable. Measured effects of operational lamp parameters are presented in the following diagram, given as 2.3.20. on the next page.

Note that the parameter, cap-design, has a considerable effect on these operating parameters. However, we are more interested in the lamp brightness which is also remarkably affected by these same parameters, as shown in the following diagram, also given on the next page as 2.3.21.

Most of the seal failures, which are the major cause of lamp failure, are caused by micro-cracks at the seal-interface between the ceramic arc-tube and the metal end-cap. Thermal on-off cycling of these lamps is probably the main contributor to propagation of micro-cracks. No significant evidence was noted due to attack of the hot sodium vapor on the micro-crystalline alumina tubulation. The luminous maintenance decreases linearly at a rate of about 5% every 3,000 hours of operating time up to about 20,000 hours.

2.3.20.-

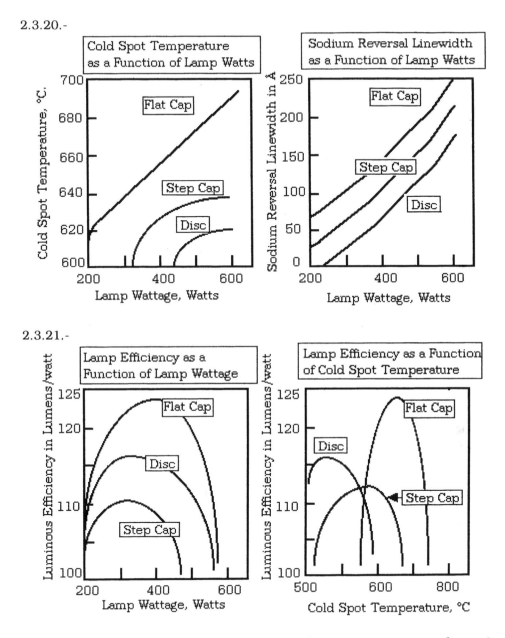

2.3.21.-

Over 2/3 of the input energy is lost as non-luminous energy, as shown in Tables 2-21 & 2-22. If these losses were to minimized, the luminous output could certainly be improved.

Nevertheless, in 1992, most of the highway and street lighting lamps employed are of the high pressure sodium vapor type. This has arisen because of at least two factors: 1) Very high output with "reasonable" color rendition, and 2) much longer operating life and maintenance of output than any other type of lamp, including arc-type mercury vapor lamps. Only where good color rendition is essential have the "color-corrected" HPMV lamps found usage for outdoor lighting purposes.

Before we leave the subject of design of high-pressure sodium lamps, we should point out that the technology made possible here is also applicable to development of a "super-output" arc lamp having improved color and lighting properties. The use of additives in the HPMV lamp produces a number of lines in the spectrum, as shown in 2.2.61. Nevertheless, such a lamp does not have the street-lighting capacity of the high-pressure sodium lamp. However, it should be possible to combine the capabilities of both types of lamps into one. A continuous spectrum between 5300-6100 Å can be obtained from the sodium spectrum. At 6100 Å, the intensity of the red wing (see 2.3.8.) is one of the strongest parts of the spectrum, but the efficiency of such lamps is only about 120 lumens/watt. However, as much green as needed for a balanced spectrum can be obtained by the addition of thallium, i.e.- TlI_3, whose emission line is at 5350 Å. If additional blue is desired, use of InI_3 produces a line at 4511 Å. By use of the ceramic Lucalox™ tube, a lamp could be produced whose output approaches 460 lumens/watt with good color rendition. This lamp would make possible improvements in street lighting and would obsolete the current high-pressure sodium lamps now in widespread use. Such a lamp would also undoubtedly find use in indoor lighting or lighting at stadiums because of its superior color and performance.

2.4.- Other Gaseous Discharge Lamps

The only other gaseous discharges that are used commercially are the rare gas discharges, i.e.- helium, neon, argon, krypton and xenon. Helium was first detected spectroscopically on the Sun in 1868 by Sir Joseph Lockyer. Argon was first separated from air in 1893 by Sir William Ramsey and Lord Rayleigh, while helium was first found in the mineral, Cleivite (from radioactive decay), in 1894. The remaining rare gases were extracted from air by these same workers in 1898.

The components of dry air are shown in the following Table, along with their known boiling points. It is easy to see that most of the components of air, other than the first three, are present only in parts per million. Thus, the so-called "rare gas", Argon, is a major component of air. Only when Georges Claude in France and Carl von Linde in Germany (137) developed techniques for liquifying air did the others become available in large enough quantities for commercial use.

Table 2-23

Composition of Normal Dry Air

Substance	% by Volume	ppm.	Boiling Point in °C.
Nitrogen	78.00		- 195.8
Oxygen	20.95		- 183.0
Argon	0.93	930,000	- 185.7
Carbon Dioxide	0.03	3000.0	- 78.5 (sublimes)
Neon	0.0018	1800.0	- 245.9
Helium	0.0005	500.0	- 268.9
Methane	0.0002	200.0	- 161.5
Krypton	0.0001	100.0	- 152.9
Nitrous oxide	0.00005	50.0	- 88.5
Hydrogen	0.00005	50.0	- 252.7
Xenon	0.000008	8.0	- 107.1
Ozone	0.000001	1.0	- 112.0

The resonance lines of all of the rare gases lie in the far ultraviolet, as was presented in Table 2-4. These atoms, which have closed electron shells, are characterized by having their first excited level relatively close to their ionization level. Although the ionization potential of Xe is only a little greater that that of Hg, its lowest (meta-stable) level is almost twice as far above the ground state. Any visible emission from excited gas atoms therefore arises from transitions between excited states.

A. Neon Discharges

The familiar red light from neon discharges consists of a large number of lines whose initial states are a group of levels lying 18.3 to 18.9 e.v. above the ground state to final states of another group lying about 16.6 to 16.7 e.v. above the ground state. When Claude first separated this gas, he used it in a long *positive-column* cold-cathode type of lamp, the so-called "Moore type" of lamp (see page 201 of this chapter). Actually, neon lamps can be made in two low-voltage varieties, the *positive-column* type, i.e.- **"neon lamps"** and a negative-column or *cathode-glow* type of discharge, called **"neon glow"** lamps. In Claude's positive-column lamps, he employed low current densities and improved cathode materials in order to obtain improved life. The short life of early lamps was found to be due to gas "clean-up". That is, disappearance of the gas was determined to be due to entrainment at the glass walls by material sputtered from the electrodes. Reducing the sputtering not only decreased the blackening observed but also led to better maintenance of internal gas pressure during operation. Claude's first neon lamps were demonstrated in Paris in 1910 (138) and were very similar to those still in use today.

At an early stage of development, Claude used a combination of Ne and Hg discharges in a low pressure positive-column lamp to achieve better color than that of Ne alone. Later, other investigators used other rare gases, colored glass envelopes were employed and much later, phosphors were used. The last type is actually a "cold-cathode fluorescent lamp" and will be discussed below in greater detail.

Data on the construction and performance of "neon tubes" has been discussed in several places in the prior literature (107,114,139,140,141). Some typical data for neon lamps are shown in the following:

2.4.1.-   Typical Operating Data for Neon Lamps

Parameter	Value
Hg pressure	6-12 mm.
Diameter	15 mm.
Length	Min. of 3.0 ft.
Current	25 milli-ampere
Voltage drop	130 volts/ft.
Electrode drop	115-150 volts
Efficiency	24 lumens/watt
Expected life	10-15,000 hrs.

Typical transformer voltages are usually about 10-15 KV for 60 foot of tubing. Thus, these devices can be characterized as low current, high voltage drop lamps, having the spectral characteristics of the rare gas (neon) discharge. The effects of current density and of neon pressure are shown in the following diagram:

2.4.2.-

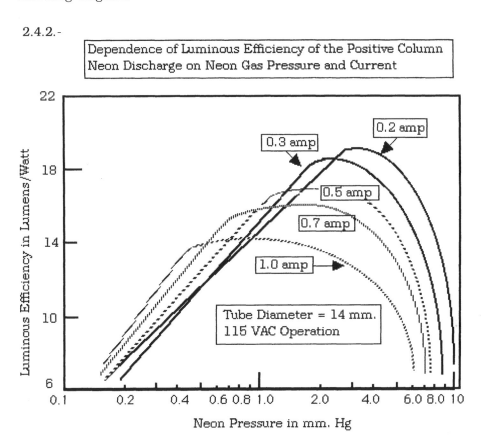

Dependence of Luminous Efficiency of the Positive Column Neon Discharge on Neon Gas Pressure and Current

Note that the luminous efficiency peaks at the lowest current, but that there is a narrow range of neon gas pressure at which the output is maximum. In actual lamps, the pressure used at the outset is usually higher than the optimum value needed for maximum efficiency so as to offset the effects of gas clean-up. It is this factor that prolongs the effective life of the lamp. Because of the decrease in efficiency with increase in current density, neon lamps are characterized by low current operation. The high voltage drop at the electrodes necessitates long and narrow lamp bodies if the efficiency is not to be degraded. The voltage drop at the electrodes may be reduced by employing externally heated cathodes. In 1928, Found and Forney (142) described a lamp 125 cm. long and 2.2 cm. in diameter, operating at 220 volts, 3.0 amperes and producing 10,000 lumens, with a life of 500 hours. The efficiencies obtained by these workers was 12-15 lumens/watt for D.C. operation and 10-12 lumens/watt for A.C. operation. The emission color was a reddish-white. For a while, this type of lamp produced considerable excitement which waned when sodium lamps with much higher efficiency were introduced.

It should be clear, then, that the usual type of neon lamp is a high voltage-very low current discharge lamp, having cold cathodes, whose operation depends upon the internal gas pressure for maximum luminous output. These are the lamps that are used for the various displays and advertisements seen operating (usually) in the night-time. A notable exception is that of Las Vegas where such lamps are operated continuously. The winding and molding of the long glass tube into letters and motifs to form a sign is, in itself, a difficult art.

Low voltage cold-cathode lamps may also be made in which the neon emission originates from the "cathode-glow" region (negative-column) of the discharge (3,143,144). These are the so-called "neon glow" lamps. They are normally made in sizes from 0.04 to 3.0 watts for operation from 110 volts A.C. lines with a simple resistive ballast. The internal operating pressure used is about 15-40 mm. and the life of these glow-lamps is extremely long, being about 3000 to 25,000 hours before failure. However, since the efficiency is usually only about 0.3 lumens/watt, such lamps are used as indicator lamps rather than as light sources and/or luminous signs. They may also be used as stroboscopic sources up to 15,000 cps. (145). If similar lamps are made with argon gas instead of neon, the emission is principally in the ultraviolet (strong line at 3948 Å) which allows the use of phosphors in some cases. However, the luminous efficiency of low-pressure, low-current discharges of the other rare gases is not as high as that of neon.

B. Other Rare Gas Discharges

Electrical discharge in the rare gases has been studied by several workers (147,148). Francis (147) concluded that a dissipation of 10,000 watts/cm.

would be necessary in helium to obtain conditions and efficiencies similar to those of Hg at 10-100 watts/cm.

The following Table, due to Suits (149) shows that the luminous efficiency does indeed increase with increasing pressure, but never becomes very great in the case of He or Ar.

TABLE 2-24

Luminous Efficiency and Brightness of Gas Discharges

Gas	Pressure (Atm.)	Efficiency in Lumens/Watt	Brightness in Candles/mm^2
Helium	1	0.40	0.09
	10	0.85	0.24
	50	1.70	1.22
Argon	1	0.4	0.055
	10	3.0	0.87
	50	10.7	5.10
	100	17.0	17.5
Nitrogen	1	2.30	0.22
	10	3.10	0.89
	50	17.0	9.10
	100	27.0	29.0

However, there does exist at least two different ways of improving both the efficiency and the emission colors in this case (146):

1. If a pulsed discharge is used, then high peak power may be achieved at low average power. In this way, the "spark" spectrum, which includes emission from multiply ionized as well as neutral atoms may be obtained (150). Such lamps, particularly that of xenon, have been studied in detail by LaPorte in France, both from the point of view of illumination and of stroboscopic applications (see below for further details of this type of lamp). Although LaPorte (150) achieved an efficiency of 40 lumens/watt in this way, this method does not seem to be an attractive

approach to general illumination. Nevertheless, pulsed xenon lamps were used for lighting the Kremlin (151).

2. At high pressures, continuum radiation from molecules is obtained (152). As in the case of Hg, excited molecules of the rare gases are stable, although they are unstable in the unexcited state (It is this fact that is the basis of the "eximer" laser). According to McCallum, this molecular emission is favored by large tube diameter and low current density.

Ultraviolet lamps based on rare gases were in use as early as 1932 (153). Later, Tanaka and Wilkinson (154) described several microwave-excited ultraviolet lamps utilizing the continuum radiation from these rare gases. Elenbaas (148) studied wall-stabilized discharges in the heavier rare gases in 1949 and found the following results for the dependance of output (as L, lumens/cm of arc length) on arc loading (as P, in watts/cm.):

2.4.3.-

Gas	L	P
Argon	9.8	> 35
Krypton	19.0	> 20
Xenon	27.0	> 13

It is seen that the luminous efficiency of such high pressure discharges is greatest for Xe, which has the lowest ionization potential. It is for this reason that only Xe lamps are of any commercial importance for illumination. It may also be noted that the heat conduction losses in such lamps decreases with increasing atomic number so that only a Xe gas-discharge is comparable to that of Hg at 10 watts/cm loading. Elenbaas also showed that this quantity is proportional to the coefficient of thermal conductivity of the gas (which indicates the same gas temperature in all cases). This is somewhat surprising and has not been very well explained.

Little information has been available on the actual gas discharge temperatures for this type of lamp arc. Elenbaas (148) found minimum potential gradients at loadings of 100, 45 and 21 watts/cm. for Ar, Kr and Xe, respectively (the corresponding figure for Hg is 30 watts/cm.). The value of the potential gradient, compared to the Hg discharge, is much lower for these gases by as much as a factor of five. Aldington (155) also found that Argon gave an efficiency of only 20 lumens/watt as compared to that of Xenon at 30 lumens/watt. He gave the following trichromatic coefficients for these high pressure gaseous discharges, viz-

2.4.4.-

Trichromatic coefficients

Gas	x-coordinate	y-coordinate
Argon	0.309	0.288
Krypton	0.338	0.323
Xenon	0.336	0.334

Thus it was agreed that a lamp based on a xenon discharge offered the best way to produce a high efficiency lamp, radiating in the visible portion of the spectrum.

C. Xenon Discharges

The high-pressure xenon gas discharge was studied intensively by Schulz (156) in Germany during and just after World War II. Subsequent commercial development was largely stimulated by his work. Development was undertaken both in England (155) and in the US where the earliest effort was at Hanovia and Westinghouse (157). By far the most effort was made, however, in Germany at Osram (158). Larche (159) found that maximum efficiency is obtained for an arc loading of 6000 watts/cm. and that the efficiency of the arc column was 5.0 candles/watt or 63 lumens/watt. Brightnesses of 1.20 x 10^6 stilb, or 12,000 candles/mm^2, were obtained with an electrode separation of 0.10 mm. At 800 watts/cm., Schulz and Steck (160) obtained 50 lumens per watt in efficiency for the same type of lamp. Bauer and Schulz (161) studied short-arc lamps with loadings up to 3500 watts/cm. and pressures up to 55 atmospheres. The arc temperature was studied by both Schirmer and Bauer (162) and the virtual electron temperature was determined to be about 8700 °K. Elenbaas (163) investigated the effect of adding a small amount of mercury to the xenon discharge, but, as might be expected because of the lower ionization and excitation potential of Hg, the emission from that lamp was due primarily to mercury, even when very small amounts of Hg was added. On the other hand, Schulz and Strub (164) found that addition of He (with its high ionization potential) increased the potential gradient without interfering with the spectrum or efficiency of the lamp. These authors also studied the effects of adding Hg or hydrogen gas to the xenon gas, while Gamber and Schultz (165) studied the addition of hydrogen gas to argon discharges. Higher arc-gradients, increased average brightness (due to a narrower arc stream), and more uniform brightness distribution may be achieved in this way with little change in the visible spectrum (the α-hydrogen line at 6563 Å is seen). However, the overall operation is affected by diffusion of hydrogen gas through the hot quartz bulb and by overheating of the upper electrode. These factors cause practical problems in both manufacture and operation of these lamps.

The outstanding feature of xenon lamps over similar mercury-xenon lamps lies in their superior color rendition. It is probably the best of any artificial light source known today (166). Comparatively, the contribution of the mercury lines makes the xenon-mercury lamps less pleasing. A typical spectrum of pure xenon lamps is shown in the following diagram:

2.4.5.-

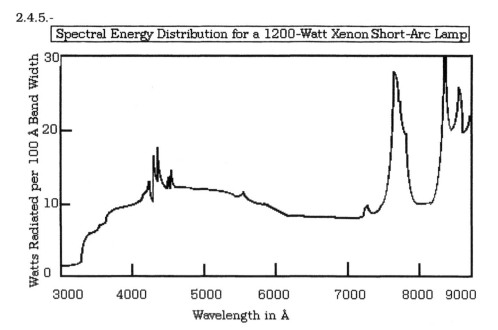

It is easily seen that the output in the visible region of the spectrum is almost entirely that of a pure continuum. The color temperature has been given variously as 5200-6400 Å. This may be due to how much of the radiation from the hot tungsten electrodes is included. Because of the ultraviolet component in their output, they have also been used as "solar-radiators" (167). The most important application has been in motion-picture projection (168). Their geometry and size makes them ideal replacements for the old carbon-arcs, and many motion-picture theaters use a 1600 watt xenon lamp (Obviously, the overall apparatus must provide for disposal of the ozone generated by the far-ultraviolet part of the radiation generated). Thus, commercial xenon lamps have been stabilized in only three different types, as shown in the following, i.e.- Tables 2-25 and 2-26. We will discuss the three types of commercial lamps immediately following:

TABLE 2-25

Electrical Characteristics of Some Commercial Xenon Lamps

1. Short-Arc Lamps

Manufacturer	Electrical Input			Spacing	Bulb Diam.	Pressure
	Watts	Volts	Amps	(in cm.)	(in cm.)	(Atm.)
Durotest	50	13.0	3.8	0.016	1.1	30
	50	22.0	2.3	0.200	1.1	30
	75	13.3	5.6	0.016	1.1	30
	75	23.0	3.3	0.200	1.1	30
	100	13.5	7.4	0.016	1.1	30
	100	24.0	4.2	0.200	1.1	30
	2000	22	91	0.300	5.4	20
	5000	25	200	0.200	6.4	6
	8000	37.5	213	0.830	10.2	6
Durotest	10000	40	250	0.830	10.2	6
Hanovia	150	22	8.5	0.385	2.0	20
	300	20	16	0.385	3.2	20
AC	800	32	28	0.650	3.9	35
DC	900	32	42	0.650	3.9	35
Osram	150	19	8	0.385	2.0	20
	450	20	23	0.250	2.9	20
	900	22	42	0.330	4.0	25
	1600	26	63	0.400	5.2	25
	2500	30	83	0.600	5.7	12

TABLE 2-25 (CONTINUED)

Electrical Characteristics of Some Commercial Xenon Lamps

2. Water-Cooled Long-Arc Lamps

Manufacturer	Electrical Input			Spacing (in cm.)	Bulb Diam. (in cm.)	Pressure (Atm.)
	Watts	Volts	Amps			
Osram	1000	90	11.5	6.5	0.8ID-1.1 OD	25
	2500	115	22.3	7.5	"	25
	6000	135	45	11.0	1.7ID-2.0 OD	25
	6000	165	37	11.0	"	25

3. Air-Cooled Long-Arc Lamps

Manufacturer	Electrical Input			Spacing (in cm.)	Bulb Diam. (in cm.)	Pressure (Atm.)
	Watts	Volts	Amps			
Osram	3000	115	23	30	2.2	120 mm.
	6000	140	41	60	2.2	120 mm.
	10000	140	73	75	3.0	80 mm.
	20000	270	75	150	3.0	80 mm.
	65000	270	250	190	5.0	40 mm.
	75000	270	250	220	5.0	40 mm

And in the following Table, the operating characteristics of the **same** lamps given in Table 2-25 are compared in terms of efficiencies and life of the lamps. The loadings used and the gradient present in these lamps are also given.

TABLE 2-26

Operating Characteristics of Some Commercial Xenon Lamps

1. Short Arc Lamps

Manufacturer	Lamp Watts	Loading			Brightness Candles/ mm^2	Effic. Lum/ watt	Rated Life in Hours
		Arc (Watts/ cm^2)	Tube (Watts /cm^2)	Gradient (Volt/ cm)			
Durotest	50	3120	13	810	800	4.0	
	50	250	13	110	70	10.0	
	75	4800	20	830	1250	6.0	
	75	375	20	115	130	12.5	

TABLE 2-26 (CONTINUED)

Operating Characteristics of Some Commercial Xenon Lamps

1. Short Arc Lamps (Continued)

Manufacturer	Lamp Watts	Loading Arc (Watts/cm^2)	Tube (Watts/cm^2)	Gradient (Volt/cm)	Brightness Candles/mm^2	Rated Effic. Lum/watt	Life in Hours
Durotest	100	6250	26	840	1700	7.5	
	100	500	26	120	170	15.0	
	2000	25000	39	125	2100	28	
	5000	6000	15	39	850	37	
	8000	9650	24	45	1250	40	
	10000	12000	30	48	1450	41	
Hanovia	150	390	12	57	75	20	300
	300	780	9	52	125	20	300
AC	800	1230	16.5	49	180	32	300
DC	900	1380	19	49	180	32	300
	1600	4000	19	65	650	34	2000
Osram	150		12			12	1200
	450	1880	17	80	450	26	300
	900	2700	18	67	550	32	300
	1600	1380	19	49	180	32	300
	2500	4170	24,7	50	610	40	1500

2. Water-Cooled Long-Arc Lamps

Manufacturer	Lamp Watts	Loading Arc (Watts/cm^2)	Tube (Watts/cm^2)	Gradient (Volt/cm)	Brightness Candles/mm^2	Rated Effic. Lum/watt	Life in Hours
Osram	1000				23	26	600
	2500				33	31	600
	3000	100	14.5			26	600
	6000	545	216	15.0	30	34	1000

TABLE 2-26 (CONTINUED)

Operating Characteristics of Some Commercial Xenon Lamps

Lamp		3. Air-Cooled Long-Arc Lamps					Rated	
		Loading						
	Watts	Arc	Tube	Gradient	Brightness	Effic.	Life
Osram	3000	100	14.5			26	600
	6000	100	14.5	2.3		23	5000
	10000	133	14.1	2.3	12,000	25	3000
	20000	133	14.1	1.8	12,000	27.5	3000
	65000	340	22	1.4	12,000	31	3000
	75000	340	22	1.4		30	3000

It should be clear from these Tables that some very large lamps are capable of very high output. Nonetheless, it is easy to see that efficiencies obtained do not compare to comparable mercury xenon lamps shown in Table 2-12 and Table 2-13.

1. Short Arc Lamps

These lamps are quite similar to the corresponding Hg-lamps. However, because of the lower gradient in xenon, the operating voltage is only 20-30 volts. The current is therefore two-three times that of mercury lamps of the same size. Because of this factor, seal and electrode design is therefore even more important than for mercury-xenon lamps. Additionally, since the electrode drop must be at least 10-12 volts/cm., the low operating voltage has an effect on the efficiency (90). Commercially, lamps are available in the range of 150-2000 watts, but experimental lamps as small as 50 watts and as large as 10 kilowatts have been made (169). The maximum efficiency that has been achieved is about 40 lumens/watt, but this is higher than predicted by Elenbaas' early measurements (148). Nonetheless, the brightness of xenon lamps of comparable electrode spacing and wattage is always lower than that of mercury lamps. However, if the arc is made very small, extremely high brightness may be obtained. The brightness is always greatest in the vicinity of the cathode. It may be noted that the loading on the quartz envelope for Durotest lamps listed in Tables 2-25 & 2-26 is greater than that normally used for mercury lamps, i.e.- 15 watts/cm^2.

The pressure used in the smaller lamps is about 30 atmospheres but that of the larger lamps is reduced to 6 atmospheres. Because of these pressures, the danger of violent explosion always exists with this type of lamp, hot or cold. As compared to mercury lamps, xenon lamps have the advantage of "instant" starting. In reality, of course, there is a difference in pressure, by a factor of 2-3, between a hot lamp and a cold one, but the output immediately after starting is about 80% of the final value and full output is reached after one minute, Because of the high pressures, these lamps require high-voltage pulses for starting. Starting and ballasting circuits have been discussed by Ramert (170).

Most xenon short-arc lamps are made for operation on DC current. Because of the presence of a gas-jet from the cathode, the stability is greatest if the anode is above the cathode (158). Indeed, stability and life are adversely affected if the lamp (and arc) is not operated in a vertical position. Buckley (171) studied the modifiability of xenon short-arc lamps when he used a third electrode for starting. However, it is true that, in general, all such lamps must be operated vertically with the anode on top of the cathode.

2. Water-Cooled Wall-Stabilized Lamps

Water-cooled xenon lamps were developed in England (155) and Germany (158), but were never made in the United States. Rompe and Ihln (172) studied this type of lamp. Typical data were given for these lamps in Tables 2-25 & 2-26. The tube diameter is very large compared to that for water-cooled mercury lamps, and "capillary" lamps cannot be made because of the high pressure used. Little use was found for this type of lamp and they have essentially disappeared.

3. Air-Cooled Wall-Stabilized Lamps

Lamps of this type, which were called "gas-arcs", were made by Aldington and Cumming (155) at Siemans in England. They are characterized by a *high pressure* air-cooled arc. Little use was found for them and they, too, have disappeared from the lamp-scene.

4. Air-Cooled Low-Pressure Xenon Lamps

The most spectacular development in xenon lamps seems to have been the *low-pressure* air-cooled "long-arc" lamp introduced by Osram in 1958 (173). It was found, in contrast to the short-arc high-pressure type, that if the internal pressure was reduced, a different type of emission resulted. These lamps were physically large, compared to other xenon lamps, and were designed to produce very high luminous output, rather than high brightness. The internal pressure of these lamps was only a fraction of an atmosphere when cold, and only about one atmosphere during operation. Thus, no danger of explosion existed, in contrast to other xenon lamps. Despite the low pressure, the current density was such that "thermal" conditions were achieved, i.e.- the electron temperature is equal to the "gas" temperature. The radiation is thus that of the white high-pressure xenon continuum (with small differences noted). The theory for this type of discharge was worked out by Schirmer (174) and the good performance predicted before such lamps were made. He found that for each pressure a value of optimum tube radius and current exists, above which the arc is of the high-pressure type. Schirmer's calculated curves went up to about 35 lumens/watt, while the commercial lamps average about 30 lumens/watt.

The electrodes in these lamps were made of thoriated-tungsten and operation was on AC current. To reduce the blackening due to evaporated and sputtered tungsten, a quartz aperture was placed before each electrode. Heat radiators were used on the electrical leads. Lamps of 6, 10, 20 and 65 kilowatt sizes were made commercially, with larger sizes being studied (174). Such lamps were designed for outdoor lighting of athletic-fields and the like. Three of the 20 kilowatt lamps were mounted on a 30 meter mast to illuminate the public square in Munich to celebrate the 800th anniversary of that city in 1958.

Of the input energy, 22% is lost by conduction, 2% as ultraviolet radiation, 12% as visible light and 64% as near infra-red from the arc, bulb and the electrodes. They are primarily designed for outdoor lighting on 100 foot masts. Operation of a 65 kilowatt lamp is truly formidable and operation in a small room would surely exemplify the Osram motto "Hell wie der lichte tag". Even 2% of the 65 kilowatts of generated radiation will produce a sunburn at a respectable distance within a few minutes. These lamps remain rather expensive, at about $600-800 each. Osram has studied the addition of mercury to these lamps (207). By properly adjusting the geometry, it is possible to operate these lamps directly from the power lines without using a ballast. However, line-voltage fluctuation severely affects light output. The Russians have also studied ballastless long-arc xenon lamps of the Osram type (217) and have made a 100 kilowatt water-cooled lamp operating at 380 volts AC. However, Lompe (174) at Osram topped this figure by demonstrating a 120 kilowatt long-arc lamp.

D. Discharges with Consumable Electrodes

The "magnetite arc" lamp was placed on the market by G.E. in 1903 (2,3,175). This resulted from the work of Steinmetz et al. This lamp was similar in principle to the "flaming carbon" arc mentioned in the previous chapter (see p. 56). The magnetite arc lamp had a non-consumable positive electrode made of copper which was cladded with iron to prevent oxidation. The lower negative electrode was made from an iron tube packed with iron oxide (magnetite) having additives of TiO_2 and Cr_2O_3. The lamp operated only on direct current and consumed 4 amperes at 75 volts. The arc was of the "flame" type and was white in color. One type had an electrode life of 350 hours and an efficiency of 11.5 lumens/watt, while another had a life of 150 hours with an efficiency of 18 lumens/watt (2).

A so-called "titanium arc' was also used at one time (176). This consisted of a cathode of titanium carbide and a copper anode. This lamp also operated as a "flame" arc and gave a higher efficiency than any competing light source of the period [30 lumens/watt for the "standard" type and up to 84 lumens/watt for experimental models (177)]. There was also a "spark" light source proposed by Eppley (178) operated in air between carbon electrodes impregnated with a mixture of MoO_3, TiO_2, and V_2O_5 to obtain a closely spaced line spectrum in the visible region. However, it was never used as a light source.

Arcs between tungsten electrodes have also been used as light sources. Early designs were suggested by Gordon in England in 1880 and by Whitney in the U.S. (USP 902,354 - 1903). The first really practical lamps were the Edison "Point-O-Light" sources developed by Gimingham and Mullard (179) and the "Tungs-Arc" lamp developed by Friederich at AEG in Germany. Several of these lamps are described by Scaupy (180). The emission is simply that of the incandescence of the hot tungsten electrode. The thermal conductivity of tungsten is, of course, much higher than that of carbon. Electrodes for tungsten-arc lamps are therefore usually mounted on thin wire supports to decrease the conduction of heat. Nitrogen, with some admixed argon, was the usual gas employed, but neon was sometimes used.

Starting of these arc lamps could be accomplished by any of the following methods:

    1. Ionization of the gas by auxiliary means such as an adjacent tungsten filament.

    2. Use of over-voltage.

    3. Touching and then separating the electrodes, usually by a thermal operated bimetallic element.

    4. First, starting the lamp as a glow discharge and allowing it to change to an arc.

The characteristics of several sizes of "Point-O-Light" lamps were given by Bourne (59). The operating voltage was 45-50 volts with a current of about 0.5-0.8 amperes, depending upon the size of the lamp. The life was about 500 hours, the color temperature about 3100 °C., the brightness about 7-15 candles/mm^2, and the efficiency varied from 19 lumens/watt (20 watt size) to 31 lumens/watt (400 watt size). These lamps are now obsolete. Arcs between tungsten electrodes in air or in nitrogen gas are still used in welding (181) with the temperature being about 6440 °K. Cobine has stated (145) that the high luminosity of tungsten-arc lamps in air is due to incandescent particles of tungsten oxide.

G.E. (182) made a lamp similar to the "Point-O-Light" lamp, but with the addition of mercury, i.e.- the S-1 Sunlamp. This consisted of two tungsten-ball electrodes which were shorted by a tungsten filament. Upon initial application of voltage, this filament drew 9.5 amperes of current at 33 volts. Electrons from the filament ionized the mercury which caused the arc to strike. Because of the lower resistance of the arc, most of the current then went through the arc (30 amperes at 30 volts). During operation, the filament contributed 7% of the luminous output, the mercury arc stream 18%, and the tungsten electrode 75%. An inductive ballast was usually employed. Although not designed as a visible light source, the luminous efficiency was 19 lumens/watt, with a color temperature of 4000 °K. The principal application was, of course, as an ultraviolet source. However, such lamps have been largely replaced by other types of mercury-vapor lamps, as discussed earlier. A similar lamp called the "Odor-Out" lamp was developed at Westinghouse, using the same principle as the old S-1 sunlamp. However, the power input was only 4 watts (0.35 amperes at 10.5 volts), using an inductor or an ordinary 40-watt incandescent lamp as ballast. The output at 1849 Å (1 milliwatt) was responsible for destroying malodorous substances, while the rest of the radiation was at 2537 Å whose intensity was about 1200 milliwatts (70).

Another type of arc-lamp was developed by the Western Union Telegraph Corp. in 1942 (3,183). They were of two varieties. The "enclosed" type was made in wattages of 2-300 watts which operated on direct current. The cathode consisted of a tube of tungsten, molybdenum or tantalum packed with zirconium oxide. The anode, usually made of molybdenum, was a flat plate spaced close to the cathode and contained a hole slightly larger than the cathode spot. Most of the emission came from the cathode spot. The gas filling was argon at about atmospheric pressure. The lamp was started by application of over-voltage (1000-2000 volts applied from a choke-coil and switch). The operating voltage was 37 volts for the 2 watt lamp and 20 volts for the larger sizes. During operation, the cathode becomes covered by a thin film of molten Zr metal formed by reduction of the oxide and operated at a temperature of about 2800 °K. The film is so thin that surface tension was sufficient to keep it in place and permitted operation of the lamp in any position. Any Zr metal lost by evaporation is replenished during operation by return of the metal as positive ions where reduction of the oxide to metal then occurred (in addition to further reduction of the oxide by the electron stream). Most of the radiation came from the incandescent Zr-metal cathode surface, but some also came from the excited gas and vapor in the arc stream. The latter has both a continuous component and the lines of excited and singly ionized Ar and Zr (mainly ArI and ZrII). The following diagram shows the visible spectrum generated by this lamp, as given on the next page:

2.4.6.-

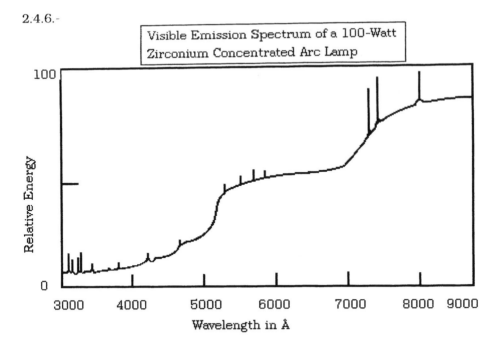

Note that several emission lines are exhibited in the infra-red region, as well as a continuum. The color temperature is about 3200 °K. For a 100 watt lamp, the cathode current density was 250 amp./cm^2, brightness 50 candles/mm^2, the efficiency 10 lumens/watt, and the life about 1000 hours. In contrast, brightness for the 2 watt lamp was 90 candles/mm^2, but the efficiency was only 2 lumens/watt and the life 175 hours. The factor associated with this is that the cathode spot size is a direct function of the current present in the arc. Thus, at low currents and a small spot size, the brightness is high but the overall output is comparatively low. If a compact high-brightness source is needed, then the 100 watt lamp (or larger wattages) would be indicated. It was observed that brightness and light output of the enclosed Zr-arc lamp decreased about 50% during the first few minutes of operation. As the lamp continued to operate, brightness then increased to 150% of the 10-minute value, accompanied by a 25-30% reduction in both spot diameter and total light output. It was concluded that the Zr film became too thick under equilibrium conditions and dissipated too much heat to the cathode walls. If non-oxidizable Pt metal parts were used in experimental lamps, then the oxygen pressure could be increased, the reduction of the oxide was retarded, and the high output was maintained. Howver, this was too costly for commercial use.

This led to another type of open Zr-arc which was operated in air. The active electrodes consisted of a Ni-rod packed with a mixture of Ni and Zr. The function of the Ni powder was to slow up oxidation of the Zr. Such electrodes were consumed during operation but rather slowly (about 0.1 inch/hr.). For DC operation, one of the electrodes was used with a copper anode (similar to the magnetite arc). For AC operation, both electrodes were identical, and the light output was doubled. These open arcs were made in two sizes: 300 watts and 1000 watts. The latter operated at 55 volts, which is considerably higher than that of the enclosed argon Zr-arc lamps of similar nature. The color temperature was 3600 °K. and the 100 watt lamp had a brightness of 200 candles/mm^2 with an efficiency of 20 lumens/watt for DC operation and 38 lumens/watt for AC operation. It was said that if Hf were to be used instead of Zr, the brightness could have been increased to 80 lumens/watt. However, the cost of Hf at that time was considered prohibitive. The open Zr-arc lamp has also been used as an infra-red source (183).

E. Discharges in Molecular Gases

Perhaps the earliest gas-discharge in which the light came from the discharge itself rather than from the heated electrodes was that due to Geissler, who invented the so-called "Geissler tubes" in 1856. At first, rarified air was used, but later other gases such as N_2, CO_2, Hg, and Ne were used by Geissler and also by Faraday and Crookes in 1879. These tubes were cold-cathode high-voltage discharges. British patents on the use of such discharges in signaling and in lighting buoys were issued to Timothy Morris, Robert Weare and Edward Monckton in 1862 and to Adolphe Miroude in 1866 (136). These lamps were not very practical, however. The major problems were low efficiency and short life.

The first practical commercial lamp of the Geissler-type was the "Moore-tube" (136,183,184). D. McFarlan Moore was a former Edison emplyee who thought that gas discharge lamps offered a way to a cool efficient light like daylight rather than the reddish relatively inefficient carbon-filament incandescent lamp. He demonstrated his first lamps in 1895. These lamps were 7-9 feet long, 2.0-2.5 inches in diameter and operated on 110 VDC. However, they had the usual life problems. Moore solved these by providing an automatic device to admit more gas as the pressure dropped below its normal value (10^{-3} atmosphere or about 0.1 mm.). The gases used in early experiments were N_2 ("soft-golden-yellow) or CO_2 (white, "like daylight"). However, since the air had been passed over phosphorous to remove the oxygen, the gas discharge was essentially that of nitrogen. Moore's first commercial installation of lamps in a hardware store in Newark, NJ in 1904 used this type of "air". Madison Square Garden was also lighted by a "long tube" in 1909. An alternating voltage of 16,000 volts was used with a continous tube about 180 feet long (1.75 inches in diameter, 0.1 ampere current, with carbon electrodes). The efficiency was 10 lumens/watt with N_2 and about half of that resulted with CO_2. This luminous output figure for N_2 was much better than the competing carbon filament lamps of the time. However, during life, the N_2-discharge lamp emission changed and argon-gas emission became more prominent as the nitrogen gas became preferentially "cleaned-up" during operation of the lamp. The use of these lamps ended when the incandescent tungsten filament lamp was introduced. In 1912, G.E. acquired the Moore Light Co. Moore himself joined the technical staff of G.E. and later did work on neon lamps. Because of their excellant color, small CO_2 Moore tubes continued to be employed as color-matching sources long after they ceased to be used for normal illumination.

A typical spectrum of this lamp is shown in the following diagram, given as 2.4.7. on the next page, which compares the "over-cast sky", i.e. "Daylight" - 6500 °K., to the emission spectrum of the lamp. It is easy to see that this lamp was quite close in its light spectrum to that of the emission of "Daylight", as defined. The complexity in the spectrum is a result of the many vibrational-rotational modes of the CO_2 molecule.

Aside from N_2 and CO_2 , the only other molecular gas which has received much attention as a "light" source is that of hydrogen. For this gas, the emission is in the ultraviolet and is of no importance as a visible source. The properties of hydrogen arcs at atmospheric and higher pressures have been described in the literature (185). Some of these lamps were of the pulsed type for generating the so-called "Lyman continuum".

2.4.7.-

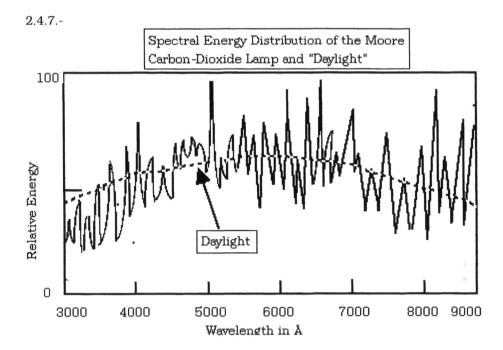

Spectral Energy Distribution of the Moore Carbon-Dioxide Lamp and "Daylight"

The theory of the hydrogen continuum has been discussed by several workers (186). For a review of the far ultraviolet region of the spectrum, see Samson (187).

F. Discharges in Various Vapors

In this section, we shall discuss discharges in vapors of several substances which are not vapors at normal temperature, as is the case for mercury. For other substances, the required vapor pressure must be maintained either by heat generated in the discharge itself (perhaps with a "fill-gas" for starting) or supplied externally. In some cases, a thermostatically controlled heater was used to vaporize the metals or compounds.

Various salts were used in this way to maintain an excited discharge at an early date. In 1915, Nearnst (102) used carbon electrodes in a closed externally-heated vessel containing $ZnCl_2$ or $ZnBr_2$. Similar studies have been made over the years, including that of Darrah (188) who used tungsten electrodes. The following Table lists some of the work that has been accomplished, where external heating was required to obtain the vapor state, so as to maintain the discharge:

Table 2- 27

Investigator	Ref.	Date	Materials Used
Stark and Küch	191	1905	Cd, Zr, Pb, Bi, Sb, Se and Te
Hoffman and Daniels	228	1932	Hg, Cd, Zn, Tl, Te, Bi & Pb
Darrah	229	1916	$SnCl_2$, $TiCl_4$, $SbCl_3$, $AsCl_3$, $SiCl_4$, BCl_3 PCl_3 $AlCl_3$ $CeCl_2$ $CaCl_2$, CaI_2, $CrCl_3$
Beese and Henry	230	1958	AlI_3, BiI_3, TlI_3, SnI_2 PCl_3 SCl_2
OS Duffendach and R Küch	231	1905	P

Note that both the metals and also certain halides were used, notably chlorides and more recently iodides. The results were disappointingly low in output. For example, the emission of Tl was green (line at 5350 Å) with only a few lumens/watt. The sulfur arc was pale blue with an efficiency of 3 lumens/watt; Arsenic produced a continuous spectrum with an efficiency of 2-3 lumens/watt (in this case, the ultraviolet spectrum was quite intense). Phosphorous also has a continuous spectrum, but the luminous efficiency was very low.

1.Tellurium Vapor Lamps

Of somewhat more interest were arcs in tellurium vapor. These have been studied at Osram by Rompe (189) and at Westinghouse by Marden, Beese and Meister (190). The ionization potential is fairly low, being about 9.01 e.v. Unfortunately, hot Te vapor is very reactive and metal electrodes cannot be used. For this reason, these investigators employed liquid "pool" electrodes in quartz tubes with a "fill gas" of Ne, Ar or He. The spectrum is continuous and is undoubtedly connected with the fact that the vapor of Te is molecular, i.e.- Te_2. The emission color becomes yellower as the loading (and hence the temperature and pressure) is increased.

The following diagram shows the spectrum obtained by Rompe:

2.4.8.-

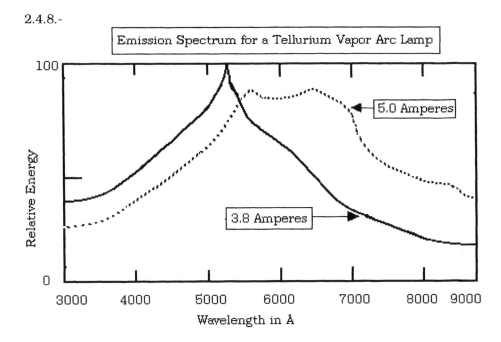

It may be noted that the infra-red intensity is low, in contrast to the high infra-red intensity of the xenon-discharge which limits its luminous efficiency. At low loadings, Marden, Beese and Meister described the color as "bluish-white" with a color temperature of 6000 °K. The efficiency was 5 lumens/watt. At higher loadings, the color becomes "ivory-white" with an efficiency of 15 lumens/watt. At very high loadings, (55 watt/cm. in a tube 8 mm. in diameter), the emission was yellow and the efficiency 40 lumens/watt (see the spectrum given above for 5.0 amperes current). These workers gave for the luminous efficiency:

2.4.9.- $L/W = 81.5 \ (1-27/P)$

where P is the loading in watts/cm. The large value of the subtractive term was attributed to the voltage drop at the pool-electrodes (stated to be 50-150 volts. Even though the value predicted by 2.4.8. is very high, i.e.- 80 lumens/watt, the development of solid electrodes has remained a major obstacle to practical use. The use of tungsten or tantalum carbide electrodes is possible as is the use of iodine vapor to prevent sputtering of such electrode materials. The use of alumina arc tubes, i.e.- Lucalox™, would also permit the use of higher vapor pressures. Discharges involving the compounds, TeI_4 or $TeBr_4$, might also be of interest. However, no attention seems to have been paid to discharges in Se. Se is intermediate to S and Te in the periodic table and has a higher vapor pressure than Te. Nevertheless, both of these elements are extremely toxic to humans and no commercial lamps have appeared to date. Perhaps this is due to the current environmental mindset in the U.S. where the cost of litigation would far outweigh any benefits that might be obtained.

2. Zinc and Cadmium Vapor Lamps

Because of their relationship to mercury, discharges in Cd and Zn vapors have also received considerable attention. The vapor pressure of Cd is higher than that of Zn and its ionization and excitation potential is also lower than those of either Hg or Zn. Many types of Cd-lamps have been described, especially for spectroscopic purposes ((191,192). The first serious attempt to produce Cd-lamps for illumination seems to have been accomplished by Pirani in 1930 (193). Since the major resonance lines are in the near-ultraviolet, high-pressure discharges were indicated. In 1937, Marden, Beese and Meister (194) at Westinghouse made both low pressure (line spectra) and high-pressure lamps in quartz tubes. For the former, the efficiency was 11 lumens/watt and for the latter 23 lumens/watt. For lamps of 1 cm. diameter and 6 cm. long, the results were:

2.4.10.-   Cd: $L/W = 36(1-26/P)$ Zn: $L/W = 26(1-37/P)$

Elenbaas (197) in 1937 also studied the voltage gradient in these vapors. His results were:

2.4.11.-   Hg: $E \cdot d^{3/2} = 3600 \ P^{0.54} \ (P-10)^{1/3} (m/200.6)^{7/12}$

Cd: $E \cdot d^{3/2} = 3850 \ P^{0.54} \ (P-15)^{1/3} (m/112.4)^{7/12}$

Zn: $E \cdot d^{3/2} = 3750 \ P^{0.54} \ (P-25)^{1/3} (m/65.4)^{7/12}$

Here, d is the arc-tube diameter and the quantities by which m is divided are the atomic weights, M, of each element (m, the added weight is in mg/cm).

Minimum gradients were observed at P = 30, 45 and 75 watt/cm., respectively. Elenbaas also showed that the constants, 10, 15 and 25 are in the same ratios as the quantity: $D^{-2}M^{-1/2} = \rho^{2/3}M^{-7/6}$, where D is the atomic-diameter and ρ is the density of the liquid at the melting point.

The low pressure Cd-discharge for use with phosphors (like the present "fluorescent" low-pressure Hg-discharge lamp) was studied by Larsen (196). Maximum efficiency was attained at 300 °C. For a lamp of 7/8" diameter, 4 feet long with 2.7 mm. Ne + 0.1 mm. Ar gases, an input of 205 watts gave line intensities of 56 watts at 3216 Å and 30 watts at 2288 Å. i.e.- 42%. This is nearly as good as the Hg vapor. Nonetheless, such lamps have not found their way into the marketplace, due undoubtedly to the problems of maintaining the rather high temperature necessary to keep the arc stream from extinguishing. Additionally, the problems of corrosion of such arc streams on the materials used for the arc-tube and electrodes have not been completely solved.

Low pressure discharges of Cd, Zn, Tl, In, Ga, Ba, Sb and Bi have been available commercially for use as spectroscopic standards (197). In some cases, such lamps have been "heated-vapor" lamps, and in other cases the lamp is excited by electrodeless discharge (198). However, none are useful for illumination.

3. Cesium Vapor Lamps

Of the other alkali metals other than Na, only Cs has received much attention. Cs has the lowest ionization potential (3.87 e.v.) of all the elements. Low pressure discharges of Cs were studied by Mohler (199) and by Steinburg (200). Commercial application of Cs-vapor devices was for a long time impeded by problems of suitable enclosures and seals. The use of ceramic tubes, similar to the high-pressure sodium lamp have not solved this dilemma.

The first useful Cs-vapor lamps were a low-pressure type developed by Beese (201) at Westinghouse about 1942-44. These were designed for use in an infra-red communication system, using the resonance lines at 8521 Å and 8943 Å. The arc tube was of glass with a special internal glaze to resist attack by the Cs-vapor. The fill gas was argon at 20-22 mm. pressure. Krypton gave

15% more infrared output but was not used at that time because of higher cost. The operating temperature 'of the lamp was 300-325 °C. and the outer envelope was evacuated. Lamps of 50, 100 and 500 watts were developed. About 25-35% of the input energy appears in the resonance lines, but life was limited to about 500 hours because of absorption of Cs on the inner walls. Impurities of O_2, N_2 and CO_2 reduced the output but small amounts of hydrogen were beneficial. Use of Zr or Ta getters were helpful in increasing the lifetime of the lamps.

The modulation characteristics of Cs lamps were studied by Frank, Huxford and Wilson (202) and by Beese (203). Acoustic resonance can be an important factor as the lamp operates, and the discharge may literally "blow itself out" due to standing acoustic waves generated within the lamp itself. A transistor circuit for stabilizing and reducing the fluctuations has been described by Benguigui (204).

High pressure high-current density discharges in Cs vapor in which nearly complete ionization occurs were studied by Mohler (199) in 1938. Although the continuum emission intensity increased as the square of the vapor pressure, the maximum attainable luminous efficiency was somewhat below that of an incandescent lamp because of the reaction which occurred between the quartz capillary tube and Cs-vapor. This limited the maximum pressure to about 17 mm. at 400 °C. In a later work (205), Mohler calculated what the efficiency could be if higher pressures could be utilized. The resistivity of a singly, completely ionized, gas is given by (199):

2.4.12.- $E/J = 1.48 \times 10^4 \ T^{-3/2} \log (0.72 \times 10^{56} \ T^2 n_e^{2/3})$

where E is the voltage gradient and J is the current density in the lamp. The electron concetration, n_e, can be calsulated from Saha's equation (33, 206), or from:

2.4.13.- $\log(n_e^2 / n_a) = \log (2g_i/g_a) + 15.385 + 3/2 \log T - 5040 \ V_i/T$

If the ionization exceeds 45%, then 2.4.12. may be used. Mohler's calculated results for Cs-vapor discharges are given in the following Table:

TABLE 2- 28

Temp. in °K.	P in mm Hg.	n_i/n_a	d in cm.	E in volt/cm	J in Amps.	Arc volts/cm	Tube kWatt/cm.	Loading Bright-candles /mm^2	Effic. - Lumens /Watt
4000	176	0.142	1.0	14.5	87	1.26	0.40	44	137
4500	176	0.35	1.0	17.8	158	2.82	0.90	103	150
5200	176	1.00	1.0	19.8	232	4.60	1.46	196	168
6000	176	3.10	1.0	22.3	300	6.69	2.12	305	180
7000	176	9.20	1.0	22.3	350	7.81	2.48	366	187
5200	760	0.55	1.0	30.0	400	12.0	3.82	250	83
6000	760	1.00	1.0	34.8	545	19.0	6.04	477	100
7000	760	3.20	1.0	43.0	815	35.0	11.10	875	100

It will be seen that a maximum luminous efficiency of about 180 lumens/watt is predicted for Cs vapor discharges at a pressure of 176 mm. at 550 °C. in tubes of 1.0 cm. diameter. This efficiency is, of course, higher than any actual source used today, and is of great interest to lamp designers. It will be noted, however, that the electrical conditions are 300 amperes and a gradient of 22 volts/cm. which corresponds to about 7000 watts/cm. of arc length, or 2100 watts/cm^2 of envelope area. The latter number is about ten times the loading of water-cooled mercury or xenon lamps (see Tables 2-12 & 2-25).

Mohler realized the difficulties in dissipating such power, even if reaction of Cs-vapor with the arc tube could be avoided, and therefore suggested the use of a pulsed or condensor discharge. At 176 mm. pressure, the discharge is not opaque enough to its own radiation to have a true color temperature, but at 760 mm pressure, the discharge would be opaque and would act essentially as a blackbody. Under these conditions, an efficiency of 95-100 lumens/watt could be expected, but the power dissipation problems would be even more severe.

These high efficiencies predicted by Mohler for Cs-vapor discharges have not been realized in practice because of the technological problems involved. Even the use of translucent high-density alumina, i.e.- Lucalox™, resulted in lamps with only 35 lumens/watt in efficiency (207). Further work is needed to fully realize the inherent efficiencies predicted for the Cs-vapor discharge.

G. - Other Types of Discharges

Gas discharges can be excited in a number of ways other than those already discussed. For example, micro-wave energy has been used for excitation of "electrode-less" discharges in many gases and vapors (including low pressure fluorescent lamps). In the Penning discharge, a magnetic field has been used to give the electrons a path length much longer than the dimensions of the enclosure so as to facilitate ionization (208). Such discharges have also been used as light sources in the far-ultraviolet (209).

1. Hollow-Discharge Lamps

This discharge makes use of the "hollow-cathode effect" (210). If a glow discharge is provided with two plane parallel cathodes, separated by a distance less than the width of the normal cathode fall space at the pressure used, then interaction occurs. The same situation will prevail for cathodes with suitably sized cylindrical or spherical cavities. The situation at such cathodes is complex and probably involves the action of excited atoms, meta-stable atoms and radiation from the discharge. Current densities of 3400 amperes/cm^2 have been achieved from such "cold" cathodes. Hollow cathode discharge lamps are appropriately known as "Schüler tubes". It is these lamps which are used as sources for Atomic Absorption Spectroscopy, a technique which has revolutionized Analytical Chemistry in the past few years by making possible analysis of impurities in the "parts-per-billion" range. Such lamps are obtainable with cathodes made from all of the metals, plus alloys of metals for "multi-elemental" analysis. The power consumption is usually low (2-3 watts) and the spectrum is that of the cathode material, since atoms liberated by sputtering are ionized in the discharge.

2. Pulsed Gas Discharges and Flashlamps

The principal application for pulsed-gas discharge lamps has been as a stroboscope for "stopping" motion or in cases where very high, transient radiant power is needed such as signal lights. Another usage has been as "flashlamps" in photography. The high-pressure Hg-Xe and Xe-discharge lamps that we have already described above may be pulsed (7,281) to produce a high-intensity output for short times, but only if they are maintained at a continuous low-level operation between pulses. However, output of these lamps is found to be limited to fairly long pulses of about 10-100 milliseconds. We are interested in much shorter pulse times such as those found in sparks.

Talbert used sparks in air for photographic purposes as early as 1852 [*Phil. Mag.* **3** **73** (1852)]. Neon lamps were also used quite early for stroboscopic purposes, as discussed earlier (211). However, it was Edgerton and associates at Massachusetts Institute of Technology who developed the stroboscopic methods now in use today (212). Many of these early lamps were similar to thyratrons, but as the demand for higher power outputs and shorter pulses became more severe, special lamp designs emerged.

A spark discharge is inherently a transient phenomenon. The usual method of applying voltage involves discharge of a charged-capacitor through two electrodes. The following Table lists some of the considerations involved in capacitor discharge:

TABLE 2-29

Definitions and Equations Involved in Capacitor Discharge

1. Circuit without Inductance

$$I = V_0 /R \; e^{-t/RC} = I_0 \; e^{-t/\tau}$$

In this equation, C = Capacitance; R = Resistance of circuit; V_0 = Initial charge on capacitor; I_0 = Maximum (Initial) value of current = V_0/R; τ = time constant = RC.

2. Circuit with Inductance (General Case)

$$\alpha = R/2L \quad \omega = 1/(LC)^{1/2} \quad \beta = (\alpha^2 - \omega^2)^{1/2}$$

$$I = (V_0/2\beta L) \; e^{-\alpha t} \, (e^{\beta t} - e^{-\beta t}) = (V_0/2\beta L) \; e^{-\alpha t} \; \sinh \beta t$$

3. Critically Damped Case (Requires $L_c = CR^2/4$)

$$I = I_0 \; (4t/\tau) \; e^{-2t/\tau} = (V_0/L)t \; e^{-\omega t} = (V_0/L)t \; e^{-\alpha t}$$

I_m = maximum current = $(2/e)(V_0/R) = 0.736 \, I_0$, when $t_m = \tau /2 = 2L/R$

4. Power Dissipation

I is maximum at $t_m = 1/2\beta \ln (\alpha+\beta)/(\alpha-\beta) = 1/\beta \tanh^{-1} (\alpha/\beta)$. It is convenient to let $L = xL_c$ or $x = 4L/CR^2$ (if x<1); $\alpha = 2/x\tau$; $\beta = \alpha(1-x)^{1/2}$. Then, $I \approx I_o [e^{-t/\tau} - e^{-2\alpha t}] - I_o [e^{-t/RC} - e^{-Rt/L}]$. As normally defined, the quality factor, Q, is: $Q = \omega L$ = volt-amperes/watts dissipated = 2π (energy stored per half cycle/energy dissipated per half cycle) = $(x-1)^{1/2}/2$.

For $x \le 1$, the current does not reverse direction and all of the energy (which is: $CV_o^2/2 = Q_o^2/2C$) is released in a single pulse. For x > 1, the discharge is oscillatory. In general, one is interested only in the power released in the first "pulse" which is:

$$W = {_0}{\int^\pi} I^2 R \, dt = (V_o^2 R/\omega_1^2 L^2) {_0}{\int^\pi} e^{-2\alpha t} \sin^2(\omega_1 t) \, dt$$

Although this last equation cannot be evaluated in closed form, we can:

1. Plot I^2 as a function of t.

2. Calculate: $Q = {_0}{\int^\pi} I \, dt$, where for the purely oscillatory case (no damping, α = o): $Q = 2Q_o = 2CV_o$ (i.e.- the charge on the capacitor is reversed). The charge at $t = \pi/\omega_1$ is $(Q - Q_o)$ when I = 0. The energy dissipation is therefore: $W = 1/2C [Q^2 - (Q - Q_o)^2] = Q/2C ((2Q_o - Q)$. These values of W can then be plotted.

3. For $\alpha \ll 0$, the ratio of energy dissipated per half cycle to that stored in the inductance is $2\pi/Q$. If the total energy is considered as that stored plus that dissipated, then: $W/CV_o^2/2 = 2\pi/2\pi + Q = (1 + [(x-1)^{1/2}/4\pi]^{-1}$. This result will hold for x >> 1.

The following diagram, given on the next page, shows the variation of voltage and current which occurs after a voltage high enough to cause "breakdown" in a gas is applied to a gap between the two electrodes. There is first an induction period or "formative time lag". This lag can be made shorter by increasing the overvoltage (defined as the excess voltage present over the breakdown voltage required).

2.4.14.-

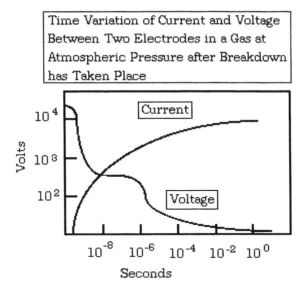

All this happens before an appreciable current begins to flow. The overall voltage then usually collapses to a lower value as the current continues to increase. During the lag time, intense (and often complete) ionization occurs in a small channel between the electrodes. Extremely high pressures and even shock waves can be generated by the high energy dissipation during this period. The acoustic "pop" accompanying a high-energy spark is well known (213). The extent of the development of the ionized channel depends upon the power supply used. If the discharge is maintained for about 1 microsecond, then an arc-like cathode becomes established and the voltage will drop again to a value typical of an arc. According to Vanyukov and Mak (214), the electron and gas temperature become equalized in approximately 10^{-7} seconds.

The initial current density in a spark may be as high as 10^5-10^7 amperes/cm^2 and the temperature as high as 50,000 °K in about 10^{-7} seconds after breakdown. As the channel diameter increases with time, the current density falls and the temperature decreases, being about 20,000 °K at 0.5 microseconds after breakdown and 12,000-15,000 °K after a few microseconds (118). Other things being equal, the temperature in heavier gases will be greater because of the slower expansion of the channel. The channel diameter at any given time depends not only on the total energy supplied up to that time, but also depends upon the way in which it has been supplied, i.e- E(t). Thus, according to Somerville (118), it has been determined that:

2.4.15- $r(t) = L\{E(t)\}^M [_0\int^t \{E(t)\}^{1/2} dt]^N$

and L, N, and M are functions of the gas and the pressure. For instantaneous release of energy, E_0, at time, t= 0 :

2.4.16.- $r(t) = L E_0^{[M + (N/2)_t N]}$

while for E(t) = W · t:

2.4.17.- $r(t) = (2/3)^N L W^{[(M + (N/2)t (M + 3N/2]}$

Experimentally, it is found that: $N \approx 0.37$, $(M + N/2) \approx 0.32$, and $(M + 3N/2) \approx 0.69$, while L descreases as much as about the 0.30 power of the pressure at one atmosphere, i.e.- L = 0.19, 0.44 and 1.10 for air, hydrogen or argon, respectively. Vanyukov and Mak (214) state that the ionized channel varies as $(V_0/p M)^{1/2}$, where V_0 is the initial capacitor voltage, p is the pressure and M is the molecualr weight of the gas. According to Marshak (215), the rate of expansion of the channel is:

2.4.18.-   dr/dt (mm./μsec.) = 0.90 E/p (volts/cm • mm)

The change in electron temperature during the spark duration obviously has an effect on the spectral output. The fact that such high temperatures can be achieved also accounts for the fact that the emission differs from that of an arc. Note that in the analytical method known as "Emission Spectroscopy", the emission lines observed for any given element are often referred to as "arc-lines" or spark-lines", depending upon which method of excitation was used to obtain greater intensity. The intensity of a short-duration spark is greatly influenced by the impedance of the circuit through which the capacitor must discharge (216). Since the voltage drop is IR, while that across an inductor is L(dI/dt), the limiting parameter for very short duration sparks is the circuit inductance. In this case:

2.4.19.- $\quad I_{max} = V_0 \, (C/L)^{1/2} \quad$ and $\quad t_{discharge} = \pi \, (LC)^{1/2}$

$$E_{stored} = c \, V_0{}^2 \, /2$$

Currents as high as 2,000,000 amperes and discharge rates of 2×10^{12} amperes/sec. have been measured. It is also found experimentally that as the energy, E, in the discharge is increased, a limiting brightness is achieved. The attainable temperature, and therefore saturation brightness increases with decreasing atomic weight of the gas employed. Vanyukov and Mak (214) gave the following results:

2.4.20.-

Gas	Temperature	Brightness
Xe	27,000 °K	1.1×10^7 candles/cm^2
Ar	35,000	1.5×10^7
N$_2$	43,000	2.1×10^7

However, the use of lighter gases such as He, Ne and H$_2$ has not led to improved brightness. It has been found that the saturation brightness is independent of pressure (if the pressure is sufficiently high).

The energy (and the pressure) required to achieve saturation becomes lower as the atomic weight of the gas is increased. Thus, although the saturation output is lower with the heavier gases, it is easier to attain. Furthermore, saturation is more easily reached at the long wavelength end of the spectrum. It therefore becomes obvious that the exact mechanisms leading to saturation output and its dependence upon physical propereties of the gas employed are not fully understood.

Spark sources of a variety of types have been studied by a number of workers (217). the luminoscity may be increased without increasing the flash duration of 10^{-6} to 10^{-7} seconds by artificially incrasing the path length of the discharge. In such "guided" or "gliding" sparks (218), the discharge occurs along the surface of a dielectric such as quartz or plastic.

All of the above sparks may be classified as "open" sparks in that the spark and spark gap does not completely fill its container. The luminous efficiency of a spark-discharge may be increased by confining it, so that the expansion of the ionization channel is impeded. A discharge in an enclosure may be said to be confined if the dimensions are such that the discharge channel expands and fills the container for most of the time required for the discharge. The nature of the electrodes may also play a role in such discharges. Reflection of the shock wave may also play an important role. This is the basis for "Flashlamps" which are typically "confined-sparks".

The most common pulsed light souces of the **confined** spark type are the rare-gas filled flashlamps. These are sometimes spherical but more often are cylindrical. The discharge is sometimes wound into a helix or other shape. Following the pioneering work of Edgerton and LaPorte, commercial flashlamps were introduced during the period following World War II (219). Several good reviews of this field are available (215, 220, 221, 222, 223) As compared to unconfined sparks, flashlamps or flashtubes give higher efficiency and brightness, but the discharge time is also longer, being 10 µsec. or longer. According to Vanyukov and Mak (214), a temperature of 94,000 °K and a brightness of 5×10^7 candles/cm^2 have been obtained for an air discharge (tube diameter - 0.40 mm. ID, 29 kVolts using a 0.011 µfd. capacitor, and 1 atmosphere pressure).

A flashlamp consists simply of a container of glass or quartz with two electrodes and a rare-gas filling of about 100-300 mm. pressure. In some cases, the cathode may be designed somewhat differently from the anode. Enlarged end-chambers may also be provided to alleviate the effects of the mechanical shock produced upon firing. In principal, a flashlamp may be employed in two ways. In one method, the lamp is simply connected across a capacitor, with perhaps a small inductance to limit the discharge current. The capacitor is charged relatively slowly through a large resistor or inductance from a DC-power supply. When the capacitor voltage reaches the breakdown voltage of the tube, discharge occurs, and the cycle is repeated until the power supply is disconnected. In practice, this method is not often used, as statistical fluctuations cause apprecible fluctuation in the breakdown voltage and hence in both the flash intensity and the time of discharge. In the more common method, the capacitor is charged below the spontaneous breakdown voltage and and the discharge is triggered at will by a high-voltage high-frequency pulse. The trigger electrode is usually a thin wire wrapped around the center of the tube, although transparent conducting coatings have also been used (296).

It was pointed out quite early by Edgerton (212) that a normal flash tube behaves essentially as a constant resistance. This results in almost complete ionization of the gas. The following diagram shows that this is only approximately true. Nevertheless, it is a useful concept.

2.4.21.-

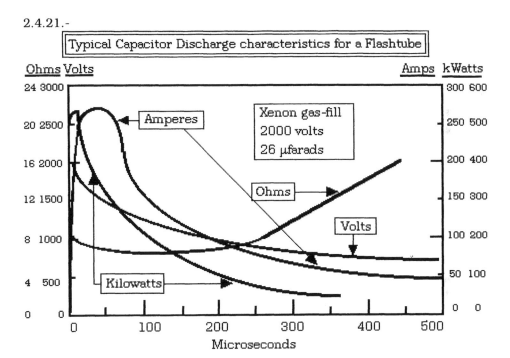

Following Murphy and Edgerton (219), the resistance is normally defined at the point where the current is a maximum. According to Olsen and Huxford (219), the light emission peak in a flashlamp occurs much later than the current peak.. These workers also found that the ultra-violet, visible and infra-red peaks occur at successively later times. On the other hand, Marshak (222) states that the flash duration is about one-half of the electrical time-constant (RC).. Robinson states that the light output persists somewhat longer than the current. The reasons for these conflicting statements has remained obscure. Perhaps it is due to the way the measurements were taken.

In general, the discharge will extinguish when the capacitor voltage has fallen below a critical extinction voltage, V_e . The power dissipated is then: $C(V_0^2 - V_e^2)/2$. This equation makes it essential to keep V_e at a low value. According to Marshak and Shchoukin (222), V_e increases rapidly if the tube diameter is reduced too far. Both the ignition voltage and the extinction voltage increase as the tube length and pressure are increased. Edgerton and Cahlender (225) have discussed "holdover" in flash tubes. This is the condition in which the tube conducts continuously at a low level and the capacitor cannot recharge.

Marshak (222) has stated that, in normal xenon flashtubes, the normal resistivity, ρ, of the discharge may be taken as: $1\text{-}2 \times 10^{-2}$ ohm-cm. for field strengths greater than about 100 volts/cm. For lower values of field, $\rho = 0.10$ $E^{-1/2}$ and the effective resistance can then be calculated in terms of the lamp dimensions.

Carlson and Pritchard (219) give the following for the width of the light pulse:

2.4.22.- $\qquad$ $i = K\, C^{0.69}\, /\, V^{0.625}$ (in μsec.)

where C is in microfarads and V is in kilovolts. If the width is defined as 50% of peak, then $K = 20$. Marshak and Shchoukin (222) measured ι at 35% of peak value and found that:

2.4.23.- $\qquad$ $\partial\, L\, dt$ (in lumen-sec.) $= 0.86\, L_{max} \cdot i$

The following diagram, given on the next page as 2.4.25., shows their results for the dependence of   on the tube diameter and the product of capacitance and tube length, $(C \cdot l)$ at $E = 100$ volt/cm.

The right hand side of the diagram shows the effect of E, the field strength, on i. These workers found that:

2.4.24.- $\qquad$ $i = A\, V_0^{-0.6}\, (C \cdot l)^p\, d^{-q}$

For $i \approx 50$ μseconds, p is ≈ 0.5, q ≈ 0.5; for $i = 500$ μsec., $p = 1.0$ and q ≈ 2.0.

In the latter case, ι ıσ proportional to the time constant, RC, where R is the

2.4.25.-

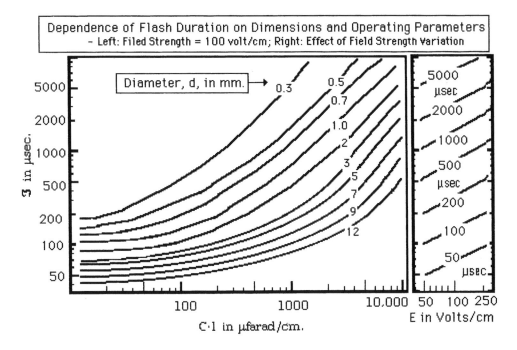

Dependence of Flash Duration on Dimensions and Operating Parameters
– Left: Filed Strength = 100 volt/cm; Right: Effect of Field Strength Variation

lamp resistance. The different behaviors for smaller values of C may arise from the fact that the discharge is no longer completely confined and does not completely fill the tube.

The effect of gas pressure on flash duration (both peak and integrated outputs), as determined by Marshakl and Shchoukin (222) is shown in the following diagram, given as 2.4.26. on the next page.

They found that the peak output, τ , reaches a maximum at pressures of the order of 200 mm.-Hg and decreases at higher pressures. The flash duration, however, increases with increasing pressure and saturates at high pressure. As a result, the integrated light putput (see 2.4.23.) and therefore the efficiency (in lumen-sec./ joule = lumens/watt) is constant for pressures above about 300 mm. The peak output is essentially a linear function of tube diameter, but the integrated light output increases less rapidly. According to Meyer (221), there is an optimum pressure and length for each tube diameter. The luminous efficiency is essentially independent of the size of the

2.4.26.-

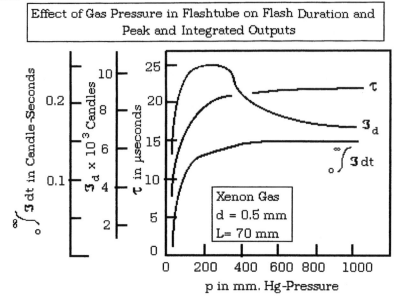

Effect of Gas Pressure in Flashtube on Flash Duration and Peak and Integrated Outputs

Xenon Gas
d = 0.5 mm
L= 70 mm

p in mm. Hg-Pressure

capacitor used or the voltage, if they exceed a certain value, i.e.- if the loading is sufficienctly high.

Marshak (215) states that the maximum efficiency will be achieved if:

2.4.27.- $\qquad$ $C V_0^2/4\pi r l \geq 1 \text{ joule/cm}^2$

This applies to the loading per unit wall area. For lower values of loading:

2.4.28.- $\qquad$ $\eta / \eta_{max} = \{C V_0^2/4\pi r l\}^{0.72}$

The efficiency will also be reduced by electrode losses if the tube is not fairly long. Marshak also gives for the peak brightness, τ :

2.4.29.- $\qquad$ $\tau = E^{0.9}/30 \{C V_0^2/4\pi r l\}^{0.72}$

where E is in volt/cm. and τ is in giga-nits (1 nit = 1 candle/m^2 = π lumens per m^2). Marshak and Shchoukin (222) state that both L_{max} and τ are roughly proportional to the atomic weight of the gas employed. However, although their data support this statement for L_{max} , the dependence is less than linear in the case of τ . Furthermore, in a later paper, Marshak (215) states the L_{max} is practically independent of the gas used, while τ (and therefore the efficiency) is proportional to the atomic weight. These contradictory statements have not been resolved. Nevertheless, there is agreement among various workers that the efficiency of a flashtube is greater when a higher atomic weight gas is used. It should be noted that this is exactly opposite to the case for the unconfined spark, as discussed above. For the confined spark, Robinson (221) gives the following data for various gases:

2.4.30.-   Xe: Kr: Ar: Ne: He = 100: 70: 50: 18: 6

The emission of Xe is also "whiter" than that of the other gases. The efficiency of Xe in a flashtube is usually about 40 lumens/watt.

There are several limitations on the ability of flashtubes to dissipate high energies while operating continuously. One, of course, is that of sputtering and evaporation of the electrodes. Another, which was discussed in detail by Marshak (224), is the ability of the molybdenum ribbons normally used in quartz seals to pass the very high currents without rupturing during operation. Finally, there is the ability of the envelope to stand up without exploding, melting or crazing (223). The average power that can be handled safely is no greater than that employed for various materials under DC operation. Thus, Marshak (224) gives the following loadings as limits for various materials used for flashlamps:

2.4.31.-

Material	Watts/ cm^2
Glass	0.5 - 1.0
Spherical Quartz	1.5 - 3.0
Tubular Quartz	10 - 15
Water-cooled Quartz	130 - 200

These limits may be exceeded for short periods of time. Since according to 2.4.27., 1 joule/cm^2 is required to achieve maximum efficiency the maximum flashing rate for tubular quartz bulbs is therefore about 10-15 per second for high efficiency. there is also a limit on the energy per flash that can be allowed. According to Meyer (221), this is 7-13 joules/cm^2 for glass and 18-39 joules/cm^2 for quartz. According to Marshak (222, 224), this limit may be written:

2.4.32.- $\qquad$ $B = C^{1-k} V^2 / 1^{1+k}$

where k = 0.5 for glass and 0.45 for quartz. B is a parameter depending upon tube design and cooling conditions. If k is 0.5, 2.4.32. may be written as:

2.4.33.- $\qquad$ $S = CV^4 / 1^3$

S is essentially independent of tube diameter for glass, but for quartz, it increases from about 0.3 to 5.3 µfarads-kV^4 / cm^3 as d increases from 1.0 to 12.0 mm, above which it is constant. This is shown in the following diagram:

2.4.34.-

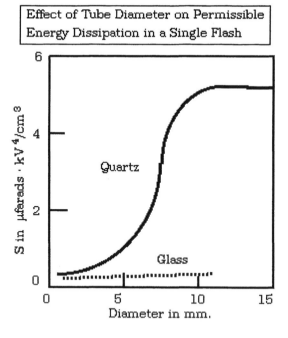

Effect of Tube Diameter on Permissible Energy Dissipation in a Single Flash

This diagram shows that the total allowable energy per flash (CV^2/2) may be increased by reducing V and increasing C. This occurs because increasing C (the capacity of the condensor used) increases the flash duration, and hence reduces the shock intensity (see 2.4.24.). Some data for commercial flashlamps are given in the following Table:

TABLE 2-30

Electrical and Optical Properties of Some Commercial Flashlamps

Tube	Envel	Oper. Volts	Max. Input		Output Peak	Integr	Eff.	Bright	Flash	Life
			Joules	Watts	Mega-Lumen	Lumen - Sec.	Lumen / watt	Candle / cm^2	Durat in μsec	# of Flashes
Strt.	Vycor	2000	400	40		25,200	63		250	10^5
Strt.	Vycor	2000	5	2		31			6	6000
Helix	Quartz	900	635	40		31,400	50		1000	1000
Strt.	Quartz	3000	10000	40		5x10^5	50		2200	50
Strt.	Glass	500	45			1900	48			
U-tube	Glass	500	100			2830	47			
U-tube	Quartz	2500	500			15000	50			
Toroid	Quartz	2700	1500			45600	46			
Strt.	Quartz	2000	60	40	50	5500	44	1.6	20	6x10^4
Helix	Quartz	450	125	5		4500	36		1.7	10^4
Helix	Quartz	6000	7x10^4	18000	450	3x10^5	4.3	0.4	5000	5000
Strt.	Quartz	6000	80000	1350	450	3 x 10^6	44	0.3	6000	5500
Strt.	Quartz	2400	15000	1250	14	6 x 10^6	42	0.2	45000	10000
Helix	Quartz	4000	24000	300	300	2x10^6	50		2700	2500

In some cases, Krypton is used, but mostly it is Xenon-gas. Life is given for maximum loading and Strt. is a straight tube configuration. In addition to tubular flash lamps (which may be bent into toroidal or helical shapes), spherical lamps with short electrode gaps have been made similar to "normal" short-arc Xe and Xe-Hg lamps. In this case, the discharge is essentially "un-confined" and the output is lower than for flashlamps (222, 224).

Flashlamps have been used for several purposes. As mentioned before, very small flashlamps have been mounted on cameras and operate off a battery. These flashlamps have essentially replaced Photo-flash lamps based on ignition of metal foils. Another use has been for Airport approaches where the flashlamps are aligned to show the air-borne approach to the landing-field. Still another usage has been on Police cars where the flashing signals are used to alert motorists. Another use that has found wide-spread application has been for "pumping" or exciting lasers (226). Special annular lamps have been built in which the discharge surrounds the Laser-Rod, and lamps capable of dissipating 25,000 joules per pulse in less than 10^{-3} seconds have been built (227). But, as of late, these are being replaced by solid-state gallium arsenide lasers which couple directly into specific energy levels of the Nd^{3+} ion in the laser crystal (YAG). This combination has raised the laser coupling efficiency by more than ten times.

REFERENCES CITED

1. G.R. Fonda, *G.E. Rev.* **32** 206 (1929); F. Koref and H.C. Plant, *Z. Tech. Phys.* **11**515 (1930); L. Prasnik, *Z. Phys.***69** 832 (1931); ibid, **72** 86 (1931); ibid, **75** 417 (1932); ibid, **77** 127 (1932); ibid, **86** 387 (1933), ibid, **99** 710 (1936); J.A.M. van Liempt, *Z. Phys.* **86** 387 (1933); Z.S. Voznesenskya and V.F. Sousten, *J. Tech. Phys. USSR* **9** 399 (1939); W. Elenbaas, *nederl. Tijdschr. Naturkunde* **6** 77 (1939); E. Bas-Taymaz, *Z. Angew. Phys.***2** 288 (1958); W. Elenbaas, *Philips Tech. Repts.* **18** 147-160 (1963).

2. W.M. Potter and K.M. Reid, *Illum. Eng.* **54** 751 (1959).

3. Iluminating Engineering Society Lighting Handbook, 3rd Ed. (1959).

4. L.E. Barbrow and J.F. Meyer, *J.Res. Natl. Bur. Stds.*, **9** 721 (1932).

5. L.G. Leighton and A Makulec, *J. Motion Pict. TV Engrs.***67** 530 (1958).

6. D.K. Wright, *Magazine of Light* (GE) p. 24, Dec. 1936.

7. British patent # 7682 by British Thomson-Houston Co. (1914).

8. H.B. Wahlin and L.V. Whitney, *Phys. Rev.* 50 735 (1936);

9. A.L. Reimann *Phil.Mag.* **41** 685 (1938).

10. J.W. Howell and H. Schroeder, *History of the Incandescent Lamp*, Maqua Co, Schenectady, N.Y. (1927).

11. A.A. Bright, *The Electric Lamp Industry*, MacMillan, NY (1949).

12. N.R. Campbell, *Phil. Mag.* **41** 685 (1922). See also: J.T. Randall and J.H. Shaylor, *Trans. Faraday Soc.* **27** 730 (1931).

13. E.G. Zubler and F.A. Mosby, *Illum. Eng.* **54** 734 (1959); See also: J.A. Schilling *Electrotech.* Zeit., **18** 485 (1959); J.A. Moore and C.M. Jolly, *GEC Journal* **29** 99 (1962).

14. J.W. van Tijen, *Philips Tech. Rev.* **23** 237 (1961/2).

15. B. Hisdal, *J. Am. Opt. Soc.* **52** 395 (1962).

16. F.J. Studer and D.A. Cusano, *J. Opt. Soc. Am.* **43** 522 (1953).

17. Photonics Spectra, p. 40, Publ. -   Jan 1991.

18. W.G. Matheson, *Natl. Tech. conf. Illum. Eng. Soc.*, Preprint # 36, Sept. 7-11 (1959).

19. M.V. Fok, *Optics and Spectroscopy,* **13** 349 (1962).

20. S.C. Peek, *Illum. Eng.* **57** 622 (1956).

21. R.G. Young, *Illum. Eng.* **57** 345 (1956).

22. A.I. Bodretsova and S.I. Levikov, *Svetotekhnika* **3** 21-4 (1961).

23. P. Schwarzkopf et al, *Refractory Hard Metals* , MacMillan, NY (1953).

24. M.R. Andrews, *J. Am. Chem. Soc.,* **54** 1845 (1932); See also: K. Becker et al, *Z. Tech. Phys.* **11** 148 (1930); loc cit., **11** 182 (1930); loc cit., **11** 216 (1930); ibid, *Tech.-Wiss. Abhandl. Osram Konzern* **2** 230 (1931); loc. cit., **2** 237 (1931); H.J. Booss, *Mettall.* **16** 668 (1962); B.H. Eckstein and R. Forman, *J. Appl. Phys.* **33** 82 (1962).

25. F. Blau, U.S.P. 2,025,565 (1935) - HfC ; M.R. Andrews, U.S.P. 2.072,788 (1937) - TaC

26. N.N. Voloshchuk and A.G. Helfgot, *Svetotekhnika* **2** 87-92 (1962).

27. D.P. Cooper, U.S.P. 2,596,469 (1952); U.S.P. 3,022,436 (1962); U.S.P. 3,022,437 (1962).

28. D.P. Cooper, G.R. Bird and L. Brewer, *XVIII Cong. Pure and Appl.Chem.*, Paper A-3-48, Montreal (Aug. 10, 1961).
29. D.P. Cooper, U.S.P. 3,022,438 (1962); D.P. Cooper, U.S.P. 3,022,439 (1962).

30. "GE Light Engine Illuminates Autos" by Jeff Waggoner, *Photonics Spectra*, p. 16, November 1990.

31. H. Kreft, *Z. Techn. Phys.* **15** 554 (1934); loc. cit. **19** 345 (1938); ibid, *Tech. Wiss. Abhandl.Osram Konzern* **4** 33 (1936).

32. W. Elenbaas, *"The High Pressure Mercury Vapor Dischage"*, Intersci. Publ., New York, (1951).

33. M.N. Saha, *Phil. Mag.* **40** 472 (1920); R H Fowler, *Phil. Mag.* **45** 1 (1923); See also- " Statistical Mechanics", Cambridge Univ. Press, London (1934); H. Einbinder, *J. Chem. Phys.* **26** 746 (1958); LS Frost, *J. Appl. Phys.* **32** 2029 (1961).

34. W. Elenbaas, *Philips Tech. Rev.* **18** 167 (1956/57).

35. JW Marden, G Meister and NC Beese, *Trans. Electrochem. Soc.* **69** 389 (1936).

36. JW Marden, G Meister and NC Beese, *Trans. Illum. Eng. Soc.*, **33** 147 (1938); ibid, *Trans AIEE* **55** 1186 (1936).

37. C Kenty, *Phys. Rev.* **61** 545 (1942).

38. EB Noel, *Illum. Engng.* **43** 1044 (1948); EB Noel and EC Martt, *Illum. Engng* **61** 513 (1956); HD Fraser, MC Unglert, WM Waldbauer and JA Walick, *Illum. Engng.* **56** 215 (1961).

39. A. Unsold, *Ann. Phys.* **33** 607 (1938); H Maecker and T. Peters, *Z. Phys.* b 139 448 (1954); G Hettner, *Z. Phys.* **150** 182 (1958).

40. F. Rössler, *Z. Phys.* **112** 667 (1939), loc cit. **122** 285 (1944); loc. cit., **133** 80 (1952); loc. cit. **139** 56 (1954); ibid, *Tech. Wiss. Abhandl. Osram Ges.* **5** 113 (1943); See also- R Borchert, *Ann. Phys.* **2** 332 (1950); W. Göing, H Meier, and H Meinen, *Z. Phys.* **140** 376 (1955); H Meier, *Z. Phys.* **149** 40 (1957); H Bartels and R Beuchelt, *Z. Phys.* **149** 594 , 608 (1957); W Neumann and H Reimann, *Ann, Phys.* **3** 211 (1959).

41. W. Finkelnburg, *Zeit. Phys.* **112** 305 (1939); ibid, **113** 562 (1939);ibid, ibid, **114** 734 (1940);**116** 214 (1941); ibid, **117** 344 (1942); ibid, **119** 206 (1943); ibid, **122** 36 (1944); ibid, **122** 714 (1944).

42. V J Francis, G G Isaacs and E H Nelson, Phil Mag. **37** 789 (1946).

43. H.K. Bourne, *Discharge Lamps for Photography and Projection*, Chapman & Hall, London (1948).

44. J. Funke and P.J. Oranje, "Gas Discharge Lamps", Philips Tech. Library, Elsevier Press, New York (1951).

45. W E Thouret, *Illum. Engng.* **55** 295 (1960).

46. BT Barnes and EQ Adams, *Phys. Rev.* **53** 545 (1938); C. Kenty, *Phys. Rev.*, **80** 95 (1950); ibid, *J. Appl. Phys,* **21** 1309 (1950); C Kenty, MA Easley, and BT Barnes, *J. Appl. Phys.* **22** 106 (1951); FA Butaeva and VA Fabrickant, *J. Tech. Phys. USSR* **18** 1127 (1948); M Dougherty and W Harrison, *Brit. J. Appl. Phys.* **Suppl #4** 11 (1955); JF Waymouth and W Harrison, *J. Appl Phys.* **27** 122 (1956); BT Barnes, *J. Appl. Phys.* **31** 852 (1960).

47. C Ellis and AA Wells- "*The Chemical Action of Ultraviolet Rays*", Reinhold, New York, (1941).

48. See for example: RO Griffith and A McKeown, *"Photoprocesses in Gaseous and Liquid Systems"*, Longmans-Green Publ. New York (1929); NH Dhar *"The Chemical Action of Light"* Blackie & Son, London (1931); WA Noyes and PA Leighton, *"The Photochemistry of Gases"* Reinhold, New York (1941); M. Luckiesh, *"Applications of Germicidial, Erythemal and Infrared Energy"*, Van Nostrand, New York (1946); LR Koller, " *Ultraviolet Radiation"* Wiley, New York, (1952); W. Summer, *"Ultraviolet and Infrared Engineering"* Wiley, New York (1962).

49. LR Koller, *J. Appl. Phys.* **10** 624 (1939); HC Rentschler, *Trans. Illum. Engng. Soc.*, **35** 960 (1940); DD Knowles and E Reuter, *Trans. Electrochem. Soc.* **78** 21 (1940); HC Renschler, R. Nagy and G. Mouromtseff, *J. Bacteriol.* **41** 745 (1941); M. Luckiesh and LL Holliday, *G.E. Rev.* **45** 223, 343 (1942); J. Janin, *Meeting Assoc. Francaise*des Eclairagistes , Dijon, France, May- 1953.

50. AW Ewall, *J. Appl. Phys. 13* 759 (1942).

51. TL Martin, *J. Franklin Inst.* **254** 267 (1952).

52. RH Clapp and RJ Ginther, *J. Opt. Soc. Am.* **37** 355 (1947); see also: H C Froelich, *J. Electrochem. Soc.* **91** 241 (1947).

53. R.C. Ropp, *Electrochem. Technol.* , **3** , 375 (1965).

54. M. LaPorte, *Bull. Soc. Franc. Elect.* **6** 298 (1956); *Le Vide* **11** 99 (1956).

55. J W Marden and M G Nicholson, *Trans. Illum. Engng. Soc.* **26** 592 (1931).

56. R O Griffith and A. McKeown, "Photoprocesses in Gaseous and Liquid Systems", Longmans Green- Publ., NY (1929); See also: M. Luckishe, "Applications of Germicidal, Erythemal and Infrred Energy", Van Nostrand- Publ., NY (1946); C M Doede and C A Walker, *Chem. & Engng. News*, p. 159 (1951).

57. See: J. Funke and P J Oranje, "Gas Discharge Lamps", Philips Technical Library, Elsevier Press, NY (1951); See also: J H Laub, *Elec. Engng.* **60** 384 (1941); R E Farnham, *Illum. Engng.* **36** 217 (1941); F G Wilde, *Illum. Engng.*

39 31 (1944); L E Barnes, *Illum. Engng.* **46** 41 (1951); W. Elenbaas, *Philips Tech. Rev.* **13** 323 (1951/52).

58. W H Kahler, *Westinghouse Engineer* **4** 144 (1944).

59. H K Bourne, "Discharge Lamps for Photography and Projection", Chapman & Hall, London (1948).

60. J. Funke and P J Orange, "Gas Discharge Lamps", Philips Tech. Library, Elsevier Press, NY (1951).

61. J E Stanworth, *J. Soc. Glass Technol.*, **23** 268 (1939).

62. H G Jenkins, *Light and Lighting* **51** 391 (1958); *Trans. Illum. Engng. Soc.* **26** 17 (1961).

63. M E Nordberg, *J. Am. Cer, Soc.* **30** 174 (1947); J E Stanworth, *Nature* **165** 724 (1950); Ibid, *Brit. J. Appl. Phys.* **1** 182 (1950); see also: T H Elmer and M E Nordberg, *6th Symp. Art Glassblowing,* (1950).

64. J W Ryde, *GEC Journ.* **4** 199 (1933); E Lax and M Pirani, *Tech. Wiss. Abhandl. Osram Konzern.* **3** 6 (1934); E Kreft and E Summerer, *Das Licht* **4** 23 (1934).

65. G H Wilson, E L Damant and J M Waldram, *J. IEEE* **79** 241 (1935).

66. J A St. Louis, *Trans. Illum. Engng. Soc.* **31** 583 (1936)

67. A.A. Bright, *The Electric Lamp Industry,* MacMillan, NY (1949).

68. J M Harris, *Illum. Engng.* **52** 363 (1957).

69. G Hermann and S Wagener, "The Oxide Cathode", Chapman & Hall, London (1951).
70. G A Freeman, *Westinghouse Engineer,* p. 116 (1960)

71. G. Heller, *Philips Tech. Rev.* **1** 129 (1936)

72. D S Gustin and G A Freeman, USP 2,241,345 and 2,241,362 (1941)

73. I Langmuir, *Phys. Rev.* **22** 357 (1923); Ibid, *J. Franklin Inst.*,**217** 543 (1934).

74. E B Noel, *Illum. Eng.* **43** 1044 (1948).

75. W S Till and M C Unglert, *Illum. Eng.* **55** 269 (1960).

76. E C Martt, K Gottschalk, and A C Green, *Illum. Eng.* **55** 260 (1960).

77. H D Fraser and W S Till, *Illum. Eng.* **47** 207 (1952).

78. E C Martt and R J Smith, *Illum. Eng.* **53** 630 (1958).

79. P J Bouma, *Philips Tech. Rev.* **6** 295 (1941); see also: G A Horten and W S Till, *Illum. Eng.* **56** 696 (1961).

80. C Bol, *De Ingenieur* **50** E91 (1935); E G Dorgelo, *Philips Tech. Rev.* **2** 165 (1937); See also: J Kern, *Tech. Wiss. Abhandl. Osram Ges.* **5** 56 (1943).

81. JW Marden, G Meister and NC Beese, *Trans. Illum. Eng. Soc.*, **35** 458 (1940).

82. E B Noel and R E Farnham, *J. Soc. Motion Pict. Engng.* **31** 221 (1938); E B Noel, *J. Appl. Phys.* **11** 325 (1940); Ibid, *Trans. Illum. Engng. Soc.* **36** 243 (1941)

83. H A Stahl, *J. Opt. Soc. Am.* **49** 381 (1959).

84. B T Barnes and W E Forsythe, *J. Opt. Soc. Am.* **37** 83 (1937); L B Johnson and S B Webster, *Rev. Sci. Inst.* **9** 325 (1938); B T Barnes, W E Forsythe and W J Karash, *G E Rev.* **42** 540 (1939); See also: *Philips Tech. Rev.* **1** 2, 62 (1936).

85. R Rompe and W Thouret, *Z. Phys.* **17** 377 (1936); loc. cit. **19** 352 (1938); Ibid, *Tech. Wiss. Abhandl. Osram Ges.*,**5** 44 (1943); Ibid, *Das Licht* **14** 73 (1944); W Thouret *Licht Technik* **2** 73 (1950).

86. W J Francis and G H Wilson, *Trans. Illum. Engng. Soc.(London)* **4** 59 (1939); J N Aldington, *Illum. Engng. Soc.(London)* **11** 11 (1946); E H Nelson, *G E C Journal* **14** 73 (1946); V J Francis and W R Stevens, *IEE (London)* **94** 423 (1947).

87. V J Francis *Trans. Illum. Engng. Soc.* **15** 315 (1950)

88. G A Freeman *Illum. Engng..* **45** 218 (1950); T C Retzer and G W Gerung *Illum. Engng..* **51** 745 (1956)

89. See for example: *Westinghouse Short Arc Bulletin* **CE-556** (April 1956)

90. A Bauer and P Schultz, *Z. Phys.* **139** 197 (1954); P Gerthsen and P Schultz, *Z. Phys.* **140** 510 (1955); A Bauer, P Schultz, and P Gerthsen, *Appl. Sci. Res.* **B5** 210 (1955); A Bauer, *Ann. Phys..* **18** 387 (1956); Ibid., *Tech. Wiss. Abhandl. Osram Ges.* **7** 39 (1958).

91. C W Jerome, *Illum. Engng.* **49** 237 (1954); Loc. Cit., **56** 209 (1961).

92. "Yttrium Phosphate-Yttrium Vanadate Solid Solutions and Vegard's Law", by R. C. Ropp and B. Carroll, *Inorg. Chem.*, **14**,2199 (1975).

93. W A Till, *Illum. Engng.* **53** 224 (1958).

94. L J Buttolph, *G.E. Rev.* **36** 482 (1933).

95. J N Aldington, *Siemans Mag. Engng. Suppl.* **#131** (April 1936); see also: W H LeMarechal and J N Aldington, *Brit. Pat. 447,428* (1936) and M E Macksoud, *USP 2,171,580* (1939).

96. E L J Matthews, *Philips Tech. Rev.* **5** 341 (1940); S Bahr and K Larché, *Z. Tech, Phys.* **21** 208 (1940); K Larché and E Summerer, *Das Licht* **10** 172

(1940); K Larché, *Tech. Wiss.Abhandl. Osram Ges.* **5** 26 (1943); J Funke and P J Orange, *Philips Tech, Rev.* **7** 34 (1942).

97. H E Strauss and W E Thouret, *Illum. Engng.* **54** 97 (1959).

98. E H Nelson, *G E C Journal* **28** 165 (1961)

99. W. Thorington, *Illum. Engng.* **54** 147 (1959)

100. C P Steinmetz, *USP 1,006,021* (1911); and *USP 1,025,932* (1912).

101. M Wolfke, *Z. Electrochem.* **33** 917 (1912)

102. W. Nernst, *Ger. Pat. 288,288* (1915) and *Ger. Pat. 288,289* (1916)

103. F Schröter, *Z. Tech. Phys.* **1** 109, 149 (1920).

104. G T Winch ands E H Palmer, *Illum. Engng.(London)* **27** 123 (1934).

105. H J Spanner *Trans. Illum. Engng. Soc.* **30** 178 (1935).

106. W. Elenbaas, *Revue d'Optique* **27** 683 (1948).

107. H. Cotton, *"Electric Discharge Lamps"* , Wiley, New York (1946).

108. V J Francis and H G Jenkins, *Rep. Prog. Phys.* **7** 230 (1940).

109. K G Schnetzler, *USP 2,240,353* (1941); and *Brit. Pat. 509,801* (1961).

110. D A Larson, H D Fraser, W W Cushing and M C Unglert, Paper presented at: *Illum. Engng. Soc. in Dallas, Sept. 9-14 -Preprint # 29* (1962).

111. A H Compton and C C Van Voorhis, *West. Lamp Co. Tech. Rep.* **No. B-12** (1919); Ibid, *Phys. Rev.* **21** 210 (1923); A H Compton, *USP 1,530,312* (1919) and *USP 1,570,876* (1926)

112. M J Druyvesteyn, *Phys, Zeit,* **33** 822 (1932).

113. G. Heller, *Philips Tech. Rev.* **1** 2, 70 (1936).

114. W. Uyterhoeven, *"Elecktrische Gasentladungslampen"*, Springer, Berlin (1938), Edwards Bros., Ann Arbor, Mich (1943)

115. G B van derWerfhorst, *Trans. Electrochem. soc.* **65** 157 (1934).

116. G. Holtz, *Ingeniew* **48** E75 (1933); H. Ewest, *Licht* **6** 243 (1936).

117. M J Druyvesteyn and N Warmholtz, *Phil. Mag*, **17** 1 (1934).
118. J M Somerville, *The Electric Arc*, Wiley, New York (1959).

119. C G Found, *G E Rev.* **37** 269 (1934); G R Fonda and A H Young, *loc. cit.* 337 (1934); Ibid, *J, Am. Opt, Soc.* **24** 31 (1934); N T Gordon, *G E Rev.* **37** 338 (1934).

120. W.E. Forsythe and E. Q. Adams, *"Fluorescent and Other Gaseous Discharge Lamps"* , Murray Hill Books, New York (1948).

121. E G Dorgelo and P J Bouma, *Philips Tech. Rev.* **2** 353 (1934).

122. W. Uyterhoeven, *Philips Tech. Rev.* **3** 157, 197 (1938).

123. M H A van de Weijer *Philips Tech. Rev.* **23** 246 (1961).

124. L Blok, *Philips Tech. Rev.* **1** 87 (1936).

125. G R Fonda, *J. Am. Opt. Soc.* **25** 412 (1935)

126. E H Nelson and S A R Rigdon, *Light and Lighting* **49** 217 (1956).

127. V Verwey and M H A van de Wiejer, *Paper # 59.22, Meeting of CIE, Brussels,* (1959); see also: M H A van de Weijer, *Philips Tech. Rev.* **23** 246 (1961/62).

128. J B de Boer and M H A van de Weijer *Light and Lighting* **55** 322 (1938).

236

129. R F Weston, *Electrical Times*, **135** 719 (1959).

130. R C Ropp, *Inorganic Polymeric Glasses* Elsevier Science Publ., Amsterdam & New York (1992).

131. K. Schmidt, *Bull. Am. Phys. Soc.* **8** 58 (1963); see also: *Illum. Engng.* **58** 8A (1963); *Proceedings 6th International Conf. on Ionization Phenomena in Gases,* Paris (1963).

132. E H Nelson, *G.E.C. Journal* **31** 92 (1964).

133. S A R Rigden, *G.E.C. Journal* **32** 37 (1965).

134. F C Lin, Paper presented, *Illum. Eng. Soc.* , Boston, Mass., Aug.- 1969.

135. F Rossler, *Ann. Phys.* **34** 1 (1939)

136. J A Burns and E H Nelson, *G.E.C. Journal* **31** 3 (1964).

137. A.A. Bright, *The Electric Lamp Industry*, MacMillan, NY (1949).

138. G Claude, *Compte Rendus* **151** 1122 (1910); loc.cit.**152** 1377 (1911); loc. cit.**156** 1317 (1913); loc. cit. **158** 692 (1914); see also: *Bull. Soc. Franc Electr.***3** 145 (1933).

139. A C Jenkins, *Argon, Helium and the Rare Gases* - G A Cook-Ed. **Vol. II - Chap. 18**, Interscience Publ., (1961).

140. H A Miller, *Luminous Tube Lighting*, George Newes- Publ., London (1945); see also: Ibid, *Cold Cathode Lighting* , Chem. Publ. Co., Brooklyn (1949).

141. F O McMillan and E C Starr, *Trans. AIEE* **48** 11 (1929); P A Kober, *Elec. Engng.* **50** 650 (1931); R R Machlett, *J. Franklin Inst.* **211** 319 (1931); H Kreft and E O Seitz, *Z. Tech. Phys.* **15** 556 (1934); Ibid, *Phys. Zeit* **35** 980 (1934); Ibid, *Tech. Wiss. Abhandl. Osram Konzern* **4** 41 (1936); M E Gemont, *Bull. Soc. Franc. Elect.* **7** 82 (1957).

142. C G Found and J D Fourney, *Trans. AIEE* **47** 747 (1928); see also: R O Ackerley, *G.E.C. Journal* **5** 216 (1934).

143. D M Moore, *G E Rev.* **23** 577 (1920); Ibid., *Proc. AIEE* **26** 523 (1907).

144. H E Watson, *Nature* **191** 1040 (1961); P. Lamaigre-Voreaux, *Bull. Soc. Franc. Elect.* **2** 18 (1961).

145. J D Cobine, *Gaseous Conductors* McGraw Hill Publ., New York (1941).
146. A Claude, *Light and Lighting* **56** 127 (1939); Ibid, *Bull. Soc. Franc. Elec.* **9** 307 (1939).

147. F Llewellyn Jones, *Proc. Phys. Soc.(London)* **48** 513 (1936); P Schultz, *Z. Naturforsch.* **2a** 662 (1947); V J Francis, *Phil Mag.* **40** 1063 (1949); K S W champion, *Proc. Phys. Soc. (London)* **65B** 169 (1953).

148. W Elenbaas, *Philips Res. Rpts.* **4** 221 (1949).

149. C G Suits, *J. Appl. Phys.* **10** 730 (1939).

150. E Bloch, L. Bloch and G Dejardin, *Comptes Rendes* **183** 171 (1926); M LaPorte *Comptes Rendes* **203** 1341 (1936); Ibid., *Comptes Rendes* **204** 1240 & 1559 (1937); Ibid. *Rev. Gen. d'Elec.* **41** 483 (1937); Ibid., *J. Phys. Rad.* **8** 338 (1937); loc cit., **6** 164 (1945); Ibid., *Bull. Soc. Franc.Elec.* **8** 185 (1948); Ibid., *LeVide* **6** 927 (1957); Ibid., *Les Lampes a Éclasirs Lumiére Blanche et Leurs Apllications*, Gauthiers-Villars - Publ., Paris (1949); see also: J. romand and B Vodat, *Comptes Rendes* **224** 1034 (1947).

151. G N Senilov, *Svetotekhnika* **9** 18-21 (1960)

152. J J Hopfield *Phys. Rev.* **35** 1133 (1930); Ibid, *Astrophys. J.* **72** 133 (1930); P Johnson, *Phil. Mag.* **13** 487 (1932); J S Townsend and M H Pakhala, *Phil. Mag.* **14** 418 (1932); S P McCallum, L Klatzow and J E Keystone, *Nature* **130** 810 (1932); Ibid., *Phil. Mag.* **16** 193 (1933); J E Keystone, *Phil. Mag.* **15** 1162 (1932); loc cit. **16** 625 (1933); A Claude, *Bull Soc. Franc. Elec.* **35** 625 (1933); S P McCallum, *Nature* **142** 252, 614 (1938); G Déjardin, *J. Phys. Rad.* **9** 142 (1938); M Laporte, *J. Phys. Rad.* **9** 228

(1938); R A Johnson B L McClure and R B Holt, *Phys. Rev.* **80 376 (1950);** W D Parkinson, *J. Opt. Soc. Am.* **41** 619 (1951); W Friedl, *Z. Naturforsch.***14a** 948 (1959).

153. J Goude-Axelos and A Claude, *Comptes Rendes* **194** 134 (1932); P Harteck anf F. Oppenheimer, *Z. Phys. Chem.* **16B** 77 (1932).

154. Y Tanaka and M Zelikoff, *J Opt. Soc. Am.* **44** 254 (1954); P G Wilkinson and Y Tanaka, *J Opt. Soc. Am.* **45** 344 (1955); Y Tanaka, *J Opt. Soc. Am.* **45** 710 (1955); P G Wilkinson, *J Opt. Soc. Am.* **45** 1044 (1955); Y Tanaka, A S Jursa and F J LeBlanc, *J Opt. Soc. Am.* **48** 304 (1958); R E Huffman, W W Hunt, R L Novak and Y Tanaka, *J Opt. Soc. Am.* **51** 693 (1961); R E Huffman, Y Tanaka abd J J Larrabee, *J Opt. Soc. Am.* **52** 851 (1962); R G Newburgh, L Heroux and H E Hintergger, *Appl. Opt.* **1** 733 (1962).

155. J N Aldington, *Trans Illum. Engng. Soc.(London)* **14** 19 (1949); H W Cumming, Light and Lighting, **43** 99 (1950); Ibid., *Trans. Illum. Engng. Soc. (London)* **16** 129 (1951).

156. P Schulz, *Reichsber. Phys.(Phys. Zeits. Suppl)* **1** 147 (1944); Ibid., Ann. Phys. **1** 95, 107 (1947); Ibid. *Z. Naturforsch.* **2a** 583 (1947); Paper-Q *CIE Meeting* (1952).

157. W T Anderson, *J. Opt. Soc. Am.* **41** 385 (1951); Ibid., *J. Soc. Motion Picture TV Engn.* **63** 96 (1954); W E Thoret and G W Gerung, *Illum. engng.* **49** 520 (1954).

158. K Larché *Lichttechnik* **2** 41 (1950); loc cit. **7** 221 (1955); Ibid., *Electrotech. Zeit.* **72** 427 (1951); loc. cit. **74** 346 (1953); Ibid., *Z. Phys.* **132** 544 (1952); K Larché, K Ittig and F Michalk, *Lichttechnik* **5** 234 (1953), Ibid, *Tech. Wiss. Abhandl. Osram Ges.* **6** 33 (1953).

159. K Laraché, *Z. Phys.* **132** 544 (1952); Ibid., *Electrotech. Zeit.* **72** 427 (1951)

160. P Schultz and B Steck, *Ann. Phys.* **18** 401 (1956).

161. A Bauer and P Schultz, *Z. Phys.* **146** 339 (1956); see also: R Rompe, W Thouret, and W Weizel, *Z. Phys.* **122** 1 (1943).

162. H Schirmer, *Z. Phys.* **136** 87 (1953); Ibid, *Tech Wiss. Abhandl. Osram Ges.* **6** 29 (1953); *Z. Angew. Phys.* **11** 357 (1959); A Bauer, *Lichttechnik* **12** 406 (1960).

163. W Elenbaas, *Appl. Sci. Res.* **B5** 205 (1955)

164. P Schultz and H Strub, *Z. Phys.* **146** 393 (1956); A Bauer and P Schultz, *Z. Phys.* **146** 339 (1956).

165. A Gambere and P Schultz, *Lichttechnik* **15** 74 (1963).

166. W A Baum and L Dunkelman, *J Opt. Soc. Am.* **40** 782 (1950); H G Früling, W Münch and W Richter, *Comptes Rendue C I E, Zurich,* Paper # 2.1DF, (1955); A Bauer and P Schultz, *Seitzber. Heidelberg Akad. Wiss.* **No. 4** 428-34 (1956/579; R Grabner and H Schlegel, *Tech. Wisc. Abhandl. Osram Ges.* **7** 26 (1958).

167. R C Hirt, R G Schmidt, N Z Searle and A P Sullivan, *J. Opt. Soc. Am.* **50** 706 (1960); D W Gibson and J Weinard. *Proc. Inst. Environmental Sci.* **1962** ; E A Boettner and L J Miedler, Appl. Opts. **2** 105 (1963).

168. H W Cumming, *Light and Lighting* **48** 158 (1955); H Ulffers, *J. Soc. Motion Picture TV Engr.* **67** 389 (1958); W B Reese, Ibid. **69** 474 (1960) ; E L G Beeson, W A Bocock, A P Castellain and F A Tuck, *J. Brit. Kinetic Soc.* **32** 59 (1958).

169. H E Strauss and W E Thouret, *Illum. Engng.* **57** 74 (1963).

170. H Ramert, *Electrotech. Zeit.* **72** 604 (1951); *Tech. Wiss. Abhandl. Osram Ges.* **6** 38 (1953); loc cit, **7** 133 (1958).

171. J K Buckley, *Illum. Engng.***58** 365 (1963).

172. R Rompe and A Ihln, *Lichttechnik* **11** 328 (1959).

173. A Lompe, *Lichttechnik* **10** 108 (1958); ibid, *Electrizitätsverwertung* **33** 283 (1958); see also: A Lompe *Tech. Wiss. Abhandl. Osram Ges.* **7** 36 (1962).

174. H Schirmer, *Z. Phys.* **156** 55 (1959).

175. C A B Halvorson, *G E Rev.* **17** 290 (1914); see also: B. Monatsch, *Illum. Engng. (London)* **3** 253, 394, 427 (1910).

176. W S Weedon, *Trans. Electrochem. Soc.* **16** 217 (1909).
177. C P Steinmetz, *G E Rev.* **17** 182 (1909).

178. M Eppley, *J. Franklin Inst.* **201** 333 (1926).

179. E A Gimingham and S R Mullard, *J. IEE (London)***54** 15 (1915).

180. F Scaupy, *Tech. Wiss. Abhandl. Osram Konzern* **1** 42 (1930); *Arch. Elektrotech.* **48** 1797 (1927); see also: D E Moore, *Trans. Illum. Engng. Soc.* **16** 346 (1921).

181. M Scholnik and T B Jones, *J. Appl. Phys.* **23** 643 (1952); J D Cobine and C J Gallagher, *Elect. Eng.* **70** 504 (1951).

182. W E Forsythe, B T Barnes and M A Easley, *G E Rev.* **33** 358 (1930); D Dooley *Phys. Rev.* **36** 1476 (1930); A H Taylor, *J. Opt. Soc. Am.* **21** 20 (1931).

183. W Buckingham and C R Deiberr, *J. Opt. Soc. Am.,* **36** 245 (1946); *J. Soc. Motion Picture Engrs.* **47** 376 (1946); loc cit., **48** 324 (1947); loc cit. **54** 567 (1950); see also: W S Huxford and J R Platt, *J. Opt. Soc. Am.,* **37** 10 (1947).

184. H. Schroeder, *"History of Electric Light"*, *Smithsonian Inst. Publ. No.* 2717 (1923).

185. C G Suits, *J. Appl. Phys.* **10** 648 (1939); W A Gambling and H Edels *Brit. J. Appl, Phys,* **7** 376 (1956).

186. G. Déjardin, *Rev. Trimester Canada* **22** 238 (1936); W Nissen, *Z. Phys.* **139** 638 (1954); G Grandsire, *Ann. Astrophys.* **17** 287 (1954); F Mastrup, *J. Opt. Sco. Am.* **50** 32 (1960).

187. J A R Samson, *Tech. Rep. No. 62-9-N (NASA)*, Geophysic Corp. of America Aug. -1962.

188. W A Darah, *Trans. Electrochem. Soc.* **29** 613 (1916).

189. R Rompe, *Z. Phys.* **101** 214 (1936); see also: *Tech. Wiss. Abhandl. Osram Konzern,* **4** 44 (1938).

190. J W Marden, H C Beese and G Meister, *J, Franklin Inst.* **225** 45 (1938).

191. J Stark and R Küch *Phys. Zeit.* **6** 438 (1905)

192. T M Lowry and H H Abram, *Trans, Faraday Soc.* **10** 103 (1914); H J S Sand, *Proc.Phys. Soc.London* **28** 94 (1916); F Bates, *Phil. Mag.* **39** 353 (1920); J R Bates and H S Taylor, *J. Am. Chem. Soc.* **50** 771 (1928); H Nagaoka and Y Sugiura, *Sci. Papers Inst. Phys. Chem. Res. Japan* **10** 263 (1929).

193. M Pirani, *Z. Tech. Phys.* **11** 482 (1930); V Göler and M Pirani, *Lichte u Lampe* **20** 67 (1931); E Lax and M Pirani, *Tech. Wiss. Abhandl. Osram Konzern,* **3** 6 (1934).

194. J W Marden, H C Beese and G Meister, *Trans. Illum. Engng. Soc.* **32** * 4 (1937)
195. W Elenbaas, *Physica* **4** 747 (1937)

196. D A Larsen, *J. Am. Opt. Soc.* **52** 306 (1962).

197. W Elenbaas and J Riemans, *Philips Tech. Rev.* **11** 299 (1950); N L Harris, *The Engineer,* **189** 143 (1950).

198. M Zelikoff, P H Wyckoff, L M Aschenbrabd and R L Loomis, *J. Opt. Soc. Am.* **42** 818 (1952).

199. F L Mohler, *J. Res, Natl. Bur. Stds.* **9** 25, 489 (1932); loc cit **10** 357, 771 (1933); loc cit **16** 227, 771 (1936); loc cit **17** 849 (1936); loc cit **21** 697, 873 (1938).

200. R K Steinberg, *J. Appl. Phys.* **21** 1028 (1950).

201. N C Beese, *J. Opt. Soc. Am.* **36** 555 (1946).

202. J M Frank, W S Huxford and W R Wilson, *J. Opt. Soc. Am.* **37** 718 (1947).
203. N C Beese, *Infrared Physics* . **1** 5 (1962).

204. L Benguigui, *Autotisme* **7** 145 (1962).

205. F L Mohler *J. Opt. Soc. Am.* **29** 152 (1939).

206. H S H Massey and E H S Burhop, *"Electrical and Ionic Impact Phenomena"* Oxford Press, London (1952); *"Atomic and Molecualr Processes"* Academic Press, NY (62).

207. F L Mohler *Illum. Engng.* **56** 138 (1961).

208. J Backus, *J. Appl. Phys.* **30** 1866 (1959).

209. R D Deslattes, T J Peterson and D H Tomboulian, *J. Opt. Soc. Am.* **53** 302 (1963).

210. A Güntherschulze, *Trans. Am. electrochem. Soc.* **44** 215 (1923); *Zeit. Phys.* **19** 313 (1923), loc cit. **30** 175 (1924); *Zeit, Tech Phys.* **11** 49 (1930); M Schüler, *Zeit. Phys.* **35** 323 (1926); K Geiger, *Zeit. Phys.* **106** 17 (1937); A Lompe, R Seelinger and E Wolter, *Ann. Phys.* **36** 91 (1939); C C Van Voorhis and A G Shenstone, *Rev. Sci. Instr.* **12** 257 (1941); M A Townsend and W A Depp, *Bell System Tech. J.* **32** 1371 (1953); P F Little and A von Engel, *Proc. Roy. soc.* **A224** 209 (1954); E Badarev, I Popescu and F Wächter, *J. Electronic and Control,* **4** 503 (1959); L M Lidsky et al, *J. Appl. Phys.* **33** 2490 (1962); T Musha, *J. Phys. Soc. Japan,* **17** 1440 (1962).

211. J R Crowley, *Illum. Engr.* **4** 189-211 (1923); E E Steinert, *G.E. Rev.* **31** 136 (1928).

212. H E Edgerton and K J Gereshausen, *Rev, Sci Instr.***3** 535 (1932); H E Edgerton, *J. Soc. Motion Pict.Engnrs.* **16** 735 (1931), loc cit. **18** 356 (1932); D C Rose, *Canad. J. Res.* **11** 780 (1934); H E Edgerton and K J Gereshausen, *Trans. AIEE,* **55** 790 (1936); ibid, *"Electronics Mag."* p. 12, Feb. -1937; H E Edgerton, K J Gereshausen, and H E Grier, *J. Appl. Phys.* **8** 2 (1937).

213. P Hoekstra and C Meyer, *Philips Tech;. Rev.* **21** 73 (1959/60)
214. M P Vanyukov and A A Mak, *Uspekhi Fiz. Nauk.* **66** 301 (1958); ibid, *Soviet Phys.- Uspekhi* **4** 137 (1958).

215. I S Marshak, *Uspekhi Fiz. Nauk.* **77** 229 (1962); ibid, *Soviet Phys.- Uspekhi* **5** 478 (1962).

216. F Früngel, *"Capacitor Discharge Engineering"* Vol. 1 & 2, Academic Press, NY (1963).

217. J W Beams et al, *J. Opt. Soc. Am.* **37** 868 (1947); L S G Kovasznay, *Rev. Sci. Inst.* **20** 696 (1949); J A Fitzpatrick, J C Hubbard and W J Thaler, *J. Appl. Phys.* **21** 1269 (1950); G D Adams, *J. Sci. Instr.* **28** 379 (1951(; R L Forgacs, *IRE Natl. Conv. Rec.* **5** Part-5 114 (1957); H Fischer, *J. Opt. Soc. Am.* **47** 981 (1957); D P C Thackery, *J. Sci. Instr.* **35** 206 (1958); P Nolan, *J. Motion Pictr. TV Engr.* **70** 632 (1961) P J Hart *J. Appl. Phys.* **33** 2983 (1962).

218. E Fünfer, *Z. Angew, Phys.* **1** 295 (1949); H Schardin and E Fünfer, *Z. Angew, Phys.* **4** 185 (1952); P Fayolle and P Naslin, *J. Soc. Motion Pictr. TV Engr.* **60** 603 (1953); H Luy and R Schade, *Z. Angew, Phys.* **6** 253 (1954); H E Edgerton and K W Cooper, *J. Soc. Motion Pictr. TV Engr.* **70** 177 (1961).

219. P M Murphy and H E Edgerton, *J Appl. Phys.* **12** 848 (1941); H E Edgerton, *J. Opt. Soc. Am.* **36** 390 (1946); S L deBriun, *Philips Tech. Rev.* **8** 26 (1946); F E Carlson, *J. Motion Pictr. TV Engr.* **48** 395 (1947); F E Carlson and D A Pritchard, *Trans. Illum. Engng. Soc.* **42** 235 (1947); J N Aldington and A J Meadowcroft, *J IEEE (London)* **9** 671 (1948); C R Bicknell *Trans. Illum. Engng. Soc. (London)* **11** 1 (1948); N Warmholtz and A M C Helmer,

Philips Tech. Rev. **10** 178 (1948/49); J W Mitchell, *Trans. Illum. Engng. Soc. (London)* **14** 91 (1949); G Knott, *Photogr. J.* **89B** 46 (1949) G D Hoyt amd W W McCormick, *J. Opt. Soc. Am.* **40** 658 (1950); H N Olsen and W H Huxford, *J. Soc. Motion Pictr. Tv Engr.* **55** 285 (1950); E B Noel and P B Davies, *Photog. Sci. Tech.* **11** (1950);

219. (Continued)- See also: W D Chesterman, D R Clegg, G T Peck and A J Meadowcroft, *Jour. IEE (London)* **98** 619 (1951); W T Whelan. *Trans. AIEE* **67** 1303 (1948).

220. J H Malmberg, *Rev. Sci. Instr.* **28** 1027 (1957); Q A Kerns, F A Kirsten and G C Cox, *Rev. Sci. Instr.* **30** 30 (1959); G Porter and E R Wooding *J. Sci. Instr.* **36** 147 (1959); ibid, *J. Photogr. Sci.* **9** 165 (1961); H Fischer, *J. Opt. Soc. Am.* **51** 543 (1961); H E Edgerton, J Tredwell and K W Cooper, *J. Soc. Motion Pictr. TV Engr.* **70** 177 (1962); S I Andreev and J Lytollis, *Elect. Comm.(ITT)*, **37** 377 (1962); S I Andreev and M P Vanyakov, *Soviet Phys. Tech. Phys.* **6** 700 (1962).

221. N W Robinson, *Philips Tewch, Rev.***16** 13 (1954); C. Meyer *Philips Tewch, Rev.* **22** 377 (1960).

222. I S Marshak and L I Shchoukin *J. Soc. Motion Pictr. TV Engr.* **70** 169 (1961); V I Vasiliev, M S Luchuk and I S Marshak, *Opt. and Spectros.* **11** 61 (1961); V P Kirsanov, V A Gavinin and I S Marshak, *Opt. and Spectros.* **13** 153 (1962); V P Kirsanov, I S Marshak and M I Epshtein, *Opt. and Spectros.* **13** 244 (1962).
223. R C McVickers, R A Dugdale and J T Maskrey, *Trans Brit.Ceramic Soc.* **60** 427 (1961).

224. I S Marshak, *Pribory i Tekhnika Eksperimenta* **3** 5-21 (1961).

225. M I Christie and G Porter, *Proc. Roy. Soc.* **A212** 398 (1952); B Meyer, *Z. Angew. Phys.* **5** 139 (1953); H E Edgerton, R. Boazoli and J T Lamb, *J. Soc. Motion Pictr. TV Engr.* **63** 15 (1954); H E Edgerton and P Y Cathou *Rev. Sci. Instr.* **27** 821 (1956); G W LeCompte and H W Edgerton, J. Appl. Phys. 27 1427 (1956); E J G Beeson and K M H Rhodes, *J. Photog. Sci.* **4** 54 (1956); E A Coulson and H E EDgerton, *J. Soc. Motion Pictr. TV Engr.* 66 616 (1957); H

245

Grabner and M Reger, *Tech. Abhandl. Osram Ges.* **7** 52 (1958); H E Edgerton and D A Cahlander, *J. Soc. Motion Pictr. TV Engr.* **70** 7 (1961); N A Kuebler and L S Nelson, *J. Opt. Soc. Am.*, **51** 1411 (1961).

226. P A Miles and H E Edgerton, *J. Appl. Phys.* **32** 740 (1961); H A Edgerton and J F Gomez, *AIEE Paper No. 62-356* (1962)

227. G Shinoda, T Suzuki and M Umeno, *Jap. J. Appl. Phys.* **1** 364 (1962).

228. RM Hoffman and F Daniels *J.Am. Che,. Soc.* **54** 4226 (1932)

229. WA Darrah, *Trans Electrochem. Soc.* **29** 613 (1916)

230. NC Beese and DE Henry, *Can, Patent* 564,555 (1958).

231. OS Duffendach and R Küch, *Phys. Zeit.* **6** 438 (1905).

246

Chapter 3

MANUFACTURE OF LAMPS AND "LAMP PARTS" USED IN LAMPS

In this Chapter, we will first survey the manufacturing protocols used to prepare what have generally been called "Lamp-Parts". Our discussion will address tungsten wire as related to lamp filaments, materials used to manufacture lamps, and raw materials used to prepare phosphors. We will then discuss methods of manufacturing both incandescent and fluorescent lamps as related to the materials needed for such manufacture (Note that we have already discussed optimal lamp design in the last Chapter). In the next Chapter, we will then discuss phosphor manufacturing practices and how phosphors for use in fluorescent lamps are made. The areas to be examined in this Chapter encompass:

1. Tungsten wire manufacture, including preparation of tungsten compounds and their chemistry, preparation and fabrication of tungsten metal powder into ductile wire, and manufacture of coils for incandescent lamps.

2. Glass used in lamps and cathode-ray tube manufacture, including glass forms prepared for specific lamps.

3. Manufacture and design of incandescent lamps.

4. Preparation of raw materials used to manufacture phosphors, including phosphates and sulfides.

5. Manufacture of fluorescent lamps.

3.1- TUNGSTEN WIRE MANUFACTURING

Up to about 1975, tungsten wire intended for use as an incandescent lamp filament was made by a compaction-sintering-densifying-swaging-die drawing process. After that time, some manufacturers changed the process to improve methods which were regarded as labor-intensive and prone to poor quality predilections. In order to clarify exactly what this means, we will first discuss the older process and then present data concerning the newer one. We do this in order to establish a firm basis of knowledge concerning the processes originally used and how the modifications to these processes came about. As is usual, the changes were made partially because of monetary considerations. Nonetheless, they serve to establish a greater understanding on our part of the extremely complicated parameters involved in the manufacturing of tungsten wire and the factors which contribute to the understanding of how tungsten wire performs as an incandescent filament.

A. MANUFACTURING PROTOCOLS USED UP TO ABOUT 1975

This area includes five aspects which need to be discussed: 1) Tungsten chemistry and preparation of chemicals used to make tungsten wire; 2) Preparation of tungsten forms suitable for wire-drawing; 3) Densification; 4) Methods of forming tungsten rods by swaging; and finally 5) Wire-drawing as related to forming ductile wire suitable for manufacturing coils. We will also refer to molybdenum chemistry where it is directly related to that of tungsten.

A-1. Chemistry of Tungsten and Chemicals Used to Make Tungsten Wire

Tungsten ores occur in nature as Wolframite (an isomorphic mixture of $FeWO_4$ and $MnWO_4$), Scheelite - $CaWO_4$, Stolzite - $PbWO_4$, and occasionally as Tungstite - WO_3. Ores such as Reinite - $FeWO_4$ and Hübnerite - $MnWO_4$ are also known. These ores are generally processed by fusion with soda-ash at high temperature to form Na_2WO_4 plus oxide salts, viz-

3.1.1.- $(Fe,Mn)WO_4 + Na_2CO_3 \rightarrow Na_2WO_4 + Fe_2O_3 + Mn_2O_3 + CO_2 \uparrow$

The sodium tungstate so-formed is soluble in water and is leached from the amalgamatcd matcrial. Hot HCl is then added to the solution to form tungstic acid which precipitates:

3.1.2.- $Na_2WO_4 + HCl \rightarrow H_2WO_4 \downarrow + H_2O$

Heating tungstic acid serves to form the oxide, WO_3. However, the oxide so-formed is not pure enough to make lamp filaments and needs to be processed further. Unlike molybdenum, the ammonium tungstate salt cannot be used for further purification steps because of the tendency of tungsten metal to react with residual NH_3 to form a nonstoichiometric **nitride** when the oxide is reduced to metal. Therefore, NaOH is used to dissolve the tungsten oxide and the resulting solution is filtered to remove any insoluble hydroxide precipitates. Organic chelating reagents may then added to the solution, depending upon the impurities needed to be removed. After filtering the precipitates formed, one obtains a solution from which tungstic acid can again be produced by the addition of HCl. This step also has a purifying effect on the H_2WO_4 formed. Since any tungsten metal intended for the wire-drawing process must be pure and substantially free of other elements (other than those deliberately added), it is common to recycle the dissolution and precipitation steps more than once to obtain a pure tungstic acid which is finally fired in air to form WO_3 Finally, the WO_3 is reduced in a hydrogen atmosphere to tungsten metal, in a manner that produces metal substantially free of oxygen.

The reduction step is critical since tungsten metal powder can be obtained in several forms, depending upon reduction conditions. One such form (where the particle size is too small, i.e.- < 1 µ) is pyrophoric. When WO_3 is reduced, it goes through several stages which include formation of the lower valence states of tungsten. The most stable valence state of tungsten is W^{6+}, but W^{5+}, W^{4+}, W^{3+}, and W^{2+} are also known. During the reduction by hydrogen, the powder produced becomes sintered into friable masses, depending upon a number of factors, as elucidated below. Several lower valence oxides are formed, i.e.-

3.1.3.- $\quad 4\,WO_3 + H_2 \quad \Leftrightarrow \quad W_4O_{11}\,\text{(violet color)} + H_2O$

$\quad\quad\quad\quad W_4O_{11} + 3\,H_2 \quad \Leftrightarrow \quad 4\,WO_2\,\text{(chocolate brown)} + 3\,H_2O$

$\quad\quad\quad\quad 4WO_2 + 8H_2 \quad \Leftrightarrow \quad 4\,W\,\text{(black to gray-black)} + 8\,H_2O$

$\quad\quad\quad\quad WO_3 + 3\,H_2 \quad \Leftrightarrow \quad W + 3\,H_2O$

In the first stage of reduction, the mixture of violet W_4O_{11} and yellow WO_3 forms a deep-blue powder, easily identifiable as such. It is this mixture which can be seen on the silica trays used to fire tungstic oxide to the metallic form. It is essential to thoroughly remove the water formed during the reduction stage since it will prevent the reaction from going to completion. Even a small amount of unreduced oxide in the tungsten powder so-produced is extremely detrimental to properties required for filament wire. Thus, a flowing atmosphere in the reduction furnace is necessary to flush out all of the water of reaction. Freedom from impurities is essential, as is a suitable physical condition required for further processing of the tungsten-powder into wire. The following Table shows a **maximum** impurity level required for tungsten wire, if it is to be used for lamp filaments.

TABLE 3-1

Upper Level for Impurities Commonly Found in Tungsten Metal Powder
Used to Manufacture Lamp Filaments

Element	Concentration in ppm.	Element	Concentration in ppm.
Mo	300	Fe	50
Si	90	Cu	30
Al	40	Oxygen	15
K	85	Nitrogen	1
Na	10	Hydrogen	1
Ca	80		

Most of these impurities come from the original ore except Si which may originate from the silica trays used during the reduction step. However, the purer that the metal powder is, the better are the wire properties as related to filament failure in incandescent lamps.

Depending upon reduction conditions, one can obtain average metal particle sizes up to 500 μ. Most methods of obtaining an ingot by pressing require a fine tungsten powder, about 2-5 μ in average size. The lowest temperature at which complete reduction is practical is about 700 °C.

In the following Table, we summarize the effect of temperature upon particle size produced. In examining this Table, it is easy to see that the reduction temperature should not go over about 850 °C. Reduction at constant temperature produces a coarser grain than is obtained when the temperature is raised gradually to a final value. Many manufacturers have a continuous furnace in which the temperature profile is controlled so that the reduction takes place in such a manner. The trays containing WO_3 to be reduced to tungsten metal are pushed through the temperature profile

250

TABLE 3-2

Particle Sizes Produced in Tungsten Powders as a Function of
Reduction Temperature

Temperature	Product Obtained	Size Range in Microns
600	WO_2 + W	0.2- 0.5
700	W	~ 1.0 µ (0.001 mm.)
800	W	~ 2.0-3.0 µ
900	W	~ 3.0-4.0 µ
1100	W	~ 10.0 µ
1500	W	~ 300-400 µ (0.3 mm)

so that each tray experiences an increasing temperature. Generally this profile starts at 550-650 °C, with a final temperature between 780 and 1100 °C, depending upon final particle size required for the end-use of the wire so-produced. WO_2 mixed with WO_3 occurs at the lowest temperature and tungsten-metal at the upper temperature stages.

Some manufacturers also produce the metal by an intermittent process wherein the reduced product is removed, cooled and then subjected to a 2nd stage of reduction after certain additives such as thoria have been made to the original reduction product. In such a process, the temperature is gradually raised from 500 to 700 °C. in about 2 to 3 hours. Then more WO_3 is mixed in and the mix is further reduced to tungsten-metal in a temperature profile from 550 to 750 to 1000 °C. over a period of 3 to 4 hours.

Particle size of tungsten metal produced has been found to depend upon:

a) Firing Time and Temperature- increasing the time or raising the temperature rapidly produces a coarser metal powder.

b) Oxide Particle Size- increasing the oxide size changes the W-metal particle size: Coarse oxide gives a coarse metal-powder (but a fine oxide can be sintered to form a coarse metal powder if one is not careful).

c) Firing temperature used for converting the tungstic acid to oxide also plays an important role in the size of oxide particles produced.

The major reason that one wishes to obtain a tungsten metal powder of small but controlled size is that the powder is to be used to form a pressed bar of controlled proportions. If the particle size is too large, then the bars produced are too fragile and cannot be handled at all. They also will contain considerable "voids". If the particles are too small, then the product obtained is likely to be pyrophoric.

As stated above, the concentration of water vapor present in the hydrogen gas has a pronounced effect on W-metal particle size. It is likely that the following reaction takes place:

3.1.4.- $W + H_2O \leftrightarrow WO_2 + 2\ H_2$

Such a reaction is cyclic and it was for this reason that the reactions given in 3.1.3. were written with a "double-arrow" since water vapor in hydrogen gas can reverse the reactions as shown. This mechanism results in the formation of large particles of W-metal and is the reason that friable masses are formed during reduction. If one wishes to make a fine metal powder of nearly constant size, it is necessary to have a rapidly flowing H_2 gas stream in the reduction furnace. Depending upon the size of the furnace, this flow may be as high as 3000-4000 liters/hour of **pre-dried** hydrogen gas, depending also upon the weight of WO_3 being reduced per unit volume of the furnace-tube per hour. This parameter also mandates the use of a thin layer of oxide if a fine metal powder is desired. Thus, another consideration is the amount of oxide present per cc. of hydrogen gas flowing at any given instant.

The following data are relevant to the "water-cycle" affecting W- size:

TABLE 3-3

Size of Tungsten-Metal Powder Produced as a Function of Gas Flow

Maximum Temperature	Grams of WO_3 Present per cc. of H_2 Gas in Reduction Tube	Diameter of Tungsten Metal Particles Produced
800 °C.	0.05 gram	0.5 µ
830	0.50	2.0 µ
900	1.00	4.0 µ

It is easy to see that the critical factors which control the particle size of the tungsten-metal powder produced include the depth of the oxide being reduced and its relationship to the amount of hydrogen-gas present at any given instant. Summarizing, the factors include:

3.1.5.- Factors Controlling Particle Size of Tungsten Metal Powder

    1. Particle size of WO_3 produced

    2. Reduction temperature

    3. Reduction temperature profile

    4. Amount of water-vapor present at any given instant

    5. Amount of oxide being reduced per unit-volume of furnace-tube, i.e.- depth of oxide layer

    6. Ratio of flowing hydrogen gas to oxide being reduced

Many manufacturers employ a rotating Kiln in which both the layer thickness of the oxide is controlled by continuous feeding, the temperature profile that the oxide experiences during reduction is controlled, and the hydrogen gas flow is controlled so as to limit the effect of water present and remove that produced by the reduction reaction.

Before we leave this subject, it is well to note that the methods used to produce tungsten-metal powder are applicable to the case of molybdenum metal as well. The major difference is that a soluble ammonium molybdate is usually employed. Because of the tendency of Mo to form polymolybdates, one usually obtains: $(NH_4)_6Mo_7O_{24} \cdot 4H_2O$. Upon firing this salt, the oxide, MoO_3, results. The use of the sodium salt has been tried, but a better metal product results from the use of the ammonium salt. The same restrictions relevant to tungsten apply for this case as well. Because Mo metal is generally used as conductive lead-in wires and foils, its metallurgical state, including grain sizes and grain boundaries within the metal, is not so critical as is the case for tungsten metal which is used for incandescent lamp filaments. One method that has been used to prepare ammonium molybdate is:

    1. MoO_3 is dissolved in ammonia. 1200 Kg. of roasted ore containing 90% MoO_3 is placed in a covered rubber-lined tank with sufficient excess NH_3 to completely dissolve the MoO_3 present. Gaseous NH_3 may also be added as required. Then air is injected to oxidize

any iron present, i.e.- $Fe^{2+} \Rightarrow Fe^{3+} \Rightarrow Fe(OH)_3$ ⇓. H_2O_2 may also be added. The resulting precipitate is then allowed to settle and the liquid is then run through a filter to remove all of the precipitate.

2.The remaining liquid is then treated with a solution of 1:4 HNO_3 and HCl (25 mol%) to a pH of 1.5. This precipitates the ammonium paramolybdate referred to above. The precipitate is now allowed to settle and then is brought to a filter press. The mother liquor contains 1-2 grams of Mo^{3+} per liter and is discarded.

3. If the pH is too high, too much Mo^{3+} will be lost, and if it is too low, H_2MoO_4 results.

4. The crystals are washed with cold deionized water and then with 2.5 mol% NH_4NO_3 solution, diluted in half with ethanol.

5. Finally, the crystals are removed from the filter and vacuum dried at about 60 °C. for 24 hours.

6. Firing at 900 - 1200 °C in air serves to form the MoO_3, which is now ready for use after it cools.

A-2. Preparation of Tungsten Forms- Pressing and Sintering Bars

Because the melting point of W is 3680 °K. (3407 °C - a temperature not readily attainable), only metal particles can be formed in the reduction step. One therefore takes this powder and presses it in a mold to form a bar, using a wax binder of some sort. Each tungsten-wire manufacturer has his own preferred binder, most using glycerine, paraffin wax or the like. The ingot size is usually about 0.50 inches square and about 24 inches long. A typical mold is made from stainless steel, and has a simple design as shown in the following diagram, given on the next page as 3.1.6.

A pressure of at least 12,000 lbs/in^2 is needed to compact the metal powder with pressures up to 50,000 lbs/in^2 being more usual. If the metal powder used is fine, the ingot so-produced may be handled, with care.

3.1.6.-

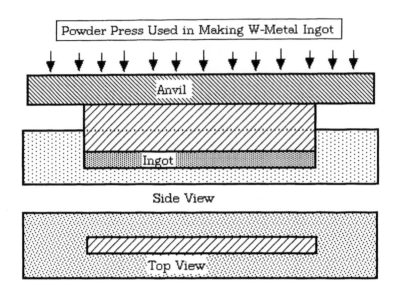

During the pressing step, an inert atmosphere is generally used since the operation tends to generate heat which might cause the formation of an oxide film on the particles. The pressing operation usually produces a porous ingot having no more than 30-40% of the density of pure tungsten metal. Although isostatic pressing has been tried, the ingot so-produced has been found to be no better in strength than that made using a mold. If the particle size of the metal particles is small enough, the pressed ingot generally has enough strength to be cautiously handled. It is then placed on a slab of W, Mo or graphite and sintered in a reducing atmosphere to fuse the metal particles into a solid bar or ingot. If a graphite plate is used to sinter the pressed bars, the sintering temperature must be kept below 900 °C. to avoid the formation of tungsten carbide, i.e.- W_2C. The usual heating cycle involves heating at temperatures between 900-1050 °C. for about 30 minutes in a flowing hydrogen-gas atmosphere and then cooling in the same gaseous atmosphere. Most manufacturers use a temperature-profile furnace and push the slabs through the hot-zone of the furnace.

The sintering temperature is low enough so that a change in surface-grain growth is not visible. An increase in strength of the sintered bar usually results, thought to be due to the reduction of an oxide surface-film on the metal particles. The reduced metal so-formed probably acts as a cement between the particles thereby increasing the strength of the sintered bar. Nevertheless, the ingots so-obtained are porous and have a density of about 60-65% of the ultimate tungsten value. As such, they are too fragile to withstand any working into a wire-form.

A-3. The "Treating-Bottle"- A Temperature Densification Step

Because of the high melting temperature of tungsten, any attempt to further densify the sintered bars will require very high temperatures to do so. In general, the use of a furnace did not appear to be practical since the insulation present in most furnaces of that time could not withstand the extremely high temperatures needed. Thus, the use of a so-called "treating bottle" evolved. This consisted of an uninsulated metal enclosure in which the top of the ingot was clamped, while the bottom was submerged in a mercury-pool. This is shown in the following diagram:

3.1.7.-

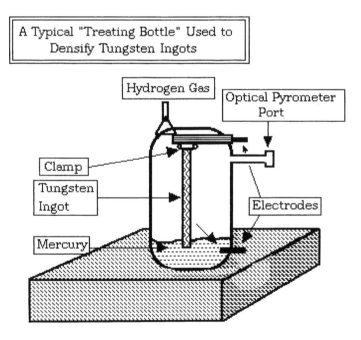

A Typical "Treating Bottle" Used to Densify Tungsten Ingots

The purpose of the mercury pool was to be able to make electrical contact while allowing one end of the ingot to be free to move as it underwent shrinkage and densification. The treatment took place under a dry-hydrogen atmosphere. Some manufacturers used a flowing atmosphere while others employed a pressurized atmosphere.

In some cases, a vacuum was employed within the bottle. Heating took place by a DC current whose amplitude depended upon the size of the ingot. To heat an 8-inch ingot close to its melting point required about 1500 amperes at 10 volts, while a 24 inch bar required about 4500 amperes. Because considerable shrinkage occurs during densification, the bar tended to deform. It is for this reason that only the top of the bar was clamped within the apparatus. While one can monitor the temperature of the ingot as it is being heated by use of an external optical pyrometer, it proved easier to monitor the current being used to heat the bar.

Grain growth begins at about 1050 °C and shrinkage accompanies this. At the same time, partial elimination of voids normally present in the bar also begins. These voids are a consequence of the pressing step and result because of interstitial spaces between particles and voids in the metal particles normally experienced during the reduction step from the oxide to the metal stage. The usual shrinkage experienced is about **17% in volume**. The whole operation takes about 30-45 minutes.

The procedure was to raise the temperature of the bar to about 92% of the fusion temperature, i.e.- $\approx$ 3300 °C. This temperature was held for about 10-15 minutes during which time grain growth continued and a coarsely crystalline microstructure was obtained. Metallurgical grain size varies considerably under differing conditions of densification, but grain size has little effect upon subsequent working properties of the metal so-produced. Ingots can be worked whose average grain size varies from 10 μ to over 10,000 μ. The average grain size produced was usually about 1500-2000 μ. per square millimeter of surface.

Since grain growth begins at 1050 °C. and continues with time, there is generally an equilibrium that occurs between density and grain growth within 10-15 minutes at any given temperature above 1100 °C. One can observe this growth by counting the grains in an etched cross-section of the treated bar. Since tungsten metal has a body-centered structure (a_0 = 3.155 ± 0.001 Å), etching takes place on the {112} planes (cleavage occurs on the {100} planes and slip occurs on the {112} planes). Generally, triangular etch-pits are seen.

(A good etching agent is 3% H_2O_2 , but a 10% NaOH + 30% potassium ferrocyanide solution may also be used). However, it did become apparent that the rate of grain-growth was modified by: 1) changes in temperature and time; 2) the size of the metal particles and 3) impurities present in the metal. Thus, it is possible to densify an ingot so that the grains present vary from about 1 to as high as 10^{10} grains (crystals) per mm^2. However, as we said, the normal figure attained is about 2500 crystals/mm^2.

The effect of temperature during the heat-treating step is complicated indeed. First, the surface is cooler due to radiation loss there and because of the presence of cooler hydrogen gas. The surface temperature may differ from the interior as much as 150 °C and the center of the bar may be much hotter than the ends. These differences give rise to peculiar types of grain growth. As the bar is heated, the center may become melted while the outside remains solid. If this bar is cooled quickly, large radial crystals will form. But if the bar is held at this high temperature for longer than about a minute, molten metal is likely to break through, leaving a hollow core. At lower temperatures, exaggerated grain growth may occur because of internal temperature gradients. Grains produced in the center of the bar may be considerably larger than those produced near the surface and edges of the bar. This exaggerated growth occurs over the temperature range of about 2500-2800 °C. Below this range, one obtains a very fine grain (but low density) while above this range uniform grain growth occurs without a large contrast in grain size (and with a relatively high density).

As shown in the following Table, given on the next page, the exaggerated grain growth mechanism has been found to be a direct function of temperature. At fairly low temperatures, one obtains rather small grains. But, as one increases the temperature, a temperature-range is reached where anomalous grain-growth occurs and the average grain size increases considerably. However, if the temperature is raised rapidly almost to the melting point, a "normal" grain-structure is obtained.

It is because of these vagaries that the use of the "treating bottle" evolved, **since it was the only method known at that time** that could raise the temperature of the bar close to the melting point of tungsten in a short period of time.

258

TABLE 3-4

Exaggerated Grain Growth Experienced as a Function of Melting Current

Sintering Temp. in °C.	% Melting Current	Grains Produced per mm^2
2550	65	1150
2850	75	2
2950	80	3
3200	88	5
3250	90	15
3300	92	1140

The particle size of the metal powder and the pressure used for compacting the metal powder particles also has a major effect on the sintering properties of the bar produced. The following observations have been made:

a. The finer the particle size, the greater is the cohesion between particles. Cohesion has a marked effect on grain growth as well. Under some conditions of time and temperature, the grain size of an ingot may be larger as finer powders are used. (However, one reaches a limit in that too fine a particle size gives a pyrophoric powder).

b. It becomes very difficult to heat an ingot made from too fine a powder through the temperature range of exaggerated growth rapidly enough to avoid very large crystal growth.

c. Coarse powder produces a fine-grained porous bar upon heat-treating and will not develop large crystals even under optimal grain-growth conditions.

d. In practice, the particle size of the metal powder that is used varies. It is best to obtain a powder whose particle distribution is narrow. This is one of the more important control parameters used to maintain high quality tungsten wire production. The best powder will have a mean diameter of 3.0 μ with a 2σ spread from 0.5 to 6.0 μ. In manufacturing, this is sometimes achieved by blending powders with differing particle distributions.

e. The pressure used to form the ingot is important only in that it affects inter-particle contacts. The temperature at which grain growth begins depends upon the particle size of the metal powder and the compacting pressure used. For a 0.6 μ powder, this temperature is 70 °C. lower than that for a 3.0 μ powder. In the following, we show the minimum temperature at which grain growth begins as a function of pressure used to compact the bar (Obviously, pressure affects how close the individual particles touch):

Pressure (tons/in ^{2})	Initial Grain Growth
10	1200 °C
25	1050
40	900

f. Impurities have an effect on the quality of grain growth. Most are volatile at relatively low temperatures and escape from the porous bar before it compacts and shrinks. Those that volatilize at higher temperatures may force grains apart and prevent normal grain sizes from forming. They may even split the bar. Therefore, intentional additions of impurities like alkaline oxides, silica or boric oxide, which volatilize between 1000-2000 °C. facilitates the escape of less volatile impurities, or leave channels for their escape. It should be noted that this purifying effect, obtained during the densification step by heat-treating, has an important influence on subsequent recrystallization and ductility of the worked metal.

g. Certain impurities are relatively non-volatile. These include oxides of thorium, uranium, calcium and magnesium. Since they are insoluble in tungsten, such particles obstruct grain growth in proportion to the number present and their relative size. Thoria, i.e.- ThO_2, is usually added to the tungsten powder before the pressing step to control grain size and/or recrystallization. The optimum value for thoria to achieve this effect is about 3-4% by weight of powder.

Other non-volatile impurities have an effect proportional to their concentration, viz-

Fe & Ca Concentration in Weight %	Number of Grains Produced in Heat-Treated Bar
0	0.54 /mm^2
0.002	1.30
0.01	59
0.05	760

h. It is possible to prepare what is substantially single-crystal tungsten wire by first forming a small rod by "swaging" and then passing the rod through a furnace, in a reducing atmosphere of hydrogen gas, at temperatures between 2400-2600 °C. at rate slow enough that the "exaggerated" growth, mentioned above, can take place. Single crystal wires with lengths up to several meters in length have been produced by this method. It is not known exactly why thoria promotes the formation of tungsten single crystals, but it has been shown that some of the thoria is converted to thorium metal during the heating operation.

Once treated, the density of the bar approaches 18.5 gm/cc., i.e.- about 95% of theoretical density. Working properties will depend upon the compactness of the bar, as indicated by the degree of shrinkage which occurs during densification, and more importantly, by freedom from certain impurities such as silicon. The treated ingot emerges from the "treating bottle" with considerable strength, but it is very brittle at room temperature. Fortunately, at higher temperatures, tungsten becomes more ductile and at about 1300 °C, it can be hammered or rolled. Reiterating, the most important parameter of control is the amount of shrinkage that occurs. Usually, if sufficient shrinkage does not occur during heat-treating, the bar cannot be worked by swaging.

A-4. Swaging of the Heat-Treated Ingot.

Tungsten metal is unique in that it is one of the few metals that becomes "work-hardened" at low temperatures. According to Valentine and Hull (1), the onset of the ductile-brittle transition of polycrystalline tungsten metal having an average grain size of about 30-300 μ occurs at 350 °K, i.e.- 77 °C. Therefore, it is necessary to subject the bar to severe working at a relatively low temperature to break up the large crystals which form when the metal is heated and the resulting grain growth renders it brittle. To do this, a specially-designed apparatus has been used, called a "swaging" machine. It is actually a rotating hammer, consisting of opposing hammers usually made of tungsten carbide, i.e.- W_2C. Their number is generally four to six and two actually strike the bar simultaneously as the swage revolves. The head of such a machine is shown as follows. Note that the individual races rotate in opposite directions.

3.1.8.-

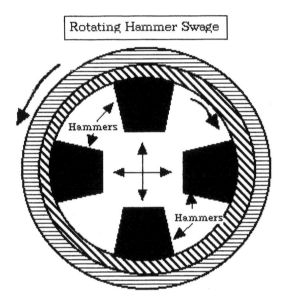

During swaging, the head of the machine becomes hot and some surface oxide forms on the bar being swaged. However, since the ingot or rod being reduced in diameter needs to be "retreated" by reheating to about 3000 °C. in a reducing atmosphere, any WO_3 formed is again reduced to the metallic state.

In general, the hammers are actuated by insertion of a bar to be swaged and the process is manual, that is- a worker "swages" one tungsten ingot at a time. Each pass of the bar through the swaging-machine reduces the diameter by about **8% with each passage**. Each ingot is heated to about 1500-1600 °C. in a furnace having a flowing hydrogen-gas atmosphere. As the ingot is withdrawn, it cools to about 1350 °C. as it is being worked. Originally, the bar was about 0.50 inches square but had been reduced to about 0.375 inches by sintering and densifying, i.e.- 9.525 mm., but can be as small as 0.350 inches square, i.e.- 8.89 mm., depending upon the Manufacturer. Obviously, more metal can

262

be swaged into a longer wire (4 times in the above example for the larger ingot). The appropriate Swaging Temperature is a function of the reduced rod-diameter as shown in the following:

3.1.9.-

Rod or Wire Diameter	Swaging Temperature
6-11 mm.	1350 °C.
5.5-2.5	1250
2.5-1.0	1175
1.0-0.05	800 - 550

After each pass through the Swager, the rod must be re-treated by heating it to 3000 °C in a reducing atmosphere for 2.0 minutes. Therefore, it is necessary to have a "Retreat Furnace" next to the swaging machine. Retreating lessens the tendency of the rod to split and gives a softer material. Care must be taken not to work the metal after cooling below its proper swaging temperature or cracks and splits will develop. The worker learns how to judge temperature from its color, and for this reason, "swaging" is a learned art.

Flexibility of the swaged rod increases as the rod diameter is reduced. However, the metal does not really become ductile until sizes below about 1.0 mm. are reached. Larger sizes, i.e.- 2-3 mm., can attain ductility by heating to a dull-red heat and then hammering or bending the metal, as in swaging. Note that the temperature of working is always below that temperature in which visible crystallization occurs, i.e.- metallurgically speaking, tungsten metal is always "cold-worked". **Thus, low-diameter ductility is characteristic of cold-worked tungsten metal.** However, the metal rod or wire must be reheated after each operation in order to reduce stresses incurred within the rod or wire because of the cold-working process.

The recrystallization temperature drops progressively as the rod or wire diameter is reduced, and the swaging temperatures used take this into account. If the metal is heated above the recrystallization temperature at any time, the ductility is immediately lost, and the metal then can only be worked at the high temperature used in the first stages of swaging. Thus, if the metal is worked too hot in later stages of swaging, the recrystallization temperature may be exceeded and the wire becomes brittle. Therefore, the primary aim of the swaging process is to break up the original equiaxial grains of the formed ingot and to reorient them and elongate them in the direction of the wire axis so that they ultimately become like a bundle of fibers. Deformation takes place by a series of slips on the {112} plane. As a result, the crystalline grains are broken up in "cold-working". In some cases, these grains are visible but eventually they are broken into sub-microscopic crystal fragments. These fragments tend to assume an equiaxial orientation with respect to the direction of working. Nevertheless, heating the wire too hot causes replacement of the fibers with crystals, whereupon ductility is lost and the wire becomes brittle. Then it is necessary to start over again at the beginning of the swaging process. The lowest temperature at which recrystallization has been observed is about 1000 °C for severely worked wires of pure tungsten. However, the presence of alumina or thoria can raise the recrystallization temperature above 2000 °C. It is primarily for this reason that most tungsten wire manufacturers use alumina additions to the tungsten trioxide (which is reduced to the metal, compacted into bars, heat-treated, and densified into ingots) before beginning the wire manufacturing process. It is well to note that the structure of drawn tungsten wire is practically unaffected by annealing at temperatures below 550-600 °C. Thus, this represents a temperature limit for reducing strain in the wire induced by the wire-drawing process.

A-5. The Wire-Drawing Process

When the original ingot has been reduced to a diameter of about 1.0 mm. (the rod so-produced from a 0.50 inch x 24 inch ingot is now about 60 feet long), one changes from swaging to wire-drawing. A typical setup is given in the following diagram, shown as 3.1.10. on the next page.

In this process, the rod is initially drawn through a die made of diamond or other hard material like W_2C. The die itself is larger on one end than the exit end so that it can accept the larger size and so that the wire being drawn becomes smaller in diameter as it passes through the die. It is common to etch one end electrolytically in NaOH solution so that it can be introduced into the die.

3.1.10.-

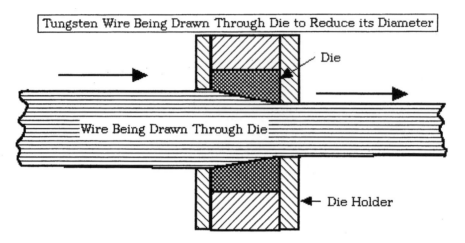

The wire-drawing process is continuous and consists of the following steps:

a. Rods are generally reduced in diameter from about 1.0 mm. to 0.5 mm. in one step to produce "heavy-wire". This product then is further reduced to about 0.25 mm. after the heavy-wire has been heat-treated at about 550 °C. to relieve strain. These steps constitute "heavy-wire drawing".

b. The 0.25 mm. diameter wire is then lubricated by passing it through a bath containing colloidal graphite, i.e.- "Aguadag™", heated just before it passes through the die, and then is wound upon a drum. Heating bakes on the lubricant, lowers both the force needed to pass the wire through the die, and lowers the stresses induced in the wire during the drawing pass through each die. Ductility of the wire is thus improved and maintained. Between each pass through a given die, the wire is heat-treated to relieve strain.

c. To measure the diameter of any given wire, one measures a standard length of exactly 20.00 cm. and weighs it. The following Table, presented on the next page, gives wire diameter, weight of 20 cm., and nominal current rating for various diameters of tungsten wire, slated for use as filaments in incandescent lamps.

TABLE 3-5

Wire Diameter	Mils	Weight of 20 cm	Nominal Current
0.180 mm.	7.08	95.65 mg.	3.40 amperes
0.017	6.69	85.47	3.12
0.016	6.30	75.99	2.86
0.015	5.90	66.20	2.58
0.014	5.51	58.12	2.34
0.013	5.12	50.31	2.10
0.012	4.72	42.80	1.865
0.011	4.33	35.60	1.625
0.100	3.94	29.30	1.400
0.090	3.54	23.80	1.195
0.080	3.15	19.20	1.015
0.070	2.76	14.50	0.827
0.065	2.56	12.40	0.732
0.060	2.36	10.60	0.651
0.055	2.17	8.90	0.569
0.050	1.97	7.30	0.488
0.045	1.77	5.90	0.416
0.040	1.57	4.70	0.353
0.035	1.38	3.60	0.290
0.030	1.18	2.65	0.231
0.025	0.98	1.82	0.172
0.020	0.79	1.15	0.122
0.015	0.59	0.65	0.080
0.011	0.43	0.38	0.050

d. It is impractical to draw wire finer than about 0.011 mm. in diameter (0.011 mm. = 0.43 mils = 0.00043 inches). However, one can make finer wire by etching. But, the etching process must be done rapidly to obtain a uniform cross-section in the wire. The best etching material is a fused mixture of $NaNO_3$ and $NaNO_2$ held at 340 °C. For example, a 0.014 mm wire is run through this bath, with a 45 second immersion to produce a uniformly etched wire of 0.007 mm. diameter.

e. An alternative method used consists of a multiple-die setup shown as follows. In this unit, a wire at 200 microns (0.200 mm.) is drawn continuously to 50 μ in six passes. Drawing speeds are 100-200 meters per minute. This method is used for wires used in high volume coils for incandescent lamps.

3.1.11.-

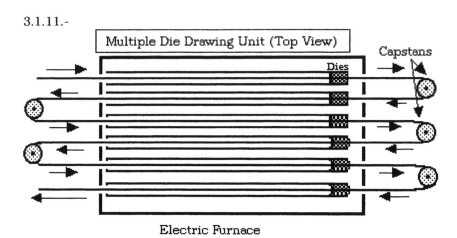

Electric Furnace

f. Once the wire diameter sought is achieved, the wire must be cleaned before further use. For sizes larger than about 0.20 mm., the "heavy-wire" is wound upon a large drum and then held in a solution of boiling KOH or NaOH for about 30 minutes. The wire is then washed by passing it through a vat of water, passing it through felt pads to remove graphite, and then rewinding it on another drum. For finer wires, the cleaning process is to spool the wire to another, passing it through boiling 40% NaOH solution, cleaning water-jets, and then felt-pads to remove any clinging graphite.

Once the wire-drawing process is complete, it is common to join wires of the same diameter together before processing further. Since tungsten metal oxidizes so easily, a special technique is required. Tungsten to tungsten joints are too brittle when joined by welding. But if nickel metal is plated on the ends of the wire, welding serves to make a good joint, both flexible and strong. Tungsten alloys with both Ni and Mo metals. Thus, strong joints can be made using either metal. But Cu and W do not alloy. It has been found that any excess metal used to join tungsten will volatilize so that its presence is not detrimental to operation as an incandescent filament (Mo melts at 2890 °K (2617 °C) whereas Ni melts at 1726 °K (1453 °C.).

A-6. The Coil-Winding Process

The primary use of tungsten-wire is as the filament in incandescent lamps. A number of methods of producing coils have evolved, depending upon the manufacturer. For the most part, the basic method has remained the same, that of winding the wire around a mandrel, usually a molybdenum wire, on a high speed automatic coiling machine. This provides a more uniform product with closer tolerances than could be reached by manual assembly. In this process, the cleaned wire, having been annealed to relieve strain, is wrapped around the mandrel in a tight spiral as a continuous single coil whose length is generally several hundred meters or more. These coils are then annealed in a hydrogen-furnace to remove the lubricant and to relieve strains induced during coiling. The initial coiling process is shown in the following diagram:

3.1.12.-

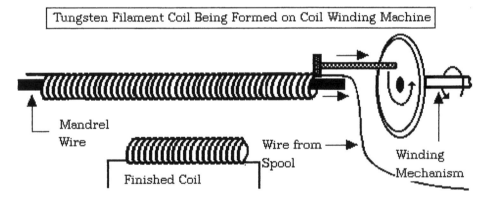

Tungsten Filament Coil Being Formed on Coil Winding Machine

Mandrel Wire

Finished Coil

Wire from Spool

Winding Mechanism

The wire with the mandrel is then cut to leave short "leads", which are then bent to form the final shape of the filament. Then, the mandrel is dissolved in an acid bath, leaving the finished coil behind. The major reason for producing a coil for use as a filament in a lamp is, as we have already stated, to lengthen the amount of fine tungsten wire present in any given lamp. There are a number of machines available to do this operation. Some manufacturers have developed their own proprietary machines, while others have purchased such a machine on the open market. Once the coil has been formed and has been cut along with the mandrel, then another Mo-mandrel wire advances and the operation begins again.

If a "coiled-coil" is to be made, a long coil is first made on a molybdenum wire mandrel by the winding mechanism. Then, the coil is reshaped around a second mandrel, usually of steel, the coiled-coil is cut to length along with the initial molybdenum mandrel, and then shaped into a final form. The second mandrel then advances, and the operation begins again. The coiled-coils are then annealed as before and the molybdenum mandrel is dissolved, leaving the coiled-coil in a finished state. An example of a coiled-coil is as follows:

3.1.13.-

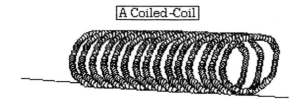

A Coiled-Coil

Coiled-coils are used in incandescent lamps because they concentrate the volume of heat generated within a smaller space so that losses by convection are minimized. These lamps also have more resistance to vibrational shock. Most manufacturers use a continuous coiling machine to produce coiled coils.

The following diagram illustrates a vertical coiling machine used to produce the first coil:

3.1.14.-

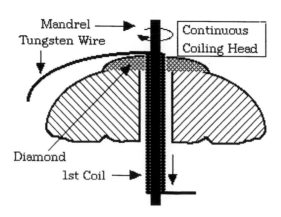

Mandrel
Tungsten Wire
Continuous
Coiling Head
Diamond
1st Coil

There are six heads per machine. The secondary double mandrel runs at about 1500-2000 rpm., depending upon coils dimensions. They are usually four-head skip-winding affairs with a constant mandrel speed where the primary coil is braked to 2/3 of the operating speed as the coiled-coil nears completion. The stop is applied by a magnetic brake, followed by the cut-off, and the coiled-coils are produced at ± 1.0 mm. length tolerance.

In the following diagram is shown some of the designs used for filaments, including that for so-called "rough-service" lamps:

3.1.15.-

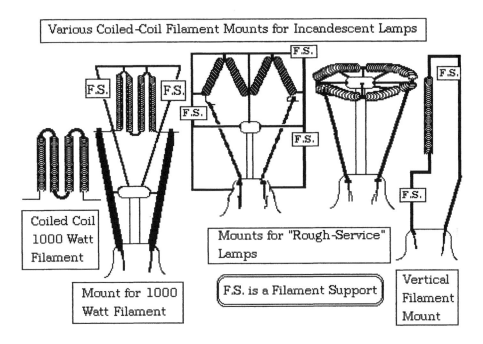

B. MANUFACTURING PROTOCOLS DEVELOPED AFTER 1975

Having described the steps used to manufacture tungsten wire up to about 1975, we are now in a position to examine "improvements" made in the process, having a firm basis of knowledge concerning the steps required to make tungsten wire. Reiterating, the steps involved:

1. Firing of WO_3 powder to obtain tungsten metal powder

2. Compaction and sintering of tungsten metal powder into a bar

3. Densifying the bar via the "Treating Bottle"

4. "Swaging" the bar into a long rod(heat-treating between passes)

5. Initial wire-drawing into "heavy wire", including annealing

6. Wire drawing into sizes as low as 0.011 mm = 0.43 mils

We will now examine each of these areas more thoroughly in order to clarify certain technological areas which have remained obscure to this point.

B-1 <u>Firing of WO_3 Powder to Obtain Tungsten Metal Powder</u>

As a result of considerable exploratory work, it became obvious that the reduction of WO_3 to the metallic form was more complicated than originally believed. Furthermore, the role of additives (and impurities) was observed to have a greater influence in the performance of tungsten-wire filaments than formerly believed. The recognition came about because some manufacturers used a potassium silicate, i.e.- $K_2Si_2O_9$, for a binder in pressing ingots before sintering and Al_2O_3 and ThO_2 as additives to control grain-size growth during sintering. Other manufacturers also added KCl, $AlCl_3$ and other compounds for the same purpose.

1. <u>Reduction Reactions to form Tungsten Metal</u>

In the following Table, given on the next page as Table 3-6, we show the reactions involved in the reduction of WO_3 to W-metal and the equilibrium constants determined as a function of temperature. In this Table, 14 reactions to form the various gaseous, liquid and solid phases of tungsten and tungsten oxides are given. These equations of formation are shown along with the corresponding equilibrium constants. Two temperature regions are of interest. The first is the region of 1000 to 1400 °K where the reduction of $WO_3(s)$ to $W(s)$ takes place. The second is the high temperature region of 2000 to 3000 °K where the final sintering and reduction occurs.

TABLE 3-6

Thermochemical Data for the Tungsten - Oxygen System

Reaction	Equilibrium Constant	$\log K_p$ at Different Temperatures, °K		
		1000	2200	3000
$W(s) + 1/2\ O_2 \leftrightarrow WO(g)$	$K_p = p_{WO}/(p_{O_2})^{1/2}$	-16.61	-4.825	-2.296
$W(s) + O_2 \leftrightarrow WO_2(s)*$	$K_p = 1/p_{O_2}$	21.281	4.961	1.512
$W(s) + O_2 \leftrightarrow WO_2(g)$	$K_p = p_{WO_2}/p_{O_2}$	-1.700	0.242	0.585
$W(s) + 1.36\ O_2 \leftrightarrow WO_{2.72}(s)$	$K_p = 1/(p_{O_2})^{1.36}$	28.444	6.788	2.148
$W(s) + 1.45\ O_2 \leftrightarrow WO_{2.90}(s)$	$K_p = p_{WO}/(p_{O_2})^{1/2}$	29.792	7.056	2.178
$W(s) + 1.48\ O_2 \leftrightarrow WO_{2.96}(s)$	$K_p = 1/(p_{O_2})^{1.48}$	30.336	7.188	2.20
$W(s) + 3/2\ O_2 \leftrightarrow WO_3(s)**$	$K_p = 1/(p_{O_2})^{1.5}$	30.596	7.231	2.213
$W(s) + 3/2\ O_2 \leftrightarrow WO_3(l)***$	$K_p = 1/(p_{O_2})^{1.5}$	29.216	7.711	3.235
$W(s) + 3/2\ O_2 \leftrightarrow WO_3(g)$	$K_p = p_{WO_3}/(p_{O_2})^{3/2}$	12.424	4.033	2.120
$2W(s) + 3\ O_2 \leftrightarrow (WO_3)_2(g)$	$K_p = p_{(WO_3)_2}/(p_{O_2})^3$	47.597	15.094	7.902
$3W(s) + 9/2\ O_2 \leftrightarrow (WO_3)_3(g)$	$K_p = p_{(WO_3)_3}/(p_{O_2})^{9/2}$	79.637	23.085	10.616
$4W(s) + 6\ O_2 \leftrightarrow (WO_3)_4(g)$	$K_p = p_{(WO_3)_4}/(p_{O_2})^6$	108.44	30.242	13.036
$3W(s) + 4\ O_2 \leftrightarrow W_3O_8)(g)$	$K_p = p_{W_3O_8}/(p_{O_2})^4$	67.323	20.100	9.529
$W(s) \leftrightarrow W(g)$	$K_p = p_{W_g}$	-37.05	-12.56	-7.104

* $3\ WO_2 \leftrightarrow W(s) + 2\ WO_3(s)$ at $T_d = 1997$ °K and higher
** $T_t = 1050$ °K, $T_m = 1745$ °K
*** $T_b = 2110$ °K

Log K_p values are given in this table for each of the reactions that occur. It should be clear that the reduction-chemistry is more complex than was evident previously. During these reactions, we assume that the $K_2Si_2O_9$ added as a compaction agent is reduced to K_2O and SiO_2, and that the alumina present is not affected below 1800 °K. It is well to observe here that thoria, i.e.- ThO_2 , is sometimes added to control grain size during sintering. The partial pressure of oxygen associated with the gas mixture during reduction can be evaluated from the equilibrium constant for water, viz-

3.1.16- $H_2(g) + 1/2\ O_2(g) \leftrightarrow H_2O(g); \quad P_{O_2} = K^{-2}_{H_2O}(p_{H_2}/p_{H_2O})^{-2}$

Here, p_{H_2}/p_{H_2O} is the partial pressure ratio of hydrogen to water vapor of the gas mixture. We shall call this ratio, "R". Note that large values of R indicate a lower ratio of water vapor to H_2 gas present. Because the reduction of $WO_3(s)$ to $W(s)$ produces water vapor, we must evaluate the effect of the water vapor oxidation-reduction potential on the equilibrium constants in the Table. This can be done by using the equations given above and substituting P_{O_2} in each of the equilibrium equations. However, in tabular form, these data are difficult to apply. Two features can be used to help visualize the various chemical processes occurring. First, oxygen is involved in many of the equations, and secondly, volatile gases form.

A plot of the volatility of the tungsten-oxygen species vs: the oxygen pressure allows one to evaluate the relative importance of each of the reactions. Such a plot is shown in the following:

3.1.17.-

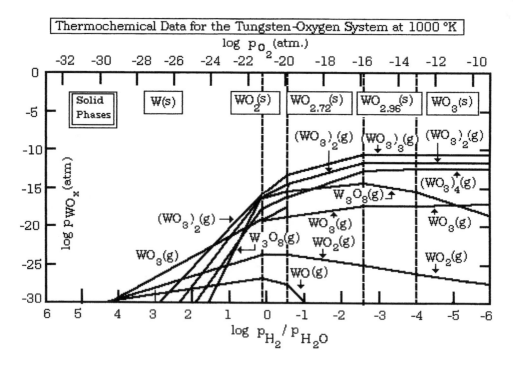

The vertical dotted lines shows the regions of existence of the several tungsten oxides involved, while the angular and horizontal lines give the volatility of these oxides as a function of the oxygen pressure (top) and the ratio of hydrogen to water vapor pressure (bottom). These data show that to avoid losses of tungsten, the initial reduction of WO_3 should be carried out at temperatures below 1400 °K. At 1000 °K, log R > 0.6 for reduction while at 1400, log R > 0. For reduction of WO_3(s) to W (s) at least 1.0% of water vapor can be present, i.e.- log R can be 2 without greatly affecting the reduction efficiency. But at 2200 °K., much purer hydrogen is required since considerable loss of W can occur through formation of volatile oxides if log R is not sufficiently large.

The amount of volatile oxides produced will depend upon the oxygen gas and water vapor content of the hydrogen gas used. Volatilization of tungsten oxides is small at 2200-2600 °K if log R > 2.0. For removal of SiO_2 and Al_2O_3, the reduction temperature and the composition of the hydrogen-water vapor gas mixture is very important. SiO(g) can be formed from SiO_2(s) at 1400 °K at an appreciable rate if log R > 4.0. Above 1800 °K, SiO (g) is formed over a wide range of log R. Al_2O_3 is inactive at 1400 °K in all gas mixtures below log R = 8.0, but reacts at 1800 °K and higher to form Al(g) and Al_2O(g). Loss of aluminum occurs by transfer of Al (s), volatile oxide, hydrides and oxyhydrides of aluminum. The best procedure for preparing tungsten ingots is to reduce the tungsten trioxide to tungsten metal at temperatures below 1400 °K.

2. Loss of Additives During Reduction

Aluminum and silicon-containing impurities can be removed in a controlled manner at higher temperatures in purified H_2 gas. A log R > 2.0 is needed to remove SiO (g) at 1800 °K. The optimum value is found at log R = 4.25. To optimize the volatility of Al, log R should be increased to 6.25. Note that we are speaking of removal of impurities by a thermal treatment prior to sintering of the ingot. This aspect is shown in the following diagram, shown as 3.1.18. on the next page.

In this diagram, we show a thermogravimetric analysis of a "doped" tungsten trioxide and the losses observed as a function of temperature of reduction by hydrogen-gas. It is apparent that part of the SiO_2 and Al_2O_3 reacted to form "Mullite", i.e.- $3 Al_2O_3 + 2 SiO_2 = Al_6 Si_2 O_{13}$.

274

3.1.18.-

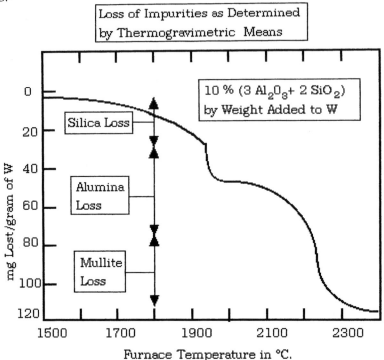

Loss of Impurities as Determined by Thermogravimetric Means

10 % (3 Al$_2$O$_3$+ 2 SiO$_2$) by Weight Added to W

Silica Loss

Alumina Loss

Mullite Loss

mg Lost/gram of W

Furnace Temperature in °C.

Note also that part of the silica and alumina volatilized before the mullite finally decomposed (T_M =1850 °C.) and was lost. This behavior occurred regardless of whether the two oxides were pre-reacted or not. It therefore becomes apparent that the initial reduction of WO$_3$ to W-metal by hydrogen gas at high temperatures serves as a purification step as well (if it is done properly).

The pronounced "bump" in 3.1.18. near to 1950 °C. is believed to be due the reaction:

3.1.19.- $3\{Al_6 Si_2 O_{13}\} + 2\ W \rightarrow 6\ SiO\Uparrow + 9Al_2O_3 + 2WO_3 \Uparrow$

Overall, these reactions may be summarized as given in 3.1.20. on the next page.

3.1.20.-

a) $2 \, SiO_2 \Rightarrow 2 \, SiO \Uparrow + O_2$

b) $O_2 + W \Rightarrow WO_3$ (vacuum)

c) $WO_3 + 3 \, H_2 \Rightarrow W + 3 \, H_2O$ (reducing atmosphere)

It is for this reason that tungsten metal powder is made using a reducing atmosphere rather than in vacuum since any oxygen present will form the trioxide unless it is reacted upon by hydrogen to form water vapor.

3. <u>Size and Morphology of Tungsten Metal Powders Produced</u>

We have already mentioned the fact that tungsten metal powder needs to be prepared with a rather small particle size and with a small range of particle sizes. A typical range of sizes is shown in the following diagram:

3.1.21.-

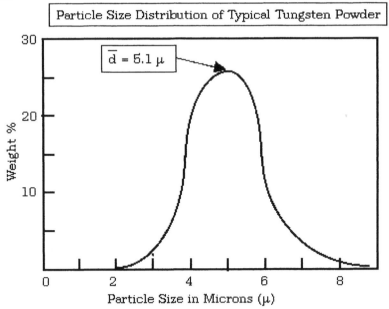

However, it has been found that additives. i.e.- "doping", have a profound effect upon both the size and morphology of the tungsten metal particles produced (Note that dopants are usually added to promote improved pressability and sinterability).

As shown in the following Table, the morphology can range from cubes to octahedra and other polyhedrons. The standard doping includes: 2% KCl, 2% $AlCl_3$ and 1% $K_2Si_2O_9$ by weight.

TABLE 3-7

Electron Microscope Examination of Tungsten Particle Morphology

Dopant	Size Range*	Morphology	Surface Appearance	Occlusions Insoluble in H_2O_2	
				Number/ W-particle	Size of Occ
Standard -see above	1- 2.5 μ	Round Cubes	Smooth-few irreg. protrusions	5	0.1 μ
3 Times Standard	1 - 5 μ	Cubes & Polyhedra	Smooth	> 10	0.3 μ
Al(NO3)3 - 2 %	1 - 5 μ	Polyhedra	Covered with many 0.1μ cube shaped protrusions	none	-----
KCl-2%	0.5-2.0 μ	Polyhedra	Smooth-some irreg. protrusions	none	-----
Standard without AlCl3	1 - 3 μ	Rounded Cubes	Covered with Geometrically Shaped Protrusions	5	0.1 μ
Standard- Reduced 2nd time	0.2 -1 μ	Cubes & Octahedra	Smooth- many Small Protrusions	none	----
Thoriated Only	0.5 - 3 μ	Cubes & Octahedra	Numerous Very Small Protrusions	1000	< .01μ
Unwashed Powder	0.5 - 3 μ	Rounded Cubes	Smooth- Few Small Protrusions	> 10	0.1- 1.0
Not doped	0.05 μ	Spheres & Bean-shaped	Smooth Surface	none	----

In all cases, it is possible to dissolve the tungsten particles in 30% hydrogen peroxide in order to determine the inert particles produced during sintering. (Note that we are not speaking of compaction and densification, but the initial process of sintering to obtain an ingot capable of being handled).

It should be clear that doping and process variations have a significant influence upon the morphology, surface topography and size of the tungsten particles being produced. This obviously has an important effect on the "voids" present when the powder is compacted into a bar.

Note in the above Table that certain additives did not produce inclusions within the tungsten particle. Most inclusions were identified as oxides of the various elements added by means of x-ray analysis.

It has been observed that certain volatile dopants added (thought mainly to be potassium), being insoluble in tungsten, exist in the vapor phase during reduction and sintering at 3000 °C. The vapor is trapped within the pores which exist because of the interstitial space between packed particles. Upon working the sintered ingot to a fine wire size, the pores are closed by plastic flow and the dopant is smeared out into rows of closely spaced submicroscopic particles aligned parallel to the working direction. Upon further annealing, the dope-particles volatilized, leaving bubbles behind. The observation of the production of bubbles is perhaps the most important one made in recent times. These bubbles have been observed to be essential to produce sag-resistance in filament wire, as well as producing an anomalously high recrystallization temperature.

B-2. Role of Annealing in the Preparation of Tungsten Wire from Ingots

To establish certain facts regarding the role of annealing during processing of tungsten into wire, four types of swaged & drawn wire were examined by transmission electron-microscopy in the as-worked condition. Similar examinations were also carried out on specimens of the same types after annealing at 2200 °C. for 15 minutes. The samples encompassed those shown in 3.1.22. After working and annealing procedures were carried out, pores were observed to have formed in each of these samples.

3.1.22.- Types of Tungsten Examined by Electron-Microscope Microscopy

         a. 4.50 mm. - swaged rod
         b. 2.20 mm. - swaged rod
         c. 2.00 mm - swaged & drawn rod
         d. 1.30 mm. - swaged & drawn rod

The pores in the swaged rods ranged from ≈ 0.05 μ to 1.00 μ in size and were randomly distributed. The larger pores were slightly elongated in the working direction.

However, the number of pores and their volume were significantly reduced by drawing from 2.20 mm. to 2.00 mm. The remaining pores were then needle-shaped. The effect of deformation (swaged & drawn rod) on the density and distribution of the bubbles produced was very evident as shown by the following diagram, given as 3.1.23. on the next page.

Annealing causes the large, elongated pores in swaged rods to become spherical or spheroidal in shape due to anisotropy of surface energy and enlarges the smaller irregular pores due to the volatilization of the dopant. In the annealed, drawn wires, bubbles aligned themselves in rows parallel to the working direction. The total bubble density, the degree of alignment, and the number of bubbles in each row were observed to increase as the size of the wire decreased, with each succeeding drawing operation.

These observations not only confirmed that working of the tungsten wire smeared out the dopant into rows of particles, but also explained the necessity of sufficient wire drawing to create a large number of small bubbles needed to create sag-resistance of wire formed into filaments. It also shows that wire-drawing is far more effective than swaging in closing up the pores, or voids.

As a result, it was concluded that:

> a. Annealing or retreating a rod (or wire) which is not completely densified is detrimental to the development of non-sag properties, because enlargement and spheroidization of the elongated pores partially destroy the work done by wire-drawing in smearing out the "dope".

> b. As a result of this work, it became more obvious that an alternative to swaging needed to be sought. A rolling process might be more effective in producing and preserving bubbles, especially if a roll with circular grooves were to be used for reduction of rods and/or ingots prior to wire-drawing.

3.1.23.-

| Transmission Electron-Micrographs of Tungsten Rods |

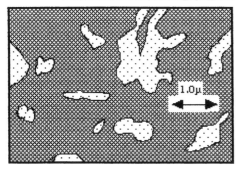

1. As Swaged - 4.50 mm. Rod

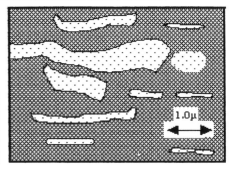

2. As Swaged - 2.2 mm. Rod

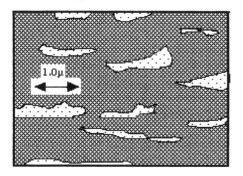

3. As Drawn - 1.3 mm. Rod

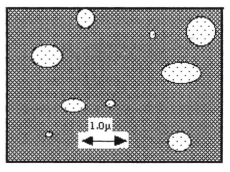

4. Swaged & Annealed - 2200 °C.
for 15 Minutes - 4.50 mm. Rod

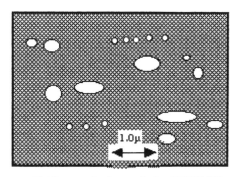

5. Drawn & Annealed - 2200 °C.
for 15 Minutes - 2.20 mm. Rod

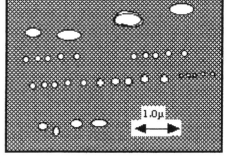

6. Drawn & Annealed - 2200 °C.
for 15 Minutes - 1.30 mm. Rod

However, let us further pursue the question of bubble-formation before we tackle the problem of advanced techniques of processing of tungsten bars into fine wire with improved "sag-resistance".

1. Doping of the Tungsten Powder and Bubble Formation

The next question that needed to be answered was that concerning the minimum number of aligned-bubbles required to produce a "sag-proof" wire. The answer was obtained by preparing a series of doped- tungsten oxides and reducing them according to a "standard" method, viz-

3.1.24.-   Standard Method for Reduction and Sintering

a. Reduction of WO_3 to W : 3 hours @ 1250 °C.

b. Wash W powder in HF

c. Sintering W Ingot after pressing: 45 minutes @ 3000 °C.

d. Swage and draw into rod (2.1 mm.)

e. Anneal @ 2100 °C. for 30 minutes

The oxides used were "doped" as shown above in Table 3-7 and then reduced to tungsten powder. Measurements were taken concerning the residue left in the powder (as determined by H_2O_2 digestion), the composition of the residue and the surface area of the particles (which measures both the apparent size of the particles and their porosity). This is shown in Table 3-8., as given on the next page.

The next step was to compact the tungsten powder, swage and draw it into wire, with appropriate sintering and annealing. Finally, a count of the number of bubbles present per 100 μ^2 was made and also an average value for grain-size present was determined. Since most, if not all, of the grains were observed to be elongated, a parameter called "M-value" was established. This value is a combination of grain-area and grain-diameter estimations.

TABLE 3-8

Analysis of Experimental Tungsten Powders

Doping History	Residue in ppm	Composition of Residue %Si	%Al	%K	Surface Area in m^2/gm.
Standard-see above	622	30	4	2	0.20
3 Times Standard	8,492	30	2	1	6.50
Al(NO$_3$)$_3$- 2 %	518		100		2.90
KCl-2%	48		100		7.30
Standard without AlCl$_3$	305	30	0.4	0.4	0.60
Std- Reduced 2nd time	209	30	4	2	0.10
Unwashed Powder	2,090	30	3	1	0.20
Not doped	12	3	2		0.20

When M was plotted against bubble-density, a strong linear correlation was seen, as shown in the following diagram:

3.1.25.-

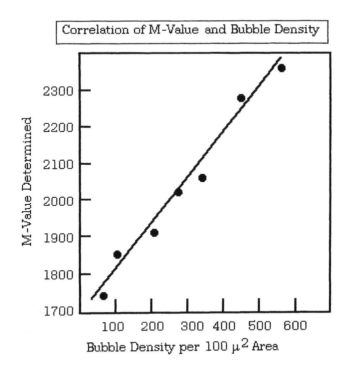

An examination of this diagram makes it clear that grain-size is linearly related to bubble density. In this manner, the recrystallized grain size is clearly related to bubble density. A bubble density of > 400 bubbles/ 100 μ^2 area apparently is required to achieve a high M-value (and satisfactory sag-resistance).

To further illustrate this aspect, a number of these specimen-rods were annealed in the range of 1600-2425 °C for times ranging from 10 to 10,000 seconds, and both bubble-densities and grainsize were determined for each of the specimens. Self-diffusion distances of some of these bubbles were calculated in microns from the diffusion equation:

3.1.26.- $D = D_0 \exp (- E /k_b T)$

which is related to Fick's Law: $\mathbf{J_N}$ = - D grad N, where $\mathbf{J_N}$ is the number of atoms crossing a unit area in unit time and D is the diffusion constant with units of cm^2/sec. E is the activation energy for the given temperature.

The following diagram, given as 3.1.27. on the next page, shows the result of the various annealing schedules employed.

Note that D_0 has been calculated as well as Q, the activation energy for diffusion. The data were found to fit all onto one curve irrespective of the annealing temperature and time. This is strong evidence that bubble formation occurs by vacancy diffusion. Since the bubble density becomes more or less constant above about 0.5 μ , which is approximately 1/2 of the grain size, it appears that the grain sub-boundaries are the source of vacancies. Moreover, this result confirms the hypothesis that a minimum bubble density is required in annealed wire for sag-resistance when used as an incandescent filament. (We shall discuss this in more detail below).

Thermodynamic considerations concerning the nucleation of a bubble by volatilization of an insoluble particle have predicted (2) that the minimum annealing temperature required for nucleation depends upon the size of the particle. However, the above experimental results have not supported this view since the bubble density appears to depend upon the diffusion distance rather than upon the annealing temperature.

3.1.27.-

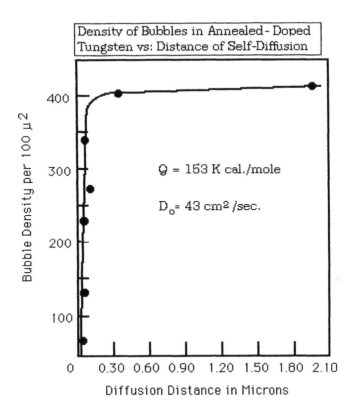

Density of Bubbles in Annealed- Doped
Tungsten vs: Distance of Self-Diffusion

$Q = 153$ K cal./mole

$D_o = 43$ cm^2 /sec.

Bubble Density per 100 μ^2

400

300

200

100

0   0.30 0.60 0.90   1.20   1.50 1.80 2.10

Diffusion Distance in Microns

This suggests that nucleation does not play a significant role for the annealing conditions used. The fact that the bubbles did not vary appreciably in size also supports this view.

Thus, much evidence has been found that the submicroscopic bubbles which form at temperatures as low as 1400 °C. in doped tungsten particles are primarily responsible for the high temperature strength and sag-resistance of tungsten filaments in lamps. Moreover, it has been established clearly that changes in grain-morphology occur as a consequence of these bubbles.

The following diagram, given as 3.1.28. on the next page, shows the grain-morphology of a doped vs: an undoped tungsten wire, both annealed under the same conditions.

284

3.1.28.-

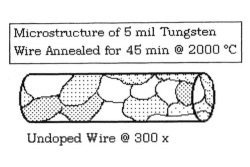

Microstructure of 5 mil Tungsten
Wire Annealed for 45 min @ 2000 °C

Undoped Wire @ 300 x

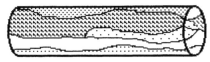

Doped Wire @ 300 x

The bubble density in the doped-wire was determined by transmission electron microscopy to be about 400 bubbles per 100 μ^2. No bubbles were detected in the undoped wire.

To confirm the postulate that bubble density is related to volatilization of inclusions remaining from added "dope", formation of bubbles by another method was sought. A very convenient way of introducing inert gas bubbles into undoped tungsten is bombardment by alpha-particles of sufficient energy, coupled with subsequent annealing to precipitate helium gas. This method has the advantage of having relatively short irradiation times and short half-lives of radioactive products so that the specimens can be handled after within 2-3 days. However, due to the nature of alpha-particles, the helium bubbles precipitate in a relatively narrow surface layer, as previously reported in alpha-irradiation experiments with beryllium (3) and aluminum (4).

When this was done, bubble density and bubble size were found to be dependant upon the temperature of annealing. For example, the first bubbles were seen under electron- transmission microscopy at about 1600 °C. (using a magnification of 1500 x). However, at magnifications of 100,000 x or higher, very small bubbles approximately 50 Å in size were found. Upon further annealing at 2200 °C for 45 minutes, the bubbles, now grown to 0.5 to 1.0 μ in size, could be seen by a light microscope at 1000 x. After an anneal of 45 minutes at 2400 °C., the bubble size had increased to 2.0-4.0 μ , but the number of bubbles had decreased considerably. However, the bubbles were mostly non-spherical and were located at or near to the grain boundaries.

Nevertheless, this work further established the role of bubbles in the annealing process. A definite change in grain morphology was noted with the production of more elongated grains becoming more prevalent, just as in "doped" tungsten wire. However, it is clear that since recrystallization is a thermal process, the number of bubbles required to retard this recrystallization process increases with increasing temperature.

It is possible to explain the dependence of recrystallized grain size upon bubble density. It is found that the recrystallized grain size is determined by the ratio, G/N, where G is the rate of growth of a recrystallizing grain in the unrecrystallized matrix and N is the rate of nucleation of the primary grains. When the bubble density is low, i.e.- high inter-bubble spacing, G is retarded due to Zener drag, but N is unaffected. If the bubble density is higher than a certain critical density, the bubbles (or a dispersed secondary phase) have a far greater effect on reducing N than G by increasing the critical nucleus size. The following equation clarifies this relationship further:

3.1.29.- $R_c = 2\,\sigma / (E_m - E_R - nF)$ $N \approx e^{-(4/3\,\pi\sigma R_c^2)} / kT$

where R_c is the radius of the critical nucleus, σ is the grain-boundary energy of the recrystallizing grain, E_M is the dislocation energy density of the unrecrystallized matrix, E_R is that of the recrystallizing nucleus, n is the number of bubbles interacting with the grain boundary per unit area, F is the Zener-drag per bubble, and T is the temperature in °K.

If G is assumed to be inversely proportional to the bubble density, a plot of G/N vs: bubble size will predict the dependence of grain size on bubble density. Such a plot is given in the following diagram presented as 3.1.30. on the next page.

Because there is a subsequently observed increase in strength of doped wire, the change can be attributed to the simultaneous formation of submicroscopic bubbles which results both in the strengthening of the matrix and inhibition of recrystallization during annealing.

3.1.30.-

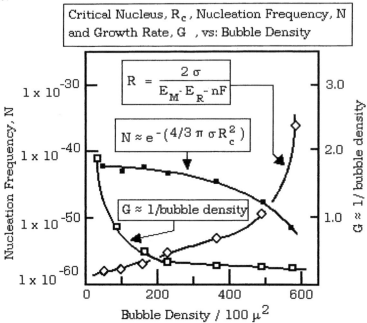

Critical Nucleus, R_c, Nucleation Frequency, N and Growth Rate, G , vs: Bubble Density

$$R = \frac{2\sigma}{E_M \cdot E_R - nF}$$

$$N \approx e^{-(4/3\pi\sigma R_c^2)}$$

$G \approx 1/\text{bubble density}$

Nucleation Frequency, N

$G \approx 1/\text{bubble density}$

Bubble Density / 100 μ^2

Indeed, measurements of tungsten wire as a function of annealing time have shown a drop of tensile strength from an original value of 380,000 psi. for a 9.0 mil wire to 140,000 psi. after annealing for 15 **seconds** at 2000 °C. (a drop of over 60%).

To further substantiate this, a similar plot of the materials shown in 3.1.22. and 3.1.23. is illustrated as follows in 3.1.31., given on the next page.

To further clarify this point, three types of 9.0 mil wire were prepared from:

        a) Fully-doped tungsten
        b) Doped, but without $AlCl_3$
        c) 2% KCl
        d) Undoped tungsten.

The annealing times ranged from 15 seconds to 8 minutes at temperatures from 1500-2400 °C. To establish the reproducibility of the results, five 4-inch specimens were annealed simultaneously at each specified time and

3.1.31.-

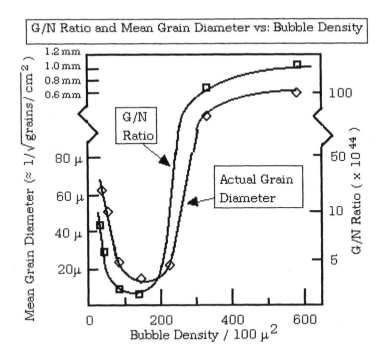

G/N Ratio and Mean Grain Diameter vs: Bubble Density

temperature. The specimens were then tensile-tested at room temperature at a strain rate of 0.05/minute. Results showed that the ultimate load (for specimens exhibiting brittle fracture: breaking load = fracture load) of these materials as a function of annealing time showed considerable deviation.

The following diagram, shown as 3.1.32. on the next page, shows results obtained at 2000 °C. which are typical of all measurements made (except that the higher temperatures promote larger bubble size formation).

288

3.1.32.-

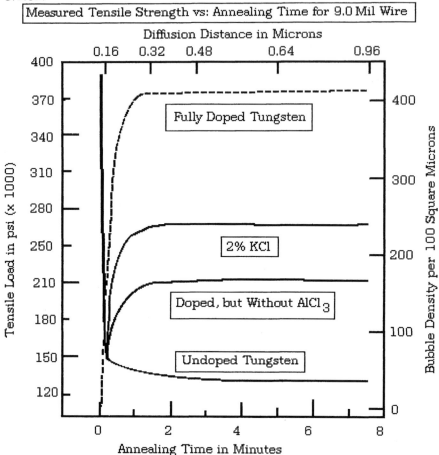

In this diagram, the initial rapid drop in strength is a consequence of either recovery from strain induced by drawing or recrystallization of the wire. The temperature at which the latter occurs depends upon the recrystallization temperature of the material. Photomicrographs of the wires annealed for 15 seconds at 2000 °C showed that the undoped wire is fully recrystallized whereas both of the other two showed only partial recrystallization. Only the fully-doped tungsten showed no recrystallization at all, and its strength increased as the annealing time was increased. Note that at times less than about 1 minute of annealing time, all specimens showed failure by brittle fracture due to lack of sufficient bubble formation. The scatter of fracture data at a given set of conditions also decreased at the longer annealing times.

The subsequent increase in strength for all wires except the undoped wire is clearly attributable to the formation of submicroscopic bubbles which results in strengthening and inhibition of recrystallization of the matrix. Included in 3.1.32. at the top is the calculated curve which relates the "self-diffusion" distance of tungsten atoms to the bubble densities (as determined by electron transmission micrographs, as calculated from the equation:

3.1.33.- $D = D_0 \exp(-Q / k_bT$ and $x = (2\ Dt)^{0.5}$

where the same values as in 3.1.25 are used. This correlation of self-diffusion distances and bubble density has established beyond any doubt that for any combination of annealing times and temperatures which result in a diffusion distance larger than approximately 0.4 μ, the density of submicroscopic bubbles doe not further increase.

If, therefore the increase in strength is indeed caused by submicroscopic bubbles, the maximum in this strength should occur when the formation of bubbles is complete. It was for this reason that the time scale and the "self-diffusion" distance scale were chosen so that they could be compared. It will be noted that for this particular set of materials, at least 2.0 minutes of annealing time are required to achieve maximum strength and that this corresponds exactly to the time when bubble density achieves its saturation. Thus, differences in strength **can** be attributed to differences in bubble density.

However, when annealing is carried out at successively higher temperatures, the materials become more brittle, resulting in lower brittle-breakage loads. This indicates that the embrittlement due to progressing recrystallization exceeds the effect of bubble strengthening, due perhaps to bubble enlargement observed at the higher temperatures.

In order to clarify this point, a series of tests were done in which the fully-doped wire was annealed for 4 minutes and tensile-tested at a strain-rate of 0.05 min^{-1}. Two samples of 9 mil wire were used, one which had been bottle-sintered, and the other which was furnace-sintered. (A vacuum furnace was used having molybdenum silicide heating elements, insulated with zirconia felt and pressed slabs of zirconia). The results are shown in the following diagram:

3.1.34.-

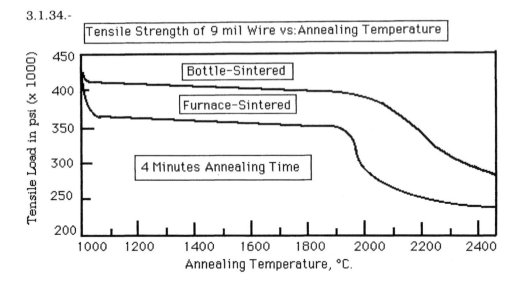

It should be clear from this diagram that annealing temperatures between 1700 and 2000 °C. are needed to promote the strongest wire, and to promote maximum growth of bubbles. However, there is a balance which must be maintained, since the formation of bubbles is related to voids left behind during the size reduction of rods to wire by drawing. Too many voids will lead to a filament which undergoes premature failure.

2. Void Migration and Filament Failure

The laws governing the motion of voids in tungsten have been been formulated mathematically in terms of known transport coefficients. What we mean by voids are those formed either by interstitials between particles of tungsten metal when compaction and sintering occurs. Void velocities can be calculated as a function of size and temperature. It will be shown that for voids less than about 5.0 μ, and temperatures less than 3000 °C., the transport mechanism is always that of surface diffusion, which dominates that of volume diffusion and vapor transport. The velocity of a void under the influence of a force, f_a, on an individual atom (or the force, F, on the entire void) can be found from the equation:

3.1.35.- $\qquad v = - 2D_s \sqrt{\Omega} f_a / r\, kT = 3/2\pi\, (D_s /kT)(\sqrt{\Omega^2/r^4})\, F$

where D_s is the surface diffusion coefficient, $\surd$ is the number of diffusing tungsten atoms per unit area of surface, k is Boltzmann's constant, T is in °K, Ω is the volume per tungsten atom in the solid and r is the void-radius. Under the influence of thermal agitation (or random walk) the motion of a void can be calculated by means of diffusion theory provided that one employs an effective diffusion coefficient, i.e.-

3.1.36.- $\qquad D_{Void} = 3/2\ \pi\ (\surd\ \Omega^2/r^4)\ D_s$

Numerical values of D_{Void} have been calculated and are given in the following Table:

Table 3-9

Effective Void Diffusion Coefficients for Void Migration via Surface Diffusion

(D_{Void} in cm^2/sec.)

T in °K	r = 25 Å	250	2500	25,000	250,000
			Values of r in Angstroms		
1500	2.35×10^{-15}	2.35×10^{-19}	2.35×10^{-23}	2.35×10^{-27}	2.35×10^{-31}
2000	1.59×10^{-12}	1.59×10^{-16}	1.59×10^{-20}	1.59×10^{-24}	1.59×10^{-28}
2500	7.93×10^{-11}	7.93×10^{-15}	7.93×10^{-19}	7.93×10^{-23}	7.93×10^{-27}
3000	1.10×10^{-9}	1.10×10^{-13}	1.10×10^{-17}	1.10×10^{-21}	1.10×10^{-25}
3300	3.60×10^{-9}	3.60×10^{-13}	3.60×10^{-17}	3.60×10^{-21}	3.60×10^{-25}

Note that these numbers indicate a rather low velocity of void migration. Nevertheless, it does occur.

The velocity of a void under the influence of a temperature gradient, dT/dx, is: given by:

3.1.37.- $\qquad v = 2\pi\ (r^3/\Omega)\ (q^*/kT)\ D_{Void}\ (1/T)\ dT/dx$

$\qquad\qquad = 3\ [(\surd\ \Omega\ )/r]\ [q^*/kT]\ [1/T]\ dt/dx$

where q^* is an energy of transport associated with thermal diffusion. For all except the smaller voids considered in Table 3-9, a temperature gradient is the more effective driving force than random walk. Numerical values are given in Table 3-10 (shown on the next page), for the case where: (1/T)(dT/dx) = 0.1 cm^{-1} .

Table 3-10

Void Velocity (cm./sec/) in a Temperature Gradient According to
the Surface Diffusion Mechanism (D_{Void} in cm^2/sec.)

| T in °K | Values of r in Angstroms | | | | |
	r = 25 Å	250	2500	25,000	250,000
1500	4.88×10^{-12}	4.88×10^{-13}	4.88×10^{-14}	4.88×10^{-15}	4.88×10^{-16}
2000	2.48×10^{-9}	2.48×10^{-10}	2.48×10^{-11}	2.48×10^{-12}	2.48×10^{-13}
2500	9.92×10^{-8}	9.92×10^{-9}	9.92×10^{-10}	9.92×10^{-11}	9.92×10^{-12}
3000	1.14×10^{-6}	1.14×10^{-7}	1.14×10^{-8}	1.14×10^{-9}	1.14×10^{-10}
3300	3.41×10^{-6}	3.41×10^{-7}	3.41×10^{-8}	3.41×10^{-9}	3.41×10^{-10}

The fact that voids migrate **up** a temperature gradient suggests that void migration is probably one of the more important factors in determining filament lifetime.

Indeed, the presence of excess voids in various regions of the filament cuts down the cross-sectional area at those regions and increases the local resistance. The current drawn by the filament from the line is, however, determined by the remainder of the filament and is, therefore, essentially the same as a homogeneous wire. As this current is "forced" through the regions of relatively high resistance, there is local overheating and concomitant temperature gradients directed **toward** them. This attracts more voids to these regions, making them even hotter and hence attracting even more voids, etc. Such a process is potentially catastrophic if there are enough mobile voids present in the wire.

3. Bubbles and Filament "Sag"

On the other hand, the presence of bubbles is akin to voids, with one important difference. Bubbles have been found to be "pinned" next to the grain boundaries and do not ordinarily migrate. Thus, they serve as "void-reservoirs". However, any movement of a bubble will cause the formation of a void in the metal structure itself. Note also that we have said that since the bubble density becomes more or less constant above about a 0.5 μ diffusion distance, which is approximately 1/2 of the grain size, it appears that the grain sub-boundaries are the source of vacancies. Any loss of bubbles therefore manifests itself as the appearance of voids which contribute to premature filament failure.

Moreover, as a filament operates at high temperature, grain growth occurs, albeit at a much slower rate when bubbles are present. Such growth is the result of movement of voids and changes the grain-boundary position and structure. Eventually, void movement overcomes the steadying influence of bubbles, slips along the {112} plane in the metal structure occur leading to "hot spots" and filament failure. This can be seen by the amount of "sag" which occurs in the operating filament as a function of time. As an example of differences in "sag" of filaments made from differing types of wire, we show the following diagram:

3.1.38.-

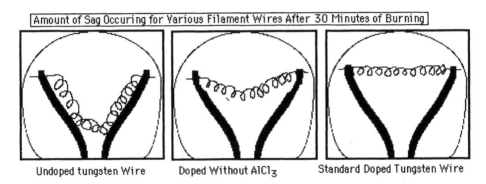

Amount of Sag Occuring for Various Filament Wires After 30 Minutes of Burning

Undoped tungsten Wire   Doped Without AlCl₃   Standard Doped Tungsten Wire

It should be clear that the amount of "sag" experienced is directly related to bubble formation in the finished wire used for filaments. The sag-test is most useful of those used in determining the quality of filaments processed from different materials. However, it is not sensitive enough to determine differences which may exist in the same type of material.

In conclusion, the above considerations on the role of void migration in filament failure have been examined mathematically. The existence of the potentially catastrophic instability given above has also been rigorously demonstrated in that the time for void migration to take place is an important part of filament lifetime. Thus, the measured lifetime of a given type of a filament should display a dependence of the form:

3.1.39.- $\qquad \tau \propto e^{39,200\,K/T}$

On the other hand, if the time for evaporation (void formation) dominates filament lifetime, the corresponding dependence on temperature would be:

3.1.40.- $\qquad \tau \propto e^{94,340\,°K/T}$

Unfortunately, even today, this factor is not known rigorously.

4. Present Status of Knowledge Regarding Tungsten Filament Life

To summarize what is known concerning failure of an incandescent lamp filament, the following is relevant:

1. It is generally agreed that undoped tungsten makes a poor lamp filament. Suitably doped tungsten, on the other hand, furnishes a satisfactory filament, long-lived and non-sagging. The exact role of the dopants has yet to be completely identified, and 80 years of lamp production, i.e.- 1910 to 1992, have only served to clarify the problem but only hinted at its solution. Still, it is known that lamps operating at temperatures of about 2600 °K. are far above the recrystallization temperature of pure tungsten. Therefore, glowing filaments of tungsten sag under the influence of gravity, with a sliding at the grain boundaries known as offsetting (5). The commercial answer to this has been to fabricate a wire whose grain boundaries form very acute angles with the longitudinal axis of the wire, so-called "overlapping grains". The large area between such grains thus reduces offsetting to a minimum.

2. P.E. Wretblad, writing in 1941 (6), attributed the overlapping grain structures to the presence of both volatile alkalies and non-volatile additives in the starting powder. The function of these dopants is not to inhibit grain growth, but rather to ensure correct orientation of the grain boundaries with respect to the wire axis, as shown in the following diagram, given on the next page as 3.1.41.

Actually, during drawing at the processing temperatures, as argued by Professor Wretblad, exaggerated grain growth occurs, but in a preferred

3.1.41.-

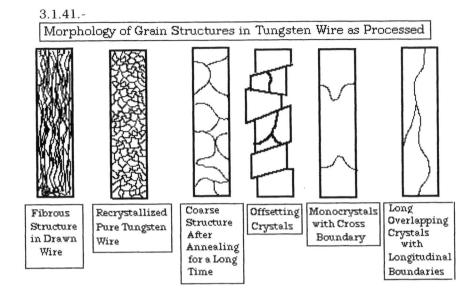

Morphology of Grain Structures in Tungsten Wire as Processed

Fibrous Structure in Drawn Wire	Recrystallized Pure Tungsten Wire	Coarse Structure After Annealing for a Long Time	Offsetting Crystals	Monocrystals with Cross Boundary	Long Overlapping Crystals with Longitudinal Boundaries

orientation. If this picture is accurate, then processing of the tungsten wire is foremost in determining how the wire will perform as an incandescent filament. Nevertheless, the role of additives prior to reduction and the exact effect of processing variables have remained obscure in several aspects. The role of vacancies generated during operation of the filament and the generation and migration of voids from grain boundary locations has still not been clarified completely and totally.

B-3. Development of Advanced Methods of Compaction and Densification

Although the prior methods of compaction and densification worked, some manufacturers sought even better methods. This came about in the United States because of the then new environmental laws passed by Congress in the late 1970's.

1. Compaction and Densification

Originally, a rolling-compaction was tried in which the tungsten metal powder was compacted by a roller. Such an apparatus is shown in the following:

3.1.42.–

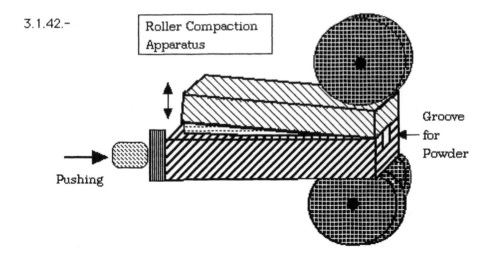

The roller-compactor shown here consisted of a 14 inch diameter roller-segment of approximately 90 ° arc and a 3/8 inch serrated mold. This apparatus produced tungsten ingots 3/8 inch square and 15 inches long. Each ingot could be handles without difficulty but had a slight bow characteristic resulting from rolling a round object over a flat surface. Ingot densities ranged from 11.9 to 12.3 gm./cc. However, when it was attempted to densify these ingots in a high temperature furnace, the ingots cracked as it was pushed into the hot zone at 2200 °C. On the other hand, if the ingots were first pre-sintered at 1100 °C., they could be densified at the higher temperature to produce ingots with densities ranging from 17.3 to 17.4 gm./cc., i.e.- 89% of theoretical density. Nonetheless, this method was not an improvement over prior ones. Therefore, another type of compaction was investigated, that of a "rocking compactor". It consisted of a rocker with a twelve foot radius of curvature with a mold 3/8 inches wide and 30 inches long. Such an apparatus is shown in the following diagram, given as 3.1.43. on the next page.

The rocker blade was 3/8 inches wide and operation required that pressure in Cylinder A and Cylinder B be programmed. The rate at which Cylinder A is pressurized is the rate that Cylinder B is decreased, and vice-versa, so that the pressure on the powder at any point along the rocker's travel will always be the same. This apparatus produced ingots having densities of 10.9 to 11.4 gm./cc., i.e.- 60.1% of the theoretical density of tungsten metal. However, this method was faster than direct mold pressing or the above roller-compactor, and in addition produced ingots that were not bowed.

3.1.43.-

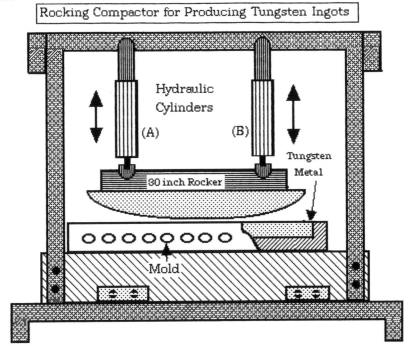

Rocking Compactor for Producing Tungsten Ingots

This led to the use of a continuous rolling compactor in which the powder was progressively compacted without the use of the top of the die, shown in 3.1.44. on the next page as follows.

This apparatus consisted of a set of four 18-inch rollers with the die having serrated edges, each roller having progressively increasing monostatic pressure, from 4000 psi to 10,000 psi. Each roller is 3/8 inch wide so that it fits into the bottom part of the die.

Tungsten powder is continuously added to the forming bar just before it encounters each roller. However, a binder and/or lubricant is required in the process, usually camphor or methacrylate.

An end-on view is shown in the diagram. The die is about 10 feet long and produces a bar about 10 feet long and 3/8 by 3/8 inches square. The bar is supported at the bottom by a strip of molybdenum metal which is removed just before the bar enters a pre-sintering furnace at 1400 °C, and then enters

298

3.1.44.-

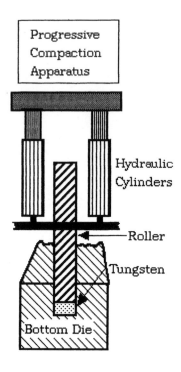

a sintering furnace where heating at 2200 °C produces the ingot. Densities up to about 17.8 gm/cc (93% of theoretical) have been achieved.

This result led to design of an apparatus to replace the swaging operation which had always been a source of frustration because of the high degree of non-control over the processing variables, i.e.- some operators could "swage" properly and others could not.

2. Advanced Swaging Techniques

To replace the swaging method of heavy-wire production required a completely different approach. A four-stand tandem-mill was developed to roll the tungsten ingot sequentially to a rod which could then be put through the wire-drawing process. This apparatus consists of sequential rolls having a groove inside of the roll which is tapered opposite to the direction of rolling so as to reduce the diameter of the rod with each pass through a specific-sized roll, as shown in the following diagram:

3.1.45.-

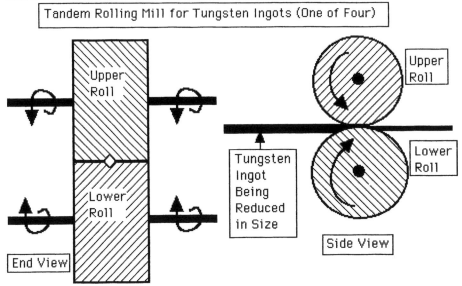

Tandem Rolling Mill for Tungsten Ingots (One of Four)

The rolls themselves were made of nodular iron and are 8 by 12 inches in size, driven by 30 HP electric motors. The inside surface of the grooves in each roll has a diamond-shaped profile to help "cold-roll" the rod. A heat-treating furnace is needed between each stand to sinter the rod at 2200 °C in flowing hydrogen gas. At first, a tungsten-rod element furnace insulated with zirconia was used, but later on, induction preheat furnaces were used. The furnace-stand combination is spaced so that the rod will cool to about 1350 °C., just as in the swaging operation. Rolling occurs at speeds in excess of 300 feet per minute. Closely fitted steel guides are fitted at entrance and exit between each mill and furnace. Inter-stand tension (or compression) can arise if speeds are not precisely set. The ingot progresses from square to oval to round at the end of the size-reduction during rolling. The following table, given on the next page, presents an example of some of the parameters of rolling. The most severe problems encountered in the rolling process are:

1) Roll wear

2) Preheat time.

The first is responsible for jam-up during processing whereas the second

TABLE 3-11
Rolling Parameters for Tandem Rolling of Tungsten Ingots

Pass No.	Output Size in Inches	Roll Gap (inches)		Mill Speed (fpm)	
		Mill 1	Mill 2	Mill 1	Mill 2
1	0.238 Sq.	0.012	0.002	350	435
2	0.191 Sq.	0.006	0.003	480	600
3	0.155 Sq.	0.001	0.008	350	445
4	0.127 Sq.	0.003	0.012	525	600
5	0.110 Ov.	0.001	0.007	375	450
6	0.102 Rd.	0.002	0.008	500	600

determines the rate of through-put of the tungsten rod processing as it progresses to wire-drawing sizes. Although continuous compaction to form a long ingot capable of being processed by pre-sintering has been successful, some manufacturers have chosen to stay with the rocker-compacted bars because of cost-savings.

The major problem observed with tungsten wire made from rolled rod was that of high iron contamination which was believed to have caused high split levels. This mandated an electrolytic cleaning process which solved both problems. The wire thus produced was entirely equal to wire made by the older method. Splits in 9 mil wire were then found to be almost nonexistent. In some cases, wire made from rolled rods was superior in that wire-breaks during coiling were almost completely eliminated.

The Preferred Process turned out to be:

1) electrolytic cleaning at finished roll size ($\approx$ 8% loss)

2) annealing of wire at 29 mil diameter.

It should be noted that molybdenum metal is now being produced by the same compaction-hot rolling procedures as outlined above. Mo is used in supports for tungsten filaments in incandescent lamps and as electrical leads in other types of lamps.

3.2.- GLASS USED IN MANUFACTURE OF LAMPS

Several types of glass are used to make glass-forms for lamps. These include both large and small incandescent lamps, fluorescent lamps of various sizes, outer bulbs for HPMV and for sodium-vapor lamps.

According to the 1986 CERAMIC SOURCE (7), the following glass formulas are used for lamps:

TABLE 3-12

Glass compositions Used for Lamps

Use	SiO_2	Al_2O_3	B_2O_3	Na_2O	K_2O	MgO	CaO	BaO	PbO	Other
Exhaust Flare	61.8	2.2		7.0	7.3		21.5			0.2 As_2O_3
Fluorescent Tube	73.6	1.4		16.2	0.4	3.4	4.8			0.2 SO_3
Sign Tubing	70.5	2.0	2.6	12.2	5.3	3.0	4.2			0.2 SO_3
Sealed Beam	78.1	2.0	14.9	4.9						0.1 Cl
Fluorescent Tube	71.4	2.2		15.0	1.7	4.0	4.6	0.8		0.2 SO_3
Lamp Bulbs	73.1	1.1		17.0		4.1	5.1		9.0	
Neon Signs	67.0			10.0	7.0					12.0 ZnO
TV Tubes	50.3	4.7		6.1	8.4	2.9	4.3	0.2	22.5	0.1F

Most of these formulations are classified as "Soda-Lime" glasses.

The steps generally used in manufacturing glass for lamps are given in the following diagram, shown as 3.2.1. on the next page.

The glass components are loaded into the front end of a large continuous furnace. Once a melt has been started, addition of components is continuous and glass is taken off the other end of the furnace.

As the molten glass accumulates in the holding area, it is drawn progressively either into a long continuous round tube (and cut to the desired length) or it

302

3.2.1.-

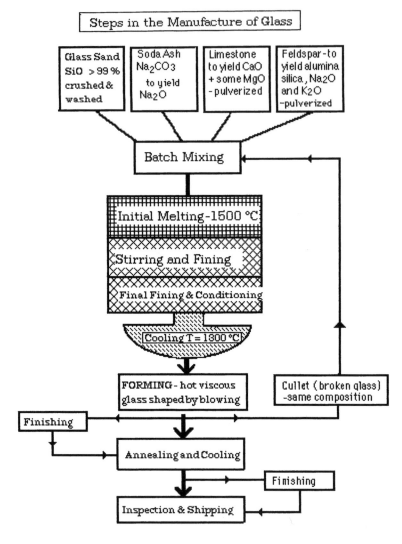

is blown into bulb shapes continuously by blow-molding. In either case, millions of items are made each month. Some lamp manufacturers have their own glass furnaces while others buy the glass forms to make into lamps. We have not shown the glass molding equipment here which consists generally of a mandrel over which the molten glass is drawn and cooled (for tubing) or the molding equipment used for bulb making. Some of the various glass forms used for manufacturing lamps are shown in the following diagram, given as 3.2.2. on the next page.

3.2.2.-

Types of Glass Forms Used for Lamp Manufacture

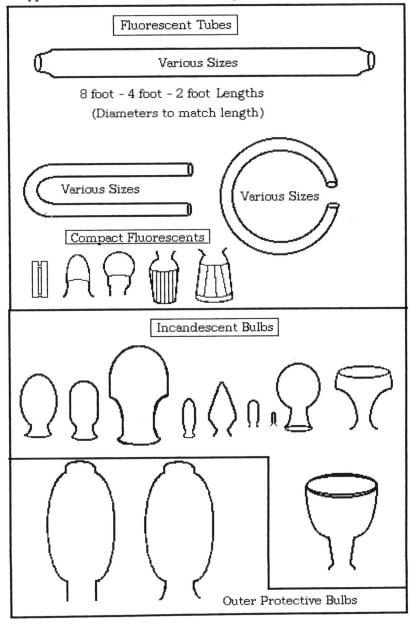

Note also that we have not provided any detail concerning the complex steps required in glass manufacture. This includes the forming operations in which the glass forms are manufactured. Various forms used for both fluorescent and incandescent bulbs are shown in the above diagram as well as their relative size. Additionally, protective shields for HPMV and high-pressure sodium discharge lamps are shown. The forms shown in 3.2.2. represent most of the glass forms used for high volume manufacturing of lamps intended for both residential and street lighting in the U.S. Additionally, blanks suitable for the manufacture of automobile headlamps are made, although we have not shown them in this diagram because of lack of space.

Most are made at high speed by blowing molten glass "blobs" into the desired shape. Immediately following the forming operation is a furnace where the glass forms are annealed and then cooled. They are then shipped to the user. Since there are about 700 different types of lamps in usage today, we have only shown a small fraction of the actual number of glass shapes that are, or have been, manufactured in the diagram given above. However, those given represent a majority of the glass forms used in the manufacture of lamps.

3.3.- MANUFACTURE OF LEAD-IN WIRES

In order to manufacture incandescent lamps, one needs to have a supply of various parts on hand, including those of lead-in wires. This section will describe some of these parts and their manufacture, including that of "Dumet™" wire.

A. Manufacture of Dumet Wire.

The manufacture of Dumet™ wire is basically simple. As we said before, Dumet wire was invented at G.E. by Colin G. Fink in 1912. The standard method involves the use of a core rod of nickel-iron, around which is wound a thin strip of brass to form a tube. this assembly is then placed within a copper tube, and the whole is heated to braze it together. This composite rod is then drawn into wire, using the same techniques used for tungsten wire. A newer process involves forming the same assembly by drawing the nickel-iron rod through a molten bath of very pure copper. The process is shown in 3.3.1. given on the next page.

3.3.1.-

| Assembly Process for Dumet Wire | Dipping Process for Dumet Wire |

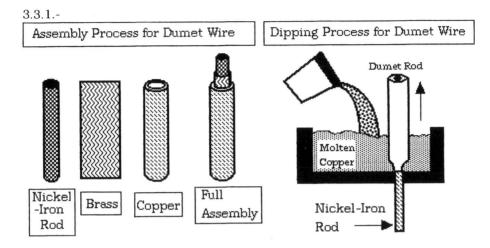

Once the composite rod is obtained, it is then drawn into fine wire, which then can be processed to form various types of lead-in wires and sealing wire lead-ins.

B. Manufacture of Lead-In Wires

The manufacture of lead-in wires involves the use of several different metals. In general, there are three sections to lead-in wires, the inner section, the press section and the outer section. This is shown in the following diagram:

3.3.2.-

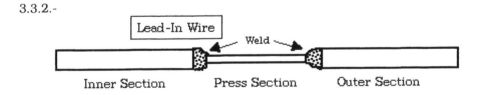

The procedure used is:

1. Inner-Section - This is the portion extending into the inside of the lamp. It conducts current and supports the filament and other parts. Standard materials used are:

Copper, copperclad steel, nickel, molybdenum, tungsten, nickel-plated iron, nickel-plated copper, chrome-copper alloy, and zirconium-copper alloy.

The temper can range from soft to full-hard. Inner sections are manufactured as straight, or with hooks to which the filament is crimped.

2. <u>Press-Section:</u> This portion is contained within the press, or lead-seal. It is used to seal the glass for a vacuum-tight seal. This is possible because its coefficient of expansion closely matches that of the glass used. By far the most common wire used for this protion is the Dumet composition. However, tungsten, molybdenum and molbdenum foil, and nickel-iron-cobalt alloy wire have also been used. This section must also be conductive.

3. <u>Outer-Section</u> - This section of the lead-in extends from the press-seal to the base of the lamp. Metals used for this portion have included: copper, nickel-plated copper, coppercald steel, manganese-nickel alloy, and nickel-plated iron.

4. <u>Other Designs</u> - The form shown above is a one-part lead. Others that have been manufactured (deopending upon the type of lamp being made) have included:

one-part lead - a straight wire
two-part lead - welded wires for direct connection
three-part lead - welded wires as shown above
four-part lead
five-part-lead.

The others are used where 4 or 5 wires are welded together to control the individual filaments used in three-way light bulbs used for residential lighting.

In general, a great variety of lead-in wires have been used to suit the purpose.

3.4.- MANUFACTURE OF INCANDESCENT LAMPS

We have already given a brief description in Chapter 2 of how incandescent lamps are manufactured (e.g.- see 2.1.1.). In the following diagram, we show a completed incandescent lamp and its parts as well as common shapes of incandescent lamps made:

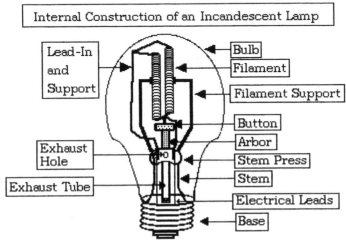

Internal Construction of an Incandescent Lamp

Lead-In and Support

Bulb
Filament
Filament Support
Button
Arbor
Exhaust Hole
Stem Press
Exhaust Tube
Stem
Electrical Leads
Base

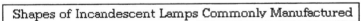

Shapes of Incandescent Lamps Commonly Manufactured

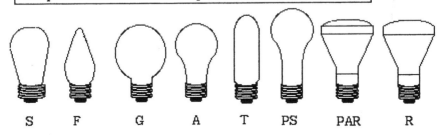

| S | F | G | A | T | PS | PAR | R |

S = straight side   F= flame   G = globular   A = arbitrary

T = tubular PAR = parabolic   R = reflector PS = pear shape

Since there are over 700 different types of incandescent lamps made, which can be expanded to several thousand because of differences in voltage, finish and base, it is impractical to describe how all lamp types are manufactured. It is for this reason that we will describe the manufacture of an incandescent lamp based on the120 volt A-type form (see 3.4.1.), which is a large volume item. Our next objective is to describe in some detail all of the intricate steps used to manufacture incandescent lamps. In this case, high speed machines are used to accomplish each of the individual steps in sequence, while maintaining an output of several thousand lamps per hour.

A. Steps Involved in Manufacture of Incandescent Lamps

The steps required to manufacture such lamps involve:

1. Filament Mount Assembly Operation: The stem-mount machine consists of a conveyer which holds the mount parts in a desired position and carries them through a series of operations in which the glass is heated and formed. The sequence of operation is:

a. A piece of glass tubing is flared on one-end.

b. It is then positioned with the flared end up.

c. Two lead-in wires as fed through cylinders are placed inside the flared-glass and come to rest in positioning guides.

d. A piece of exhaust tubing (about 3/16 inch diameter) is also placed inside the flared glass and forced to a stop at about one half of its length.

e. Filament support wires are then inserted into this assembly which is then heated to soften the glass. Thereupon, clamps move in and form a flat solid section called the "press" which now contains the supports and lead-in wires sealed into the glass.

f. Just before the "press" solidifies, and a puff of air through the exhaust tube provides a hole in the "press" through which the lamp can be evacuated of air at a later stage of manufacture of the incandescent lamp.

g. Following this operation, the end of the exhaust tube is heated and shaped into a button of glass about 1/4 inch in diameter and

1/8 inch thick, and a molybdenum support wire is inserted into the button before it cools.

h. The top ends of the two lead-in wires are shaped and separated at a distance of about a millimeter **less** than the length of the tungsten filament coil.

i. The coil is lifted into position just as the glass-lead wire assembly arrives opposite to it and the coil is mechanically clamped to the two lead-in wires at both the top and bottom and the other filament supports are attached to the filament.

This completes the filament mount assembly operation. These operations are accomplished on a highly automated machine which completes mounts at a rate of several thousand per hour. Many manufacturers have designed their own version of such machines which are regarded as proprietary and whose designs are not readily available.

2. Bulb Washing: In general, the glass bulbs need to be washed before use. This is done by several means, depending upon the Manufacturer. Mostly, it consists of mounting the bulbs vertically in a moving rack which moves the bulbs over a series of water-jets spraying upwards so as to wash the inside of the glass. A detergent may be employed as may an acidic solution (usually a mixture of HCl and HF). A final rinse completes the operation.The bulbs are then dried in a Lehr equipped with infra-red heating elements (sometimes just large high-intensity lamps), moving through at a rate sufficient to thoroughly dry the surfaces thereof. Air jets during this phase may also be employed. All of these operations are accomplished on a large machine having stations equipped for each individual step of the process so as to produce several thousand washed (and/or coated) glass bulbs per hour.

3. Bulb Coatings If the bulb is to be coated (and most are), the coating is applied either as an electrostatic coating (a "puff" process for silica coating) or as a frosting. For the former, the silica particles are charged by passing them through an electric field in an air stream so that they will adhere to the freshly prepared glass surface. The bulbs then pass through to a Lehring process where applied heat makes the coating

more adherent. If the glass is to be frosted, a strong HF solution is applied for several seconds so as to "frost" the glass by dissolving part of the surface, i.e.-

3.4.2.- $SiO_2 \text{ (glass)} + 6 \text{ } HF_\text{aqueous} \Rightarrow H_2SiF_6 \text{ (aqueous)} + 2 \text{ } H_2O$

A final rinse finishes this operation, and the bulb proceeds to the drying station. If a colored coating is desired, a lacquer composed of either nitrocelluose or ethyl cellulose dissolved in xylol is used, in which silica plus an inorganic pigment is suspended. The lacquer is prepared by milling the vehicle (which contains dispersing agents plus xylol) with both pigments for several hours. The lacquer is sprayed onto the glass surface and then dried. The organic part must then be removed by passing the coated bulbs through a Lehr whose temperature is sufficient to burn off the organics present without leaving any carbon residues. The bulbs are then allowed to cool before encountering the sealing-in stages of the operation.

4. Sealing In Operation : The sealing and exhausting process is generally accomplished in two operations on a machine consisting of a rotating turret having two circular receiving mechanisms, one on the top of the other and separated by about 10-12 inches, with the top one being smaller in diameter than the bottom one.

a. The mount is automatically positioned in the top section in a hole slightly larger than the exhaust tube (see 2.2.1.) and held fairly rigid.

b. A lamp bulb glass blank is dropped down over the mount. Note that the inner surface of the bulb may have already been coated.

c. A flame is applied at the neck of the bulb in the region of the mount flare and the flare-bulb neck are fused together into one piece.

d. At the end of the sealing-in process, the lower end of the bulb is reshaped in a mold while still hot to facilitate a better fit with the base later on.

e. Subsequently and automatically, the assembled mount-bulb is transferred to the bottom circular mechanism where the exhaust tube is positioned on a vertical rubber compression fitting, which in turn is connected to the exhaust manifold and fill-gas tanks through a series of valves.

f. First, the internal air within the bulb-mount is removed down to about 1.0 μ pressure. At this point, the "getter" (usually a mixture of cryolite and red-phosphorous) is introduced into the bulb while it is being flushed out with gas, usually nitrogen. This is done by squirting a small amount of a lacquer in which the "getter is suspended. The bulb is then re-exhausted, repeating this cycle several times. Finally, the bulb is refilled with the final fill-gas at a pressure of about 2-3 mm. and then "sealed-off".

g. Three gases are used in incandescent lamps. Of these, argon is the most important. It is used with a trace of nitrogen in perhaps 98% of all gas-filled lamps. The major characteristics that make it desirable are its relatively high molecular weight and its low heat conductivity, which result in a slower rate of tungsten evaporation and in smaller heat losses from the lamp. The trace of nitrogen makes the mixture less susceptible to short-circuiting and arcing. Nitrogen is still used for lamps of high wattage such as projection lamps and where distances between filament sections or lead-in wires is small. Krypton gas is also used, and would be used more widely if its costs were not so high. Recent advances in the isolation of Krypton at lower cost from air have made it one of the major gases for production of high intensity incandescent lamps. Krypton not only increases output but has a beneficial effect on lifetime of the lamp as well.

h. Sealing-off consists of heating the exhaust tube until it collapses and is fused and closed, meanwhile maintaining the proper gas pressure within the bulb. The assembly has now become a "lamp", but without a base. The lamp is carried to the basing reel by conveyer.

5. <u>The Basing Operation</u> : Metal bases (mainly brass) are usually purchased, although some manufacturers make their own. The basing operation includes:

a. Bases are filled automatically with a thermosetting cement on a circular turret. The cement, viscous in form, is forced under pressure through an orifice that deposits it around the inner periphery of the base just below the start of the threaded section.

b. The basing reel consists of a large circular rotatable turret with an automatic vertical positioning mechanism equally spaced around the circumference. Directly above these is a mechanism for holding the base-bulb assembly in a base-up position.

c. The threading of one lead wire through the center of the base, and the other around the bottom of the bulb is next accomplished by the machine operation.

d. Flames are positioned so as to provide varying degrees of heating as the turret rotates so as to cure the cement.

e. The turret indexes circumferentially so that several sequential operations occur. These consist of cutting off excess of the lead wires, soldering or welding the side lead wire to the side of the base at the point where the bulb neck and base meet and soldering the other lead wire to the base eyelet at the bottom of the base.

f. The last few indexing positions are used to "flash" the lamp. Flashing a vacuum lamp is done to ignite the reducing chemical, i.e.- "getter" to combine with the remaining oxygen in the lamp. For filled-gas types of lamps, the filament is lighted by a voltage considerably less than full voltage, in a repeated fashion, in order to condition it for use later.

The operations described above complete the manufacturing operation for incandescent lamps. Most manufacturers have developed high speed machines in which the parts are loaded and in which the various operations are carried out. Note that two separate pieces of glass are used to form the "mount". These are sealed together to form a glass rod support for the filament. Support wires are added together with electrical lead-in wires, and all of these are sealed into the glass. All of these steps are accomplished on high-speed machines which can make up to one lamp every 5 seconds of operating time. Note also that most lamps are "frosted" on the inner surface of the glass bulb. This process involves either acid-etching of the surface by HF, or coating the inside surface.with a silica coating to produce a diffusing effect on the light generated by the lamp. Very few, if any, lamps are manufactured today having a bare glass bulb in which the hot filament is readily visible. Most consumers prefer a lamp whose light is diffused by some sort of internal coating on the inside of the glass bulb. The silica coating may even be colored, as described above.

The **major** problems encountered in the manufacture of incandescent lamps are as follows:

   1. Glass Bulb Preparation:

      i.- Bulb Washing- water temperature, time & rinse

      ii.- Drying- use of IR heat or filtered air blast plus heat

      iii.- Bulb Handling- small household lamps get an air-blast prior to applying silica coating on machine. Reflector and Large Lamp types do not.

   2. Coatings:

      i. Pigment lacquers- viscosity problems arise caused by heat and weather, drying and lehring problems, furnace time and temperatures, air-blast impurities introduced and bulb surface impurities.

      ii. Silica coatings- particle size of silica and its moisture content, air-pressure to be used on "puff" (electrostatic) process, moisture content of compressed air used, Lewhring temeperature and index time, pressure of air-flush used.

314

3. Filaments and Getters:

i. Getters- maintaining proper cryolite and red-phosphorous ratios, viscosities of lacquer containing "getter", effects of basing, soldering, flashing and seasoning on "getter" function.

ii. Filaments- operating temperature, effects of stem insertion, exhausting, basing soldering on filament seasoning, vibration check for filament failure (i.e.- the "drop-test", oxidized Dumet obtained on press, bad filament lead clamps, poorly positioned filament "stand-offs"

4. Miscellaneous Problems:

i. Reflector Type- control of sand fill prior to flashing of aluminum coating, surface cleanliness due to sitting prior to coating of the glass bulb (most of these types of lamps are not made on high speed machines. The cleaned bulbs are placed within a vacuum-coater, filled with sand to limit the reflective coating, "flashed" to vaporize the aluminum as a coating and then are removed and placed on the seal-in machines at a later date), purity of reflective coating and its adherence to glass, "flashing" time and voltage, and time elapsed before actual manufacture into lamps (the glass and coating may further ozidize and pick up moisture from the ambient air). The "sand-fill" also causes problems in that the reflector glass must be washed with warm water prior to putting them on the manufacturing line of the lamp manufacturing machine.

B. Types of Incandescent Lamps Manufactured

The major consideration of every Lamp Manufacturer is, of course, the costs of making such lamps. Lamp sales have become so competitive that every fraction of a cent saved per lamp translates into thousands of dollars of profit per year.

The types of lamps that have been manufactured include:

    1. <u>Residential Incandescent Lamps</u> :
       Wattages of - 25, 40, 60, 75, 100, 200, 500, 1000 & 1500.

    2. <u>Photoflood & Reflector Lamps</u> :
       Wattages of - 75, 150, 200, 300, 500, 1000, 2500 & 5000.

    3. Subminiature Lamps:

    4. Automobile Headlamps:

    5. Specialty Incandescent Lamps:

1. <u>Incandescent Lamps for Home Use</u> - the major differences between these lamps lies in the size and shape of the filaments used. As stated above, a wide variety of bulb shapes and coil-shapes have been used and continue to be used for a wide variety of lighting purposes. Most of the lamps sold are of the "frosted" variety, i.e.- the glass surface is etched with hydrofluoric acid, but some have a pink-tinted internal coating of silica.

2. <u>Photoflood and Reflector Lamps</u> : These lamps are in general larger in wattage than those used for general lighting and require larger sizes of tungsten wire for operation. As such, lifetime is less important than that of light output. Many lamps have internal reflectors of sputtered aluminum metal to totally reflect internal light generated by the filament in a forward direction.

<u>3. Sub-Miniature Lamps</u>:

These lamps have been used in many areas including indicator lamps and lamps arranged to form signs and letters for advertising. Because the lamps are very small, they are made in a slightly different process. Most filaments consist of a a single coil and many are a small piece of straight tungsten wire which operates of milliwatts of power.

The following diagram, given as 3.4.3. on the next page, shows typical construction of these lamps.

3.4.3.-

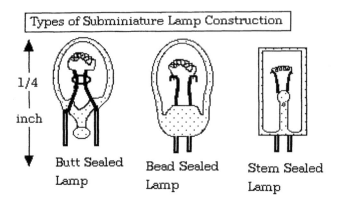

Types of Subminiature Lamp Construction

1/4 inch

Butt Sealed Lamp   Bead Sealed Lamp   Stem Sealed Lamp

Note the very small size of the lamps. Even smaller lamps have been made for use in digital quartz watches.

The electrical characteristics of these lamps are presented in the following Table. The long life is a resultant of the low filament temperatures.

TABLE 3- 13

Electrical Characteristics of Subminiature Lamps

Lamp No.	Volts	Amperes	Candle Power- Ave.	Average Life (Hours)	Filament Temp. (° K)
680	5.0	0.060	0.030	100,000	1850
683	5.0	0.060	0.050	100,000	1950
713	5.0	0.075	0.088	25,000	2100
328	5.0	0.115	0.150	40,000	2125
327	28.0	0.180	0.340	3,000	2275
		0.340	0.340	7,000	2200

In addition to the bayonet terminals shown in 3.4.3., these lamps are also available with screw-in bases.

4. Automobile Headlamps :

All automobile manufacturers have provided headlamps for driving at night since the inception of the automobile. The first light sources were very crude affairs, being oil lamps adapted for the purpose. At that time, the incandescent lamp was still under development, although the horse and buggy mode of transportation continued to use such lamps. However, the greater speed of the automobile made these lamps nearly useless because the wind usually blew out the flame, even though a glass chimney was in place. As shown in the following diagram, given as 3.4.4. on the next page, acetylene lamps were next on the list.

This source was somewhat better but the amount of light produced was still nearly non-existent. It was not until about 1910 that incandescent tungsten filament lamps had sufficiently improved in quality so as to provide sufficient light for night driving. One factor was the fact that early cars had magnetos to provide the spark for ignition and batteries were not used at all. Most of the incandescent bulbs operated at 110 VAC, not 6 VDC. Batteries were installed in cars about 1912 and low voltage incandescent lamps began to be used as headlamps.

The first lamps used were vacuum lamps which provided about 30,000 candlepower in the forward direction. However, the light beam was diffuse and not well directed. In 1915, the use of gas-filled lamps began, due to the fact that they had shorter coiled filaments and, more importantly, were used with a glass lens over the front of the headlight. This use continued until 1924 when the first two-filament lamps for automotive use were introduced. These lamps had the advantage that an upper headlight beam could be used for general driving at night, but the light beam could be shifted to a lower beam so as not to blind oncoming traffic. However, this type of headlight had the tendency to shift the light beam out of initial focus, as vibrations affected the lamp position so as to make the beam very diffuse.

It was not until 1928 that a fixed focus headlight was introduced. In this headlamp, the bulb was placed at a point where all of the light was collimated. By 1934, a pre-focused lamp assembly began to used in which the lamp had a bayonet socket which turned into a socket and was held in position at the exact focus of the reflector at the back of the assembly. This was the pre-focused headlamp, as shown in the following diagram, given on page 319 as 3.3.5.
However, it was not until 1939 that the pre-focused lamp- reflector unit was replaced, this time by an entirely new type of headlight, the Sealed Beam headlight. This lamp was manufactured quite differently in that the dual

3.4.4.

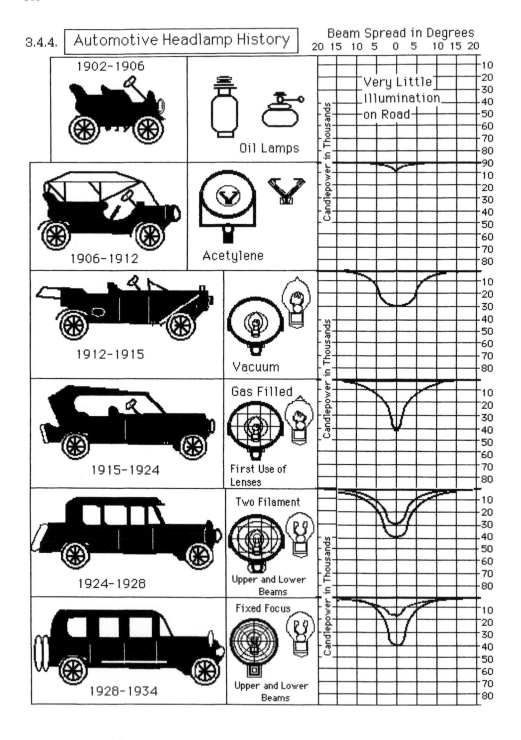

| Automotive Headlamp History | Beam Spread in Degrees |

3.4.5.-

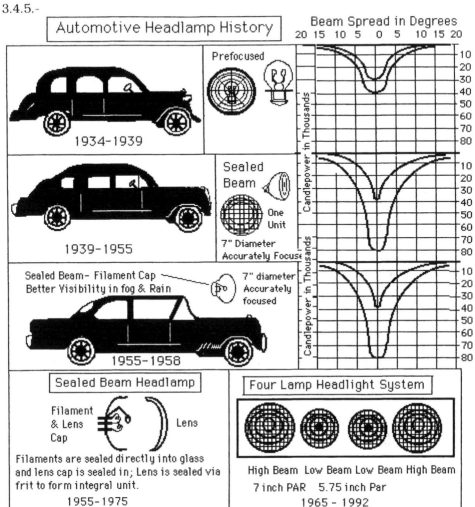

Automotive Headlamp History

Beam Spread in Degrees
20 15 10 5 0 5 10 15 20

Prefocused

1934-1939

Sealed Beam

One Unit

7" Diameter
Accurately Focused

1939-1955

Sealed Beam- Filament Cap
Better Visibility in fog & Rain

7" diameter
Accurately focused

1955-1958

Sealed Beam Headlamp

Filament & Lens Cap

Lens

Filaments are sealed directly into glass and lens cap is sealed in; Lens is sealed via frit to form integral unit.
1955-1975

Four Lamp Headlight System

High Beam Low Beam Low Beam High Beam
7 inch PAR   5.75 inch Par
1965 - 1992

filaments were mounted directly into the glass of the reflector base which was then sealed to a fixed glass lens. Only then did the light output of theseheadlamps exceed 50,000 candlepower. By 1955, a cap had been added over the low beam filament to improve the seeing distance when fog or rain was present.

Dual headlamps were used and aiming-pads were molded into the outer surface of each glass lamp to facilitate mechanical aiming of the light beams. In 1965, a four-lamp system was introduced which used two high beam (75 watt filament- 7" diameter) sealed beam lamps and two low beam (60 watt- 5 3/4" diameter) lamps arranged in a horizontal row. The higher wattage used allowed the generation of as much light in the low beam configuration as was available formerly in the high beam mode.

However, it was not until about 1975 that quartz-iodine sealed beam headlamps were introduced. This was done to facilitate the introduction of even brighter automotive headlamps at about the same wattage. The major difference was, of course, the loading that could be achieved at little increase in cost. The increase in candlepower approached 150% that of the old sealed-beam lamps, particularly when krypton gas replaced the older gas mixture formerly used. In this case, the quartz tube was mounted within the bottom half of the sealed-beam lamp, replacing the filaments and the lens part was then sealed to form the integral unit.

5. <u>Specialty Incandescent Lamps</u>

These lamps include both the so-called "lumiline" lamps and "showcase" lamps. These lamps were merely long incandescent lamps designed to fit into small spaces:

3.4.6.-

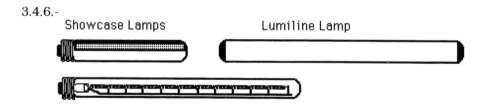

Also included in this category are the quartz-iodine lamps which have of late begun to find their way into all types of incandescent lamp bulbs. This is shown in the following diagram, given as 3.4.7. on the next page.

Although the quartz-iodine lamp runs hotter, it does not consume more power, its lumen output is nearly 50% higher than corresponding incandescent filaments, and its life is considerably longer. See Chapter 2- pages 84-87 for a more complete discussion of lamp design.

3.4.7.-

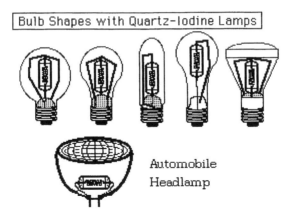

Bulb Shapes with Quartz-Iodine Lamps

Automobile
Headlamp

In general, manufacture of such lamps starts with a quartz tubing of selected length and diameter. The filament is made up as given above for incandescent lamps except that flat ribbons of molybdenum are used. An exhaust tube is attached to the main body of the quartz tube and the filament assembly is placed within the quartz tubing. The tube is then heated on each end and a "press" is made on each end. The flat ribbons of molybdenum form an excellent seal with the quartz with no other additives required. The sealed tubing is then evacuated via the exhaust tube, about 5-6 mg. of iodine (I_2) is added to the lamp and the tubing is backfilled with about 5 mm. of gas, typically nitogen and argon.

More recently, krypton gas has been used to produce a lamp superior in both output and lifetime. the exhaust tube is then sealed off and the completed quartz-iodine lamp is then mounted in its final outer protective glass bulb. Because this type of lamp runs hotter than the usual incandescent lamp, the outer bulb is necessary to protect the quartz tube from thermal gradients that may be present in the ambient atmosphere.

3.5.- PREPARATION OF RAW MATERIALS FOR PHOSPHORS

In this section, we will describe methods of manufacture of raw materials used to fabricate both fluorescent lamp and cathode-ray tube phosphors. Such materials include:

3.5.1.-   CaHPO$_4$ Alkaline Earth Carbonates SrHPO$_4$ ZnS
           BaHPO$_4$ ZnCdS$_2$ CdNH$_4$PO$_4$ • 2H$_2$O   MnNH$_4$PO$_4$ • H$_2$O

Of these, CaHPO$_4$ is the most important material for manufacture of fluorescent lamp phosphors and ZnS is the primary one used for preparing cathode-ray tube phosphors.

The general method of preparation of such materials involves precipitation from solution. Specifically, one prepares a soluble solution of both the cationic and anionic parts of the material desired and then pumps one into the other to cause an insoluble precipitate to form. In Industry, a 2000 gallon glass-lined tank is most often used, with smaller sized tanks of the same class used for holding and storage of purified solutions. Probably the most important criterion for all raw materials used to prepare phosphors is that they must be as **pure** as possible, and certainly must **not contain** impurities whose total concentration exceeds about 1.0 part per million (ppm). In a prior work (8), the nature of such impurities which are detrimental to phosphor quality were defined as "killers". These included the following cations:

3.5.2.- Cationic Quenchers of Luminescence

Divalent			Trivalent					Quadrivalent
Ni	Cu		Ti	V	Cr	Ru		Os
Pd	Co		Zr	Nb	Mo	Rh		Ir
Pt	Fe		Hf	Ta	W			Re

Most of these are transition elements with d-electrons and unpaired spins. However, they form insoluble precipitates with either sulfide or many organic chelating compounds, whereas the alkaline earth sulfides and/or chelates are soluble. Thus, these analytical reagents can be used for solution-purifying.

A general method of both solution purification and formation of insoluble precipitates, including those of alkaline earth orthophosphates, is shown in the diagram given on the next page as 3.4.3., along with the equipment needed in the process. In general, one dissolves a chloride or nitrate of the appropriate cation in one processing-tank and uses a solution of an ammonium salt of the anion desired in the other.

3.5.3.-

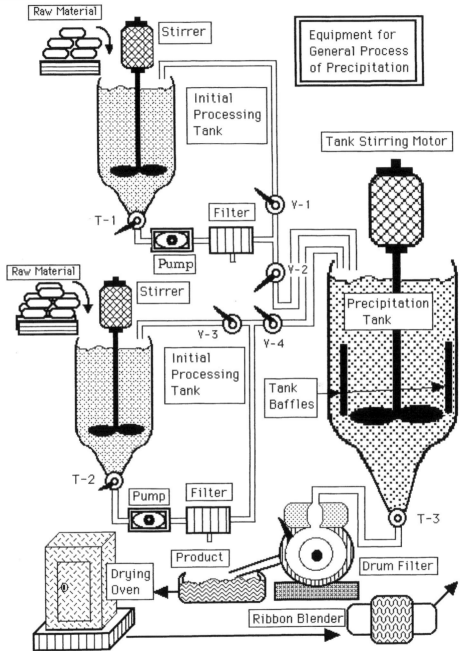

Raw Material

Stirrer

Equipment for
General Process
of Precipitation

Initial
Processing
Tank

Tank Stirring Motor

V-1

Filter

T-1

Pump

V-2

Precipitation
Tank

Raw Material

Stirrer

V-3

V-4

Initial
Processing
Tank

Tank
Baffles

T-2

Pump

Filter

T-3

Drum Filter

Drying
Oven

Product

Ribbon Blender

Although ammonium polysulfide solution, i.e.- $(NH_4)_2 S_x$, is generally used to form insoluble transition metal sulfides, precipitation of the impurities by addition of certain organic chelating agents can also be used. Both solutions are first stirred after the precipitating agent is added, and then allowed to remain quiescent so that the precipitate formed settles to the bottom of the tank. Then both T-1 and T-2 valves are opened and each solution is circulated through its corresponding filter to remove the precipitated impurities from the solution. At the same time, the pump circulates the solution back into each corresponding tank, valves V-1, V-2, V-3 & V-4 having been turned so as to be able to facilitate this operation. This continues until the solution is free of the precipitate. At the same time, samples of the solution are analyzed to determine when to stop the recirculating cycle.

The following lists the organic chelating agents that are generally employed in this process, viz-

3.5.4.-   Organic Agents Used for Solution Purification

Ammonium 1-Pyrrolidinedithio Carbamate
Ammonium Nitrosophenyl Hydroxylamine
8-Hydroxyquinoline
Dimethyl Glyoxime
Sodium or Ammonium Polysulfide

The next part of the process is the critical one. As has been detailed in a prior work (7), **any** precipitate to be used to prepare phosphors must be as close to exact stoichiometry as possible. If one adds one solution to the other to cause precipitation to take place, it is necessary to add a slight excess of the one to cause a complete precipitation. Since it is known (7) that the presence of an excess of the cation, ever so slight, is detrimental to phosphor quality, one **always** adds to excess the anionic solution to the cationic one.

To illustrate this fact, consider the following. The solubility product of the hypothetical compound, MX, is given by:

3.5.5.- $K_{SP} = [M] [X]$

where [M] and [X] are the molar concentrations of the cation and anion respectively. When "X" is added to slight excess, then **all** of the cation possible is precipitated as the compound, MX (Note that all of the solvated ions cannot be completely precipitated since the K_{SP} **never** equals zero). However, since the precipitation process is a phase-change, i.e.- a change from solvated ions in a liquid to a solid suspension, a surface charge called the "zeta-potential" always prevails. Therefore, if one ion is in excess in solution **after** the precipitate has formed, it will **adsorb** on the surface of the precipitate particles, due to attraction by the surface charge. It is for this reason that it is better to have an anionic excess during precipitation since anions are usually optically transparent and cations are not (7). It is also for this reason that ammonium salts of the anion are used since ammonia is volatile at rather low temperatures, whereas the use of a soluble metal salt of the anion desired would result in surface-contamination of the particles by both the metal ion and the anion.

In general, the solution is stirred as the precipitation takes place. Particle size produced is a function of both solution concentration and temperature. Precipitation at room temperature usually produces small particles. This is due to the fact that the solubility product is never zero and a small amount of both cations and anions will always be present in solution in equilibrium with the precipitate. Even the most insoluble compound known, i.e.- ZnS, has a solubility product of 10^{-23} (which is not zero). This results in soluble cations and anions left in the solution. However, at higher temperatures in solution, the solubility product usually increases slightly. This fact is used to grow larger particles, since a dynamic equilibrium between soluble ions and the solid phase exists in all cases where a precipitate has formed.

Most ammonium salt solutions are subject to loss of volatile NH_3 when heated to elevated temperatures. Thus, the ammonium anion solution cannot be heated above about 40-45 °C without loosing appreciable amounts of ammonia as vapor. In cases where the pH is important during precipitation, the ammonia solution is kept below 45 °C. while the cation solution can be as high as 95 °C. Once the addition of anion is complete, the solution with its suspension of formed particles is stirred to promote growth of larger particles. The final step in the precipitation process involves pumping the suspension into a drum filter where the so-called "mother-liquor" is separated from the precipitate. The powder is then put into an oven at 110-125 °C. to dry and then is placed in a ribbon-blender to break up any lumps that may have formed during the drying process.

In the following sections, the specific conditions to be used in manufacturing the various products are detailed. These include:

> $BaHPO_4$, $SrHPO_4$,
> $CaHPO_4$, $CdNH_4PO_4 \bullet 2H_2O$,
> and alkaline earth carbonates.

A.- MANUFACTURE OF $BaHPO_4$ BY PRECIPITATION

1. Barium Nitrate Solution = 261.34 MW - Concentration needed is 2.43 molar or 1.40 lbs/gallon. 112 lbs. of $Ba(NO_3)_2$ produces 100 lbs. of $BaHPO_4$. Using 80% efficiency (actual measured is 87%), we need 140.0 lbs. of $Ba(NO_3)_2$, or 243 mols. The procedure is as follows:

a. Fill 300 gallon glass-lined tank with 200 gallon of 70 °C. deionized water. Start stirrer.

b. Add 140 lbs. of approved lot of $Ba(NO_3)_2$ to tank and stir until material dissolves.

c. Add 400 ml. of $(NH_4)_2S_x$ to tank with stirring. Stir 15 minutes. Allow to stand 2 hours at 65 °C. to digest and form larger particles.

d. Fill 5 gallon pail with Solka-Floc™ and slurry with hot deionized water. Pour slurry into filter-press and wash floc until wash-water is clear.

e. Filter the $Ba(NO_3)_2$ solution by recirculation until the solution is clear. If the solution still has a greenish cast, add more Solka-floc™ and continue recirculation until a pale yellow solution is obtained. Flush lines with approximately 10 gallons of hot deionized water to clear lines.

f. Add 170 ml. of 30% hydrogen peroxide with stirring. The solution should turn white.

g. Clean filter press, and re-coat with Solka-Floc™ as before.
h. Filter $Ba(NO_3)_2$ solution when it turns from a yellowish cast to whitish cast, recirculating to obtain a clear solution.

i. When clear, pump purified $Ba(NO_3)_2$ solution into 700 gallon precipitating tank.

j. Test 50 ml. of purified solution by adding 5 ml. of ammonium sulfide to solution. If solution turns darkish, it must be reprocessed before further use.

k. Clean Filter Press for further use.

2. Diammonium Phosphate Solution = 132.056 MW: For 243 mols of Ba^{2+} solution, a 1.05:1.00 ratio is required. This is 255 mols, or 74.3 lbs. of $(NH_4)_2HPO_4$. In 60 gallons of deionized water, the concentration is 3.05 molar.

a. Fill a 90-gallon glass-lined tank with 60 gallon of 60°C. deionized water. Start stirrer.

b. Add 75.0 lb. of approved lot of $(NH_4)_2HPO_4$ to tank with stirring. Maintain at 40-45 °C while stirring to dissolve salt.

c. Add 410 ml. of ammonium sulfide and continue to stir about 10 minutes. Stop stirrer and allow to digest for 6 hours, maintaining temperature between 35-40 C.

d. Re-coat filter press with Solka-Floc™ as given above. Wash the Floc before attempting to use coated filter-press.

e. Filter the $(NH_4)_2HPO_4$ solution to remove the blackish precipitate. Continue to recirculate the solution until it turns to a clear pale yellow solution. If it remains greenish, re-coat the filter-press and refilter the solution.

f. Clean filter press and flush lines back into tank. Re-coat filter press with Solka-Floc™ as before.

g. Add 165 ml. of 30% hydrogen peroxide to oxidize excess sulfide to sulfur. The solution will turn whitish, or it may become almost clear. Remove the whitish precipitate by recirculation through the filter-press and continue until the solution is completely clear. Maintain temperature at 35 - 40 °C. Check purity by analysis.

h. Check pH of solution. If lower than 7.80, adjust to pH $\geq$ 8.2 with NH_4OH, adding reagent slowly.

i. Clean filter-press for next use.

3. Precipitation of $BaHPO_4$

a. Start the stirrer in the purified $Ba(NO_3)_2$ solution tank and heat to 70-75 °C.

b. Test the pH of the $(NH_4)_2HPO_4$ solution. It should be about pH = 8.3. If it is lower than about 8.0, add NH_4OH to adjust pH to 8.0 - 8.5. Maintain temperature between 35-40 °C.

c. Add the $(NH_4)_2HPO_4$ solution at a rate of about 6.0-10.0 gallon per minute, while maintaining its temperature between 35-40 °C. This will take about 6-10 minutes to complete the precipitation.

d. It is important to direct the $(NH_4)_2HPO_4$ solution at a point in the precipitation tank where it is instantly dispersed within the $Ba(NO_3)_2$ solution. If it encounters the surface of the solution, an immediate precipitate forms and floats upon the surface of the solution. This results in a precipitate of indeterminate particle size. The $(NH_4)_2HPO_4$ solution should be directed against one of the tank baffles where it can be instantly dispersed. Increase stirring speed if necessary to accomplish this action. When the $(NH_4)_2$ HPO_4 solution has been completely added, wash the tank with hot deionized water and add to precipitation tank while stirring.

e. Allow to stir for 15 more minutes. Test clear liquid for excess phosphate with previously prepared barium nitrate solution. A precipitate will form if excess is present.

f. Allow $BaHPO_4$ precipitate to settle and siphon off supernatent liquid. Add enough hot (85-90 °C) deionized water to re-suspend precipitate. Allow precipitate to settle and remove supernatent water.

g. Repeat this washing operation for 4 more times for a total of 5 washes.

h. Using drum-filter, remove precipitate and place in stainless steel trays. Dry in oven at 125 °C. for 10 hours.

i. Remove trays from oven, allow to cool and place in tared plastic-lined drums.

j. Weigh complete product and record weight plus yield.

This completes the production of the $BaHPO_4$ precipitate which is then ready for use. During precipitation, the pH in the $Ba(NO_3)_2$ solution changes from about 3.0 to about pH 7.0 during the precipitation, as shown in the following diagram, given as 3.5.6. on the next page.

The critical parts of the procedure have been determined to be:

1. pH of the $(NH_4)_2HPO_4$ solution

2. Stirring after addition of $(NH_4)_2S_x$ should be minimal

3. The $(NH_4)_2HPO_4$ solution should always be added to the $Ba(NO_3)_2$ solution, not vice-versa. The amount of excess

3.5.6.-

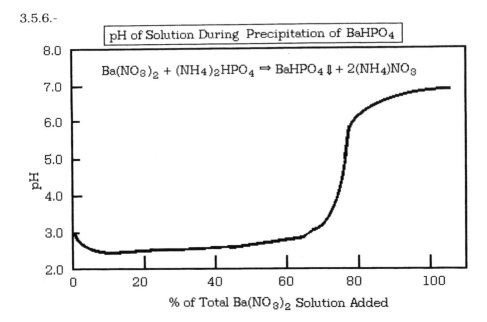

phosphate is not critical and can be as high as 1:1.25. However, the precipitate must be thoroughly washed to remove as much of excess phosphate present as possible.

4. The rate of addition of the $(NH_4)_2 HPO_4$ solution has some effect on particle size. Faster rates tend to form smaller average sizes and vice-versa. A average size of about 5.6 μ is considered normal.

B. PRECIPITATION OF SrHPO₄

The preparation of $SrHPO_4$ is quite similar to that of $BaHPO_4$ except for one factor. Whereas the latter exists in one form, the former exists in one of **two** forms, depending upon the temperature of precipitation (9). At temperatures below about 25 °C, β- $SrHPO_4$ is obtained and above 40 °C., α-$SrHPO_4$ is formed. The particle size of the precipitate also is more temperature dependant than for $BaHPO_4$, as will be shown below.

1. Strontium Nitrate Solution = 211.63 MW   - Solubility @ 60 °C. is 938 gm/liter or 7.82 gm. /gallon of water. 115.0 lb. will produce

100.0 lb. of $SrHPO_4$. Using 80% efficiency, 144 lbs. are required. This is 308.6 mols, or 0.41 molar if 200 gallons are used.

    a. Steps "a" through "k" are followed as for $BaHPO_4$, except that 144 lbs. of $Sr(NO_3)_2$ are used.

2. Diammonium Phosphate Solution = 132.056 MW: For 308.6 mols of Sr^{2+} solution, a 1.05:1.00 ratio is required. This is 324.1 mols, or 94.3 lbs. In 60 gallons of deionized water, the concentration is 1.43 molar.

    a. Steps "a" through "k" are followed as for the $BaHPO_4$ procedure except that 94.3 lbs. of $(NH_4)_2HPO_4$ are used.

3. Precipitation of $SrHPO_4$

    a. The critical part of the procedure involves the decision concerning which form of $SrHPO_4$ is desired. The temperature is controlled either at < 25 °C. to produce β-$SrHPO_4$ or at > 40 °C. to produce α-$SrHPO_4$. In general, steps "a" through "i" are followed, except for temperature control of the solutions.

    b. Below 25 °C, β-$SrHPO_4$ is formed and above 40 °C. α-$SrHPO_4$ obtains. Between 25-40 °C., a mixture of the two results. The major difference between the two allotropic forms is the particle size produced. This is shown in the following table, viz-

Physical Properties of the Two Forms of $SrHPO_4$

Form	Aver. Particle Size	Surface Area (BET)
β-$SrHPO_4$	19.1 μ	20.6 m^2/gm.
α-$SrHPO_4$	8.2 μ	5.4 m^2/gm.

It should be clear that the major difference between the two forms is that of surface area. Whereas β-$SrHPO_4$ has a larger average size than that of α-$SrHPO_4$, it is composed of much smaller particles agglomerated into one larger one. This accounts for the surface area. The larger surface area particle produces a brighter phosphor, in general, than does the α-SrHPO$_4$, when used in a solid state reaction to produce the phosphor. Only when the phosphor, Sr$_2$P$_2$O$_7$: Sn, is to be prepared does the α-SrHPO$_4$ produce a superior product.

c. The effect of precipitation temperature on particle size is considerable, as shown in the following:

Precipitation Temperature	Allotropic Form	Aver. Particle Size	σ of Particle Distribution
95 °C.	α-SrHPO$_4$	3.9 μ	1.56
80	α-SrHPO$_4$	9.7	1.59.
60	α-SrHPO$_4$	13.8	1.62
20	β-SrHPO$_4$	17.6	1.56

Thus, the precipitation temperature to produce a particle size of 5-6 μ should be about 85 °C.

d. However, if the Sr^{2+} solution is added to the phosphate, i.e.- the reverse addition, the pure SrHPO$_4$ does not result. At 80 °C., 12% of the product is the hydroxyapatite, i.e.- Sr$_5$OH(PO$_4$)$_3$ and 88% is SrHPO$_4$. At 95 °C., the product consists of 15% hydroxyapatite and 85% SrHPO$_4$. Thus, the reverse addition should never be used in the preparation of SrHPO$_4$.

e. The final pH of the solution should be adjusted to about pH = 6.8-6.9 with ammonia, as necessary, stirring about 15 minutes before the precipitate is allowed to settle and before the mother liquor is withdrawn and washing of the precipitate begins.

C. MANUFACTURE OF $MnNH_4PO_4 \bullet H_2O$ (10 gallon batch size)

Since this compound is used as a source of activator for some phosphors, its preparation will be described. Obviously, large amounts are not usually required and precipitation takes place in a smaller glass-lined vessel.

1. MnCl$_2$ Solution- MW = 161.86: 14 liters of de-ionized water are added to a 10 gallon tank and heated to near 100 °C. Add 40.0 mols (17lbs 7 oz.) of $MnCl_2 \bullet 4H_2O$ with stirring. Maintain temperature near to 100 °C. Since the Mn^{2+} solution cannot be purified by the usual reagents, it is used as the analytical reagent grade purity.

2. Diammonium Phosphate Solution = 132.056 MW: For 40.0 mols of Mn^{2+} solution, a 1.10:1.00 ratio of PO_4/Mn is required. This is 45 mols or 13 lbs. 1 oz. of $(NH_4)_2HPO_4$ which is dissolved in 17.2 liters of warm ($\approx$ 45 °C.) deionized water. The solution is purified as before, using steps "a" through "c" given above, except that only 30 ml of ammonium sulfide is used. It is important that **all** of the sulfide be removed by oxidation with hydrogen peroxide , and the sulfur precipitate is filtered out before this solution is used. Therefore, the solution is cooled to about 25 °C. before attempting to oxidize the sulfide to a sulfur precipitate. The solution is filtered through a 0.45 µ filter before use.

3. Precipitation : The phosphate solution is heated to 55 °C and is added slowly to the Mn^{2+} solution which has been maintained near to 100 °C. Addition time should be about 15 minutes or about a rate of addition of 1.2 liters/minute. After addition of the phosphate solution is complete, continue stirring while adding ammonia to raise the pH to about 6.8. The precipitate is then allowed to settle before the mother liquor is drawn off.

4. Washing: Wash the precipitate with warm deionized water by filling the tank near to the top as stirring continues. Allow to settle and repeat the procedure. Test the wash water for chloride by adding silver nitrate solution and continue washing until the

test is negative. Allow to settle for the final time, and filter into a prepared large Büchner filter.

5. Drying: Dry the precipitate for at least 16 hours at a temperature no higher than 110 °C. so as to prevent loss of the ammonium part of the prepared salt. Too high a temperature results in loss of ammonia and loss of certain physical properties important in subsequent phosphor preparation.

D. MANUFACTURE OF $CdNH_4PO_4 \cdot H_2O$ (20 gallon batch size)

This compound is also used as an additive for some phosphor preparations as well as a base for $Cd_5Cl(PO_4)_3$:Mn phosphor. Since the Cd^{2+} solution cannot be purified by the usual methods given above, it is necessary to resort to other means. It is well to note here that this method can also be used for manufacture of $MnNH_4PO_4 \cdot H_2O$.

1. $Cd(NO_3)_2$ Solution- MW = 308.46 : Dissolve 7934.6 gm. (17 lbs. 8 oz. = 25.72 mols) of $Cd(NO_3)_2 \cdot 4H_2O$ in approximately 3.0 gallon of cold deionized water. This solution is then gassed with NH_3 until the solution is clear (The solution first turns whitish due to formation of $Cd(OH)_2$ and then changes to a clear solution of $Cd(NH_4)_4^{2+}$ ion, due to the reaction:

$$NH_3 + 2H_2O \Rightarrow NH_4^+ + OH^- + Cd^{2+} \Rightarrow Cd(OH)_2\Downarrow + NH_3 \Rightarrow Cd(NH_4)_4^{2+}$$

When this cation precipitates as the phosphate, the actual allotropic compound has the stoichiometry: $(CdHPO_4)_3 \cdot 3\ NH_3 \cdot 3H_2O$ (9). The final pH will then be about 8.5. Add 39 grams of tannic acid to solution and let sit for 18 hours. Filter this solution through an Ertel pad and then through a 0.45 µ filter to remove all traces of the precipitate.

2. Diammonium Phosphate Solution = 132.056 MW: A ratio of 1.4 PO_4 : 1 Cd is needed. This is 36.0 mols or 4755.7 gm. (10 lbs. 8 oz.). Add this weight of $(NH_4)_2HPO_4$ to 13.8 gallon of deionized water which has been heated to 40 °C. with stirring to form a 2.61 molar solution. Add 55.0 gm. of tannic acid while stirring, stop the

stirring and allow solution to sit for 16 hours. The solution will cool while the tannic-acid precipitate of impurities is forming and settling. At the end of the settling time, filter the solution through an Ertel™ pad and then through a 0.45 µ filter to remove all traces of the precipitate.

3. Degassing the Cadmium Solution: In order to obtain a stoichiometric compound, i.e.- $(CdHPO_4)_3 \cdot 3 \ NH_3 \cdot 3H_2O$ or $CdNH_4PO_4 \cdot H_2O$, the excess ammonia present (which was added in order to cause the tannic-acid precipitate to form at pH = 8.5) must be removed before final precipitation takes place. Therefore, the cadmium solution is heated to 85-95 °C where heating causes NH_3 gas to boil off as the solution is stirred. This process continues until a faint trace of a white precipitate $[Cd(OH)_2]$ appears. The solution is then transferred to a 20 gallon glass-lined tank and diluted to a 6.0 gallon total volume.

4. Precipitation of $CdNH_4PO_4 \cdot H_2O$: The degassed cadmium solution is heated to 75-80 °C with stirring while the phosphate solution is being heated to 40 °C. The latter is then pumped into the Cd^{2+} solution at a rate of 1.0 liter per minute. The solution must be well-stirred during this time to break up any clumps of precipitate that may form, and the phosphate solution is directed at the baffles present in the tank to facilitate dispersion of the phosphate solution during addition. After the precipitation is complete, test the solution for pH and adjust the pH with NH_4OH to about pH = 6.8, continuing to stir the solution during this time. Allow to stir for at least 15 minutes and then let the precipitate to settle.

5. Washing of the Precipitate: Siphon off the mother liquor and wash the precipitate by re-suspension with 40 °C deionized water, filling the tank to the brim. Allow the precipitate to settle and repeat the sequence for five times.

6. Drying of the Precipitate: Filter off the precipitate and oven dry for 16 hours at 110 °C. Do not let temperature rise above this limit since loss of ammonia from the salt will occur. At the end of this

time, remove from oven, let cool, record weight and yield, and place in tared plastic-lined drums.

E. Manufacture of Alkaline Earth Carbonates

These materials are used in the manufacture of phosphors and as emission materials for electronic filaments in cathode-ray tubes and the like. Their preparation follows the general method given above in that purification is achieved by either sulfide additions (with subsequent oxidation and removal of excess sulfur) or by use of organic precipitants, as stated above. In general, either nitrate or chloride salts can be used. The latter have the advantage that washing can be controlled by testing for chloride as it proceeds.

    1. <u>Alkaline Earth Solution: $BaCl_2 \bullet 2H_2O$ - MW =244.31;</u>
       <u>$CaCl_2$ - MW = 110.99; $SrCl_2 \bullet 6H_2O$ - MW = 266.64</u>

The formulas given above are the usual commercial forms. They are all very soluble in both hot and cold water. A 2.08 molar solution is made up in a 20 gallon tank by adding one of the following to 10.0 gallons of deionized water:

Material	Weights to be added		Carbonate Yield @ 95%
$BaCl_2 \bullet 2H_2O$	19,244.30 gm.	42 lb. 7 oz.	32 lb. 9 oz.
$CaCl_2$	8,742.68	19 lb. 4 oz.	16 lb. 8 oz.
$SrCl_2 \bullet 6H_2O$	21,003.23	46 lb. 4.5 oz.	24 lb. 6 oz.

Obviously, if a mixture is needed, e.g.- 1Ba:2Sr:6Ca, one would multiply the above weights by 1/9, 2/9 and 6/9 respectively. A larger batch such as that required for a 2000 gallon tank would follow the same procedure, i.e.- multiply by 100.0. More concentrated solutions can be used, but the above molarity has been found to produce carbonates with the required particle size when it (they) is (are) used to prepare phosphors and/or emission mixtures (which are milled with other additives such as titania, TiO_2, and/or silica, SiO_2).

a. Purification of Solution: - Steps "a" through "k" are followed as given for the BaHPO$_4$ manufacture except that 40 ml. of (NH$_4$)$_2$S$_x$ and 17 ml of 30% hydrogen peroxide is used during this stage. The solution is filtered through an Ertel™ pad and then through a 0.45 µ filter to remove all remaining impurities. Solka-floc™ can be added just prior to filtration to prevent the precipitated sulfides from deglommerizing and passing through the filter. The solution temperature is maintained at 70 °C.

2. Ammonium Carbonate Solution- MW = 157.11 : The usual commercial form is: NH$_4$HCO$_3$ • NH$_2$CO$_2$NH$_4$. A 1.4 CO$_3$: 1 M^{2+} ratio is required since the ammonium carbonate in solution at 50-60 °C. is not stable, and tends to decompose by losing gas as NH$_3$ and CO$_2$ to the open air. However, because of the solubility limit, i.e.- NH$_4$HCO$_3$ • NH$_2$CO$_2$NH$_4$ is soluble only to the extent of 510 gm. per liter at 50 °C, higher temperatures are used to promote higher yields of carbonate, even though some of the carbonate in solution is lost before the precipitation can take place. The following amounts of ammonium carbonate are added to 9 gallons of deionized water at 50 °C in a separate 10 gallon glass-lined tank.

Mols Carbonate	Weight of Carbonate	
110.27	17,325 gm	38 lb. 3 oz.

a. Purification of Solution: - Steps "a" through "k" are followed except that 41 ml. of ammonium sulfide and 17 ml of 30 % H$_2$O$_2$ is used. Filtering of the impurity precipitates is accomplished with an Ertel™ pad and a 0.45 µ filter as before, taking care that the precipitated sulfides do not deagglomerate. Solka-floc™ may be used as a filter aid.

b. Adjustment of pH: The pH is measured and adjusted to about 7.5 by addition of ammonium hydroxide solution.

3. Precipitation of Carbonates: The purified alkaline earth solution is heated to 70 °C. with stirring. The purified carbonate solution is heated to 50 °C. with stirring and then pumped into the alkaline

earth chloride solution at a rate of 280 ml./ minute so that the overall precipitation takes about 2 hours. Addition is made so that clumps of precipitate do not form and so that dispersion of the carbonate solution occurs quickly. At the end of the addition, the lines are flushed with deionized water and the precipitate is allowed to settle.

4. Washing: Wash the precipitate with warm deionized water by filling the tank near to the top as stirring continues. Allow to settle and repeat the procedure. Test the wash water for chloride by adding silver nitrate solution and continue washing until the test is negative. Allow to settle for the final time, and filter into a prepared large Büchner filter.

5. Drying: Dry the precipitate for at least 16 hours at a temperature no higher than 110 °C. so as to prevent loss of the carbonate part of the prepared salt. Too high a temperature results in loss of carbon dioxide and loss of certain physical properties important in subsequent phosphor preparation.

6. At the end of this time, remove from oven, let cool, record weight and yield, and place in tared plastic-lined drums.

Mixed alkaline-earth carbonates are used as emission materials on tungsten filaments for both various vapor lamps and cathode-ray tubes. They are generally applied as the carbonates from a slurry (non-aqueaous) directly to the filament. When the filament is heated, the carbonates decompose to the oxides, and the oxides thereby greatly increase the emissivity of the metal filament. The mixture can be prepared by milling the separate carbonates together for several hours, or by co=precipitation. The final mixture used depends upon the individual manufacturer, each having his own proprietary composition. However, it is usually composed of:

	Mols/ Mol of Carbonate
56% $BaCO_3$ by weight	0.450
13% $CaCO_3$	0.206
32% $SrCO_3$	0.344

If coprecipitated carbonates are desired, then one coprecipitates the mixture according to the mol ratios given above. Thus, one adds the following weights to 10.0 gallons of deionized water:

Material	Weights to be added	
$BaCl_2 \bullet 2H_2O$	8,655.30 gm.	19 lb. 1 oz.
$CaCl_2$	1,798.0	3 lb. 15 oz.
$SrCl_2 \bullet 6H_2O$	7,226.1	15 lb. 15 oz.

The same procedure is followed as given above to produce the co-precipitated carbonates which are then dried before use.

F. Manufacture of $CaHPO_4$

This compound is used to manufacture the halophosphate phosphor used in fluorescent lamps and is made in tonnage quantities each month. $CaHPO_4$ is polymorphic like the strontium orthophosphates in that it forms "brushite", i.e.- $CaHPO_4 \bullet 2H_2O$ at precipitation temperatures below about 35 °C. and "monetite" , i.e.- $CaHPO_4$, above about 65-70 °C. More importantly, it is possible (10) to convert brushite to monetite from a slurry suspension by merely reheating the suspension. Since the solubility product of $CaHPO_4$ is $K_{SP} = 3.4 \times 10^{-14}$, it is evident that about 10^{-7} mols/liter of both Ca^{2+} and $HPO_4^=$ will be present in a saturated solution containing both of these ions. The solution equilibrium is a dynamic one:

3.5.7- $Ca^{2+} + HPO_4^= \Leftrightarrow CaHPO_4 \downarrow$

Brushite precipitates as very small crystals of poor crystalline quality. Monetite prepared directly from solution at 75-80 °C is somewhat more crystalline. However, if one first prepares brushite, one can cause the brushite to redissolve and form monetite of high crystalline quality. This can be done either on a continuous basis or a batch basis. The former method removes the mother liquor, and resuspends the precipitate to form a slurry, whereas the latter method accomplishes the conversion directly in the mother liquor.

The following diagram shows the various types of $CaHPO_4$ that are possible to obtain by such a conversion process:

3.5.8.-

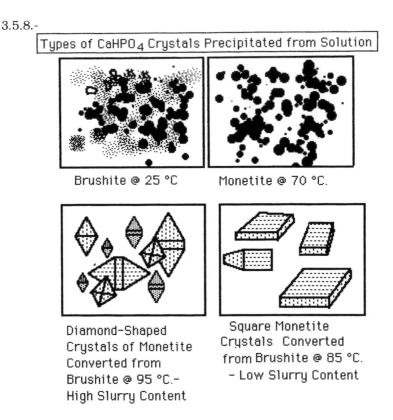

Types of $CaHPO_4$ Crystals Precipitated from Solution

Brushite @ 25 °C.   Monetite @ 70 °C.

Diamond-Shaped Crystals of Monetite Converted from Brushite @ 95 °C.- High Slurry Content

Square Monetite Crystals Converted from Brushite @ 85 °C. - Low Slurry Content

Obviously, a mixture of these two types of crystals can be obtained under intermediate conditions by controlling the slurry proportions.

Because the manufacture of $CaHPO_4$ is a high volume procedure, some manufacturers have used a continuous manufacturing procedure. Such a system is shown in the following diagram, as given as 3.5.9. on the next page.

In this process, a series of processing and holding tanks are used to purify both the $CaCl_2$ and the $(NH_4)_2HPO_4$ solutions, whereupon each solution is stored ready for use. Sulfide is added on a continuous basis

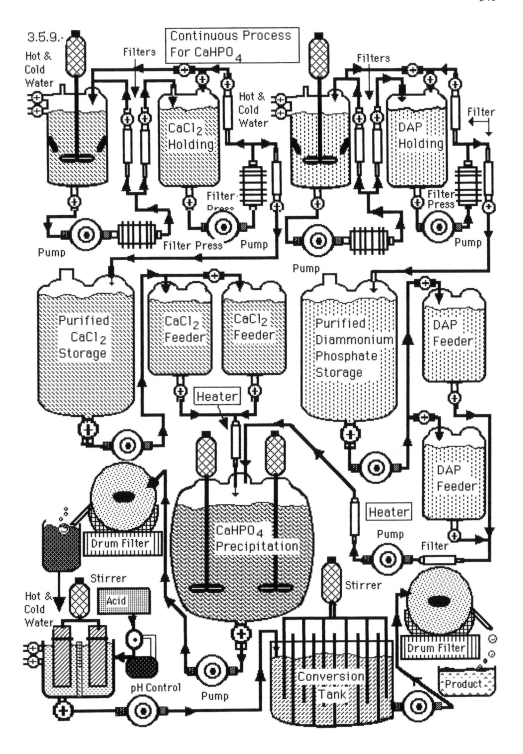

3.5.9.-

Continuous Process For CaHPO$_4$

and since the solutions are not allowed to sit so as to coagulate the impurity sulfides, the solution is pumped on a continuous basis through a filter press and 0.45 μ filters until analysis determines that all of the impurities have been removed. Each pure solution is then pumped to a storage tank.

These solutions are then pumped into a precipitation tank to form the $CaHPO_4 \cdot 2H_2O$ on a continuous basis. Then, the slurry is fed to a drum filter where the mother liquor is separated. The wet cake is then re-slurried at a specific ratio of solid to water (sometimes containing a small amount of soluble phosphate) and the pH of the slurry is adjusted. Thereupon, it is pumped to a conversion tank where the particles undergo a change from brushite to monetite of predetermined size and proportions. As stated above, a thick slurry will form the diamond-shaped crystals, whereas a thinner slurry forms the square crystals.

The procedure to be used is as follows:

1. <u>Calcium Chloride Solution = 110.99 MW</u> : The molarity of the solution desired is 2.43 mols/liter or 2.25 lbs/gallon of water. The 300 gallon tank is filled with 200 gallons of 20-25°C. water and 450 lbs. of $CaCl_2$ are added with stirring. 450 ml of $(NH_4)_2S_x$ is added to tank with stirring and the solution is pumped through the filter press and 0.45 μ filter into the holding tank where it is allowed to stand for 4 hours. Then, the solution is re-circulated through the filter press and the 0.45 μ filter until analysis indicates that the solution contains less than 1.0 ppm. of total impurities. The pure solution is then pumped into the $CaCl_2$ storage tank. Meanwhile, another batch of $CaCl_2$ solution has been started and is being processed. Once the storage tank is full, the feeder tanks are filled.

2. <u>Diammonium Phosphate Solution = 132.056 MW:</u> For a 2.43 molar solution of Ca^{2+} , the phosphate solution needs to be 2.56 molar, a ratio of 1.05:1.00. Therefore, a 200 gallon batch requires 2.82 lbs/gallon for a total of 563.9 lbs of $(NH_4)_2 HPO_4$. The procedure is as given above in 1. except that the batch has 400 ml of $(NH_4)_2S_x$ is added to the tank with stirring. Once the solution

has been analyzed to be of sufficient purity, it is pumped to its storage tank. Meanwhile, another batch is being processed. Once the storage tank is full, the feeder-tanks are filled as well.

3. Initial Precipitation of Brushite: The precipitation tank has a total volume of about 500 gallons. Two of the feeder tanks are selected and their pumps are adjusted to a rate of 1.50 gallons/minute. The line-heaters are adjusted so that the solutions are heated to 25-30°C. and the tank-heater is adjusted to maintain this temperature. Once the precipitation tank is full, the pump at the outlet of this tank is adjusted to 1.50 gallons/minute to draw off the precipitate and mother liquor from the bottom of the tank into the drum filter. The wet cake is washed on the drum filter and is fed into the reslurrying tank (shown on the lower right hand side of 3.5.9.).

4. Conversion of Brushite to Monetite Crystals: A slurry of about 3.75 lbs./gallon is made up by adjusting the flow of hot (85 °C) water into the slurry-tank. Hydrochloric acid is used to adjust the pH of the slurry on a continuous basis to about 3.6-3.8 pH. while heating the solution to near boiling. The slurry is then pumped into the conversion tank which is stirred slowly while maintaining the slurry temperature near to boiling so that the slurry resides there about 5-10 minutes. A characteristic drop in pH will occur wherein the pH will drop to about 2.6-2.8 indicating a conversion from brushite to monetite. The slurry is then pumped into the final drum filter where the wet cake is washed free of residuals and finally is dried in an oven at 125 °C. for 16 hours.

It is important to observe, on a continuous basis, the type of crystals being formed during conversion. Increasing the solids content of the slurry will produce diamond-shaped crystals while decreasing it will tend to form elongated cubes rather than flat square crystals. The particle size should be about 8.0 μ on the average to produce a phosphor whose covering power inside a fluorescent lamp is such that all of the ultraviolet radiation is absorbed by the phosphor. This covering power is a function of both particle size and shape as well as how the particles "fit" together to form a layer within the lamp.

It is for the latter reason that most manufacturers have chosen to prepare flat square platy crystals to form the particulate film. This is referred to in the Trade as "covering power". It is now well established that particles less than about 3.0 µ and greater than 35-40 scatter the ultraviolet and generated visible light within the film to such a degree that a loss of output of the fluorescent lamp results. Thus, the optimum particle size has been determined to be:

3.5.10.-   3.0 µ < optimal particle size < 35-40 µ

It might seem incongruous to discuss particle size and shape in relation to preparation of a raw material used to prepare a phosphor. Nonetheless, it has been established that the Particle Habit of the $CaHPO_4$ used to form the phosphor determines the final Particle Habit of the phosphor so produced. What this means is that if flat square crystals are used, the final shape of the phosphor particles will be flat squares. This fact has been firmly established by observing the particle habit produced as a function of the solid state reactions ocurring.

For the halophosphate phosphor which is universally used in all fluorescent lamps, the solid state reactions involve:

3.5.11.- Solid State Reactions Leading to Formation of Halophosphate

<u>Overall Reaction:</u>

$$2\ CaHPO_4 + CaCO_3 + CaF_2 + CaCl_2 + Mn_2O_3 + Sb_2O_3 \Rightarrow$$
$$Ca_5(F,\ Cl)(PO_4)_3 : Sb: Mn\ (Halophosphate)$$

<u>Partial Reactions</u>

a. $2\ CaHPO_4 \Rightarrow \beta\text{-}Ca_2P_2O_7\ (monoclinic) + H_2O \Uparrow$
b. $CaCO_3 \Rightarrow CaO + CO_2 \Uparrow$
c. $Ca_2P_2O_7 + CaO \Rightarrow \beta\text{-}Ca_3(PO_4)_2\ (triclinic)$
d. $\beta\text{-}Ca_3(PO_4)_2 + CaF_2 + CaCl_2 \Rightarrow Ca_5(F,Cl)\ (PO_4)_3\ (hexagonal)$

In the above reactions, we have left out the side reactions that occur which involve the two activators, Mn^{2+} and Sb^{3+}. A more detailed discussion of the $Sr_5(F, Cl)(PO_4)_3$: Sb: Mn phosphor was presented in a prior work (8).

As shown in the following diagram, electron micrographs taken of the reaction product between the flat square platelets of $CaHPO_4$ and various components of the above reactions show that the particle habit of the $CaHPO_4$ is retained in fired product of the solid state reaction.

3.5.12.-

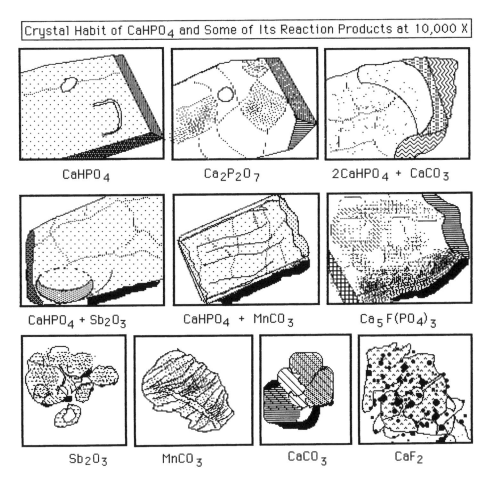

Crystal Habit of $CaHPO_4$ and Some of Its Reaction Products at 10,000 X

$CaHPO_4$ $Ca_2P_2O_7$ $2CaHPO_4 + CaCO_3$

$CaHPO_4 + Sb_2O_3$ $CaHPO_4 + MnCO_3$ $Ca_5F(PO_4)_3$

Sb_2O_3 $MnCO_3$ $CaCO_3$ CaF_2

The reason for this behavior lies in the relative reaction temperatures of the various components. According to Wanmaker et al (11), Reaction "a."

in 3.5.11. starts as low as 380 °C. Since it is merely a "lattice-rearrangement" with loss of water vapor- that is, the basic structure of the PO_4-tetrahedra shifts slightly- the overall crystal habit of the $Ca_2P_2O_7$ does not change much from the flat-platy form of the original $CaHPO_4$ crystal. However, grain boundaries can be noted in the fired product. If $CaCO_3$ is present, the reaction is that of "c." to form the tribasic calcium phosphate, i.e. ß - $Ca_3 (PO_4)_2$. This compound was identified by x-ray analysis. However, the "a." reaction takes place first because $CaCO_3$ does not react until a temperature above about 650 °C. is reached(12). By that time, the ß- $Ca_2P_2O_7$ crystal habit has already become stable. Thus, the CaO formed in Reaction "b." diffuses into the structure and changes the lattice structure, but not the outer crystal habit form. Note also that the crystals of the other components are not the same as the $CaHPO_4$ and the reaction between each one has little effect on the final crystal habit of the fired product.

This thermal behavior of the reacting components has important consequences concerning halophosphate phosphor preparation in that the crystal habit and form of the fired product can be controlled to a significant extent. As mentioned above, the properties of the particulate film on the inner surface of the glass tube of the fluorescent lamp plays a significant role in how the lamp performs, both in initial brightness and maintenance of that brightness over the life of the lamp. Indeed, by eliminating particles smaller than about 3.0 µ, the initial brightness was increased by over 10% and the output maintenence was significantly improved. Lamps are coated according to a "coating weight" which is specified in mg./cm^2 of surface. What is required is to cover the surface of the glass to form a particulate film of a particular depth. How the particles interlock on the surface is critical since too large a particle will leave open spaces between particles. A coating weight of about 6.5 mg./cm^2 is usual, but this depends upon the density of the crystals as well as their crystal habit. Although the diamond-shaped crystals will interlock to form a dense coating, the platy- square crystals have been observed to form a coating which has a better maintenence during the life of the lamp. If the particles are too small, the scattering of ultraviolet light generated by the mercury discharge increases radically and results in a lower brighness lamp. This is the reason for preparing crystals between the size limits given above in 3.5.10.

The above description has presented a method for producing $CaHPO_4$ on a continuous basis. If a batch method is required, a somewhat different method is used. In this process, the concentrations of both Ca^{2+} and PO_4^{3-} are lower and the conversion takes place directly in the mother liquor. Product yields range from 30 yo 95% of theoretical yield, based on calcium, depending mainly on the pH to which the dihydrate suspension is adjusted (conversion pH) before reheating the suspension to near boiling. It is possible to produce very highly crystalline, closely-sized, luminescent grade $CaHPO_4$ possessing characteristic form and habit.

1. <u>Calcium Chloride Solution = 110.99 MW</u> : The molarity of the solution used in this case is 0.333 mols/liter or 0.309 lbs/gallon of water. A 2000 gallon tank is filled with 1400 gallons of 20°C. water and 432 lbs. of $CaCl_2$ are added with stirring. 450 ml of $(NH_4)_2S_x$ is added to tank with stirring and the solution is pumped through the filter press and 0.45 µ filter into the holding tank where it is allowed to stand for 4 hours. Then, the solution is re-circulated through the filter press and the 0.45 µ filter until analysis indicates that the solution contains less than 1.0 ppm. of total impurities. The pure solution is then pumped into a $CaCl_2$ precipitation tank. Alternately, $CaNO_3$ may be used at 0.50 mols/ liter concentration.

2. <u>Diammonium Phosphate Solution = 132.056 MW:</u> For a 0.333 molar solution of Ca^{2+}, the phosphate solution needs to be 0.334 molar, a ratio of 1.003:1.000. Therefore, a 2000 gallon batch requires 300 gallons of 1.72 lbs/gallon for a total of 515.5 lbs of $(NH_4)_2HPO_4$. The procedure is as given above in 1. except that the batch has 400 ml of $(NH_4)_2S_x$ is added to the tank with stirring. Once the solution has been analyzed to be of sufficient purity, it is held prior to precipitation.

3. <u>Initial Precipitation of Brushite:</u> The $(NH_4)_2HPO_4$ solution is pumped into the $CaCl_2$ solution at 20 °C, with stirring at a rate of about 1.50 gallons/minute. The line-heaters are adjusted so that the solutions are heated to 25°C. and the tank-heater is adjusted to maintain this temperature. Once the precipitation tank is full, stirring continues.

4. Conversion of Brushite to Monetite Crystals: Hydrochloric acid is used to adjust the pH of the slurry to about 3.1 pH with stirring. The slurry is then heated to near boiling ($\approx$ 104 °C) for 10-15 minutes. The conversion can be judged to be complete when the pH of the slurry drops to a value lower than the original pH adjustment. Usually, this will be in the range of 2.4- 2.8 pH. The slurry is then pumped into the final drum filter where the wet cake is washed free of residuals and finally is dried in an oven at 125 °C. for 16 hours. The yield will be in the range of 75-85% of the theoretical yield.

5. Yields and Concentration of Solutions : It is possible to increase the initial concentrations of the solutions used for precipitation. For example, 1480 lb. of purified $CaCl_2$ is added to 1400 gallons of pure water and the volume is adjusted to 1600 gallons total volume. One can then add directly to this solution 1757 lbs of **solid** $(NH_4)_2HPO_4$ which dissolves and then causes brushite to form in solution. The temperature in this case is best held at about 30-35 °C. Once the precipitation is judged to be complete (the pH will rise above 7.00), the slurry continues to be stirred while the pH is adjusted to pH = 3.1 with HCl. (If $Ca(NO_3)_2$ is used, then one uses HNO_3 for pH adjustment). About 4000 ml. of HCl will be required. The tank temperature (and the slurry) is then increased to a boiling stage and held for about 1.0 hour, or until a characteristic drop in pH to 2.5-2.6 is observed. The total time required for heating and conversion is about 4.0 hours. The product in this case will consist of flat, platy diamond-like crystals of about 7-9 μ in size, i.e.- . Yield expected is about 1590 lbs. or 87.7% of theoretical.   If a truly diamond-shaped crystal is desired, i.e.- the procedure is modified as follows. The precipitaion of brushite proceeds as given above and stirring is continued for 1/2 hour before conversion. The brushite precipitate is then allowed to settle in its own mother liquor for about 2.0 hours or until the mother liquor above the settled presipitte is clear, meanwhile maintaining the temperature within the tank between 30 and 35 °C. Then approximately one-half of

the mother liquor is removed from the tank, taking care that none of the precipitate is lost during this process. The next step requires that the precipitate is resuspended by stirring and the conversion proceeds as given above. Obviously, less acid will be required to adjust the pH and the time needed to reach the boiling stage will also be less.

G. Manufacture of CdO_2 and ZnO_2

These materials are used in phosphor manufacture and are made by the following method. Either or both the cadmium and zinc salts can be used in this procedure.

1. $Cd(NO_3)_2$ Solution- MW = 308.46 : Dissolve 7934.6 gm. (17 lbs. 8 oz. = 25.72 mols) of $Cd(NO_3)_2 \cdot 4H_2O$ in approximately 3.0 gallon of cold deionized water. This solution is then gassed with NH_3 until the solution is clear (The solution first turns whitish due to formation of $Cd(OH)_2$ and then changes to the $Cd(NH_4)_4^{2+}$ ion, due to the reaction as given above. Add 39 grams of tannic acid to solution and let sit for 18 hours. Filter this solution through an Ertel pad and then through a 0.45 μ filter to remove all traces of the precipitate.

2. Precipitation of CdO_2 Using Hydrogen Peroxide

In this case, the $Cd(NH_4)_4^{2+}$ solution is not degassed as in the preparaton of the cadmium ammonium phosphate procedure given above. The solution is transferred to a 20 gallon glass-lined tank and diluted to a 6.0 gallon total volume. It is then cooled to a temperature **below** 15 °C and the 30% H_2O_2 is pumped in at a rate of about 125 ml. / minute until 5.0 gallon of peroxide has been added. A very fine precipitate will result. The solution mixture is allowed to stand overnight and the temperature will gradually rise to 85-90 °C. due to decomposition of the excess peroxide. In addition, the particle size of the CdO_2 will increase due to recrystallization to the point where it can be easily filtered. Some foam will be observed.

3. Washing of the Precipitate: Siphon off the mother liquor and wash the precipitate by re-suspension with 85-95 °C deionized water, filling the tank to the brim. Allow the precipitate to settle and repeat the sequence for five times.

4. Drying of the Precipitate: Filter off the precipitate and oven dry for 16 hours at 100 °C. Do not let temperature rise above this limit since loss of oxygen from the salt will occur. At the end of this time, remove from oven, let cool, record weight and yield, and place in tared plastic-lined drums.

The zinc salt may also be made by the same methods except that 5902.20 gms (13 lbs. 1 oz.) are used since the molecular weight of the zinc salt is: $Zn(NO_3)_2 \bullet 4\ H_2O$ - MW = 229.45

H. Manufacture of ZnS and $Zn_{1-x}Cd_x S_2$

These materials are used primarily in the manifacture of cathode-ray tube screens, including color-television. Two methods are commonly used to manufacture these materials, that of the chloride process and that of the sulfate process. We will address the latter method, although one should keep in mind that the other process is also being used. Although zinc sulfide phosphors were tried early on in fluorescent lamps, they were found to decompose under the influence of the 2527 Å radiation present in the low pressure mercury discharge. Even when used as a coating on the outer bulb of a high pressure mercury discharge lamp having 3650 Å radiation (and not directly exposed to the mercury discharge), these phosphors deteriorate fairly rapidly during the initial life of the lamp. Thus, their main use has been as cathode ray phosphors, and more recently as optical components for the infra-red region of the spectrum.

Typically, the following raw materials, as given in the following table, are required for manufacture of both of these materials, viz-

Raw Materials Required for Manufacture of Zn & Cd Sulfides

ZnO	NH_4OH	$NaSO_3$	Na Diethyl-Dithiocarbamate
CdO	HCl	CaO	DiMethylglyoxime
H_2S	$AgNO_3$	$CaCl_2$	$Na_2S_2O_3$
H_2SO_4	Cupferron	Ludox™	HNO_3
$CaCl_2$	$Cu(NO_3)_2$	$BaCl_2$	NaCl
HNO_3	$MgCl_2$	Ludox™	$Na_2S_2O_3$

In general, both ZnS and CdS are made by the same methods and use the same equipment given above in 3.5.3. Some manufacturers prefer to use a $ZnCl_2$ solution since the presence of Cl^- is known to be necessary in the final phosphor composition to obtain bright and efficient cathode-ray phosphors. Others prefer to use $ZnSO_4$ solutions in order to be able to control the amount of chloride ion added to form the final phosphor.

As shown in the following diagram, given as 3.5.13. on the next page, the procedure involves use of ZnO and sulfuric acid. The steps involved are:

1. Dissolve ZnO in H_2SO_4 to form solution of $ZnSO_4$ (4 hrs).

2. Purify solution by addition of Dimethylglyoxime - let sit 8 hours, filter out sludge (8 hrs.).

3. Repeat purification with Cupferron- filter out sludge (16-20 hrs.).

4. Repeat purification with Sodium Diethyl-Dithiocarbamate - filter out sludge (6-20 hrs.).

5. Analyze solution for approval (4 hrs.).

6. Precipitate ZnS by H_2S gassing for 8 hours under pressure ($\approx$ 15 psi.) while stirring slowly.

7. Let settle and decant mother liquor (2 hrs).

3.5.13.- | Process For Precipitation of ZnS |

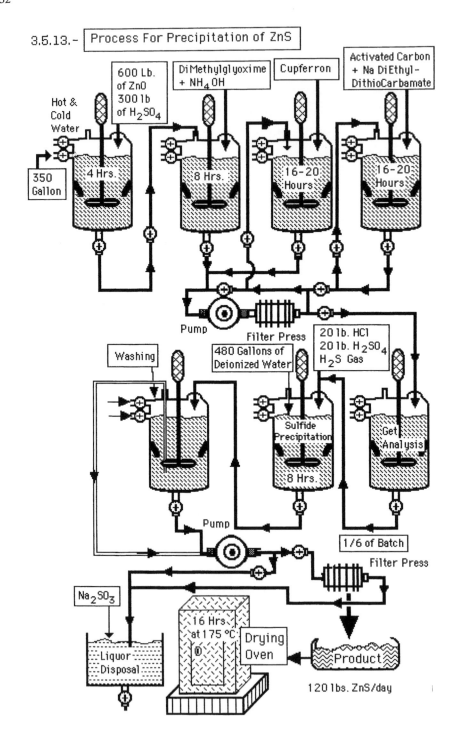

8. Wash precipitate 4-5 times with hot deionized water, letting settle in between washes and decanting supernatent liquid (8-12 hrs.).

9. Treat **all** mother liquor and wash water with sodium sulfite to oxidize sulfide to thiosulfate.

10. Pump sulfide precipitate slurry into drum filter to separate.

11. Filter and dry the wet cake obtained for 16 hours @ 175 °C.

12. The same procedure applies for CdS preparation.

Note that the original volume of the purified solution is about 360 gallons and that 1/6 of this is diluted to 480 gallons before sulfide precipitation takes place. This means that 6 sulfide batches can be made from the original purified batch and that another batch can be started and purified while precipitation is taking place. Times required to complete each step are also given in the above. About 120 lbs of ZnS can be prepared each working day.

The sludges obtained during each separate purification step are shown in the following:

Reagent Used	Ions Precipitated
DiMethylglyoxime	Ni^{2+} , Ni^{3+}
Cupferron	Cu^{2+}
Na Diethyl-Dithiocarbamate	Fe^{3+}, Co^{2+}, Pb^{2+}, Mn^{2+}

The above procedure represents but one way to prepare ZnS and CdS. Some manufacturers prefer to make various combinations of Zn^{2+} and Cd^{2+} solutions and coprecipitate $Zn_{1-x}Cd_xS_2$. Others prefer to make each separately and then combine them before firing to form the phosphor. Still others will add $AgNO_3$ and co-precipitate all three as $Zn_{1-x-y}Cd_x S_2: Ag_y$ (where Ag_y is the activator). It should be clear, then, that each manufacturer has his own special method of manufacture but that the products are very similar. Although 3.5.13. shows a tank for each separate step, this is not necessary.

354

In the following diagram, we show a typical equipment arrangement which is sufficient for manufacture of both ZnS and CdS, viz-

3.5.14.-

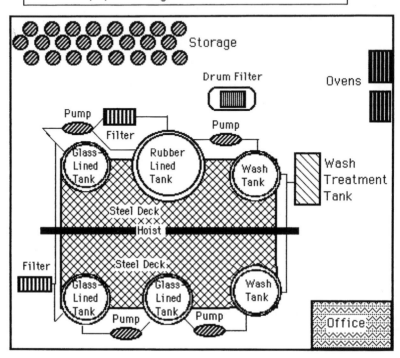

The actual piping-hookup is not shown here. There are three glass-lined tanks which are used for purification of the solutions and precipitation of the sulfide and one rubber-lined tank which is used for acid dissolution steps. Sometimes, the rubber-lined tank is also used for sulfide precipitation. Usually, a drum filter is used for the product and the filter presses are reserved for the impurity sludges produced.

This completes the description of manufacture of raw materials used to prepare phosphors for use in both fluorescent lamps and cathode-ray tubes. Although we could discuss the preparation of YVO_4: Eu^{3+} as a raw material by precipitation methods, we prefer to wait until the next chapter so as to be able to compare the product of the solid state method to that of the precipitation method.

3.6. - MANUFACTURE OF FLUORESCENT LAMPS

Before we turn to the manufacture of phosphors in the next Chapter, we need to discuss briefly how fluorescent lamps are made. There are many varieties of such lamps, including differences in size, wattage and shape. The manufacture of all of these lamps takes place on highly automated machines which have a through-put of thousands of individual lamps on a daily basis. We do not intend to discuss the design of such machines since this could be the subject of a separate and lengthy monograph. Most fluorescent lamp manufacturers produce between two to six million lamps per month, or about 90,000 to 272,000 lamps daily. This rate is about 4,000 to 11,000 lamps per hour on a 24 hour basis, or 1 to 3 per second on a continuous basis. Such productivity is obviously accomplished through the use of more than one manufacturing and finishing line.

The steps involved in the manufacture of fluorescent lamps can be delineated as follows:

A. Washing Fluorescent Lamp Bulbs
B. Preparation of "Phosphor Paint"
C. Bulb Coating and Drying
D. Lehring
E. Manufacture of Flared Stems
F. Stem and Flare Mounting
G. Exhausting of Lamps
H. Mercury Dosing and "Tipoff".
I. Base Mounting
J. Lamp Testing

Each of these steps require a particular protocol which has been developed by each Manufacturer according to his own needs and desires. Our description will assess these steps in a general manner without trying to include all of the various approaches that have been made for each individual process.

The following diagram, given as 3.6.1. on the next page, shows a flow-chart of each of these individual steps in the manufacture of fluorescent lamps. Again, as we have stated, this represents just one way that such lamps can be manufactured.

356

3.6.1.-

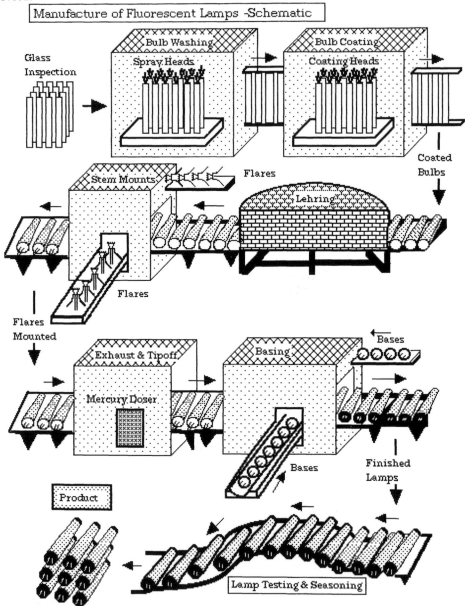

Manufacture of Fluorescent Lamps -Schematic

Glass Inspection

Bulb Washing
Spray Heads

Bulb Coating
Coating Heads

Coated Bulbs

Stem Mounts

Flares

Lehring

Flares

Flares Mounted

Exhaust & Tipoff

Mercury Doser

Basing

Bases

Bases

Finished Lamps

Product

Lamp Testing & Seasoning

A. Washing Fluorescent Lamp Bulbs

In general, the glass bulbs are visually inspected to determine amount of breakage present, if any. Any bulbs with end-cracks are obviously rejected. The quality of glass can be easily checked by running a bare bulb through the Lehr, along with the coated bulbs. The bare bulb is then retrieved and is washed internally with a dilute solution of HCl. The wash solution is then analyzed for Na^+ content. If an inferior glass batch has been used to form the tubing, sodium ions internal to the glass structure will diffuse out from the interior to the surface when the glass tube is subjected to the Lehr temperature of about 650 °C. By analyzing the amount of Na^+ recovered, one can obtain an empirical result concerning the relative quality of the glass. The amount of Na^+ found in the wash water should not exceed about 0.05 mg/cm^2 of glass surface.

By far the greatest volume of the fluorescent lamps manufactured involve either 48 inch or 96 inch long glass tubes. They are generally received with "shoulder-formed" ends, viz-

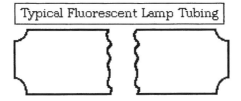

Typical Fluorescent Lamp Tubing

They are manually placed in a vertical position within a "washing" machine where they progress from station to station while hot water is sprayed downward internally to wash the surface clean of any particles present. This is followed by an upward hot water spray which allows the wash water to drip from the bottom of the tubing. Sometimes, an acidic wash is used as a next step in which about 2% by weight of HCl and 1% HF by weight is present in the wash water. A final rinse is then necessary. Hot air jets are then used to dry the bulbs before they emerge from the washer to be cooled to room temperature by a blower. All these operations take place within a closed space so that outside contamination by dust or other particles cannot occur.

It is important to use clean, filtered water as well as filtered air for the washing and drying processes. The temperature of the wash water needs to be controlled at about 180 °F. and the presence of algae is discouraged by the use of an anionic detergent present in the wash water, which is filtered and then recirculated through the washing machine on a continuous basis. Once the clean bulbs exit the washing machine, they are ready to be coated, or "painted". In this process, a "paint" is first prepared by forming a suspension of phosphor in one of several vehicles.

B. Preparation of Phosphor Paint

A typical organic-based "paint" consists of an ethylcellulose base dissolved in a mixture of xylol and butanol. It is prepared by mixing the following ingredients:

1. Organic Solvent Based Paints :

a.- A volume of n-butanol and xylol (xylol is a mixture of xylenes, i.e.- para-, meta- and ortho- dimethylbenzenes) is prepared consisting of about 60% butanol and 40% xylol. To this is added about 5.0% by weight of ethylcellulose and about 2.5% by weight of dibutyl phthalate (the latter is a plasticizer). This mixture is stirred and rolled about 24 hours until the lacquer is smooth and all of the solids have dissolved. The lacquer is then filtered to remove any remaining particles. This lacquer can also be purchased from:

> Pierce & Stevens Chemical Corp.
> 724 Ohio St.
> Buffalo, NY 14203

> o r   Raffi & Swanson
> Wilmington, Mass. 01887

b. The "paint" is prepared by placing 1.0 gallon of the ethylcellulose lacquer in a 4-gallon mill containing about 28 lbs. of 3/4 inch "French-flint" pebbles (contact Paul O. Abbe, Little Falls, NJ for details). To this is added 50.0 ml. of Armeen CD™ (which is a primary aliphatic amine which acts as a dispersing agent and is available from the Armour Co., Chicago, Ill.). Add 12.0 lbs. of phosphor and run mill for 1.0 hour. Check suspension for texture and adherence by dip-coating a glass slide and letting it dry at room temperature. Mill for 15 minutes longer if necessary, as judged by phosphor particles which may protrude from the surface of the film. Ultrasonic dispersion is also an accepted practice.

c. The viscosity and specific gravity of the paint must be adjusted to obtain the desired density of coating as well as uniformity of density from end to end. Lacquer lowers specific gravity but raises viscosity, while xylol lowers both viscosity and specific gravity. Viscosity is measured by a Zahn-cup (usually #3), specific gravity by use of hydrometers, while coating density is measured by light transmission orthoganal to the axis of the glass tube. Usually, the paint must be adjusted so as to double the original milled volume to give a viscosity of about 15-30 Zahn-seconds, at a specific gravity of about 1.100-1.400. These values may depend somewhat upon the ambient conditions of humidity and temperature.

An optimum value useful for both Cool White and other colors has been found to be:

$$28 \text{ Zahn-seconds}$$
$$1.320 \text{ Sp. G.}$$

These are values for the physical parameters of the paint used to coat both 40T12 and 96T12 lamps (Such lamp designations are based on first the wattage, configuration and diameter of the lamps, i.e. 40 = 40 watts, T = tubular, & 12 = 12/8 inch in diameter. The actual lengths needed for these wattages are: 48 inches and 98 inches, respectively). A phosphor coating weight of about 7.5 mg/cm^2 will result after the lamp has been lehred.

d. It is important to minimize evaporation losses and to avoid operator exposure to fumes whenever possible. An exhaust system is to be used at all times in the paint preparation process. Armeen CD is corrosive to the skin and must be washed off with vinegar immediately. Since the paint is very flammable, all sources of flame and sparks must be carefully avoided. All sources of dirt-contamination must also be avoided that might cause damage to the phosphor coating as application of the paint proceeds.

e. In another ethylcellulose paint variation, the following ingredients have been used:

Xylol	= 78.0 gallons
Ethylcellulose Lacquer	= 28.0 gallons
Butanol	= 24.0 gallons
Armeen CD	= 2.50 gallons
Alon-C	= 15.5 lbs
Phosphor	= 1500 lbs.

This mixture is milled for 8.0 hours and the viscosity is then adjusted as given above. Alternately, ultrasonic mixing and dispersion may be used in place of the milling step.

Because of the flammability and possible damage to human respiratory systems, **water-based paints** have been more recently used. Preparation of such paints involves the following:

2. Water Based Paints

a. Polymethacrylate Lacquer

i. To make this lacquer, 300 liters of pure water and 15.0 liters of an ammonium polymethacrylate solution (~ 7% solids by weight), Vulcastab-T™ (Code 05971 from ICI Industries, 33 Brazennoze St., Manchester, England) are fed into a reaction tank. The basic solution is steam-heated until about 150 liters are left in the tank and/or the pH reaches pH = 7.00. Boiling is continued longer if the pH remains above this limit. Water is then added to make a total of 300 liters of solution, which is then pumped through a filter to remove insolubles and then into a storage tank. The reaction tank is cleaned by stirring 50.0 liters of wash water to clean it, and this volume is added to the storage tenk. Once the solution is cool, its viscosity is measured and water is added to adjust it into an acceptable range. This procedure usually results in a total volume of around 425-450 liters of premix. To make the paint, 40 liters of premix is mixed with 80.0 kilogram of phosphor in an 80 liter tank for 15 minutes. This volume is then moved to a 400 liter ball-mill containing 120 kilograms of flint stones, and the mill is rolled at 18.5

rpm. for 1.0 hour. 40.0 liters of water are then added and milling is continued for 2.0 minutes. The paint is then pumped into a reservoir and is ready for use. A wetting agent, called Lissapol NXA™ (Code 06544 - available from ICI Industries, England) is automatically added as 1.0 drop per 5.00 liters of dispensed paint as it is drawn from the paint tank.

ii. Another method for preparing the same type of water-based paint involves the following procedure. A 7% solution of Vulcastab-T™ (Code 05971) is mixed with pure water at a rate of 6.25 gallons and diluted at a 1:7 ratio with water to obtain a 1.0% solution. At this point, the viscosity is about 50 seconds using a DuPont #70 cup. About 1.0 cc. of a wetting agent, Supronic B75/25 (available from Glovers Chemical Ltd., Wortley Low Mills, Whirehall Road, Leeds, England) is added (having been diluted 150/1 with water before use to about a 0.67% concentration). This premix solution is then ready for use.

A phosphor slurry is prepared by placing 50.0 kilogram of phosphor in a 50 gallon tank along with 7.5 kilogram of citric acid crystals, 15 liters of a 10% hydrogen peroxide and enough water to fill the tank. This mixture is stirred to disperse the phosphor and then is filtered to remove the phosphor which remains as a wet cake (which has been acid-washed). The filtration product is resuspended at a concentration of 1.00 gram per 1.00 ml. of water to form the slurry, along with an adherence agent such as barium calium borate (about 10 grams per liter of final volume). This suspension is then mixed at an equal volume ratio with the premix solution. It is this final paint suspension which is then used for coating of the lamps.

iii. Note that in the two methods given above for use of polymethacrylate lacquer, the first neutralizes the system by boiling off the ammonia present, while the second method uses an acid wash of the phosphor to present an acidic surface to the basic ammonium polymethacrylate in order to obtain a neutral paint suspension.

iv. In most of these paints, adherence agents are usually added to improve the coherence and sticking of the phosphor layer to the glass

362

after lehring. These agents have generally consisted of one of the following:

Alon-C (which is a 300 Å -sized alumina, Al_2O_3)
Barium calcium borate
Titania, TiO_2 (micron-sized)
Diammonium phosphate, $(NH_4)_2HPO_4$ (which is soluble in water-based paint)

One of these agents is usually added prior to the final milling of the paint. Ultrasonic dispersion is also used by some Manufacturers.

b. Hydroxypropylmethylcellulose Paint

Another paint used has been based upon Methocel™- HG grade, a modified cellulose available from Dow Chemical Co. The paint is prepared by making a stock solution consisting of 420 gms. of Methocel-HG to 100 liters of pure water. This initial solution is rolled for about 1-2 hours until a clear solution is obtained. This is then filtered to remove any gel particles that may have formed. Then, 5500 ml. of the solution is diluted with 4500 ml. of water to which has been added 10 cc. of Igepol™ CO610, a dispersing agent (also available from Dow Chemical Co.) and 20 gram of any dye to be added (It is quite common to identify various phosphor colors in specific lamps by adding a soluble organic dye, which remains in the paint until lehring is done to bakeout the organics in the paint), to form the final composition of the premix.

i. To the 10 liters of premix is added 7.00 kilograms of phosphor, 100 gm. of Alon-C, and 10.0 ml. of Igepol CO610. This suspension is stirred and then rolled for about 90 minutes. Following this, the suspension is then ready for use as a paint.

ii. In another variation of this procedure, Carboxymethylcellulose is substituted for the Hydroxypropylmethylcellulose used in the paint making process.

C. Coating of Lamps

The objective in the coating operation is to obtain a uniform internal coating of the paint containing the phosphor particles by "squirting" a predetermined volume of paint onto the inner surface of the verticle glass tube and allowing it to run down the sides. It is essential that the coating be uniform from the top to the bottom of the glass tubing. Also, the "density of coating" needs to be uniform from end to end of the finished lamp. It is the density of coating, or the "powder weight" which determines the final quality of the lamp as it operates throughout its rated lifetime of operation. Most manufacturers strive to obtain a powder weight for Cool White Halophosphate phosphor between 6.00 to 8.00 milligrams of powder per square centimeter of internal glass surface, with 6.8 to 7.5 mg/cm^2 being the average. This, of course, depends upon the crystal density of the phosphor being used, as noted elsewhere (e.g.- see "Manufacture of $Cd_5Cl(PO_4)_3$:Mn Phosphor" in the next chapter).

The coating chamber consists of a continuous moving rack in which the clean, vertical glass tubes move through the chamber. One of the first positions heats the tubing to about 75-80 °F by use of a heated air stream (about 1000 cfm. for a 96 inch tubing), at a controlled humidity index level so as to dry, or keep dry, the internal surface of the glass. A dehumidifier is generally used in this room. At the next position, the paint is squirted at the exact top of the tubing and is allowed to run down the sides of the tubing so as to completely cover the inside of the tubing with the paint. The excess drips from the end of the tube and is recaptured and sent back to the paint position. One of the biggest problems encountered is the production of "thin-ends" at the bottom of the tubing, especially if 96 inch tubing is being coated. Many manufacturers have solved this problem by "top-coating" using a specific volume of paint which runs downwards and also "bottom-coating" by squirting a smaller volume of paint upwards from the bottom so that the two volumes meet. During coating, it is essential to keep the temperature of the lacquer constant (even though it is being stirred constantly and recirculated from drippings of the tubes) so as to obtain reproducible coatings.

The next position is called a "cold-set" position since the air temperature is lowered to about 60 °F so as to stop most of the paint flow at this point. At this point, the coated tubes exit the coating chamber where they proceed to the Lehr.

D. Lehring of Lamps

The coated-bulbs next proceed to the entrance of the Lehr (which is nothing more than a low temperature furnace equipped to take glass tubing. The tubes rotate around to a horizontal position just before they enter the Lehr, and are caught on each end by a series of rollers which support and spin the glass tubes as they pass through the Lehr. One end-roller has an air-jet positioned which blows air at a set velocity through the tube as the organic binder gets hot enough to burn out, leaving behind just an adherent coating of phosphor. At the front end of the Lehr, the tubes are kept apart as they pass into the interior of the Lehr. The maximum temperature that the coated glass tubes reach is about 650 °C., a temperature high enough to cause the glass to sag if they were not rotating continuously.

In some Lehr designs, the tubes are allowed to roll, caught by their ends, down a slight incline built within the lehr, kept apart by means of asbestos or other ceramic flexible strips so that the tubes do not touch each other (if they did happen to touch one another, "microcracks" would develop in the glass wall of the tube and the tube would likely break somewhere further in the process). The air-jet velocity starts at about 600 cfm. and advances to about 2000 cfm. at the point where the burning process is most vigorous. In the case where rollers are used in the Lehr, the air-jets move with the tubes, whereas in the rolling tube version, the air-jets are stationary.

The lehring operation causes complete "burn-out" of the organic binder from the phosphor when the inside temperature reaches about 600 °C. It is important that no carbon particles are left behind during burn-out. It is for this reason that the air-jet velocity is varied to remove the combustion products from the tubes. The "smoke" and fumes must be directed to an air-scrubber to protect operating personnel. For a complete burn-out, the temperature profile inside the Lehr must be carefully monitored and controlled. This is usually accomplished by a set-point recorder which continuously monitors temperature through a series of thermocouples strategically placed within the Lehr.

Once the coated bulbs reach the end of the Lehr, they exit and are allowed to cool while moving on a carrier-belt to the next step of the process.

E. Sealing Machine Operation

Once the coated bulbs exit the Lehr, air-jets are played across the outer surfaces to cool the glass bulbs. The bulbs are picked up by a series of rollers while still in a horizontal position and fed into an "end-sealing" machine. At the first station of this high-speed machine, two brushes, one on each end, are used to remove excess phosphor about 1/4 inch from the end so that the ends can be sealed to form the basic lamp (It is important to use a vacuum to pick up any excess powder that results from this operation since the excess may cause damage to the phosphor coating during subsequent operations if left inside of the lamp). Once this is done, the bulbs can have a stem attached to each end, to form the basic lamp. It is important to have the vacuum system to pick up the powder removed by brushing since many final rejects can result if this operation is not done properly.

1. Stem Manufacture

Fluorescent lamp stems are comprised of a flare with a pressed-stem sealed within, and contain a coiled-coil filament very similar to that used in incandescent lamps. Indeed, these stems are made in a like manner. The procedure is as follows:

a. A piece of glass tubing is flared on one-end. The flare must be just slightly larger than the end of the glass tubing used to form the fluorescent lamp.

b. It is then positioned with the flared end up and two lead-in wires, as fed through placement-cylinders, are aligned inside the flared-glass and come to rest in positioning guides.

c. A piece of exhaust tubing (about 3/16 inch diameter) is also placed inside the flared glass and forced to a stop at about one half of its length. This is done for only one of the stems used in the fluorescent lamp, i.e.- for 1/2 of the stems.

d. Filament support wires with properly placed "Dumet" connectors are then inserted into this assembly which is then heated to soften the glass. Thereupon, clamps move in and form a flat solid section called the "press" which now contains the supports and lead-in wires sealed into the glass.

e. Just before the "press" solidifies, and a puff of air through the exhaust tube provides a hole in the "press" through which the lamp can be evacuated of air at a later stage of manufacture of the fluorescent lamp.

f. The top ends of the two molybdenum lead-in wires are shaped and separated at a distance of about a millimeter **less** than the length of the tungsten filament coil.

g. The coil is lifted into position just as the glass-lead wire assembly arrives opposite to it and the coil is mechanically crimped to the two lead-in wires at each end of the filament. The coils will have already been coated with an emission mixture of carbonates.

h. To coat the filaments, it is necessary to make up a lacquer containing the emission mixture. The most usual electrode coating composition is:

45.60% $BaCO_3$

9.025% $CaCO_3$

40.375% $SrCO_3$

5.00 % ZrO_2

The lacquer is made up by mixing 5.0 gallons of amyl acetate with 2.0 gallons of ethyl acetate and 1.0 gallon of a nitrocellulose binder, S-1667 which contains 5.7% solids (available from Raffi & Swanson). This mixture is stirred together and filtered to remove any gel particles. To this volume is added about 16.0 lbs of the above mixture of carbonates, and the miz is milled for 24 hours at about 170 rpm. in a 10 gallon mill containing 1/2 inch flint pebbles.

i. The emission mixture lacquer is then adjusted with amyl acetate to a viscosity and specfic gravity between:

20-30 centipoises and 1.700 to 1.780 Sp.G.

depending upon the type of fluorescent lamp being manufactured.

2. Stem Mounting

Two versions of the completed stem assembly are shown in the following diagram. One stem is used as the exhaust port while the other forms a seal at the other end of the tube, viz-

3.6.2.- Stem Designs used for Mounting to Form Fluorescent Lamps

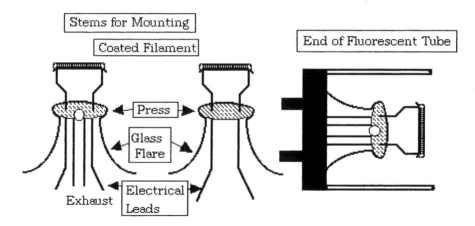

Just as in incandescent lamp manufacture, the ends of the coated tubes (now cleaned of powder left from the coating operation) are heated until the glass is softened (about 650 °C.) Stems are fed into the sealing machine from each end, one having an exhaust tubing already attached, and the other without. Each stem is belt-mounted and auto-indexed to co-incide with the end of each coated tube as it progresses through the stem-sealing machine. The flare of each stem comes into contact with the tube end while the tube is being heated first by a flame. Then the flare is heated (since it has thinner walls than the tubing). As both become hot and have softened, they are pressed together by machine action to form the sealed glass joint. The flame-sealing operation is then adjusted to a lower temperature so as to give an annealing effect to the glass joint. This operation is critical since "leakers" can result if the glass-seal is not perfect. The seals are then allowed to come to room temperature while each lamp proceeds toward the Exhaust machine, meanwhile being rotated as the seals cool.

F. Exhaust Machine Operation

At this point the coated tubing now begins to look like a fluorescent lamp. At the exhaust machine, each lamp is exhausted down to a 0.05 micron vacuum. The lamp is then refilled with argon or nitrogen gas to atmospheric pressure and this cycle is usually repeated at least twice. On the third pump-down, the lamp enters an oven set at 350-400 °C. where degassing of the lamp occurs. Another pumpdown then occurs and mercury is introduced into the lamp. The amount is usually about 70.0 milligrams for a 40T12 lamp and varies more or less, depending upon the lamp size. The lamp then exits the oven and is subjected to an increasing series of voltages, applied through the electrical leads. These voltages start at 0.50 volts AC and are increased by 1.5 volt stages so as to break down the emission mixture when the filament becomes heated above 1200 °C. All of these steps take place as the lamp progresses from stage to stage. Near the end of the exhaust cycle, the lamp is again filled with argon to atmospheric pressure. All of these evacuations take place through the exhaust part of the stem. Finally, the internal pressure of the lamp is adjusted to about 2.3-2.5 mm. of argon, before the exhaust tube is tipped off by heating in a flame. The lamps now proceed through a basing machine cycle.

G. Basing and Seasoning Machine Operation

This final part of the operation is very similar to that of the incandescent lamp basing operation.

> 1. Bases are filled automatically with a thermosetting cement on a circular turret. The cement, viscous in form, is forced under pressure through an orifice that deposits it around the inner periphery of the base just below the start of the threaded section.

> 2. The basing reel consists of a large circular rotatable turret with an automatic vertical positioning mechanism equally spaced

around the circumference. Directly above these is a mechanism for holding the base-bulb assembly in a horizontal position.

3. The threading of one lead wire through one pin of the base and the other around the other pin is next accomplished by the machine operation.

4. Flames are positioned so as to provide varying degrees of heating as the lamp rotates so as to cure the cement.

5. The lamp indexes circumferentially so that several sequential operations occur. These consist of cutting off excess of the lead wires, and soldering or welding the lead wires to the side of the base by application of solder to the joint.

Once this is done, the lamps proceed to a "starting" turret where they proceed around the turret while a voltage is applied to start the lamp. This usually requires an overvoltage to do so on the first start. The lamp operates for about 60 seconds, is cut off, and then restarted. After this cycle repeats itself about three times, the lamp is allowed to operate about 5 minutes before it is stopped. finally, the lamp proceeds to a storage area where it is boxed for shipment.

There are several causes for rejection of the finished lamps. These include:

"Non-starters"
"Leakers"
Coating Defects
Breakage due to handling after manufacture

Sometimes, jam-ups can occur in the Factory-Line and this too causes breakage and loss of production. Quality checks involve both visual and photometric evaluation. To do the latter, one requires a photometric sphere in which the lamp is placed and the total luminous output is directly measured after 1.0 hour of operation. This is called "zero- hour" lumens. The test lamp is then operated for 100 hours longer to obtain an output in lumens. This luminosity is generally about 95% of the initial zero-hour value.

H. Types of Fluorescent Lamps Manufactured

The list of fluorescent lamps that have been manufactured is long indeed. We will delineate some of these in order to separate them into lamp types and their electrical operating circuits. Such lamps include:

Preheat lamps

Rapid start lamps

Instant start lamps

High output lamps

Very high output lamps

The differences in these lamps lies primarily in the electrical contacts and the power used in the lamp. The following diagram, given as 3.6.3. on the next page, shows some of the electrical connections of ballasts used for various types of fluorescent lamps.

Note that the difference lies in how the ballast is connected to the power Line, as well as the number of individual chokes and coils.

In the Preheat Start, the filaments are heated to ionize the mercury and then the voltage is switched across the lamp to start the lamp and its attendent mercury vapor arc-stream.

In a Rapid Start lamp, the electrodes are also preheated to start the arc, and two contacts at each end are required for the heating circuit. However, there is no switch and a small heating current flows through the elctrodes while the lamp is operating.

Instant Start lamps are started directly by application of sufficient high-voltage to strike the arc without any preheating of the electrodes. Thus, Instart Start lamps require only one contact on each end to start and operate the lamp (see 3.6.3. on the next page).

3.6.3.-

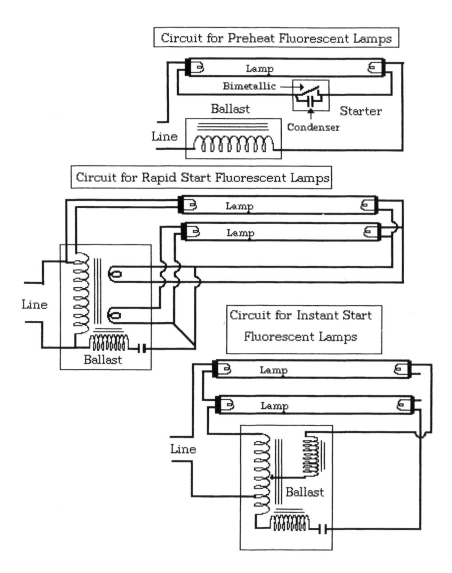

High Output (HO & VHO) lamps involve a special electrode design so that more power can be used to operate the lamp. Whereas the ordinary 96T12 lamp uses 97.7 watts (63 VAC and 1.550 amps.), the 96T12HO lamp uses 120 watts (150 VAC and 0.800 amps). The 96T12VHO lamp uses 258 watts (172 VAC and 1.500 amps). This is achieved by use of the following type of filament mount (two to a lamp), viz-

Electrode Design for VHO Lamps

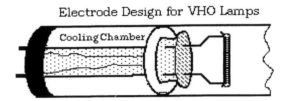

Note that a "cooling chamber" is present behind the filament electrode. This prevents the mercury vapor pressure from becoming too high during operation at these high wattages. Also, a special mixture of rare gases is used to provide long life for the electrodes and to give good luminous output. A comparison of these lamps is shown in the following:

Lamp Type	Wattage	Rated Lumens
96T12	97.7	5,560
96T12 HO	120.0	8,500
96T12 VHO	258.0	15,000

These values are typical of these types of lamps.

Most of the VHO type of lamps areused for industrial lighting or in Supermarkets and the like. 40T12 lamps are used in Restaurants and private homes. In general, no VHO varieties are made in the 40 watt range of fluorescent lamp.

REFERENCES CITED

1. A P Valentine and D Hull, *J. Less Common Metals*, **17** 353-361 (1969).

2. R.A. Swalin, Private Communication (1967).

3. CE Ells and W Evans, "The Effect of Temperature During Irradiation on the Behavior of Helium on Beryllium", *AECL Report #898*, Chalk River, Ontario, Canada (Oct.- 1959).

4. CE Ells, *Acta Met.* **11** 87-96 (1963).

5. C Agfe & J Vacek, *Tungsten and Molybdenum*, NASA Transl. - Washington, DC (1963).

6. P.E. Wretblad, *Powder Metallury*, Publ.- Amer. Soc. Metals-J Wulff -Ed., p. 433 (1942).

7. Published in *Ceramic Source '86* Vol. **1** p.303- compiled by the Amer. Cer. Soc., Inc. (1986)

8. R C Ropp, *"Luminescence and the Solid State"*, Elsevier Sci. Publ., New York & Amsterdam, (April 1991)

9. MA Aia, USP 3,113,835 (1963).

10. "X-ray Powder Diffraction Patterns for Some Cadmium Phosphates", by R.C. Ropp, R.W. Mooney and C.W.W. Hoffman, *Anal. Chem.*, **33**, 1240 (1961).

11. WL Wanmaker, AB Hoekstra, and MG Tak, *Philips Res. Rpts.* **10** 11 (1955).

12. "Phosphor Industry and X-ray Diffraction Analysis" by C.W.W. Hoffman and R. C. Ropp, *Encyclopedia of X-Rays and Gamma Rays* - Ed. by *H. Clark*, Reinhold Publ. Co., New York (1963).

Chapter 4

Chapter 4

MANUFACTURE OF PHOSPHORS USED IN MERCURY DISCHARGE LAMPS

In this Chapter, we will survey the commercial and manufacturing protocols for preparation of phosphors used in fluorescent lamps and in the next chapter will present methods of preparation of phosphors for cathode-ray tubes. To do so, we need to discuss briefly some of the background theory concerning phosphors, including the principles which govern luminescence.

4.1.- <u>A Theory of Luminescence</u>

Although luminescence has been known in both organic and inorganic systems for more than 100 years, a thoroughly descriptive account of the collective mechanisms that prevail has not been made. Perhaps, this is due to the significant disparities that exist between the two systems. For example, organic phosphors can luminesce even when dissolved in a solvent whereas the inorganic ones cannot. Furthermore, organic phosphors involve specific molecules while inorganic phosphors involve certain lattice structures. Nevertheless, it should be obvious that the luminescence mechanisms operating in organic compounds embody the same principles as those that have been recognized for many years for the earlier known inorganic phosphors. More explicitly, it is likely that organic phosphors have "activator centers" quite similar to those already defined for inorganic phosphors. However, the theory usually quoted for organic phosphors deals with "singlet" and "triplet" excited states of the organic molecule. While these states undoubtedly do make a major contribution to the electronic transitions occurring during the emission process, they are not the "raison d' etre" for the mechanism occurring. In inorganic compounds, it is necessary to create an "activator-center" capable of emitting photons without loss of energy to other competing processes within the lattice. However, in organic compounds which do emit photons when stimulated by ultraviolet light, the nature of the emitting center has never been completely defined. For this reason, we intend to outline a reasonable description of the processes occurring during luminescence using inorganic compounds as the example. However, one should keep in mind that the same type of processes undoubtedly occur in the organic phosphors as well.

The typical inorganic phosphor consists of a cation combined with an anion to form a given compound. The cations occupy certain lattice sites in the crystal structure and their total charge exactly equals that of the anions which occupy other sites within the same crystal lattice structure. To obtain an inorganic phosphor, one must substitute a different, optically-active, cation for the lattice-cation within the structure. Similarly, it is possible to substitute an optically-active anion for the lattice-anion.

In a prior work (1), we presented the cations, anions and activators which can be used to form inorganic phosphors, and which are capable of absorbing electromagnetic energy and transforming that energy into visible light at room temperature. This is the definition of a "phosphor". To facilitate our discussion, the following diagrams are repeated from the earlier work:

4.1.1.-

The Periodic Table as Related to Phosphor Composition

H	(2+)	Cations That Can Be Used to Form Phosphors							(3+) (4+)			He
Li^+	Mg											Ne
Na^+	Mg	(3+) (4+)						(2+)				Ar
K^+	Ca	Sc	Ti				Zn	Ga	Ge			Kr
Rb^+	Sr	Y	Zr				Cd	In	Sn			Xe
Cs^+	Ba	La	Hf				Hg	Tl	Pb			Rn
Fr^+	Ra	Ac	104									

La^{3+}				Gd^{3+}				Lu^{3+}
Ac^{3+}				Cm^{3+}				Lw^{3+}

Note that these cations are those that possess closed electron-shells and so are **not** optically active. Furthermore, these cations must be in the valence state shown above when the final crystal is formed, or the electron shell will have unpaired spins and will not be a closed shell. For example, the +4 cations, i.e.- M^{4+}, also have +2 states. As we have pointed out before, the cations, Ti^{2+} , Zr^{2+} , Hf^{2+} , Ge^{2+} , Sn^{2+} , and Pb^{2+} will not function solely as cations in our phosphor since they are optically active (they are in fact electronic states of the "activator" class). If they compose 100% of cations present in the structure, no luminescence will be forthcoming because of quenching due to intra-center interactions with the lattice processes. We have already pointed out that it should be possible to prepare the phosphor: Sn^{4+} SiO_4 : Sn^{2+} (1), if Sn^{2+} is maintained below 5 mol%.

The same rules also apply to anions which can be utilized to form phosphors. That is, they must also be optically transparent if the "activator" is to function properly. Those that can be utilized to produce phosphors are shown in the following diagram:

4.1.2.-

The Periodic Table as Related to Phosphor Composition

H		Anions that Can be Used to Form Phosphors			(3-)	(4-)	(3-)	(2-)	(1-)	He
					BO_3				F	Ne
					AlO_3	SiO_4	PO_4	SO_4	Cl	Ar
					GaO_3	GeO_4	AsO_4	SeO_4	Br	Kr
					InO_3	SnO_4	SbO_4	TeO_4	I	Xe
	La					PbO_4	BiO_4			Rn
	Ac	104								

La																
Ac																

Since a wide variety of anionic groups of atoms can exist in the solid state, the above diagram only shows one of a class of anions, e.g.- SiO_4 represents all of the known silicates including $Si_2O_7^{6-}$, $Si_3O_9^{6-}$, SiO_3^-, $Si_4O_{11}^{5-}$ etc. The same is true for the other anions shown above.

We will not dwell upon the anions which also function as activators except to say that they must be combined with cations which are optically inert to form an efficient phosphor. A good example is $MgWO_4$:W, where W indicates the tungstate, WO_4^{2-}, which functions both as the anion and the activator, i.e.- the phosphor is "self-activated" (However, the charge on the optically active central ion, i.e.- W in this case, may not be the same as the rest of those comprising the lattice, i.e.- W^{+6}). Self-activating anions are shown in the following diagram:

4.1.3.-

The Periodic Table as Related to Phosphor Composition

H		Anions That Are Optically Active - "Self-Activation"															He
																	Ne
		(4-) (3-) (2-) (1-)															Ar
		TiO$_4$ VO$_4$ CrO$_4$ MnO$_4$															Kr
		ZrO$_4$ NbO$_4$ MoO$_4$															Xe
	La	HfO$_4$ TaO$_4$ WO$_4$ ReO$_4$															Rn
	Ac	104															

Note that most of these involve anions classified as transition element oxides having various degrees of coordination. As before, each of these anions represents a class of anions which are optically active. However, not all of the class may be optically-active. Consider the tungstate anion. Only the so-called "orthotungstate", i.e.- WO_4^{2-}, is optically active (and only when combined with specific cations). The poly-tungstates, for example, have never been observed to form luminescent compounds, e.g. - $M_5W_{12}O_{41}$.

The factors leading to formation of an efficient self-activated phosphor are complicated indeed. The phosphor, $MgWO_4$:W, has been used for many years. It is prepared by calcining (firing at elevated temperature) MgO and WO_3 together. But if one uses CaO or BaO, one does not obtain an efficient blue-emitting phosphor like $MgWO_4$:W. The "phosphor", $CaWO_4$:W, when irradiated by ultraviolet light, only emits when cooled to 78 °K (liquid nitrogen temperature = LN_2) and $BaWO_4$:W does not emit at all under the same conditions. In other words, the crystal lattice structure of $CaWO_4$:W does not permit emission to occur at room temperature but it does occur if the lattice is sufficiently cooled. This behavior relates to an increased amount of vibronic coupling that occurs for $CaWO_4$:W at room temperature as compared to

MgWO$_4$:W. Only when the former is cooled to LN$_2$ is the amount of vibronic coupling lowered to the point where emission can occur.

Additionally, we have a limited number of **cations** which can be substituted into a given crystal structure and composition so as to function as the "activator-center". These are shown in the following diagram. Such cations are based on the fact that only those valence states which conform to the electronic configuration of d^{10}s^2 which has the ground state of ^{1}S$_0$ (Russell-Saunders coupling) will function as activators.

4.1.4.-

The Periodic Table as Related to Phosphor Composition

Cations That Can Be Used as Activator Centers

Half-filled Shells (1+) (2+) (3+)

(1+) (2+) (3+)

(1-) (0)

H																	He
																	Ne
																	Ar
			Cr	Mn	Fe				Cu	Zn	Ga	Ge	As			Kr
				Tc					Ag	Cd	In	Sn	Sb			Xe
	La			Re					Au	Hg	Tl	Pb	Bi			Rn
	Ac	104															

(3+)

	Ce	Pr	Nd		Sm	Eu	Gd	Tb	Dy	Ho	Er	Tm	Yb	
	Th		U											

Note that the electronic states given in this diagram include valence states which range from +3 through 0 to -1. All of these states have the ^{1}S$_0$ ground state, i.e.: S= 0, L = O, and J = 0. While some of these states might seem electronically difficult to attain in a crystal, they are not completely unknown states, having been studied by other authors as well (2).

One might wonder how one would stabilize an activator cation in the **metallic** state, e.g.- Zn^0. If one calcines the compound, ZnO, in a reducing atmosphere, the following defect reactions, as given in 4.1.5. on the next page, take place.

4.1.5.- Defect Reactions that Occur for ZnO in a Reducing Atmosphere

1. $ZnO + x H_2 = Zn(O_{1-x}, x V_o) + x H_2O + (2 e^-)_x + x Zn^{2+}_i$
2. $x Zn^{2+}_i + (2 e^-)_x = x Zn^o$
3. $x Zn^o + x V_O = = x Zn^o_{O}^{=}$
4. $ZnO_{1-x} + x Zn^o_{O}^{=} = ZnO: x Zn^o_{O}^{=}$

Note that the defect formation behavior is one where the hydrogen gas reduces the host lattice and forms an oxygen vacancy **plus two free electrons.** These electrons then reduce the **interstitial** Zn^{2+} cation to the metal valence state, i.e.- Zn^o. (The Zn^{2+} cation became an interstitial in order to satisfy the charge compensation mechanism of the lattice when the oxygen anion was removed by the reducing atmosphere to form water). The Zn^o atom is too large to remain as an interstitial, and therefore occupies the oxygen vacancy already created. (Note that the radius of Zn^o is 1.48 Å while that of the oxygen vacancy is 1.50 Å). This is one of the few cases where the cation in its metallic form functions as an activator. The energy efficiency of this phosphor is high, as is its quantum efficiency, QE.

We need to explain why only those cations with the ^1S_O ground state will function as activators. Consider the Sn^{2+} activator substituted in a $Sr_2P_2O_7$ lattice. The actual composition will be:

4.1.6.- $(1-x)Sr_2P_2O_7 \bullet xSn_2P_2O_7$

The $Sr_2P_2O_7$ lattice consists of Sr^{2+} ions and a double-tetrahedron, $P_2O_7^{4-}$, arranged in an orthorhombic lattice structure. However, $Sn_2P_2O_7$ itself is tetragonal. Since 95-97% of the lattice is $Sr_2P_2O_7$, the Sn^{2+} cation must fit into the main lattice structure, substituting for Sr^{2+}. Fortunately, both cations are the same size, i.e.- Sr^{2+} = 1.13 Å and Sn^{2+} = 1.12 Å and have the same valence state. Therefore, one can fit into the same space in the lattice as the other without problems of lattice strain due to size or that of defect charge compensation. Nonetheless, Sn^{2+} is a "foreign" atom in the structure. The whole structure is vibrating at room temperature and a specific phonon spectrum is present consisting of both optical and acoustic branches. However, at the site occupied by the Sn^{2+} cation, the lattice phonon spectrum is modified because of the presence of this "foreign" cation. If the electronic structure of Sn^{2+} happened not to be ^1S_O, i.e.- S = 0, L = 0, J = 0, vibrational coupling of the phonon states to the electronic states of the Sn^{2+} site would occur, i.e.- "vibronic coupling". Any transfer of lattice-absorbed energy o r direct absorption by the Sn^{2+} site itself would not result in excitation because the energy would be stolen by the phonon (vibrational) states, thus obviating any possible luminescence.

This is the phenomenon that we have referred to as the "The Ground State Perturbation Factor", which occurs for the electronic states of any cation and anion within the structure (1). These processes involve specific electronic and vibrational coupling modes. Briefly, they include phonon processes with vibronic coupling between electronic states of the activator and the phonon energies of the crystal host lattice. Following absorption at the activator site, phonon creation and destruction operators at the site populate adjacent Stark levels in the excited center which are further broadened by vibronic coupling. The isolated cation can be thought of as interacting with its crystalline environment via the crystal field, which is time-varying because the lattice is vibrating. However, this does not occur for those cations whose ground state involved $^1S_0$.

Let us examine why this is so. Firstly, an ion with a $^1S_0$ state has 2J + 1 Stark states or a **single** Stark state in the ground state configuration. Therefore, the phonon perturbation operator cannot populate adjacent levels via vibronic coupling of the localized vibrational modes at the activator site, because **none exist**. Secondly, with the absence of a total momentum operator (J = 0), the ground state electron wavefunctions are spherically symmetric, making vibronic coupling practically non-existent at the site. In this way, we can obtain an activator center, substituted within a host crystal, which is not coupled to the phonon modes of the lattice in its ground state. However, once it becomes excited, our cation is subject to the full perturbation of vibronic coupling. Therefore, we have achieved our goal of eliminating the ground state perturbation factor. All other cations have electronic states with either spin coupling factors or angular momenta which can couple via vibronic processes to the phonon modes of the host lattice. This results in excessive loss of excitation energy before the excitation-emission process can take place, i.e.- the excitation energy is lost as phonon modes (heat) to the lattice.

In the ZnS type of phosphor used in cathode-ray tubes, the mechanism is quite different. These phosphors are covalent and semi-conductive in nature, and include the zinc and cadmium sulfides and/or selenides. The criterion for selection of a semi-conducting host for use as a phosphor includes choice of a composition with an energy band gap of at least 3.00 ev. This mandates the use of an optically inactive cation, combined with sulfide, selenide and possibly telluride. The oxygen-dominated groupings such as phosphate, or silicate or arsenate, etc. are not semi-conductive in nature. And, none of the other transition metal sulfides have band gaps sufficiently large to be use as a covalent host for visible emission. We find that the ZnS and ZnSe host structures are excellent semi-conductors and exhibit very high efficiencies of photoluminescence and cathodoluminescence, especially when activated by copper, silver and gold. There is a large body of prior literature wherein the valence state of these activators is considered to be: Cu^+ , Ag^+ and Au^+. These are known stable states of these elements. The electron configuration of the elemental state is: $d^{10}s^1$. Note that the d^{10} electron configuration would be optically inert. Therefore, it is likely that the monovalent ion has the ground state of:

4.1.7.- $\qquad Ag^+ = d^9s^1 = {}^2S_{1/2}$

However, in our scheme of activator ground states, this electron configuration would be subject to substantial ground state perturbation, particularly when used in dielectric host crystals. Indeed, they do not function very well as activators when incorporated into such hosts. However, they do function well in the semi-conducting hosts, ZnS (CdS) and ZnSe. The lowest energy excited state for this electron configuration is: ${}^4D_{1/2}$ but the transition is **not dipole allowed,** i.e.- the oscillator strengths are in the range: 10^{-3} to 10^{-7}.

A much more plausible explanation considers the silver activator to have the Ag^- electronic configuration. The ground state would then be:

4.1.8.- $\qquad Ag^- = d^{10}s^2 = {}^1S_0$

But the Ag^- center would have to substitute on the sulfide site, viz:

4.1.9.- $\qquad ZnS: Ag^-{}_S=$

382

where the same mechanism as already given for the formation of the ZnO: Zn phosphor applies. Indeed, the ZnS phosphors are "self-activated" and exhibit very high luminescence efficiency under both photoexcitation and cathode-ray excitation. The band gap of ZnS (or fundamental absorption edge as it is sometimes called) is about 3.71 ev (or 3350 Å). Silver introduces an energy level at about 0.31 ev in the band gap (3.40 ev below the conduction level).

The "self-activated" phosphor likely has the formula:

4.1.10.-   ZnS : x $Zn^0 S^=$

In both of the above cases, the ionic radii of the activator states and the sulfide site are:

4.1.11- $S^= = 1.84$ Å
                $Zn^0 = 1.48$ Å
                $Ag^- = 1.62$ Å (estimated)

Thus, the electronic states for these activators as discussed in 4.1.4. would conform to our rules for functioning efficient activators. However, it is well to note that the mechanism given in 4.1.9. has a **negative monovalent** ion substituting at a negative **divalent** site. This would imply that an additional negative charge is required for charge compensation in the crystal. The emission band of silver occurs at 4500 Å (2.76 ev). The "self-activated" ZnS phosphor emits at about 4450 Å. Therefore, delineation between these two types of sites is nearly impossible. Furthermore, the probability is high that the "self-activated" center forms first in this crystal (when a sulfide atom is lost as sulfur + 2 electrons) and that the Ag^- center forms after that. The best explanation for the formation of silver centers seems to be that when a sulfide atom is lost, as in 4.1.12a., a corresponding zinc atom becomes interstitial. This would give rise to the following defect reactions, shown in 4.1.12. on the next page.

The ion, Ag^-, has been studied by Kleeman (3) in single crystal alkali halides. He was able to equate Ag^- with other isoelectronic ions like In^+ and to show that three transitional absorption bands exist which he attributed to: $^1P_1$, $^3P_1$ and $^3P_2$. These are the allowed states according to Russell-Saunders coupling rules. Emission in KCl occurred near to 4080 Å, but efficiently only at 78 °K.

4.1.12.- DEFECT REACTIONS FOR FORMATION OF THE ZnS:Ag PHOSPHOR

a. $ZnS + xH_2 \Rightarrow Zn(S_{1-x}, x V_s) + x H_2S \Uparrow + (2e^-)_x + Zn^{2+}_i$

b. $Ag^+ + 2e^- \Rightarrow Ag^-$

c. $Zn^{2+}_i + e^-_{lattice} \Rightarrow Zn^+_i$

d. $Zn(S_{1-x}, x V_s) \Rightarrow (Zn, Zn^+_i) S_{1-x} : Ag^-_S =$

Nevertheless, because the ZnS crystal is a semi-conductor, it is nearly impossible to determine the exact mechanism and state of the various activators in this class of phosphors. Therefore, the above discussion must be regarded as speculative and not completely proven, even though it conforms to our rules for phosphor design.

Let us now turn to the manufacture of phosphors. We will discuss:

A. Manufacture of the so-called "Halophosphates" first because this class of phosphors comprises 85-90% of the phosphor weight used in low pressure mercury vapor fluorescent lamps.

B. Manufacture of phosphors of differing emission colors which are used in blends with Halophosphates.

C. Manufacture of phosphors used in specialty fluorescent lamps

We will present luminescent characteristics in terms of:

a. Excitation - Emission Bands
b. CIE Color Coordinates, i.e. - $\bar{x}$ and $\bar{y}$
c. Peak Emission Wavelength
d. Quantum Efficiencies = QE
e. Absorbance, A , = (1-R)
f. Cation to Anion Ratios Used
g. Activator Concentration

The Excitation-Emission bands will be shown in terms of wavenumbers because this method allows a better display of the bands involved, particularly that of the excitation band. The specification of the CIE coordinates, x & y, will allow one to determine relative color of the phosphor being described. QE will be given for all phosphors as will be the values of A. The ratios of cation/anion and the mol% of activator added will serve to define the amounts of raw materials required to produce an efficient phosphor which exhibits a good performance in a fluorescent lamp.

Although quantum efficiency has been defined as the ratio of photons emitted to those absorbed, most of the values given will be relative ones. That is, a standard material like sodium salicylate (which has a constant intensity absorption band from about 2000 Å to 3900 Å and a QE of 50.0%) will be compared to the unknown phosphor, i.e.-

4.1.13-

QE is usually defined as:

$$QE = \frac{Photons_{Emitted}}{Photons_{Absorbed}}$$

But is measured as:

$$(QE)_U = (QE)_S \frac{\{(Ed\lambda_{excitation})_S\}\{Ed\lambda_{\cdot emission})_U (1-R)\}_S}{\{(Ed\lambda_{excitation})_U\}\{Ed\lambda_{\cdot emission})_S (1-R)\}_U}$$

S = Standard Material
U = Unknown Material

where absorbance, A, is defined as (1-R) and is equated to excitation. Definitively speaking, this is not exactly true since a material can absorb radiation without emission. What is actually required for exact specification in the above equation is the integration of all reflectance values over the entire absorption band. But, this is a very onerous task. However, if one uses absorbance values close to the maximum of the excitation band, this problem is resolved. In general, luminosity values will not be given, although they are available for each phosphor. Luminosity can be calculated from :

4.1.14.- $\quad \eta = \int e\,(\lambda)\,\overline{y}\,(\lambda)\,d\,\lambda\,/\int e\,(\lambda)\,d\,\lambda$

where e (λ) is energy of the emission band as a function of wavelength, λ. If one multiplies this value by 680, one obtains the maximum luminous output of the phosphor of that color.

4.2.- Manufacture of Lamp Phosphors

Commercial manufacture of phosphors requires specific methods to do so. Steps required to **prepare** phosphors were set forth in a prior work (1) and are quite similar to those needed for the **volume manufacture** of phosphors. In following sections, we will detail the preparation of each individual phosphor. This will include the individual steps, the type of equipment requisite for manufacture, and the conditions mandatory for commercial manufacture of a phosphor suitable in luminous output, i.e.- "brightness", and stability (maintenance as it is called in the Trade, i.e.- how the luminous output decays with lamp operating time) for use in a fluorescent lamp.

A. GENERAL METHODS OF MANUFACTURING PHOSPHORS

The manufacture of **all** phosphors requires the following specific measures. They include:

4.2.1.- Steps Required in the Preparation of Phosphors

        a. Selection and Assay of Materials
        b. Preparation of Phosphor Mix
        c. Blending Steps
        d. Firing Steps
        e. Post-firing Steps
        f. Washing Steps (if any)
        g. Evaluation and Record Keeping Steps

All of these steps will be delineated as they apply to manufacture of each specific phosphor. But first, we need to explain in more detail exactly how these actions are carried out.

    a. Selection and Assay of Materials

    This step is one of the more important since the solid state reaction to prepare the host matrix depends upon the nature of the components used to do so. For example, suppose we wished to prepare a pyrophosphate matrix. We could do so by a number of methods including the use of H_3PO_4 added to an oxide, viz-

4.2.2.- $$CaO + H_3PO_4 \Rightarrow Ca_2P_2O_7 + H_2O \Uparrow$$

However, what we will actually get is a mixture of compounds including $Ca_3(PO_4)_2$ and $Ca_5OH(PO_4)_3$. But if we use $CaHPO_4$, the reaction proceeds smoothly, viz-

4.2.3.- $$2\,CaHPO_4 \Rightarrow Ca_2P_2O_7 + H_2O \Uparrow$$

Therefore, we will specify that only certain compounds are to be used to prepare the formulation so as to have a control of the solid state reactions that take place. As shown below in Table 4-1, specific compounds are delineated for halophosphate phosphor manufacture, since they are the ones normally employed in the process. Needless to say, all materials chosen for this purpose must contain no more that a few parts per million (ppm) of transition metals present, since it is these that act as "killers" of luminescence as has been detailed previously (1). This factor is one of the most important ones which affect the final brightness attained in the phosphor to be manufactured.

By assay, we mean the amount of reacting component present in the original material used. That is, there is always a small amount of adsorbed water present in any material that we use. This must be accounted for, since we put together a phosphor formulation based upon the amount of fired product obtained and its contained molar ratios. For example, the molecular weight (MW) of $CaHPO_4$ is 136.06 while that of $Ca_2P_2O_7$ is 254.10, and water is 18.02. Thus, we expect to get 254 grams of $Ca_2P_2O_7$ from 272 grams of $CaHPO_4$ (2 x 136.06) = 272.12). However, if we fire the former compound to form the pyrophosphate, we find that 100.0 grams of the former do not produce 93.4 grams of product but some lower number. If that number is, say - 91.6 grams, then the assay of is 98.1% $CaHPO_4$ present and 1.9% adsorbed water. This must be done for all components in the mixture used.

b. Specific amounts of each component are weighed out. It is essential to use free flowing powders that will blend well together and not to have clumps and lumps of one or the other.

c. The mixture is blended together by use of a tumbler or other device. It is common to use a hammermill, which is a device having free-swinging hammers rotating upon a stainless steel screen, to effect mixing by a slight grinding action to obtain a free-flowing powder. The mix is usually hammermilled twice, blending in between and after each hammermilling step.

d. Firing in usually accomplished in a box-type of furnace in air or a long enclosed furnace, i.e.- "tube-furnace", capable of controlling and maintaining a desired internal atmosphere. The heating elements are usually "Globar™", which are composed of silicon carbide. Trays or boats containing the mix are pushed in and out of the tube- furnace on a prescribed time schedule, or they may be pushed through on a specified cycle. If a box-furnace is used, one usually uses 7 " silica crucibles or a 2-liter volume crucible for production purposes. Silica boats or trays can have variable volume and can be as large as 6 inches high by 12" wide and 16" long. Silica crucibles used for production usually have the ratio of height to diameter shown in the following diagram, viz-

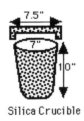

Silica Crucible

The mix is placed in the tray or crucible and tamped by vibration until it fully fills the container. Then, the crucible or tray is placed inside of a **hot** furnace which has been previously set at the proscribed firing temperature and kept there for the specified time. Following this time, the hot crucible is removed from the furnace and allowed to cool in air to room temperature. If the phosphor is susceptible to oxidation, it may be kept within the tube furnace under the firing atmosphere used and allowed to cool there.

e. Post firing steps include removal of the phosphor (cake) from its crucible, examination of the cake under ultraviolet light to facilitate removal of inert material (silica scale from the crucible and the like), and finally crushing the friable cake into small lumps and fragments. These are usually broken down further by a screening operation in which the lumps are shaken on a 200 mesh screen to obtain a lump-free powder. Finally, the powder is screened on a 325 mesh screen to remove all particles larger than about 44 μ. Sometimes, the powder is hammermilled without the screen being present. This prevents grinding of the phosphor particles while breaking up lumps to form a fee-flowing and uniform material.

f. Washing of the powder may be specified. One suspends the powder as an aqueous suspension in which the amount of water per lb. of powder is specified. Many phosphors are acid-washed by adding the requisite amount of specified acid to the suspension. The washed powder is than allowed to settle, the water is removed by suction or the like, and the powder is then re-suspended by adding water while stirring. Following this, the powder is again allowed to settle. This cycle is repeated at least 3-4 times. Alternately, a small amount of base such as NH_4OH may added to neutralize the acid before the rinsing steps are done. Finally, the powder is separated as a wet cake and placed in an oven set at about 110-125 °C to dry overnight.

g. The "brightness" of the phosphor is measured by comparing it to a standard phosphor while both are illuminated by an ultraviolet light. Many times, this is done visually by spotting one on the other. Or, a simple device consisting of a UV source and a detector is used to compare response to the UV source. The particle size is usually measured as well.

h. Record keeping consists of recording the lot numbers of the components used to manufacture the phosphor as well as the brightness, the particle size of the phosphor and the number of lbs. produced.

All of these steps are normally carried out for each batch of commercial phosphor. However, because the manufacture of halophosphate phosphors occurs in tonnage volumes, special equipment for handling this volume is necessary.

B. MANUFACTURE OF HALOPHOSPHATE PHOSPHORS

We will begin with a description of the manufacturing protocol used to make the so-called "halophosphate" phosphors, since these represent at least 85% of the phosphors employed of the internal fluorescent coating present in the low pressure fluorescent lamp. By "halophosphate", we mean a class of phosphors exemplified by:

4.2.4.- $M_5X (PO_4)_3 : Sb :Mn$

which is activated by both Sb^{3+} and Mn^{2+}, and where M can be any of the alkaline earth cations, except Mg^{2+} (which does not form an apatite). (Actually, in these phosphors, the Sb^{3+} cation is the "sensitizer" and Mn^{2+} is the "activator"). X can be any of the halides, including Cl^-, Br^- and I^-. These compounds are hexagonal, have the apatite structure, and form extensive solid solutions among themselves. A more complete discussion of the apatite compositions was given in a prior work (1).

It is well to note that a very large body of literature exists concerning various types and compositions of phosphors. Our criterion for discussion herein will be based solely upon those phosphors which are **commercially manufactured**. Thus, we will first discuss solely the phosphor used in greatest volume to manufacture fluorescent lamps involves the calcium halophosphates, i.e.- $Ca_5(F,Cl)(PO_4)_3$: Sb:Mn. The barium analogs do not have equivalent efficiencies, i.e.- "brightness", compared to the calcium based phosphors and the strontium apatites do not have quite the emission color considered to be useful in lamps, except for the $Sr_5F(PO_4)_3$:Sb phosphor, which is a blue-emitting phosphor used in Deluxe blends.

When most manufacturers began to make phosphors for fluorescent lamps, they found that certain colors were needed because the Public was used to incandescent lamps and the outdoors. This gave rise to specific colors used in fluorescent lamps which were designated as "Cool White", "Warm White", etc., viz-

4.2.5.- <u>Halophosphate Phosphor Colors Usually Manufactured for Fluorescent</u>
<u>Lamp Use</u>

> a. "Warm White" = WW
> b. "White" = W
> c. "Cool White" = CW
> d. "Daylight" = D

These are the four major colors that are used in fluorescent lamps today. (Actually, Daylight is a blend of phosphors approximating 6500 °K emission). The emission colors in halophosphate phosphors can be shifted from a bluish white to a reddish white by change in the ratio of the two "activators added. A number of variations have also appeared in the marketplace such as "Deluxe Warm White", etc. These are lamps which have extended colors present in the emission band (mostly in the deep red part of the spectrum) which are used to promote better color rendition of indoor objects. In the following table is shown a general formulation that has been used for these phosphors. Following that is a diagram (4.2.6.) showing the the excitation-emission bands of the Cool White phosphor which is typical of the whole series. Note that an emission band exists for each activator and that the excitation band has its peak close to that of the 2537 Å emission line of mercury (which is the dominant emission line in the low pressure mercury discharge lamp).

TABLE 4-1
Generalized Formulas for Halophosphate Phosphors

Emission Color	Formulation in Mols					
	Lumin.*	$CaHPO_4$	$CaCO_3$	CaF_2	NH_4Cl	Sb_2O_3	$MnCO_3$
Yellow Halo = YH	0.620	3.00	1.20	0.50	0.50	0.05	0.18
Warm White = WW	0.552	3.00	1.20	0.50	0.50	0.05	0.17
White Halo = WH	0.542	3.00	1.20	0.50	0.50	0.05	0.12
Cool White = CW	0.493	3.00	1.20	0.50	0.50	0.05	0.08
Blue Halo = BH	0.386	3.00	1.20	0.50	0.50	0.05	0.05

*Luminosity

The peak intensity of each activator has been determined to be dependent upon the relative ratios of the two activators incorporated into the phosphor.

4.2.6. -

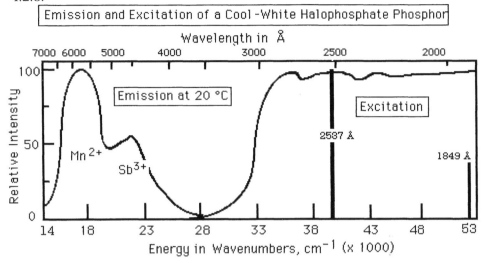

It is this factor which makes the Halophosphate phosphor so useful in fluorescent lamps. For the YH phosphor, the major emission peak is due to primarily to Mn^{2+} whereas in the BH phosphor the emission is dominated by that of the Sb^{3+} center peaking at 4550 Å. Note that the major difference in the above formulations is the amount of Mn^{2+} activator added. This has a singular effect upon the properties of the emission of this series of commercial phosphors in that emission colors ranging from reddish white to a bluish white can be obtained, with high quantum efficiencies as well. In contrast, the only phosphor that is suitable in the strontium apatite system is the "Strontium Blue" phosphor. For this reason, we will not discuss any barium analogs.

As a matter of record, the following values have been determined for the Cool White composition, when excited by 2537 Å irradiation:

4.2.7.- Cool White phosphor: QE = 0.82 (1-R) = 0.90

Generally, these values are used as a criterion in the development of any new phosphors to be used in fluorescent lamps.

The following diagram shows the various lamp colors used in the trade plotted upon a CIE color chromaticity coordinate plot:

4.2.8.-

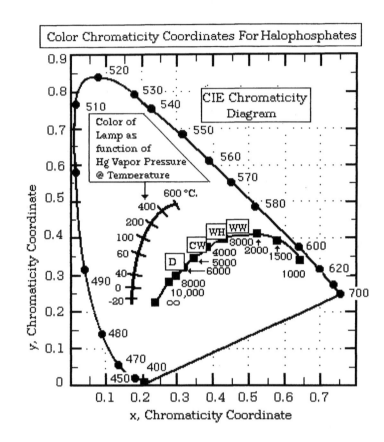

Color Chromaticity Coordinates For Halophosphates

Note that the four lamp colors are located close to the "black-body" line of temperatures. Thus WW is usually specified as a 3500 °K white, WH = 4000 °K, CW = 4500 °K, and D (Daylight) = 6500 °K. BH is used in blends to make the D lamp color, whereas YH is used in WW blends as well.

Also shown is the color shift due to shift in relative intensities of mercury spectral lines as a function of lamp operating temperature. Since most low pressure mercury discharge lamps operate near to 40 °C, this is not a significant factor in fluorescent lamps. But it is significant in HPMV lamps where phosphors are also used. Thus, if a phosphor is to be used in HPMV lamps, one can estimate the color-shift which will occur when the phosphor is operated at the temperature within the lamp.

Nevertheless, one can see that the calcium halophosphate system admirably fits the criteria required for illumination by fluorescent lamps. It was for this reason that this phosphor system was chosen over other potential systems to replace the $ZnBeSiO_4$: Mn phosphor system then being used in 1948. (Another major reason for replacement was the discovery of the virulent toxicity of beryllium to the human organism).

1. Methods and Formulas Used for $Ca_5(F,Cl)(PO_4)_3$:Sb:Mn Phosphors

Manufacture of halophosphates is carried out using special equipment. The following formulas are those used by several manufacturers to manufacture halophophate phosphors of various colors and emissions. The data presented in the following generalized formulas shows some of the formulations that have been used.

4.2.9.- Generalized Formulas for Cool White Halophosphate Phosphors

One Manufacturer

$$Ca^{2+} + Mn^{2+} + Sb^{3+} + Cd^{2+} = 4.85$$
$$PO_4^{3-} = 3.00$$
$$Cl^- + F^- = 1.00$$

or:

	Mols
$CaHPO_4$	= 3.00
$CaCO_3$	= 1.04
CaF_2	= 0.50
NH_4Cl	= 0.50
Mn_2O_3	= 0.08
Sb_2O_3	= 0.05
$CdCO_3$	= 0.05

Another Manufacturer

$$Ca^{2+} + Mn^{2+} + Sb^{3+} + Cd^{2+} = 4.70$$
$$PO_4^{3-} = 3.00$$
$$Cl^- + F^- = 1.50$$

or:

$CaHPO4$	= 3.00
$CaCO_3$	= 0.90
CaF_2	= 0.50
NH_4Cl	= 0.50
Mn_2O_3	= 0.08
Sb_2O_3	= 0.05
CdO	= 0.04

Another Manufacturer (in Europe)

$$Ca^{2+} + Mn^{2+} + Sb^{3+} + Cd^{2+} = 5.01$$
$$PO_4^{3-} = 3.00$$
$$Cl^- + F^- = 0.51$$

or:

$CaHPO4$	= 3.00
$CaCO_3$	= 1.67
CaF_2	= 0.21
NH_4Cl	= 0.10
Mn_2O_3	= 0.08

Another Manufacturer (in Europe)	Mols
Sb_2O_3	= 0.04
$CdCO_3$	= 0.05

If one wishes to manufacture these phosphors, one will find that small changes will need to be made in order to obtain specific colors for use in blends. This arises because of small differences in raw materials used to make these phosphors. Actually, each halophosphate, i.e. - WW, CW, W, etc., has its own optimized formulation which varies from Manufacturer to Manufacturer. Therefore, those we have given must be regarded as representative of those actually used. We have given three formulas that have been, or are being, used to manufacture halophosphate phosphors. However, the above formulas are not directly useful unless assay and weights are specified.

It has been found that the following formulation, shown in Table 4-2, is more useful. This specific formula has a 4.85/3.00 cation to anion ratio. The mix will produce a phosphor color similar to that of a Cool White. Cd^{2+} has been added to improve the phosphor brightness in lamps.

Table 4-2

A Formula Used for the Manufacture of Halophosphate

Material	Mols/Mol	Assay	Grams	Lbs./Lb. of Mix
$CaHPO_4$	3.00	≈ 98.8 %	413.23	0.6312
$CaCO_3$	1.04	≈ 99.0%	105.14	0.1606
CaF_2	0.50	≈ 99.5%	39.23	0.0599
NH_4Cl	1.00	-----------	53.49	0.0817
Sb_2O_3	0.08	≈ 99.9%	23.34	0.0357
$MnCO_3$	0.10	≈ 99.8%	11.52	0.0176
$CdCO_3$	0.05	≈ 98.8%	8.73	0.0133

2. Equipment Used for Manufacture

The equipment required for volume manufacture is shown in the following diagram, given on the next page as 4.2.10. The individual parts and steps involved include:

a. A double cone- blender or "V" - blender of sufficient size.

4.2.10.- Process for Manufacture of Halophosphate Phosphor

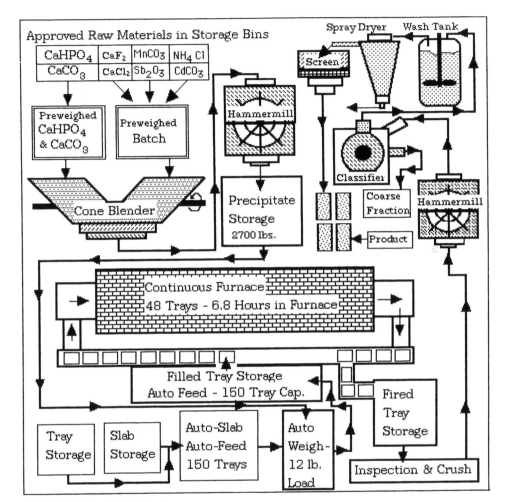

b. Two hammermills, one for making the initial mixture of compounds prior to firing and the second for breaking up the lumps that may have occurred after the mixture has been fired to form the phosphor composition.

c. A continuous furnace capable of reaching about 1350 °C maximum. Usually "globars™" are used as the heating elements in the furnace. The furnace may be equipped with an automatic tray pusher which transports filled trays from a filled tray storage to a continuous belt where each tray is fed into the furnace at programmed intervals. At the other end is a fired tray storage.

d. Inspection of the fired trays is manual, and the output of each tray is combined and fed into the second hammermill which is operated so as not to damage the particles while producing a powder.

e. The output of the hammermill is fed into the particle classifier where the coarse and very small fractions are separated from the major part of the powder. This powder is then put into a storage bin. About a 5% loss of material is usually experienced.

f. The powder is washed in batches where the first wash is 1% HCl, followed by a 1% NH_4OH-NH_4NO_3 wash, with two water rinses in between the acid-base washes. These washes will remove any left-over free oxide salts not completely reacted during firing, and also leaves an oxidized surface on the particles. About 1% of the phosphor weight is lost in this process.

g. The slurry is then fed into the spray dryer and the dried powder emerges onto a continuous 325 mesh rotating screen which breaks up any lumps that may be present and produces a free flowing powder.

h. The product is then placed in plastic-lined drums, weighed, and the yields recorded.

Note that the batch size is usually about 2700 lbs to start and that about 490 lbs is lost due to firing by loss as water vapor, NH_3 and CO_2 (and other volatiles). Thus, the overall loss is 18.2% and about 2200 lbs of fired product results. Further losses due to post-firing processes usually do not exceed 10% and the product ends up to be about 1980 lbs, or a 73-75% yield of finished phosphor, compared to the gross weight of the components originally blended before firing.

3. Process Used for Halophosphate Manufacture

The process that is used for processing of the halophosphate phosphors is given in the following steps, using the layout of equipment shown in 4.2.8.

a. Selection and Assay of Raw Materials

Most manufacturers make both $CaHPO_4$ and $CaCO_3$ which is used to prepare phosphors. In the previous chapter, methods of manufacture of these raw materials were set forth. The most important criteria for the $CaHPO_4$ used in production of halophosphate phosphors are the particle size (which should be about 7-9 μ in average size), the crystal habit (most manufacturers use flat, platy diamond-like crystals of $CaHPO_4$ since these are the easiest to make), and the purity of the crystals (which should contain no more than about 5-9 ppm of total transition metals). The size of the $CaCO_3$ is of less importance since it normally reacts with the $CaHPO_4$ present and has little effect upon the nature of the halophosphate crystals so produced. Usually, the other components are purchased after the purity has been ascertained. Note that assays are usually obtained for all raw materials. The next step is that of preparation of the mix to be fired to form the phosphor.

b. Preparation of Phosphor Mix

The usual batch size is 2700 lbs. This requires the amounts of raw materials as given in the following Table on the next page. Many times, both $CaHPO_4$ and $CaCO_3$ are pre-weighed and then added to cone-blender. A pre-weighed batch of a mixture of all of the other components may also be used, particularly if the same phosphor, i,e, Cool White, is manufactured in successive batches.

$CdCO_3$ is used in this case since if the use of CdO may result in losses of cadmium due to the fact that reaction with NH_4Cl will form $CdCl_2$, which is volatile. This will lead to a lower luminous output in lamps.

c. Blending Steps

For volume production, the mix is usually hammermilled twice, blending before and after each hammermilling. The mix may then

Table 4-3

Raw Materials to be Added for 2700 lb. Batch

CaHPO$_4$	1704 lb. 4 oz.
CaCO$_3$	433 lb. 11 oz
CaF$_2$	161 lb. 11 oz.
NH$_4$ Cl	220 lb. 10 oz
Sb$_2$O$_3$	96 lb. 6 oz.
MnCO$_3$	47 lb. 8 oz.
CdCO$_3$	35 lb. 14 oz.

be stored prior to firing. This is done because the hammermilling steps take several hours to accomplish and can be done while the firing of the already prepared mix is taking place. Sometimes, silica trays used for firing are filled and kept in storage, except when the type of halophosphate phosphor manufactured is changed.

d. Firing Steps

This part of the manufacturing procedure is the most critical of all of the steps accomplished. What we are speaking of is shown in the following diagram which shows a specific furnace used for the firing step in the manufacture of halophosphate phosphors, viz-

4.2.11.-

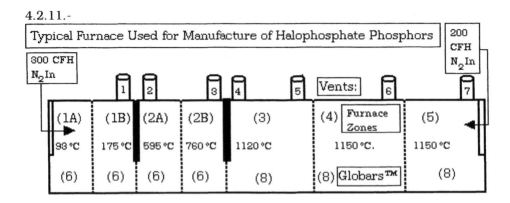

An automatic loading device is used to load trays with the powder mix. Note that the load of mix is usually about 12 lbs./tray. Filled trays are then loaded upon a conveyor belt and then pushed into the furnace automatically, following a preset schedule of loading and pushing each tray. The total time that each tray is in the furnace is 6.8 hours, but this can be varied, depending upon the type of halophosphate phosphor being manufactured. After firing, the trays are allowed to cool and then emerge from the furnace and go to a fired tray storage for further processing.

Actually, each type of halophosphate is fired according to a temperature profile within the furnace that has been determined to be optimum for that color-type of halophosphate, shown in the following Table:

Table 4-4

Optimum Firing Temperatures for Several Phosphors

Warm White		White		Cool White	
1 A	595 °C	1 A	110 °C	1 A	110 °C
1 B	760	1 B	550	1 B	535
2 A	1120	2 A	635	2 A	635
2 B	1150	2 B	765	2 B	825
3	1150	3	1135	3	1155
4	1150	4	1135	4	1160
5	1150	5	1140	5	1155

e. Post-firing Steps

Once the phosphor cakes are cool, they are inspected under ultraviolet light and all of the inert material present is scraped off and discarded. The cake is crushed and the lumps and powder is processed through a hammermill in which the screen has been removed so as not to grind the phosphor particles. A free-flowing powder results. Next, the halophosphate is processed through a particle classifier wherein all particles less than about 3.0 μ are removed. One must take care not to grind any of the particles while they are being classified since grinding destroys the crystallinity of the particles and hence the brightness of the powder. Both the coarse and fine fractions are discarded.

f. Washing Steps

Nearly all halophosphates are washed. Each manufacturer has his own procedure. One method involves washing the powder in 1% HCl in which about 300 gallons of deionized water are added to the tank and about 1% by volume of concentrated HCl is added. The phosphor powder is added with stirring to form a slurry at a concentration of about 1.5 lbs./gallon. The slurry is stirred about 15 minutes and then the particles are allowed to settle out. The mother liquor is removed by decantation and a water wash is applied by resuspension to form a slurry. then about 1% NH$_4$OH is added to neutralize any remaining HCl, stirring about 15 minutes.

The particles are then allowed to settle, the wash water removed, and two water washes are applied as before. Finally the slurry is spray-dried and the resulting powder is screened through a rotating 325 mesh screen to remove all particles larger than about 44 μ. The product is then put into plastic-lined drums.

g. Evaluation and Record Keeping Steps

Most batches are evaluated by making test lamps. However, one also evaluates the "brightness" of the phosphor is measured by comparing it to a standard phosphor while both are illuminated by an ultraviolet light. Many times, this is done visually by spotting one on the other. Or, a simple device consisting of a UV source and a detector is used to compare response to the UV source. The particle size is usually measured as well.

A record of the lot numbers of the raw materials is kept and the weights used to make the batch of phosphor. Finally, the yield of phosphor is recorded.

The steps outlined above give a good perspective of the steps involved in manufacturing these phosphors. However, we need to better define the processes in order to produce the best phosphor that we can.

4. Recommendations for Optimization of Halophosphate Production

In order to gain a better understanding of all the intricacies involved in the processes which are used to manufacture halophosphate phosphors, we need to delve into the individual procedures more thoroughly, particularly from the standpoint of optimizing the process to produce as bright a phosphor as possible. Therefore, we will present a detailed study that has been made of the individual processes from the standpoint of making specific recommendations concerning the optimization of methods for each operation.

a. Selection and Assay of Materials

1. It is essential that all materials be assayed correctly. This means that each material needs to be fired individually at a temperature that ensures that complete reaction is obtained. The best way to do so is to use Thermogravimetric Analysis where the decomposition of each compound can be observed directly.

2. The use of alternative materials other than those specified can lead to unwanted side reactions. A good example is the substitution of CdO for $CdCO_3$. When the mix containing the former is hammermilled, it will react with NH_4Cl to form $CdCl_2$ which is volatile and may lead to loss of Cd^{2+} in the final phosphor. CdO can only be used if the hammermilled mix is fired immediately.

3. The use of the flat, platy diamond-like crystals of $CaHPO_4$ leads to a halophosphate whose molar composition is closer to the formulation stoichiometry than that produced form pure diamond-like crystals. The latter usually has some excess $Ca_2P_2O_7$ present which must be removed by washing in acid. This means that the washing step must be carried out more carefully. The fluorescent lamp produced from either phosphor having a specific crystal habit rivals the other in both output and maintenance. (Some manufacturers prefer one over the other because of perceived advantages in "covering power"). However, the

CaHPO$_4$ used for Warm White should not exceed 8.8 μ and preferable should be between 6.6 and 8.0 μ to obtain the most consistent results in lamps. Needless to say, the crystals of CaHPO$_4$ should contain little hydroxyapatite. If it does, the maximum amount tolerated is about 1-3%.

b. Preparation of Phosphor Mix

1. The CaCO$_3$ ratio used in any halophosphate formulation should not exceed 1.23 mols/3.00 mols of CaHPO$_4$. Apparently, a high CaCO$_3$ ratio exceeding 1.23 mols affects the incorporation of the Sb^{3+} activator into the phosphor during firing (by formation of the cubic Romeite composition as shown below).

c. Blending Steps

1. The use of CaCl$_2$ as a source of Cl$^-$ is not recommended since it is hygroscopic and leads to attack on all metal surfaces into which it comes into contact. During hammermilling, it may stick to the internal metal surfaces and result in a phosphor composition which is not uniform in chloride content when the mix is fired.

2. One of the reasons for using a double-hammermilling procedure is to ensure that the final mix before firing becomes uniform. If any material tends to stick to surfaces, reprocessing, as a second hammermilling step, tends to level out the composition and makes it more uniform from tray to tray.

d. Firing Steps

1. The major reactions involved which form the apatite crystal phase has been found to involve the following:

Approx Reaction Temp.

$$2\ NH_4Cl + CaO \Rightarrow CaCl_2 + 2\ NH_3 \uparrow + H_2O \uparrow \quad \sim 150\ °C.$$

$$2\ CaHPO_4 \Rightarrow Ca_2P_2O_7 + H_2O \uparrow \quad \sim 250\ °C.$$

$$CaCO_3 \Rightarrow CaO + CO_2 \uparrow \quad \sim 600\ °C.$$

$$Ca_2P_2O_7 + CaO \Rightarrow Ca_3(PO_4)_2 \qquad \sim 1050 \ °C.$$

$$Ca_3(PO_4)_2 + CaF_2 + CaCl_2 \Rightarrow Ca_5(F,Cl)(PO_4)_3 \quad \sim 1050 \ °C.$$

The above temperatures of reaction are approximate, since they are affected by the actual ratio of compounds in the mix.

2. It was noted that operation of the furnace and its attendant temperature zones had a profound effect upon the firing and that side-reactions occurred which involved the activator compounds in the mix, particularly that of Sb^{3+}.

The following diagram shows one furnace setup that was studied:

4.2.12.-

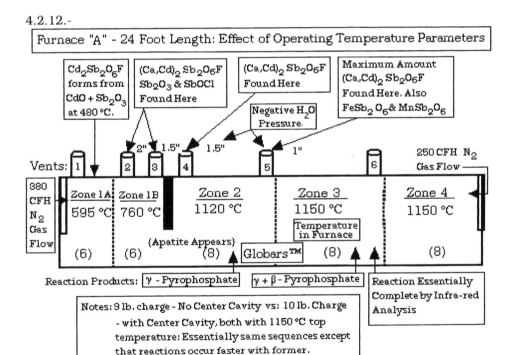

3. In this furnace (called Furnace-A), the lengths of the individual temperature zones are shown and their operating temperatures. Also shown are some of the observations made as a result of change in firing procedure. This furnace was 24 feet long and had N_2-gas as an internal

atmosphere, flowing into both ends of the furnace. The ratio of preheat to firing zone was 25% of the total length. The number of Globars™ in each section of the furnace is also shown. There are a total of 6 vents and a total of 5 firing zones in this furnace.

4. Sampling was done through the individual VENTS of the furnace. Originally, these were designed into the furnace to control the flow of nitrogen gas, but it was soon found that buildup of deposits during firing occurred. By analyzing these deposits, it was determined that the composition was primarily that of various antimony crystal structures which deposited at the cooler vents of the furnace. These compounds included:

4.2.13.- $Cd_2Sb_2O_6F$ Sb_2O_3 SbOCl $CaCdSb_2O_6F$

         $SbCl_3$ Sb_2O_5 $SbOCl_3$ $MnSb_2O_6$

                                          $FeSb_2O_6$

The complex oxide, i.e.- $CaCdSb_2O_6F$, was determined to be a "fluoro-romeite", a cubic compound formed by the following reaction:

4.2.14.- $3/2\ CdO + 1/2\ CaF_2 + Sb_2O_3 + 3/4\ O_2 \Rightarrow (Ca_{0.5}Cd_{1.5})Sb_2O_6F$

Note that oxygen gas is required in the reaction to form the compound which contains oxidized antimony, i.e.- Sb^{5+}. Obviously, this state is not luminescent, even if the antimony were to be incorporated into the apatite structure. This leads to loss of the Sb^{3+} state and a phosphor with lower output when incorporated into a fluorescent lamp.

5. Another furnace, designated as "Furnace-B", is shown in the following diagram which is given on the next page as 4.2.15.

This furnace, which was 25 feet in length, had a ratio of pre-heat to firing zone of 40%. The preheat temperatures shown were specified to allow the lower temperature reacting components to fully react before the trays encountered the higher temperatures where the apatite structure was formed. There are 7 vents present in this furnace and the

4.2.15.-

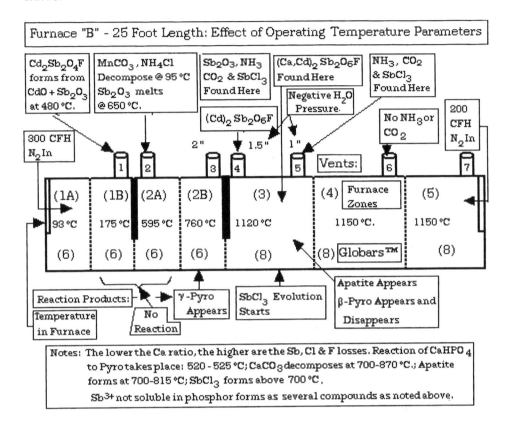

Furnace "B" - 25 Foot Length: Effect of Operating Temperature Parameters

$Cd_2Sb_2O_4F$ forms from $CdO + Sb_2O_3$ at 480 °C.

$MnCO_3$, NH_4Cl Decompose @ 95 °C Sb_2O_3 melts @ 650 °C.

Sb_2O_3, NH_3 CO_2 & $SbCl_3$ Found Here

$(Ca,Cd)_2 Sb_2O_6F$ Found Here

NH_3, CO_2 & $SbCl_3$ Found Here

Negative H_2O Pressure.

$(Cd)_2 Sb_2O_6F$

No NH_3 or CO_2

200 CFH N_2 In

300 CFH N_2 In

2"   1.5"   1"

Vents:

(1A)   (1B)   (2A)   (2B)   (3)   (4) Furnace Zones   (5)

93 °C   175 °C   595 °C   760 °C   1120 °C   1150 °C.   1150 °C

(6)   (6)   (6)   (6)   (8)   (8) Globars™   (8)

Reaction Products: γ-Pyro Appears $SbCl_3$ Evolution Starts   Apatite Appears β-Pyro Appears and Disappears

Temperature in Furnace   No Reaction

Notes: The lower the Ca ratio, the higher are the Sb, Cl & F losses. Reaction of $CaHPO_4$ to Pyro takes place: 520 - 525 °C; $CaCO_3$ decomposes at 700-870 °C.; Apatite forms at 700-815 °C; $SbCl_3$ forms above 700 °C.
Sb^{3+} not soluble in phosphor forms as several compounds as noted above.

number of globars™ in each section is also shown. Notice that there are 7 firing zones in this furnace as well. The types of compounds found in the vents as well as the results of the ongoing solid-state reactions determined to be taking place are also shown. Condensations regarding some of the conclusions drawn are also given.

6. Note that both $\gamma - Ca_2P_2O_7$ and $\beta - Ca_2P_2O_7$ appear as separate phases during the course of solid state reaction. It is these which have to be removed during the washing procedure (if any exist after firing). It should be noted that the formation of the antimony-based compounds given above leads to a lower degree of incorporation of Sb^{3+} into the phosphor and hence to a lower brightness fluorescent lamp. For this reason, it is essential to minimize their formation within the furnace. Note also that $\gamma - Ca_2P_2O_7$ appears at about 760 °C. in the mix being

fired. Ordinarily, $CaHPO_4$ alone reacts as low as 535 °C. to form this compound (8).

7. It was discovered that control of the vent-openings had a significant effect upon the course of the solid state reactions occurring. By opening a vent in Zone-1A of Furnace-A and a corresponding vent in Zone-2A of Furnace-B, the gases produced by reaction there by NH_4Cl and $CaHPO_4$ were removed faster, thus preventing buildup of "fluoro-romeite" by oxidation there and loss of the Sb-activator. Control over the solid state reactions was further increased by raising the temperatures in the pre-heat zones to above 585 °C. This increased the initial rate of reaction in the furnace and decreased the amount of antimony loss in the final phosphor composition.

e. Post-Firing Processing

1. Hammermilling to break up lumps after Inspection should be done so as not to produce a fine fraction of particles by breakage. The particle-classifier must also be closely controlled so as not to produce a separate "fine" fraction of particles.

2. Washing of the powder is also a critical step. Once the 1% HCl wash has been applied and the powder filtered on the Drum-Filter, it is then re-slurried in the tank and treated with a 1% NH_4OH-NH_4NO_3 solution. It is essential to remove all of the HCl-wash water and wash the powder on the drum filter since the HCl wash liquid contains soluble Ca^{2+} (which may be present due to an incomplete solid state reaction between components in the original formulation). When the powder is then washed with the basic solution, hydroxyapatite (non-fluorescent) is then reformed on the surface of the phosphor particles, leading to a less luminous output when coated within a fluorescent lamp. Normal losses in the washing step are about 1% by HCl dissolution. Some of the soluble Ca^{2+} can remain within the Drum Filter and contaminate subsequent batches during the washing operation. It has been estimated that as much as 0.5 lb. of hydroxyapatite can result from the final 1980 lbs of phosphor, i.e.- an 0.025% or 250 ppm. contamination. It should be clear that this operation is critical to the production of a high brightness phosphor and highly efficient fluorescent lamps.

f. Final Conclusions Drawn Concerning Optimized Firing Methods

The most significant finding concerning the firing of Halophosphate phosphors is that fluorescent lamps repeatedly show increased gains in lumen output when the mixture is fired at higher temperatures in the preheat zones and lower temperatures in the full reaction zones.

1. Using 525 °C in the preheat zones and 1150 °C in the last two zones of each furnace causes a faster rate of reaction in the preheat zones, and a slower rate of reaction in the last two zones of either furnace. Gains up to 40 lumens/watt resulted in fluorescent lamps.

2. Increasing the powder charge from 9-10 lbs./tray to 12 lbs./tray also resulted in lumens gains in lamps, as well as increasing the overall efficiency of the firing process.

3. Differences in Warm White firing occurred primarily because of the higher activator concentrations. By opening Vents #1 & #2 in both furnaces during firing, gains in lumen output in lamps of up to 50 lumens/watt occurred for Warm White, but losses were incurred for Cool White formulations. Venting is critical for Warm White and Cool White firing but not so much for White halophosphate compositions.

4. In the following Table, given on the next page, we present some of the test results observed for firing of the Cool White phosphor in both of these furnaces.

One can follow the general trend in the solid-state reaction by comparing the amount of residual materials determined to be present. In comparing the two furnaces, there was usually more $\beta-Ca_2P_2O_7$ from Furnace-A, but less free Sb^{3+} and residual chloride in the final washed phosphor. Most of the reaction occurred in the 1120 °C. zones where the loss of free Sb^{3+} and Cl^- occurred.

If a larger tray charge is used, the initial decomposition rate is faster, due presumably to the larger heat capacity of the load. Most of the final reactions occur in the 1150 °C zones of the furnace.

The use of a 12.0 lb. charge in the trays was found also to promote faster reactions in the preheat zone and a brighter phosphor.

TABLE 4-5

Extent of Reaction Observed in Various Zones of Furnaces A & B during Firing of Cool White Phosphor Formulations

% $Ca_2P_2O_7$

Vent	Zone Temp. f= front (°C) r= rear (°C)	% React. of Total	% Residual CO_2	% Apatite	γ −	β−
1	A- 595- f	0	100	0	0	0
	B- 175 - r	0	100	0	0	0
2	A- 760 - f	5	100	0	0	0
	B- 595 - r	10	80-90	0	0	0
3	A- 760 - f	20	40-90	0	0 -8	0
	B- 760 - r	20	50-80	0-10	0 -5	0-Tr
4	A- 1120- f	60	10 - 90	50	5-15	7
	B- 1120 -r	50	30-70	60-100	7	5
5	A- 1120- f	95	0-20	40-100	8	4
	B- 1120 -r	90	0-25	20-100	Tr	10
6	A- 1150- f	95	Trace = Tr	80-100	Tr	3
	B- 1150-r	90	5	70-100	Tr	1
7	A- no vent	....	....	....	......	....
	B- 1150 -r	100	0	100	0	1.5
End	A- Cooling	95-100	0	95-100	0	2
End	B- Cooling	100	0	100	0	1

5. The same analysis of reactions occurring for the Warm White formulation led to essentially the same conclusions as those reached for the Cool White formulation. The specific data are given in Table 4-6, as shown on the next page.

Note that more residual β-$Ca_2P_2O_7$ resulted for the 3500 °K formulation as compared to the 4500 °K phosphor mix. Again, it was noted that the condition of the vents had considerable effect upon the extent and degree of solid state reaction occurring at any time.

TABLE 4-6

Extent of Reaction Observed in Various Zones of Furnaces A & B during Firing of Warm White Phosphor Formulations

					% $Ca_2P_2O_7$	
Vent	Zone Temp. f= front (°C) r= rear (°C)	% React. of Total	% Residual CO_2	% Apatite	γ –	β –
1	A- 595- f	0-5	80-100	0	0	0
	B- 175 - r	0	100	0	0	0
2	A- 760 - f	5	75-90	0	0	0
	B- 595 - r	5	90	0	0	0
3	A- 760 - f	30	40-90	20	0 -8	2
	B- 760 - r	10	40-80	T r	0 -5	2
4	A- 1120- f	60	20-70	60	5-12	7
	B- 1120 -r	5-60	30-80	80	15	5
5	A- 1120- f	95	0-20	80-100	8	4
	B- 1120 -r	85	10-25	20-80	6	10
6	A- 1150- f	85-95	Trace = Tr	80-95	0	3
	B- 1150-r	90-100	5	70-90	0	1
7	A- no vent	....	....	....	......	....
	B- 1150 -r	100	0	100	0	1.5
End	A- Cooling	100	0	100	0	2.5
End	B- Cooling	100	0	100	0	1.5

6. It was also noted that the crystal habit of the $CaHPO_4$ used had significant effect upon the final content of the phosphors so produced. This is shown in the following table, given as Table 4-7 on the next page.

Note that most of the analyses show small but significant changes in the CW phosphor composition when the flat, platy square crystals of $CaHPO_4$ are used. This results in more lumen output in the lamp. There are also less residual compounds like β–$Ca_2P_2O_7$ which is actually an unreacted part of the original formulation.

TABLE 4-7

Effect of Type of CaHPO$_4$ used on Final Composition of Phosphors

			CaHPO$_4$ Used		Washed Phosphor		
Phosphor	Type	Remarks	ppm Na$^+$	%Sb	%Cl	%Cd	%Mn
Cool White	Batch	Low Crystallinity	220	0.71	0.46	1.18	0.91
	Squares	High "	5	0.80	0.40	1.08	0.85
	Diamond	High " (7.7µ)	< 3	0.71	0.38	1.13	0.84
	Diamond	High " (15.5µ)	< 3	0.58	0.48	1.14	0.90
	Prefired	γ -Ca$_2$P$_2$O$_7$	< 3	0.68	0.41	1.32	0.92
CW in "A"	Batch	Low Crystallinity	220	0.55	0.50	1.20	0.96
CW in "B"	Batch	Low Crystallinity	220	0.64	0.56	1.32	0.97
CW in "A"	Squares	High "	4	0.77	0.41	1.04	0.95
CW in "B"	Squares	High "	4	0.74	0.61	1.28	0.97
WW in "A"	Batch	Low Crystallinity	220	0.87	0.58	1.16	1,25
WW in "B"	Batch	Low Crystallinity	220	0.69	0.59	1.21	1.26
WW in "A"	Squares	High "	4	0.76	0.53	1.38	1.27
WW in "B"	Squares	High "	4	0.71	0.49	1.36	1.22

As a result of this work, it is safe to say that a gain of 200-300 lumens resulted in fluorescent lamps compared to those phosphors and lamps previously being manufactured.

Once these results were implemented in phosphor production, lamp manufacturers were able to market lamps with more lumen output and so illuminate inside areas at the same lumen level while using fewer lamps than were required previously to do so. This gain was also implemented by the number of lamp colors that were subsequently marketed.

g. Types of Halophosphate Lamps Being Manufactured

The following table, given on the next page, shows the great diversity of emission colors available in the halophosphate phosphor system. All of the lamps contain blends of various phosphors. the standard lamps, i.e.- WW, W, CW and D, contain mostly halophosphate phosphors blended with other phosphors and are used for indoor illumination.

TABLE 4-8

Lamp Colors Manufactured Using the $Ca_5(F,Cl)(PO_4)_3$: Sb:Mn Phosphor System plus Blends with Other Phosphors

Lamp Color	Color Temp.	Lamp Color	Color Temp.
Daylight	6500 °K	Cool White	4100 °K
Super Deluxe Daylight	7000	Cool White	5100
Super Deluxe Daylight	7000	Deluxe Cool White	4200
		Deluxe Cool White	4100
White	3500	Super Deluxe Cool White	4200
Lite White (vegetables)	4100		
Universal White	4000	Warm White	2600
Universal White	4300	Warm White	3000
Soft White	4100	Deluxe Warm White	3000
		Super Deluxe Warm White	3000
Natural White (meat)	2600	Warm Tone	2800
Cool Green (vegetables)	6800		

The other lamps are mostly blends of phosphors other than halophosphates and and are used for specialty lighting such as that found in Supermarkets where meat counters and vegetable stands require special output illumination to emphasize certain reflective colors in the green and red portions of the spectrum.

One should also recognize that this listing is not all-inclusive and that new types of lamp colors are being introduced as Lamp Manufacturers perceive demand in the Marketplace. The current demand has been for fluorescent lamps that use less power and deliver more (or the same) lumens than prior lamps.

In this regard, we have also described compact fluorescent lamps which are currently replacing incandescent bulbs. The reason that they have not done so in entirety is, of course, their relatively high cost in comparison to incandescent lamps. However, you may be sure that when the cost does diminish due to volume manufacture, the incandescent lamp will become like the dinosaur of past ages.

5. MANUFACTURE OF $Sr_5F(PO_4)_3$:Sb PHOSPHOR ("Strontium Blue")

The only other halophosphate phosphor being manufactured (other than the calcium halophosphates) is the $Sr_5F(PO_4)_3$:Sb phosphor. The procedure follows that already given above for the Calcium Halophosphates. Indeed, the same equipment and furnaces are utilized to manufacture this phosphor. The formulation that is generally used is:

4.2.16.- Typical formulation Used for $Sr_5F(PO_4)_3$:Sb Phosphor

Compound	Mol Ratio	Assay	Grams	Lb/Lb
$SrHPO_4$	3.00	98.2%	560.9	0.667
$SrCO_3$	1.46	99.2	217.3	0.258
SrF_2	0.45	99.8	56.64	0.067
Sb_2O_3	0.02	99.9	5.84	0.007

This formulation is blended, hammermilled twice with blending in between, loaded into 2000 ml silica crucibles (with Covers) and then fired at:

1100 °C. for 4 hours.

At the end of the firing period, the crucible is removed from the furnace, the phosphor is cooled, inspected and then washed as given above in a 1% HCl solution.

The phosphor so-produced is blue emitting with an emission peak at 4800 Å. as shown in the following diagram, given as 4.2.17. on the next page.

The QE of this phosphor is 0.72, the luminosity is 0.45, absorbance is 0.88, and the average emission wavelength is near to 4800 Å. A 4.95 cation to 3.00 anion ratio is used to produce the optimum phosphor.

If this phosphor is to be fired in a continuous furnace like those described above, the formulation to be used differs slightly from the above. This can be seen in the following, given as 4.1.18. on the next page.

4.2.17.-

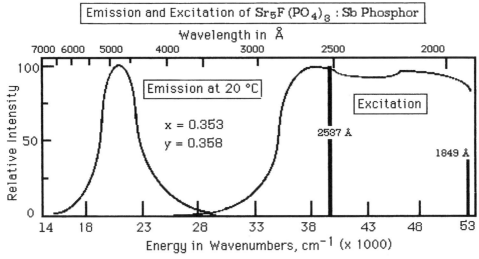

4.2.18.- <u>Formulation Used for Continuous Firing</u>

Compound	Mol Ratio	Assay	Grams	Lb/Lb
$SrHPO_4$	3.00	98.2%	560.9	0.667
$SrCO_3$	1.38	99.2	205.4	0.244
SrF_2	0.50	99.8	62.94	0.075
Sb_2O_3	0.04	99.9	11.72	0.014

The mix is processed as before. The settings for the continuous furnace "B" are as follows:

Zone 1 = 560 °C   Zone 4 = 1050 °C   Zone 7 = 1100 °C
Zone 2 = 650 °C   Zone 5 = 1050 °C
Zone 3 = 875 °C   Zone 7 = 1050 °C

Vent settings: # 1,2,5,6, & & closed Vents 3 &4 = 2 inch hole Nitrogen gas flow = 200 cfm. Tray loading is 8 lbs. per tray.

This phosphor is used as one of the components in Deluxe Cool White blends.

C. MANUFACTURE OF PHOSPHORS USED IN LAMP BLENDS

Nearly all of the other phosphors manufactured are used to prepare blends for use as the phosphor coating in fluorescent lamps. We will describe their manufacture according to the type of activator involved since each type requires a differing approach to its manufacture. A good example is that of Sn^{2+} activated phosphors which mostly require a double firing. The mix is first fired in air to develop the matrix (which is non-luminescent). It is then refired in a reducing atmosphere to develop the luminescent properties of the phosphor.

In general, all phosphors are manufactured according to the steps outlined as given in 4.2.1. above. Reiterating, these steps may be abbreviated as:

4.2.19.- General Procedures for Phosphor Preparation

 a. Assay all materials
 b. Weigh out components
 c. Mix by blending and then by hammermilling all components together.
 d. Load silica trays or crucibles with mixed powder.
 e. Fire according to a proscribed firing schedule, maintaining whatever firing atmosphere is considered essential within the furnace being used.
 f. Cool fired trays and phosphor in specified atmosphere until they reach room temperature. Many times, this will be air.
 g. Inspect fired phosphor "cake" under ultraviolet radiation and scrape off all inert material. Crush cake.
 h. Place in trays and refire as specified in the procedure. Cool as before.
 i. Inspect and crush phosphor cake. Sift through 325 mesh stainless steel screen.
 j. Wash powder by suspension in an aqueous solution if so specified. Filter powder to separate wash solution and to form wet cake.
 k. Dry overnight in oven.
 l. Measure relative luminous efficiency, weigh amount produced and record.
 m. Place in plastic-lined drums for future use.

These steps apply to the manufacture of all phosphors.

1. Manufacture of $Sr_2P_2O_7$: Sn Phosphor

This phosphor is a blue-emitting phosphor which is used in various blends with halophosphate phosphors for specific types of lamp colors. Its emission peak occurs at 4620 Å, its quantum efficiency is 0.84 and its absorbance at 2537 Å is 0.96, i.e.- 96 out of 100 ultraviolet photons are absorbed and 4% are reflected. Quantum efficiency is one measure of the luminous efficiency of a phosphor. In our case, this phosphor emits 86 photons of blue light for every 100 photons of ultraviolet light absorbed. The wavelengths of these emitted photons range from about 4000 Å to 5500 Å with the majority being about 4620 Å in wavelength. The emission-excitation spectrum of this phosphor is shown in the following diagram, viz-

4.2.20.-

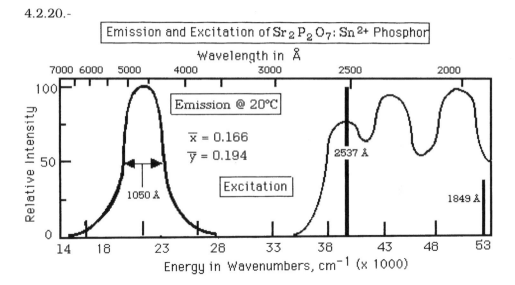

Also given are the CIE color coordinates and the half-width of the emission band. The substitution of Ca^{2+} for Sr^{2+} has no observable effect upon the emission band in quantities less than about 0.40 mols/mol of phosphor. However, the composition used generally has about 12.5 mol% of calcium substituted for strontium, due to the better performance of this composition in lamps.

The formula to be used to prepare this phosphor is given as follows:

416

4.2.21.- Sr$_2$P$_2$O$_7$: Sn Formulation

Mols per Mol Phosphor	Weight in Grams	Lbs./Lb.
1.75 mol SrHPO$_4$	321.30	0.8399
0.25 mol CaHPO$_4$	34.01	0.0889
0.05 mol SnO	6.73	0.0172
0.05 mol (NH$_4$)$_2$HPO$_4$	6.60	0.0174
0.26 mol NH$_4$ Cl	13.91	0.0364

The ratio of cations to anions is 2.00: 2.00 and the activator content is 0.05 mol Sn^{2+} per mol of (Sr,Ca)$_2$P$_2$O$_7$. This phosphor is unique in that most tin-activated phosphors need to be fired in air to form the host crystal (in this case, the matrix is ß-Sr$_2$P$_2$O$_7$ which has the tin incorporated therein), which is then refired in a reducing atmosphere to form the Sn^{2+} valence state in situ. Only the Sn^{2+} valence state has the proper electronic configuration to function as an activator. {It is possible to form a non-luminescent stannate compound which is then subjected to a reducing atmosphere to form the final phosphor. A good example of this is Mg$_2$SnO$_4$:Mn which is inert until refired in a N$_2$-5% H$_2$ gas mixture. Thereupon, this phosphor exhibits a green emission band due to the Mn^{2+} activator. However, the Sn^{2+} cation does not emit but acts as a sensitizer for the green emission which is typical of Mn^{2+} in a cubic crystal environment).

The use of a 2 liter volume silica crucible has been found to produce a brighter phosphor at lower cost. If such a crucible is to be used, the formulation required is:

4.2.22.-

Mols per Mol Phosphor	Weight in Grams	Lbs./Lb.
1.75 mol SrHPO$_4$	1044.2	0.8399
0.25 mol CaHPO$_4$	76.55	0.0889
0.05 mol SnO	15.14	0.0172
0.05 mol (NH$_4$)$_2$HPO$_4$	14.85	0.0174
0.26 mol NH4 Cl	31.30	0.0365

The formulations given above were chosen so that the mix could be fired in air to directly produce the phosphor without having to undergo the two-step firing process. In the latter process, a furnace capable of providing a reducing atmosphere is required, whereas the first-stage firing process can be accomplished in any furnace, and open to the air. The firing stage includes the following steps:

1. Load the mix into a 7.0 inch or 2 liter silica crucible, packing the powder mix by vibration until the crucible is full.

2. Cover the crucible with a loose-fitting silica lid. The general shape of the commercially-available crucible is:

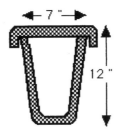

3. The crucible is placed in a furnace set at 1175 °C and removed exactly after 2.0 hours of firing.

4. The mechanism of reaction to form the phosphor involves first the reaction of the phosphates to form the pyrophosphate crystal incorporating the tin- activator:

4.2.23.- $\qquad$ $2 \ SrHPO_4 \ \Rightarrow \ \beta - Sr_2P_2O_7 \ + H_2O \ \Uparrow$

This reaction takes place at about 850 °C. The $(NH_4)_2HPO_4$ decomposes at about 150 °C and reacts with the SnO present to form either stannous or stannic pyrophosphate. The possible reactions are:

4.2.24.- $\qquad$
$$(NH_4)_2HPO_4 \ \Rightarrow \ 2 \ NH3 \Uparrow + H_3PO_4$$
$$2 \ SnO + 2 \ H_3PO_4 \ \Rightarrow \ Sn_2P_2O_7 + 3 \ H_2O \Uparrow$$
or $\quad SnO_2 + 2 \ H_3PO_4 \ \Rightarrow \ SnP_2O_7 + 3 \ H_2O \Uparrow$

5. It is not certain whether the tin oxidizes first to SnO_2 and then is reduced to Sn^{2+} or remains in the divalent state while the solid state

reactions are taking place. Nor is it certain how much of the total tin present reacts. However, Plateeuw and Mayer (2) have reported on the vapor pressure of SnO over the temperature range of 1280-1400 °K.

Their equation gives the vapor pressure of SnO at any temperature:

4,2.25.- $\log p \text{ (mm)} = -13,160/T + 10.775$

At 1120 °C., the vapor pressure is 21.6 torr. Thus, SnO is a vapor at the firing temperature of the phosphor. These authors also state that SnO is stable at these temperatures.

According to Remy (3), NH_3 decomposes to N_2 and H_2 above about 1000 °C. Thus, the purpose of the NH_4Cl present in the formulation is to provide a reducing atmosphere during firing in air, thus obviating a second firing in a reducing atmosphere to form the luminescent state of the tin activator. Undoubtedly, the NH_4Cl also functions as a flux during firing to form the phosphor.

Note that this is the only phosphor formulation known to date that can be prepared by this method. All other tin-activated phosphors need to be double-fired, once in air and the second time in a reducing atmosphere. The reason seems to lie in the fact that only the tin-activated pyrophosphate matrices are stable when reheated in air close to their firing temperatures. Koelmans and Cox(4) have shown that orthophosphate phosphors such as $(Ca,Sr)_3(PO_4)_2$:Sn lose up to 50% of their original brightness when reheated in air to 1100 °C. However, the $Sr_2P_2O_7$:Sn phosphor shows no loss at all when it is reheated in such a manner. This phosphor is normally produced as a fairly soft cake which is then crushed and sieved to remove the largest particles, using a 325 mesh stainless steel screen. Mean particle size produced is usually about 12- 15 μ.

This phosphor is used in Daylight blends with Warm White phosphors and a small amount of a deep-blue emitting phosphor.

2. Manufacture of (SrMg) $_3(PO_4)_2$:Sn Phosphor

Several other tin-activated phosphors are manufactured for use in use for blends with halophosphate phosphors. The methods required involve prefiring to form an inert composition which is then refired in a reducing atmosphere to develop luminous emission.

The procedure for preparing this phosphor follows the steps given in 4.2.19. plus the steps used for firing in silica crucibles. The formula to be used is:

4.2.26.- Formulation Used

Compound	Mol Ratio	Assay	Grams	Lb/Lb
$SrHPO_4$	2.00	94.6%	388.16	0.775
$SrCO_3$	0.50	97.6%	75.63	0.151
$3MgCO_3 \bullet Mg(OH)_2 \bullet 3H_2O$	0.28 MgO	44.6% MgO	25.31	0.051
$CaCO_3$	0.08		8.01	0.016
SnO	0.03		4.04	0.008

The reactions involved are:

4.2.27.- $2\ SrHPO_4 \Rightarrow Sr_2P_2O_7 + H_2O \Uparrow$

$Sr_2P_2O_7 + SrCO_3 \Rightarrow Sr_3(PO_4)_2 + CO_2 \Uparrow$

The mix is prepared and then fired:

a, First Firing (in Air): 4.0 hours @ 1150 °C

b. Second Firing (in 95% N_2 - 5% H_2): 2.0 hours @ 1175 °C

The gas flow will depend upon the cross-sectional area of the furnace used. A total gas flow of about 15.0 liters/minute is generally sufficient. After the first firing, the optically-inert cake is inspected, crushed and sieved through a coarse screen to produce a free-flowing powder before the second firing is carried out. In general, ß- $SrHPO_4$ produces a better phosphor than the α- $SrHPO_4$ high-temperature form. Once the product is reduced to its final luminescent state, the powder is inspected under ultraviolet light and sieved via 325 mesh to remove all particles larger than about 44μ in size. The emission-excitation spectra are shown in the following diagram:

420

4.2.28.-

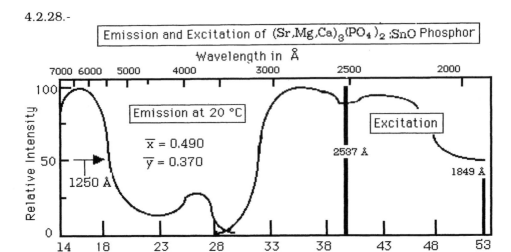

| Emission and Excitation of $(Sr,Mg,Ca)_3(PO_4)_2 :SnO$ Phosphor |

The CIE color coordinates are also given. The excitation spectrum peaks at about 2900 Å and the emission peak occurs at about 6280 Å with a 50% bandwidth of about 1250 Å.. The QE is 0.92, the absorbance 0.97, the ratio of cations to anions is 2.90 : 2.00 and the activator content is 3 mol%. Particle sizes usually produced range from 9 to 12 μ.

This phosphor is used in Deluxe blends where a deep red color rendition is required. It has been determined that the above specified addition of the 0.08 mol of $CaCO_3$ to the formulation results in a phosphor with a gain in luminous output in lamps of over 10% without changing the emission color. Additionally, an improvement in long term maintenance of over 20% is seen when such a phosphor is used in blends. If MgO is used in place of the $3MgCO_3 \cdot Mg(OH)_2 \cdot 3H_2O$ specified above, the output of the phosphor is lowered by about 10%.

This phosphor has also been employed as the main component in Gro-lux™ blends along with a blue-emitting phosphor to manufacture a lamp to stimulate plant growth. However, it was found not to be stable during bakeout when applied in the standard ethylcellulose lacquer to form the internal phosphor layer in lamps. To correct this, it is necessary to add about 0.05% $(NH_4)_2HPO_4$ to the lacquer before applying the phosphor-lacquer suspension to the lamp and then baking out the lacquer. This procedure prevents loss of phosphor brightness, particularly when SO_2 is employed to prevent glass surfaces of the lamp bulbs from seizing during lehring (bakeout). Up to 100 lumens (2.5 lumens/watt in a 40 watt fluorescent lamp) can be obtained in this manner.

In some cases, the performance of this phosphor can be improved in lamps by a washing technique:

4.2.29.- <u>Washing Procedure for (SrMg) $_3$(PO$_4$)$_2$:Sn Phosphor</u>

    1. Weigh out 400.0 grams of powder and place in a one quart ceramic mill. Add 200.0 ml. of water.

    2. Roll for 2.5 hours.

    3. Drain into a 1.0 liter beaker and add two 200 ml. water washes to remove all powder from the mill.

    4. Stir 10 minutes. Make up a HNO_3 solution containing 974 ml, of water and 26 ml. of concentrated HNO_3. Add 200 ml. of this solution to 1000 ml. beaker containing powder being stirred. Stir 10 more minutes.

    5. Let powder settle, mark both powder level after settling and top of water level. Siphon off water.

    6. Add water to top mark and stir 10 minutes. Let settle and repeat this procedure twice more.

    7. Let settle and filter. Dry wet powder in 110 °C. oven overnight.

This procedure also helps in protecting this phosphor from loss of output during the lamp making procedures.

    3. <u>Manufacture of (SrZn)$_3$(PO$_4$)$_2$:Sn Phosphor</u>

A related composition involves a slightly different combination of components which is fired in a manner similar to that given above for the (Sr,Mg)$_3$(PO$_4$)$_2$: Sn phosphor. This composition involves the following formulation:

4.2.30.- Formulation Used for $(SrZn)_3(PO_4)_2$:Sn Phosphor

Compound	Mol Ratio	Assay	Grams	Lb/Lb
SrHPO4	2.00	94.6%	388.16	0.776
SrCO3	0.55	97.6%	83.19	0.164
ZnO	0.24	98.8%	19.77	0.040
Al2O3	0.083	99.8%	8.48.	0.016
SnO	0.019	------	2.56	0.004

The mix is prepared as specified above for $(SrMg)_3(PO_4)_2$:Sn phosphor.

This composition is fired identically to those conditions already specified for $(SrMg)_3(PO_4)_2$:Sn phosphor except that the firing temperatures are slightly higher, i.e.-

First Firing (in Air): 4.0 hours @ 1250 °C

Second Firing (in 95% N_2 - 5% H_2): 2.0 hours @ 1175 °C

The cake is processed as given above before the refiring step.

The emission-excitation spectra are given in the following diagram:

4.2.31.-

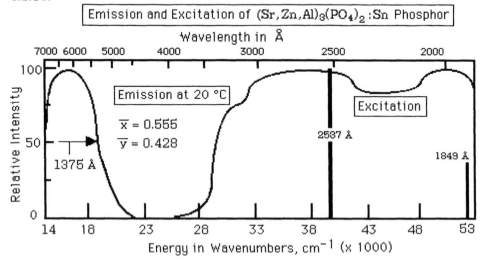

Emission and Excitation of $(Sr,Zn,Al)_3(PO_4)_2$:Sn Phosphor

The QE of this phosphor is 0.80, the absorbance 0.92 and the ratio of cations to anions is 2.89: 2.00. Mean particle size is 32 μ. Note that the CIE colors are shifted from that of the $(SrMg)_3(PO_4)_2$:Sn phosphor. However, this phosphor has not been used in fluorescent lamps due to its lower efficiency. It was originally developed for use in HPMV lamps because at temperatures above about 300 °C, its response to HPMV ultraviolet radiation (primarily 3650 Å) increases dramatically.

Note the extra-"hump" in the excitation curve given above. At 300°C, this part of the excitation response increases radically, becoming a major excitation of the phosphor. Its major use was for HPMV lamps. The coated HPMV lamp was called "Silver-White" and was marketed under that name to the Trade. However, it is now obsolete, having been replaced by both vanadates for red rendition, and by additive HPMV lamps.

4. Manufacture of $(Ca,Zn)_3(PO_4)_2$: Sn Phosphor

The procedure for preparing this phosphor follows the steps given in 4.3.19. including those of firing in air, followed by refiring in a reducing atmosphere.

The formula usually employed is:

4.2.32.- Formulation for $(Ca,Zn)_3(PO_4)_2$: Sn Phosphor

Compound	Mol Ratio	Assay	Grams	Lb/Lb
$CaHPO_4$	2.00	98.6%	275.98	0.758
$CaCO_3$	0.55	97.6%	56.40	0.155
ZnO	0.22	98.8%	18.12	0.049
$3MgCO_3 \bullet Mg(OH)_2 \bullet 3H_2O$	0.04 MgO	44.6% MgO	3.63	0.010
$SrCO_3$	0.04	99.2%	5.95	0.016
SnO	0.03		4.04	0.011

The solid state reactions that occur are very similar to those already given above except that ZnO is involved in the reactions.

The mix is prepared and then fired in two stages, as shown in the following firing schedule:

4.2.33.- Firing Schedule for $(Ca,Zn)_3(PO_4)_2$: Sn Phosphor

First Firing (in Air): 4.0 hours @ 1150 °C

Second Firing (in 95% N_2 - 5% H_2): 2.0 hours @ 1175 °C

The gas flow again depends upon the cross-sectional area of the furnace used. A total gas flow of about 15.7 liters/minute is generally sufficient with about 1800 ml./minute of H_2 gas. The inert cake is inspected, crushed and sieved through a coarse screen to produce a free flowing powder which is then refired. the final step is to inspect the powder under ultraviolet light and to sieve the powder through a 325 mesh screen to remove all particles larger than 44μ. In general, ß- $CaHPO_4$ produces a better phosphor than the α-$CaHPO4$ high-temperature form. The emission-excitation spectra are shown in the following diagram:

4.2.34.-

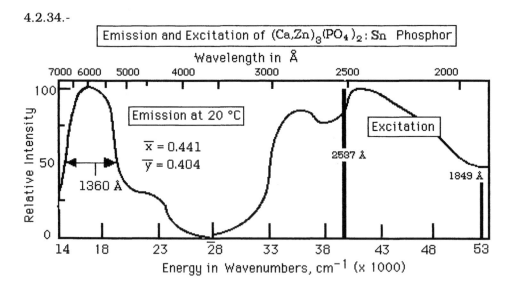

The QE of this phosphor is 0.86, absorbance is 0.84, and the ratio of cations to anions is 2.88 : 2.00 with an activator content of 3 mol%. The half-width of the emission band is 1360 Å. The emission band is quite wide and peaks near to 6100 Å with the maximum of the excitation band near to 2480 Å. Average particle size ranges from 9 - 12 μ.

This phosphor is also used in deluxe blends with halophosphate phosphors but does not have quite the red rendition of the $(SrMg)_3(PO_4)_2$:Sn phosphor. Therefore, its use has not been as extensive as that of the latter.

5. Manufacture of $Zn_2 SiO_4$:Mn Phosphor

This phosphor is a green-emitting phosphor whose peak wavelength occurs at 5280 Å. The emission is due to the Mn^{2+} activator which is probably not the actual energy absorber since its electronic configuration is: $d^5 = {}^6S_{5/2}$. As stated previously (1), this cation is a good activator but requires a sensitizer for efficient emission. Thus, the most probable formula for this phosphor is:

4.2.35.- $\qquad\qquad Zn_{2-x} SiO_4 : (Zn^0{}_{(O=)}): Mn_x$

where the neutral Zn^0 cation occupies an oxide site in the silicate tetrahedron of the lattice structure. Its electronic structure is: $d^{10}s^2$ with a ground state of $^1S_0$.

A formula used by one manufacturer to prepare this phosphor is:

4.2.36.- Formulation for $Zn_2 SiO_4$:Mn Phosphor

	Mols/ Mol Phosphor	Weight in	
		Grams	Lbs./Lb.
ZnO	2.00	162.68	0.6934
SiO_2	1.08 (as SiO_2)	64.88	0.2766
$MnCO_3$	0.06	6.90	0.0294
PbF_2	0.00053	0.130	0.00055

This mix is blended thoroughly following all the steps given in 4.2.15. The assay of SiO_2 will vary according to the nature of the silica being used. In general, it must be a pure grade with no more than about 5 ppm total transition metal elements present. A large 2-liter silica crucible is generally used for firing:

4.2.37.- Firing Schedule for $Zn_2 SiO_4$:Mn Phosphor

Fire 6.0 hours in air @ 1235 °C using a covered crucible.

The fired cake is then processed as given above in 4.2.19.

Still another formulation that has been used for this phosphor is given as follows:

4.2.38.- Another Formulation for Zn $_2$ SiO$_4$:Mn Phosphor

	Mols/ Mol Phosphor	Weight in	
		Grams	Lbs./Lb.
ZnO	1.82	148.03	0.6910
SiO$_2$	1.00 (as SiO$_2$)	60.08	0.2804
MnCO$_3$	0.05	5.75	0.0268
PbF$_2$	0.0007	0.172	0.00055
As$_2$O$_3$	0.0010	0.197	0.00080

This mix is blended as before and fired 11.0 hours @ 1270 °C. in a covered crucible. The cake is then processed as stated before.

The excitation-emission spectra of this phosphor are shown in the following diagram:

4.2.39.-

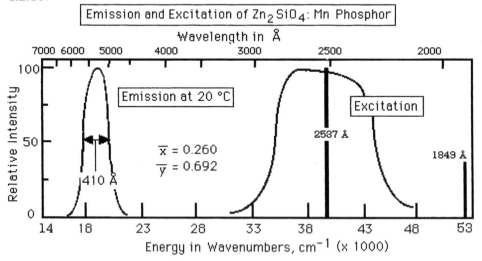

Average particle size produced is usually 6-9μ. The QE is 0.82, the absorbance is 0.93, the activator content is generally 5-6 mol% and the ratio of cations to

anions is about 2.00 : 1.08. The phosphor is generally used in blends with halophosphates and/or other phosphor colors.

In some cases, one requires a lamp phosphor whose decay is longer, that is- the time required for the phosphor to reach 1/e of its original intensity once the excitation source intensity ceases.

In that case a formulation that has been used to produce this effect is given in the following, viz-

4.2.40.- Formulation for a Long Decay Zn_2SiO_4 :Mn Phosphor

	Mols/ Mol Phosphor	Weight in	
		Grams	Lbs./Lb.
ZnO	2.00	162.68	0.6732
SiO_2	1.16 (as SiO_2)	69.69	0.2884
$MnCO_3$	0.06	6.90	0.0269
PbF_2	0.00816	2.00	0.0083
As_2O_3	0.00189	0.373	0.00154

This mix of materials is fired:

3 hours @ 1300 °C. in a covered silica tray.

It has been found that the addition of Pb^{2+} increases the luminous efficiency of the phosphor, as does the addition of As^{3+}. However, the latter introduces a large increase in the decay of the phosphor, probably by introduction of vacancies which form mandatory upon the introduction of a trivalent cation into a divalent cationic lattice. This factor has also been found useful for cathode-ray usage of this phosphor where the As^{3+} is increased considerably over the formula given above.

We have presented all of the above formulations to show the formula variations that have been used by various manufacturers. Actually, the Zn_2SiO_4 :Mn phosphor is one of the more important phosphors used for lamp blends. It has been known since the 1880's and occurs as a natural ore, "Willemite", from which ZnO and Zn metal are extracted. Willemite fluoresces under irradiation by ultraviolet light (primarily 3650 Å) and this was used early-on for detection of ore bodies. Indeed, in the late 1880's, the only phosphors available were the ZnS phosphors and artificial willemite. Willemite was designated as the first cathode-ray phosphor, i.e.- P-1 (according to JEDEC notation).

However, there are some applications where this phosphor is used as the major component in the lamp. In this case, it has been found that the long-term maintenance is very poor. It has been found that a thermal treatment of the phosphor after it has been prepared serves to produce a material that exhibits good maintenance in a fluorescent lamp.

The heating schedule to be used is as follows:

a. Heat linearly at a rate of 7.1 °C/ minute to a temperature of 1065 °C. (This takes about 150 minutes).

b. Hold this temperature for 90 minutes and then decrease temperature to 1038 °C. over a period of 30 minutes (This is a rate of 1.0 °C/minute).

c. Raise the temperature to 1150 °C at a rate of 3.7° C per minute for the space of 30 minutes and hold this temperature for 30 minutes.

d. Following this time, drop the temperature rapidly to 1065 °C and maintain this temperature for 45 minutes. At the end of 45 minutes, the temperature is raised to 1095 °C and maintained for 45 minutes. The total thermal treatment time has been 505 minutes.

e. At the end of this time, the temperature in the furnace is allowed to drop at a rate of 100 °C/ minute. Since the furnace will cool in an exponential manner, this will take about 12 hours for the crucible to reach 400-450 °C.

f. At the end of the 12 hour period, the crucible is removed from the furnace and allowed to cool naturally.

g. The thermally treated phosphor is then sieved through a 325 mesh screen to obtain a soft powder.

h. A lamp coating lacquer is prepared in a normal manner using the thermally treated phosphor and 0.04% by weight of Sb_2O_3 of the phosphor weight used is added and milled into the lacquer suspension before coating to form the luminescent layer, followed by subsequent lehring.

It has been found (5) that the Zn_2SiO_4 crystal undergoes a dimorphic transition above about 1095-1115 °C. However, the transition is reversible and it is necessary to quench the high temperature crystal modification by fast cooling. Although it is possible to remove the hot phosphor crystals from the furnace and to quench them directly upon a cold steel plate, the above controlled temperature cycling through the transition temperature zone seems to work better and produces a Zn_2SiO_4 crystal phosphor which exhibits a much improved luminous output and maintenance of that output when used as the major component in a fluorescent lamp.

Although several newer green-emitting phosphors have been developed, the Zn_2SiO_4 : Mn phosphor remains the "work-horse" for use in blends, due primarily to its lower cost of raw materials and ease of manufacture.

6. Manufacture of $CaSiO_3$: Pb: Mn Phosphor

This phosphor is a orange emitting phosphor sensitized by Pb^{2+} with Mn^{2+} as the activator. The formulation usually employed is given as follows:

4.2.41.-   Formulation Used for $CaSiO_3$: Pb: Mn Phosphor

Compound	Mol Ratio	Assay	Grams	Lb/Lb
$CaCO_3$	1.00	0.986	101.51	0.538
SiO_2	1.20	0.769	78.12	0.414
PbF_2	0.006	0.973	1.512	0.0080_6
$MnCO_3$	0.06645	-------	7.638	0.0405

This formulation is weighed out, mixed together and then hammermilled.

The firing conditions to be used are:

a. Place mix in a covered 7" silica crucible and fire 6 hours in air at 605 °C. Let cool, inspect cake and crush into powder.

b. Wet mill for 4 hours in a ball mill, using about 400 grams of powder per liter of deionized water. Filter and dry powder at 110 °C. for 12 hours.

c. Place milled powder in a 7" silica crucible and refire in air at 1105 °C

At the end of the firing, the cooled cake is inspected, crushed and sieved through a 325 mesh screen to remove all > 44μ particles.

Still another formulation that has been used is the following mixture:

4.2.42.- Formulation Used for CaSiO$_3$: Pb: Mn Phosphor

Compound	Mol Ratio	Assay	Grams	Lb/Lb
CaCO$_3$	0.823	0.986	83.54	0.493
CaF$_2$	0.0084	0.992	0.661	0.0039
SiO$_2$	1.00	0.769	78.13	0.461
PbO	0.0035	0.973	0.803	0.0047
MnCO$_3$	0.054	-------	6.207	0.0366

The components are blended, hammermilled twice together with blending in between and fired:

6.0 hours in air at 1200 °C in a covered silica tray

The phosphor cake is then processed as given above.

The emission-excitation characteristics are given in the following diagram, labelled as 4.2.43. on the next page

Note that emission bands due to both Mn^{2+} and Pb^{2+} are present. There are two excitation bands, due probably to crystal-field splitting of the $^3P_1$ state.. The QE is 0.78, the absorbance is 0.88, the average particle size produced is about 8-9 μ, and the emission is classified as "deep-red".

4.2.43.-

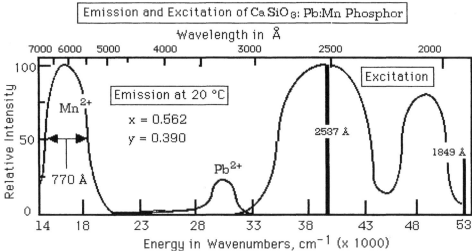

This emission is typical of that of Mn^{2+} in an octahedral $(^4T_2)$ coordination and may be compared to that of Mn^{2+} (green) in a tetrahedral $(^4T_1)$ coordination. The ground state of Mn^{2+} is $^6S_{5/2}$ which transforms in a crystal field to $^6A_1$ where all the electron spins are parallel. In the excited state, a single spin-reversal occurs. Such excitation transitions are parity-forbidden but spin-allowed. Thus, the excitation transition of Mn^{2+} alone is a quadrupole-quadrupole type with intensities of $<10^{-7}$. For this reason, Mn^{2+} cannot absorb energy efficiently but can become excited by energy transfer from a more efficient activator like Pb^{2+} which has a dipole allowed transition. However, once Mn^{2+} has become excited, its transition, being spin-allowed to the ground state, occurs with high efficiency. It is for these reasons that Mn^{2+} is a good emitter, but is a poor absorber of radiation. Thus, a good absorber like Pb^{2+} is required to produce an efficient red emission from manganese.

That the above statements are true can be seen by examining the following diagram which shows the excitation-emission characteristics of $CaSiO_3$: Pb., as shown in the following diagram, given as 4.2.44. on the next page.

In this case, the major emission peak is due to Pb^{2+} and occurs at 3300 Å. The excitation peaks are nearly identical to those given above for the $CaSiO_3$: Pb: Mn phosphor.

432

4.2.44.-

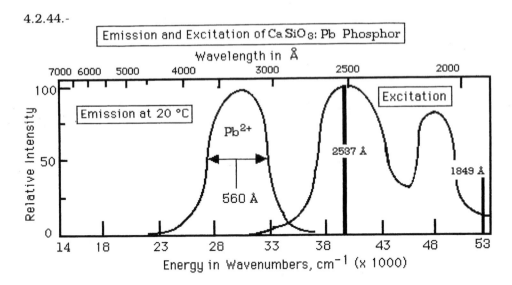

Emission and Excitation of CaSiO₃: Pb Phosphor

Thus, it should be obvious that excitation occurs in the Pb^{2+} center but that the red emission comes from the Mn^{2+} center. Additionally, it is possible, by controlling the ratio, $Pb^{2+} : Mn^{2+}$, to produce a phosphor in which the emission band of each activator are equal to each other in intensity.

The above phosphor is prepared according to the formula given as follows. In this case, the only activator is Pb^{2+} and the preparation is identical to that given above.

4.2.45.- Formulation Used for CaSiO₃ Phosphor

Compound	Mol Ratio	Assay	Grams	Lb/Lb
$CaCO_3$	1.00	0.986	101.51	0.562
SiO_2	1.20	0.769	78.12	0.434
PbF_2	0.006	0.973	1.512	0.004₆

This formulation is blended, hammermilled twice with blending in between and then is fired:

      a. Place in a covered 7" silica crucible and fire 6 hours in air at 650 °C

b. Wet mill for 4 hours in a ball mill, using about 400 grams of powder per liter of deionized water, and then filter to produce a wet cake.

c. Place wet cake of milled powder in a 7" silica crucible and refire in air at 1100 °C. This has the effect of firing in steam and helps promote the crystallinity of the final phosphor

The fired cake is cooled, inspected and sieved through a 325 mesh screen to remove particles larger than about 44 μ.

The QE of this phosphor is 0.75, the absorbance is 0.85 and the particle size is about 15 μ in average size. However, it has been supplanted as an ultraviolet emitter by some of the newer divalent europium activated phosphors which are more stable and have higher intensities of emission.

Still another formulation that has been used to make this phosphor is given as

4.246.- Formulation Used for $CaSiO_3$: Pb: Mn Phosphor

Compound	Mol Ratio	Assay	Grams	Lb/Lb
$CaCO_3$	1.000	0.986	101.51	0.485
CaF_2	0.010	0.992	0.786	0.0038
SiO_2	1.20	0.769	93.75	0.448
PbO	0.035	0.973	9.612	0.046
$MnCO_3$	0.033	-------	3.793	0.0182

This mix is blended as before and hammermilled twice, with blending in between. It is then fired in covered trays in air at:

4 hours at 1200 °C.

The QE of this phosphor is 0.80, A is 0.88, x= 0.588 and y = .406, the Pb^{2+} peak is 36% of that of the Mn^{2+} peak (in contrast to the 20% peak shown in 4.2.34.), and the peak wavelength of Mn^{2+} is 6600 Å (in contrast to the 6300 Å peak shown in 4.2.34.) Average particle size is 10-12 μ and the luminosity is 0.38. Thus, this formulation results in a "redder" phosphor than that of 4.2.43.

The CaSiO$_3$: Pb: Mn phosphor is not used very frequently in fluorescent lamps any more, having been replaced by other broad-band red-emitting phosphors such as the tin-activated phosphors described above. However, it still finds extensive use in sign-tubing, i.e.- see Chapter 2- page 188., and hence is still manufactured, but in smaller quantities than in the past.

7. Manufacture of MgWO$_4$:W Phosphor

This phosphor is a blue-green emitting phosphor with an emission peak at 4780 Å. As described above, it is "self-activated", and is usually prepared by calcining MgO and WO$_3$ together. Thus, the activator is the optically active tungstate anion (whose exact electronic structure and valence is not definitively known). The formula to be employed is:

4.2.47.- Formulation Used

Compound	Mol Ratio	Assay	Grams	Lb/Lb
3MgCO$_3$ • Mg(OH)$_2$ • 3H$_2$O	1.58 MgO	44.6% MgO	142.78	0.375
WO$_3$	1.00	97.4%	238.04	0.625

This mix is hammermilled twice, blending before and after each hammermilling step. The mix is then loaded into 2 liter silica crucibles and fired:

6 hours at 980 °C. in air

The mol ratio can vary between 1.46 to 1.80 MgO per 1.00 mol of WO$_3$. It is usual to make a small batch first to determine the optimum ratio since the brightest phosphor to be obtained depends upon the actual particle size of the two components being fired (which determines the rate of reaction between the two components being fired). Once this ratio is established, then manufacture of a larger batch can proceed.

Once the phosphor is cool, it is inspected and then dry-milled for about 2 hours. It is essential not to overmill the product so as not to lower the brightness of the phosphor. It has been found that the process used to prepare this phosphor is sensitive to over-grinding and so that a minimum of milling is used so as to disperse lumps and produce a free-flowing powder. Finally, the powder is screened through a 325 mesh screen to remove all particles larger than about 44μ. Sometimes, the powder is acid-washed (~ 5% HCl solution) to remove excess MgO present.

The excitation-emission bands are shown in the following diagram:

4.2.48.-

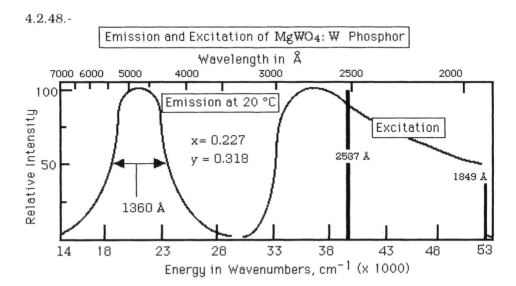

The QE is 0.90, A = 0.94, luminosity is 0.39, and the average particle size is about 1 -3 μ. This phosphor is used in fluorescent lamps primarily because of the position of its emission band at 4780 Å. It is utilized in Deluxe Cool white blends and Super Deluxe Cool White blends for fluorescent lamps, both of which are high volume sales items.

8. Manufacture of CaWO₄ :Pb Phosphor

This phosphor is blue-emitting and has an emission band at 4330 Å. As we stated before, $CaWO_4$ does not emit at room temperature but the addition of Pb^{2+} as an activator produces a room-temperature emitting phosphor.

One formula that has been employed is given as follows on the next page as 4.2.49., viz-

4.2.49.- Formulation Used for Calcium Tungstate Phosphor

Compound	Mol Ratio	Assay	Grams	Lb/Lb
$CaSO_4 \bullet 2 H_2O$	1.347	32.8% CaO	233.3	0.491
WO_3	1.000	97.4%	238.0	0.501
$Pb(NO_3)_2$	0.011	98.7%	3.68	0.00177

The mix is hammermilled twice, blending before and after each hammermilling. It is important to obtain complete blending by hammermilling so that the solid state reaction will go to completeion under the set firing conditions.

The mix is then fired:

2 hours in a covered tray at 1050 °C.

Milled for 2 hours

Refired in a covered tray for 2 hours at 1050 °C.

After firing, the cake is inspected, crushed and screened through a 325 mesh screen to remove particles larger than 44μ. It has been found that **all** tungstate phosphors are sensitive to milling, and if this is done, care must be exercised that loss of brightness does not occur during the dry milling step. (In general, wet-milling is never used because of the propensity of tungstates to react with water, thereby lowering their luminous output).

The excitation-emission characteristics are shown in the following diagram, given as 4.2.50 on the next page.

The QE is 80%, A = 0.92, the luminosity is 0.20 and the average particle size obtained is about 4-5 μ.

Actually, this phosphor has been manufactured using several formulations and methods. One manufacturer makes a $CaWO_4$:W phosphor (it actually has a small amount of Pb^{2+} added as an activator), a $CaWO_4$:Pb phosphor, and a $(Ca,Mg)WO_4$: Pb phosphor.

4.2.50.-

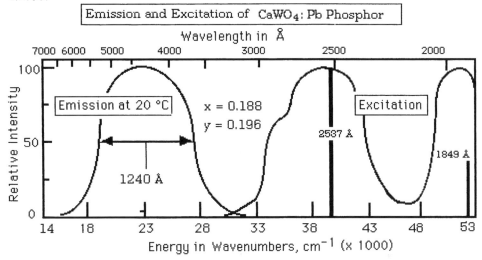

The formulations used for these variations are:

4.2.51.- Formulation for $CaWO_4$:W Phosphor (Phosphor-A)

Compound	Mol Ratio	Assay	Grams	Lb/Lb
$CaCO_3$	1.005	98.6%	102.02	0.230
WO_3	1.000	97.4%	238.04	0.699
PbO_2	0.00166	98.7%	0.4023	0.001182

The mix is hammermilled twice as given before and fired:

5 hours at 985 °C. in a covered silica tray

After firing, the cake is processed as given above.

Still another formulation that has been used is:

4.2.52.- Formulation for $CaWO_4$: Pb Phosphor (Phosphor-B)

Compound	Mol Ratio	Assay	Grams	Lb/Lb
$CaCO_3$	1.347	98.6%	136.74	0.362
WO_3	1.000	97.4%	238.04	0.631
PbO_2	0.011	98.7%	2.666	0.007063

This mix is hammermilled as before and fired in air in covered trays at:

3 hours at 1070 °C in covered silica trays

Dry mill for two hours (about 300 gm/quart mill)

6 hours at 1040 °C in covered silica trays

After the first firing, the phosphor is inspected, crushed and sieved through a 200 mesh screen to break up lumps so as to obtain a uniform powder. It is then refired in air, cooled, inspected and then sieved through a 325 mesh screen to remove particles larger than 44 μ.

Still another formulation that has been used is as follows:

4.2.53.- Formulation for (Ca,Mg)WO $_4$:Pb Phosphor (Phosphor-C)

Compound	Mol Ratio	Assay	Grams	Lb/Lb
$CaCO_3$	0.98	98.6%	99.48	0.221
$3MgCO_3 \cdot Mg(OH)_2 \cdot 3H_2O$	0.30 MgO	44.6% MgO	108.38	0.241
WO_3	1.000	97.4%	238.04	0.529
PbO_2	0.010	98.7%	3.680	0.008185

In this case, part of the Ca^{2+} cation is substituted by Mg^{2+} to form a $(Ca,Mg)WO_4$:Pb composition:

The mix is processed as before and fired in silica crucibles, viz-

Firing Conditions Used

6 hours at 1040 °C in air in a covered silica tray

Mill 2 hours (300 gm. in a quart mill)

3 hours at 900 °C. in air in a covered silica tray

The cake is then processed as before. The reasons why these variations in formulation were made can be seen by examination of their spectral properties, viz-

4.2.54.- Comparison of Spectral Properties of Phosphors "A", "B" & "C".

Property	Phosphor "A"	Phosphor "B"	Phosphor "C"
Peak Emission:	4330 Å	4460 Å	4610 Å
1/2 Bandwidth:	1140 Å	1220 Å	1240 Å
Color Coord.			
x =	0.238	0.182	0.182
y =	0.206	0.191	0.221
QE =	0. 76	0.80	0.84
A =	0.92	0.92	0.92
Luminosity =	0.20	0.21	0.26
Cation to Anion:	1.00666:1.0001	1.358:1.000	1.290:1.000
Act, Conc. :	0.00166	0.011	0.010

It should be clear that Phosphor "B" is bluer than any of the other phosphors. Phosphor "C" can be classified as being more of a "green-blue" emitter. It is for this reason that the latter one, i.e. - Phosphor "C", is not being manufactured any more. These phosphors are used in Deluxe blends in fluorescent lamps.

As a matter of general interest, it has been determined that the addition of SiO_2 in an amount about 0.10% by weight of the mix improves the phosphor by about 10% in luminous output over that not containing silica. It is added during the milling step before the second firing. It is believed that the SiO_2 acts as a flux, thereby producing a more uniform phosphor. However, none of the phosphor manufacturers seem to be using this procedure at the present time.

9. Manufacture of $Ba_2P_2O_7$: Ti Phosphor

This phosphor was originally developed for use in Daylight blends. However, it has found use in many other blends as well and remains one of the more useful blend phosphors.

The formulation used for its manufacture is:

4.2.55.- Formulation for Manufacture of $Ba_2P_2O_7$:Ti Phosphor

Compound	Mol Ratio	Assay	Grams	Lb/Lb
$BaHPO_4$	2.0000	98.2%	475.17	0.858
BaF_2	0.0532	99.2%	9.403	0.0170
TiO_2	0.5840	99.0%	47.133	0.0852
AlF_3	0.0556	99.5%	4.693	0.0848
BaO_2	0.00276	98.5%	9.558	0.0173
$(NH_4)_2HPO_4$	0.0560	-------	7.395	0.0134

These ingredients are hammermilled twice, blending before and after each hammermilling. A 2 liter silica crucible is used for firing. (If a smaller volume than this is used, the phosphor so-produced will be inferior in luminous output in a fluorescent lamp). The firing schedule is:

Fire in air for 4.0 hours at 985 °C.

Refire in air for 4.0 hours at 985 °C.

After the first firing, the phosphor is inspected and rolled out. 10.00 gram of BaO_2 per 1000 gram of prefired phosphor is then added with thorough blending to form a uniform mixture. This mixture is then refired as given above. Alternately, 15.00 grams of AlF_3 can be added in place of the BaO_2.

The excitation-emission spectra are given in the following diagram, given as 4.2.56. on the next page.

The emission band is very wide in relation to those of other phosphors and peaks at 4940 Å. The QE is 0. 88, A is 0.92, the luminosity is 0.45 and the average particle size produced is about 4-6 μ. The cation to anion ratio is : 2.00 : 2.50, and the activator concentration is 0.553 mol%.

The solid state reactions are believed to follow the scheme shown in 4.2.57. as given also on the next page.

4.2.56.-

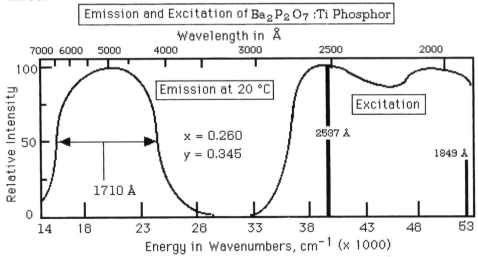

4.2.57.- Reactions Leading to the formation of the $Ba_2P_2O_7$:Ti Phosphor

$2\ BaHPO_4$	$\Rightarrow$	$Ba_2P_2O_7 + H_2O \Uparrow$
$2\ BaF_2 + 4\ (NH_4)_2HPO_4$	$\Rightarrow$	$Ba_2P_2O_7 + H_2O \Uparrow + 2\ HF$
$2\ BaO_2$	$\Rightarrow$	$2\ BaO + O_2$
$Ba_2P_2O_7 + 2HF + TiO_2 + O_2$	$\Rightarrow$	$Ba_2P_2O_7 : TiO_3F + H_2O$

This phosphor is used in Daylight, Super Deluxe white, and Deluxe Cool White blends for fluorescent lamps.

10. Manufacture of $CdBO_3$:Mn Phosphor

This phosphor is a red emitting phosphor wherein the Mn^{2+} center is in an octahedral lattice site i.e.- ($^4T_2$) coordination, as discussed above for the $CaSiO_3$:Pb:Mn phosphor. The actual phosphor composition is probably:

$$Cd_{1-x}BO_3 : (Cd^O_{(O\ =)}): Mn_x$$

where Cd^O occupies an oxide site, as discussed for the Zn_2SiO_4 crystal phosphor. Note that this valence state is specified because it is the one which possesses the proper $^1S_0$ ground state to function as an activator for the Mn^{2+} emitter.

One formulation used has been:

4.2.58.- Formulation for Manufacture of CdB_2O_4 :Mn Phosphor

Compound	Mol Ratio	Assay	Grams	Lb/Lb
$CdCO_3$	1.00	98.6%	174.87	0.571
H_3BO_3	1.20 as B_2O_3	57.3% as B_2O_3	129.49	0.423
$MnCO_3$	0.015	99.4%	1.735	0.00567

This mix is hammermilled and blended as described above and then is fired in covered silica trays in air.

4.2.59. - Firing Schedule for CdB $_2O_4$

> 6 hours in air at 770 °C
> Mill 2 hours
> Refire: 6 hours in air at 750 °C

After the first firing, the cakes are inspected and rolled out. The powder is then milled in as an aqueous suspension for 2 hours, using about 400 gram of powder per 600 ml. of water (enough to make a thick slurry). After filtration and drying of the wet cake, the dry powder is then refired, inspected again and finally sieved through a 325 mesh screen to remove particles larger than 44 μ.

Another formulation that has been used is:

4.2.60.- Formulation for Manufacture of CdB_2O_4 :Mn Phosphor

Compound	Mol Ratio	Assay	Grams	Lb/Lb
$CdCO_3$	1.00	98.6%	174.87	0.632
H_3BO_3	0.94 as B_2O_3	57.3% as B_2O_3	101.43	0.367
$MnCO_3$	0.003	99.4%	0.3469	0.000125

This mix is blended and hammermilled as before and then is fired:

4.0 hours in air in open silica trays at 770°C

The phosphor cake is then processed as before except that it is not milled but screened through a 325 mesh screen.

The excitation-emission spectra are shown in the following diagram:

4.2.61.-

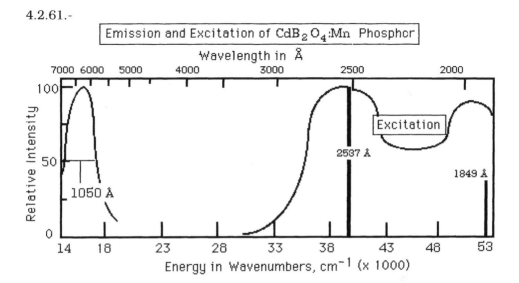

The QE is 0.72, A = 0.66, the luminosity is 0.31, and the average particle size produced is 6-9 μ. The peak emission occurs at 6180 Å and the cation to anion ratio is: 1.00 to 1.20.

This phosphor has become obsolete today because of the EPA regulations concerning cadmium disposal and the fact that several other phosphors emit in this same red region and have much better maintenance in fluorescent lamps. However, when lamp phosphors were first being developed, this phosphor was one of the few deep-red emitters available. It was used with the $ZnBeSiO_4$:Mn phosphor in early fluorescent lamps.

11. Manufacture of $Cd_5Cl(PO_4)_3$:Mn Phosphor

This phosphor is another example wherein the sensitizer is probably Cd^0, as presented above, i.e.- the actual composition is expected to be:

4.2.62.- $\quad\quad\quad Cd_{5-x}Cl(PO_4)_3$: $(Cd^0{}_{(O=)})$: Mn_x

where the Cd^0 occupies an oxide site and sensitizes the Mn^{2+} emission. Although this phosphor is based on cadmium, it is still manufactured to some extent because of its emission color and the fact that it will produce a 3400 lumen lamp with good maintenance.

The formulation generally used for its production is as follows. Note that the preparation of the three compounds used in the formula was described in Chapter 3. (Only $CdCl_2$ is commercial and is not manufactured in house).

4.2.63.- Formulation Used for $Cd_5Cl(PO_4)_3$:Mn Phosphor

Compound	Mol Ratio	Assay	Grams	Lb/Lb
$CdNH_4PO_4 \bullet 2\,H_2O$	2.06	0.985	716.09	0.7211
CdO_2	0.90	0.995	130.62	0.1315
$MnNH_4PO_4 \bullet H_2O$	0.04	0.982	7.5752	0.007628
$CdCl_2$	0.75	0.991	138.74	0.1397

Note that an excess of cation in the form of $CdCl_2$ is used. The reason for this is that the $CdCl_2$ acts as a flux and also is volatile at the firing temperature. An excess is used to compensate for the volatile part so as to ensure that the stoichiometric chloride needed for the hexagonal apatite composition is actually incorporated into the structure. This mix is hammermilled twice, blending in between hammermillings. It is then fired in air, in 2 liter silica crucibles with a lid, for:

3.0 hours at 895 °C.

If a 1000 ml. silica crucible is used, the firing time is 2.0 hours. At the end of the firing period, the crucibles are removed and allowed to cool in air. They are then inspected and the cakes crushed and screened through a 200 mesh screen.

Since the formulation includes a large excess of $CdCl_2$ (~2.5 times that needed to provide the stoichiometric amount in the lattice), it is necessary to wash out the excess by suspending the powder in hot water. The water temperature must be > 80°C and the wash water is tested with an $AgNO_3$ test solution until no free chloride ion is shown to be present in the wash. Usually, three washes are sufficient. The following data shows how the chloride is utilized during firing:

<u>Total Chloride Present = 17.57% $CdCl_2$ by Weight</u>

Present in Lattice = 4.01% Cl^- ≈ 10.36% $CdCl_2$

6.72% Lost Through Volatilization as $CdCl_2$

0.39% $CdCl_2$ Removed by Washing

After the water washes, the phosphor is subjected to an acid wash accomplished as follows:

a. Add concentrated HNO_3 equivalent to 10 ml. of concentrated acid per lb. of phosphor in suspension, while continuing to stir. The acid may be diluted before addition to the tank.

b. Three hot wash washes are then applied, followed by the addition to the last water wash of 5.0 ml. of concentrated NH_4OH per lb. of phosphor present.

c. Then an addition of 0.05% by weight of NH_4HPO_4 per lb. of phosphor present is added, and the suspension is then filtered. The wet cake is then dried for 16 hours at 125 °C.

Since the phosphor is already suspended, many manufacturers then coat this phosphor with colloidal silica by adding Ludox™ to the final wash. The procedure is to add 5.0 ml. of Ludox™ per lb. of phosphor present in the final wash suspension and to stir for 15 minutes before final filtration of the suspended powder. After the powder is dry, it is classified to remove all particles smaller than about 8.0 μ in size and larger than about 80 μ

Alternately, it may be classified to remove the smaller particles and then screened through a 200 mesh screen to remove particles larger than about 80 μ.

The excitation-emission spectrum of this phosphor is:

4.2.64.-

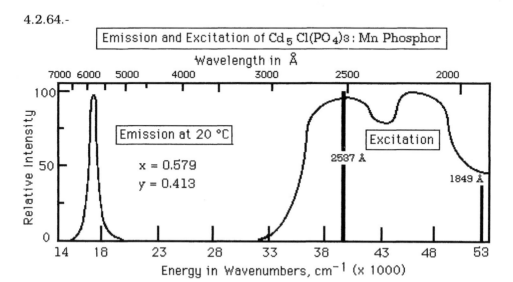

The QE of this phosphor is 0.88, A = 0.97, the cation to anion ratio added is: 3.75 : 2.06 (or 5.46 : 3.00), the luminosity is 0.815 and the peak emission occurs at 5850 Å. Average particle size produced is usually 35-40 μ.

In order to produce a fluorescent lamp having good maintenance, it is essential to about double the amount of phosphor used from about 6.0 mg/cm² to 11.0 mg/cm² of internal glass surface. The reason for this is that the high density of the crystal, i.e. d = 5.33 gm/cm³ (compared to 3.18 gm/cm³ for calcium halophosphate phosphor) mandates higher coating weights in order to have an equal number of particles present per square centimeter of glass surface. To do this requires a phosphor having a closely sized particle distribution.

In the manufacture of this phosphor, it is important to:

1. Avoid any contamination by sulfate from any source

2. Use high bulk density $CdNH_4PO_4 \bullet H_2O$

3. Classify powder to eliminate all particles below 6-8 μ and above 80 μ.

4, Wash to eliminate all free $CdCl_2$

5. Acid wash with HNO_3 to clean particle surfaces before coating

6. Coat the surface with both phosphate and silica

These aims are easily accomplished using the methods set forth above and in Chapter 3 for preparation of $CdNH_4PO_4 \bullet H_2O$ and CdO_2.

When properly accomplished, a 40 watt fluorescent lamp has been obtained having an initial luminous output of 3430 lumens at 0.0 hours and 3175 lumens at 100 hours of operating time. This phosphor can also be used in both Cool White and White blends in lamps.

12. Manufacture of Mg_4FGeO_6:Mn Phosphor

This phosphor is very interesting from the standpoint of the valence state of the activator. The emission comes from the Mn^{4+} state which does not resemble that of the band emission of Mn^{2+} shown previously in many other phosphors. Rather, it consists of five fairly narrow bands which are almost line spectra. Efficient, i.e.- "strong', emission can be obtained only in two hosts, magnesium fluorogermanate and magnesium arsenate as phosphor matrices.

The fluorogermanate phosphor is prepared as given in the following formulation, shown as 4.2.65. on the next page. In this case, the ingredients are weighed out and blended. Then they are hammermilled twice, with blending in between each hammermilling.

Because GeO_2 is relatively expensive, care must be taken not to lose any of the powders being mixed by loss as dust in the baghouse which is usually used to control dust being generated by this process.

448

4.2.65.- Formulation for Manufacture of Mg_4FGeO_6:Mn   Phosphor

Compound	Mol Ratio	Assay	Grams	Lb/Lb
$3MgCO_3 \bullet Mg(OH)_2 \bullet 3H_2O$	3.51 as MgO	44.6% MgO	317.19	0.6959
MgF_2	0.50	0.993	31.38	0.0688
GeO_2	1.00	0.986	106.08	0.2327
$MnCO_3$	0.01	0.995	1.1552	0.00253

The firing schedule used is:

Fire in covered silica crucibles in air for 2 hours at 1100 °C

When cool, the cakes are inspected, crushed and then hammermilled without the use of a screen, taking care not to grind the particles. The powder is then reloaded in the same crucibles and fired uncovered:

40 hours at 1100 °C.

At the end of this period, the phosphor is cooled, inspected and then crushed before screening it through a 325 mesh screen to remove particles larger than about 44μ. The emission and excitation characteristics are shown in the following diagram:

4.2.66.-

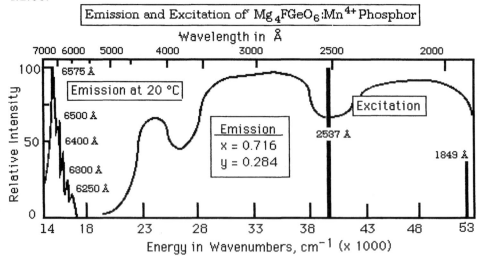

The QE is 0.82, A = 0.96, the luminosity is 0.11, and the major emission peak occurs at 6550 Å., with four other "lines" at 6250 Å, 6300 Å 6400 Å and 6500 Å. The average particle size produced is about 12-14 μ.

The structure of this phosphor is not certain but it may resemble that of the apatites in composition (However, it is well known that Mg does not form an apatite structure in the phosphate analogs). Note also that this phosphor responds to near ultraviolet and blue excitation as well as to 2537 Å excitation, i.e.- from about 4880 Å to nearly 2000 Å.

Mn^{4+} is a $3d^3$ cation whose ground state is $^4A_2$ in the fluorogermanate phosphor (9). The lattice site is one of octahedral symmetry and absorption takes place into the $^4F_2$ and $^4F_1$ states. Emission occurs from the split $^4F_2$ state to the ground state. Since this ion is one whose electronic structure is not symmetrical, i.e.- the free-ion ground state is $^4F_6$, we would not expect it to exhibit efficient luminescence because of vibronic coupling at the lattice site. Isoelectronically, it is like Cr^{3+} which in Al_2O_3 , i.e.- "ruby", exhibits spectra similar to that of Mn^{4+} in the fluorogermanate. We have already discussed Cr^{3+} spectra in Al_2O_3 in a prior work (1). Both of these activators do exhibit luminescence but only Mn^{4+} is efficient enough to be of use in lamps. It is likely that the Mn^{4+} occupies a Ge^{4+} site in the germanate anion. The fluorogermanate phosphor is used both in HPMV lamps and in blends for use in fluorescent lamps.

13. Manufacture of $Mg_5As_2O_{11}$: Mn Phosphor

The formulation that has been generally used for manufacture of the magnesium arsenate phosphor is given in the following. Note that it also is activated by tetravalent manganses, i.e.- Mn^{4+}.

4.2.67.- Formulation for Manufacture of Mg_4FGeO_4:Mn Phosphor

Compound	Mol Ratio	Assay	Grams	Lb/Lb
$3MgCO_3 \bullet Mg(OH)_2 \bullet 3H_2O$	5.50 as MgO	44.6% MgO	497.02	0.5520
As_2O_3	2.00	0.985%	401.71	0.4461
$MnCO_3$	0.015	0.995 %	1.7329	0.001924

This mix is blended as before by hammermilling twice, blending before and after each hammermilling step. Extreme care must be exercized while these components are being handled due to the great toxicity of the arsenic oxide being used. All chlothing worn needs to be discarded at the site after use. A good baghouse is used to control any dust being generated during the process steps. All waste needs to have special handling, according to EPA rules.

The phosphor mix is then fired in air using closed 2 liter silica crucibles for:

6.0 hours at 1220 °C.

Hammermill - no screen

Refire 16 hours at 1220 °C. in air in covered crucibles

The phosphor cake is inspected, crushed and then hammermilled without a screen so as not to break any crystals during the procedure. It is then refired as given above.

The emission spectra of the arsenate phosphor differs slightly from the fluorogermanate phosphor as shown in the following diagram:

4.2.68.-

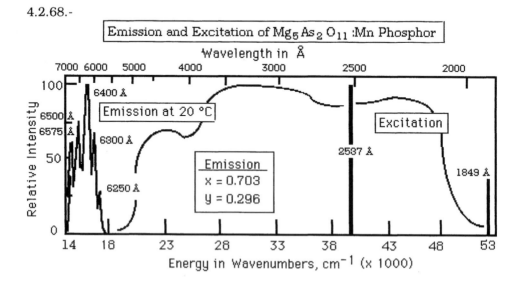

Note that the major peak occurs at 6400 Å but all five "lines" are still present. This makes the arsenate somewhat brighter and slightly less saturated than the germanate phosphor.

The QE of this phosphor is 0.80, A = 0.96, the luminosity is 0.15, and the major emission peak occurs at 6400 Å. Note that the excitation band again extends from about 48800 Å to nearly 2000 Å. The average particle size is about 12-14 μ.

The actual composition of the matrix has not been determined and the formula given above is that expected if all the arsenic is present as As^{5+}. If As^{3+} is the state of the arsenic atoms present in the anion, then the phosphor formula would be: $Mg_5 (AsO_4)_2$:Mn . It may be, since Mn^{4+} is the state of the activator, that a combination of the two arsenic electronic states are present. At any rate, the phosphor is best made as described above regardless of its actual composition.

This phosphor is also obsolete in the U.S. due to recent EPA restrictions on the disposal of arsenic and is no longer used, although the fluorogermanate phosphor continues to find use in Deluxe blends for lamps. It is still being manufactured in Europe.

14. Manufacture of $Y_2 O_3$:Eu Phosphor

This phosphor was originally developed by the author as a red-emitting phosphor for color-television in 1959 (10). However, it is also used in fluorescent lamps both as a red phosphor for blends and in the so-called "tri-phosphor" high output lamps whose emission consists of a narrow-band blue (Eu^{2+}), green line emitter (Tb^{3+}) and a red (Eu^{3+}). The phosphor is prepared by first co-precipitating Y^{3+} and Eu^{3+} as oxalates and then firing to obtain the mixed oxides in solid solution.

The oxalates are prepared by the following steps:

1. Rare Earth Solution: A HNO_3 solution is prepared by adding 23.5 gallons of concentrated HNO_3 to 250 gallons of deionized water in a 500 gallon glass-lined tank. This forms a 1.200 molar solution. To this is added 51.3 kilograms (113 lb. 1 oz.) of Y_2O_3 and 13.1 kg. of Eu_2O_3 (10 lb.

452

2 oz.) with stirring. The oxides should be mixed beforehand and should be added slowly since the dissolution reaction is exothermic and the solution tends to boil over if the oxides are added too fast. The solution is allowed to cool while the oxalic acid solution is being prepared.

2. Oxalic Acid Solution: To 250 gallons of deionized water in a 300 gallon tank is added 167.71 kg. (369 lbs. 9 oz.) of oxalic acid with stirring. The oxalic acid is soluble to the extent of 94 kg at 20 °C and 1135 kg. at 100 °C. Therefore, the solution needs to be about 40-50 °C in order to dissolve all of the oxalic acid. When the dissolution is complete, 65 gallons of ethyl alcohol are added to this volume. The solution is then heated to 80 °C. just before precipitation.

3. Precipitation: To the rare earth solution (which now should be about 25-30 °C in temperature) is added 10.0 gallons of 30% H_2O_2 with stirring. The solution is stirred for 15 minutes and is then heated to about 80 °C.

The hot (80 °C.) oxalic acid solution is then pumped into the Y^{3+} and Eu^{3+} solution at a rate of about 5.0 gallons per minute while stirring. There is an induction period before any precipitate will be seen. The total time of precipitation will be about 1.0 hour. The precipitate is allowed to settle and the mother liquor is removed by decantation. One water wash is given to the precipitate by resuspension. The suspension is then filtered on a drum-filter using a water wash wherein the cake is sprayed on the filter. The wet precipitate is then dried for 16 hours at 125 °C.

The weight of oxalate is then recorded, the oxalate is assayed and the particle size distribution is measured. The yield is then calculated and recorded.

4. Notes on Precipitation Procedure:

i. The optimized phosphor will contain 0.115 mols of Eu^{3+}/mol of Y_2O_3: when the precipitated oxalates are fired.

453

ii. The ethyl alcohol is added to the oxalic acid solution to increase the yield of the oxalate precipitate. The yield should be in excess of 98%. In the absence of ethyl alcohol, yields as low as 80-85% have been experienced.

iii. The purpose of the hydrogen peroxide is to form a peroxy-oxalate when the rare earth precipitate is formed. When this is then fired, the excess oxygen present prevents the oxidation of Eu^{3+} to Eu^{2+} by the carbon monoxide produced during the solid state reaction:

4.2.69.- $\quad\quad$ $(Y,Eu)_2(C_2O_4)_3 \;\Rightarrow\; (Y,Eu)_2O_3 + 3\,CO_2 \Uparrow + 3\,CO \Uparrow$

A redder emitting phosphor is obtained at a lower Eu^{3+} content than that obtained when H_2O_2 is not used (10). (Actually, we do not know the exact composition of the oxalate produced when hydrogen peroxide is added, but the effect is the same. It may be that the europium ions present are kept in their highest oxidation state just before precipitation).

iiii. Precipitation is done at 80 °C in order to obtain the trihydrate, i.e.-

4.2.70.- $\quad\quad$ $(Y,Eu)_2(C_2O_4)_3 \bullet 3\,H_2O$

If the precipitation is done at lower temperatures, higher hydrates are obtained and the decahydrate obtains ar 20 °C or lower. All of these produce a phosphor whose luminous output is much lower compared to the trihydrate.

The phosphor is prepared by firing the precipitate:

6 hours at 1275 °C in an open silica tray

At the end of the firing period, the trays are allowed to cool and then are dumped. The powder so-produced is usually rather soft. If it is not, the cake is inspected, crushed and then sieved through a 325 mesh screen to remove particles larger than about 44 μ.

The phosphor is then weighed, and yields calculated. Since this phosphor is one of the most expensive ones produced (at about $ 80/lb.) it is important to save as much product as possible.

Excitation- emission properties are given in the following diagram:

4.2.71.-

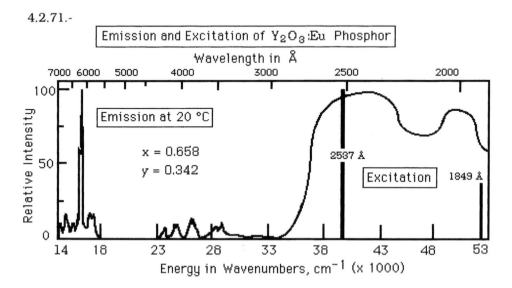

The emission consists of a set of lines with the major one peaking at 6130 Å. The QE is 0.85, A = 0.97, luminosity is 0.43 and the average particle size is 12-15 μ.

Note that the excitation band is broad and that there are low-lying excitation peaks between about 3500 and 4500 Å. The former is the charge transfer band of Eu^{3+} and the latter are other excitation levels in the ^5D multiplet which cascade down to the lowest excitation level, i.e.- $^5D_0$, from which comes the emission. The major line transition is: $^5D_0 \Rightarrow {}^7F_2$, i.e.- an allowed "forced-dipole" transition, but the other lines are caused by crystal field splitting of this state and transitions to the other levels of the ^7F_J multiplet.

The most important factor affecting luminous performance in a fluorescent lamp is that the original Y_2O_3 (or Eu_2O_3) must not contain more than about 2-4 ppm. of Dy^{3+}.

15. Manufacture of YVO$_4$: Ln^{3+} Phosphors

Rare earth vanadate phosphors used in lamps include:

$$YVO_4:Eu^{3+}, \quad YVO_4:Dy^{3+} \text{ and mixed } Y(P,V)O_4:Eu^{3+}$$

They have been prepared by at least two methods, one a solid state method and the other a precipitation method(11). We did not describe the precipitation process in the last chapter because the process actually produces a phosphor having about 40% of its final brightness after firing. This phosphor is one of the two systems known to the author in which a phosphor is produced directly without a calcination step (The other is EuPO$_4$ and/or TbPO$_4$).

a. The Solid State Method

In this method, yttrium oxalate is fired together with vanadium pentoxide to form the phosphor according to the reaction:

4.2.72.- $(Y,Eu)_2(C_2O_4)_3 + V_2O_5 \Rightarrow 2\,YVO_4 : Eu^{3+} + 3\,CO_2 \Uparrow + 3\,CO \Uparrow$

The reaction does not go to completion unless there is excess V$_2$O$_5$ over the stoichiometric quantity required in the reaction. Therefore, one uses a large excess and then removes it from the reaction product by dissolution with NaOH. The reason for this approach is that the solid state reaction is sluggish and would require a much longer firing time than is actually used. In most cases, the particle size would also be much too large for use in HPMV lamps and/or color television tubes.

The steps involved in the manufacture of this phosphor include:

1. The co-precipitated oxalates of Y^{3+} and Eu^{3+} are obtained in the exact same manner as given above, except that the quantity required is 0.05 mol of Eu^{3+} per mol of Y^{3+}. (Again, the quantity of Dy^{3+} actually present in the raw materials must be limited to < 2-3 ppm.). For a 250 gallon solution, the quantity to be added is: 5.696 Kg. of Eu$_2$O$_3$ (the other weights remain the same).

2. The typical phosphor formulation used is:

Compound	Mol Ratio	Assay	Grams	Lb/Lb
$(Y,Eu)_2(C_2O_4)_3 \bullet 3\ H_2O$	1.000	0.5559	417.56	0.6933
V_2O_5	1.400	0.985	184.73	0.3067

The usual batch size is about 1100 lbs. This mix is hammermilled twice, blending in between the hammermilling steps. (Note that V_2O_5 dust is toxic and that this step need to be accomplished within an enclosed space, having suitable dust collection facilities).

3. Initial firing is done in silica trays at a load of about 5 kg. per tray. The mix is loaded by vibrating the trays to ensure that air is completely removed (so as to prevent loss of powder when the tray is heated). The firing schedule is:

6 hours at 1220 °C in air

At the end of the firing schedule the trays are allowed to cool in air.

4. The trays are then dumped, the cakes are crushed and then hammermilled. The powder at this point will be a dark reddish color due to the excess V_2O_5 present.

5. Washing:

A 2.00 molar solution of NaOH is prepared by adding 165.0 lbs. of NaOH pellets to 250 gallons of deionized water with stirring., The water will heat up as the pellets dissolve. When dissolution is complete, the solution is heated to 80 °C. The powder is added in increments as the caustic solution is being heated while continuing to stir the suspension.

The dissolution reaction is: $3NaOH + V_2O_5 \Rightarrow 3\ Na^+ + VO_4^{3-}$

The suspension is stirred for 30 minutes after the temperature has reached 80 °C and all of the powder has been added. The suspension is allowed to settle, the mother liquor is siphoned off , and two water washes are applied. The suspension is then filtered to obtain a wet cake of powder. The cake is then dried at 125°C. for 16 hours.

6. Refiring

Some manufacturers then refire the powder as obtained:

6.0 hours at 1200 °C. in air using silica trays

Other manufacturers weigh the powder, add 1.00 % NaCl as a flux, blend the powder thoroughly, and then refire. The trays are then allowed to cool and the cakes are inspected and crushed.

7. 2nd Washing Step:

A 1.0% solution of oxalic acid is prepared by adding 25 lbs. of oxalic acid to 250 gallons of deionized water with stirring. The powder produced above is then added to form a suspension, which is stirred about 30 minutes. The suspension is allowed to settle, and the mother liquor is removed by suction. The powder is then washed again by resuspension in deionized water. A final wash is given while it is being filtered on the drum filter. The wet cake is then dried for 12 hours at 125 °C.

8. The dry powder is then screened through a 325 mesh screen to remove all particles larger than 44 μ.

Actually, the above description is only an outline of all of the steps involved in production of this phosphor. The following diagram, shown as 4.2.73. on the next page, gives all of the details involved in the manufacturing process.

Note that there are at least three hammermilling steps involved as well as two washing, filtering and drying steps. It should be clear that this is a rather complicated procedure.

Because of the toxicity of V_2O_5 being used, the same precautions as outlined for the arsenate phosphor given above need to be implemented. The rules concerning disposal of V_2O_5 are not so stringent as those of arsenic disposal. Nevertheless, extreme caution regarding dust generated must be exercized.

458

4.2.73.-

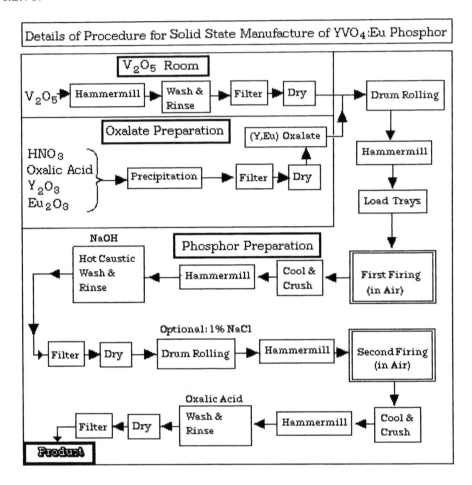

Details of Procedure for Solid State Manufacture of YVO$_4$:Eu Phosphor

V$_2$O$_5$ Room

V$_2$O$_5$ → Hammermill → Wash & Rinse → Filter → Dry → Drum Rolling

Oxalate Preparation

(Y,Eu) Oxalate

HNO$_3$, Oxalic Acid, Y$_2$O$_3$, Eu$_2$O$_3$ → Precipitation → Filter → Dry

Hammermill → Load Trays

Phosphor Preparation

NaOH → Hot Caustic Wash & Rinse ← Hammermill ← Cool & Crush ← First Firing (in Air)

Optional: 1% NaCl

Filter → Dry → Drum Rolling → Hammermill → Second Firing (in Air)

Oxalic Acid

Filter ← Dry ← Wash & Rinse ← Hammermill ← Cool & Crush

Product

The excitation-emission spectra are shown in the following diagram, given as 4.2.74. on the next page.

The QE of this phosphor is 0.82, A = 0.92, the peak emission wavelength is 6210 Å , and the usual particle size produced is usually about 9-12 μ. Note that the excitation band is unusually broad. This is the excitation band of the vanadate anion, as we discussed previously (1). Eu^{3+}, having a charge transfer band of its own, produces an efficient phosphor in this host because the vanadate charge transfer band can couple to that of the Eu^{3+} activator.

4.2.74.-

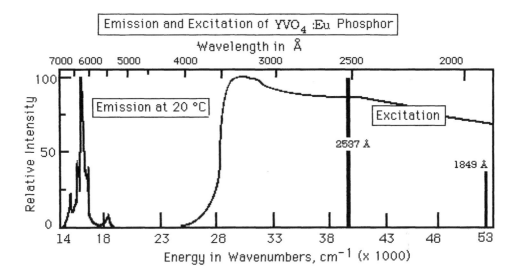

b. The Precipitation Method.

In this method, V_2O_5 is dissolved as is the Y_2O_3 and Eu_2O_3. It is essential to use HCl since the requisite reaction does not occur as well with HNO_3. A solution of H_2O_2 and NH_4OH is then used to titrate the mixed solutions of VO^{2+} and Ln^{3+} to cause the precipitate to form. We will address four items: solution chemistry of vanadium; a general precipitation method for $YVO_4{:}Eu^{3+}$ phosphor; manufacture of $YVO_4{:}Dy$ phosphor; and manufacture of $Y(P,V)O_4{:}Eu$ phosphor by precipitation.

i. SOLUTION CHEMISTRY OF VANADIUM

The solution chemistry of vanadium is quite complex. Vanadium has valence states of V^{2+}, V^{3+}, V^{4+} and V^{5+} but only the last three states are stable in solution, and exist solely as oxygen-bound complexes. What this means is that only certain ions are found in solution and the number of soluble vanadium ions is limited to: VO^+, $VO_2{}^+$ as cations, and the anions, $VO_4{}^{3-}$ and $V_{10}O_{28}{}^{3-}$, the first two in acid solution and the last two in basic solution.

If one wished to form $YVO_4{:}Eu$ by precipitation, one could start with V_2O_5 and dissolve it in NaOH solution. One would obtain a stable orthovanadate solution:

4.2.75.- $\qquad$ $3 NaOH + V_2O_5 \Rightarrow 3 Na^+ + 2 VO_4{}^{3-}$

However, the lanthanide oxides must be dissolved in an acid since they are not soluble in a base. When the pH of a basic solution of a soluble vanadate is lowered (as would happen upon the addition of the acidic Ln^{3+} solution), the $VO_4{}^{3+}$ ions tend to polymerize and one obtains a polymeric anionic aggregate:

4.2.76.- $\qquad$ $10 VO_4{}^{3-} + 12 H^+ \leftrightarrow V_{10}O_{28}{}^{6-} + 12 OH^-$

$\qquad$ $V_{10}O_{28}{}^{6-} + 6 H^+ \leftrightarrow 5 V_2O_5 \downarrow + 3 H_2O$

The first reaction is reversible but the latter is not. Thus, upon the addition of the acidic Ln^{3+} cation solution, one obtains a mixed precipitate, due to the presence of the decameta-vanadate and vanadium pentoxide. The formation of these compounds cannot be prevented because of the nature of the solution chemistry involved. Additionally, the precipitate contains considerable adsorbed Na^+ ions. These are nearly impossible to remove from the precipitate by washing.

A much easier method (12) involved the dissolution of V_2O_5 in HCl (it does not dissolve in oxidizing acids). The reaction involves:

4.2.77.- $\quad$ $V_2O_5 + 6 HCl \Rightarrow 2 VO^{2+} + Cl_2 \uparrow + 3 H_2O + 4 Cl^-$

The acidic lanthanide solution can be added to this solution. Upon the addition of the titrate solution, a series of complex reactions result as shown in the following:

4.2.78.- <u>Reactions Leading to Precipitation of Orthovanadates</u>

$\qquad$ $2 VO^{2+} + H_2O_2 \Rightarrow 2 VO_2{}^+ + 2 H^+$ $\qquad$ $E^0 = \approx 1.0$ volt

$\qquad$ $2 VO_2{}^+ + Cl^- \Rightarrow VO_2Cl \downarrow$

$\qquad$ $10 VO_2{}^+ + 8 OH^- \Rightarrow V_{10}O_{28}{}^{6-}$ $\qquad$ pK = 6.75

$\qquad$ $3 V_{10}O_{28}{}^{6-} + 6 OH^- \Rightarrow 10 V_3O_9{}^{3-} + 6 H^+$ $\qquad$ pK = 7.85

$\qquad$ $V_3O_9{}^{3-} + 3 OH^- \Rightarrow 3 VO_4{}^{3-} + 3H^-$ $\qquad$ pK = ?

Note that the VO^{2+} ion must be oxidized to VO_2^+ before the reactions to form vanadate in solution can proceed. It is not known whether a VO_2Cl precipitate forms in suspension. If it does, it most certainly redissolves so that the succeeding reactions can take place. The important part is that if a Ln^{3+} solution is present in the beginning, a $LnVO_4$ precipitate is obtained with high yield. As we have demonstrated previously (13), these reactions take place sequentially in solution.

As shown in the following diagram, one can follow the complex reactions taking place in the solution by measuring both pH and conductivity:

4.2.79.-

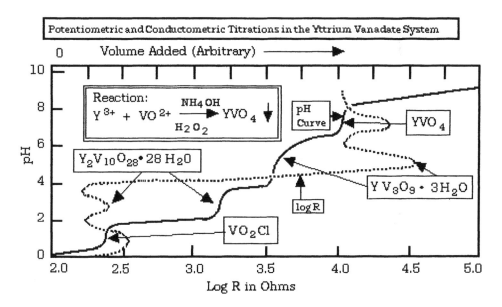

Note that all of the reactions except the initial oxidation one can be followed. Each reaction has a specific range in which it occurs:

4.2.80.- $2\ VO_2^+ + Cl^- \Rightarrow VO_2Cl \downarrow$ pH = 0.80

$10\ VO_2^+ + 8\ OH^- \Rightarrow V_{10}O_{28}^{6-}$ pH = 3.00

$3\ V_{10}O_{28}^{6-} + 6\ OH^- \Rightarrow 10\ V_3O_9^{3-} + 6\ H^+$ pH = 5.20

$$V_3O_9^{3-} + 3\,OH^- \Rightarrow 3\,VO_4^{3-} + 3H^- \qquad\qquad pH = 8.0$$

This method is applicable to all lanthanide orthovanadates and we have already shown that $LaVO_4$ is dimorphic (12) using this method of preparation. The YVO_4: Eu phosphor, obtained as a precipitate, has about 40% of its final luminous output before firing.

ii. PRECIPITATION METHOD FOR MANUFACTURING YVO₄:Eu PHOSPHOR

The precipitation method of manufacture involves the following steps:

1. Rare Earth Solution

A 2.00 molar solution of Ln^{3+} is employed.. This requires 0.93023 mols of Y_2O_3 (210.06 gm./liter) and 0.06977 mols of Eu_2O_3 (24.560 gm./liter). The working volume is 520 liters or 138.70 gallons.

Weigh out the following amounts of rare earths for dissolution:

$$Y_2O_3 = 109.23 \text{ kg.} = 240.6 \text{ lbs. } (520 \times 210.86 \text{ gm/mol})$$
$$Eu_2O_3 = 12.77 \text{ kg.} = 28.10 \text{ lbs. } (520 \times 24.560)$$

Mix these oxides together thoroughly before use. The following totals are the volumes required for dissolution of the oxides:

286 liters, or 75.6 gallons, of concentrated HCl
234 liters, or 61.8 gallons, of deionized water

To a 150 gallon glass-lined tank, add 168.0 liters, or 45 gallons, of concentrated HCl to the tank containing about 10 gallons of deionized water. The resulting diluted solution will heat up. Maintain stirring and add small increments of the mixed oxide while stirring until about 1/4 of the total has been added. The solution will heat up proportionally to the amount of oxides added and how fast they are added. Do not allow the solution to boil.

When the rate of dissolution has slowed down and some of the rare earth oxides remain in suspension, add approximately 20-25 liters (5-6 gallons) of concentrated HCl and stir until dissolved. Add more oxide and continue the process until all of the mixed oxides are completely dissolved and a total of 286 liters of concentrated acid plus 234 liters of deionized water has been added. If the Ln_2O_3 dissolves before all of the acid has been used, dilute volume to final volume using deionized water.

2. Vanadyl (VO^{2+}) Solution

A 2.10 molar solution of vanadyl ion is required. Thus, 1.05 mols of V_2O_5 (191.0 gm/liter) must be dissolved in 1.00 liter of concentrated HCl. Since the working volume is 520 liters or 138.7 gallons, we will require 99.32 kg. or 218.7 lbs. of V_2O_5 (520 x 191.00 @ 95.2% assay).

The total volume of 138.7 gallons of concentrated HCl is placed in a 150 gallon glass-lined tank having an enclosed cover equipped with a vent structure and scrubber. The total weight of 218.7 lbs. of V_2O_5 is added to the tank with stirring. A suspension will first form. Heat the solution in the tank while continuing to stir until all of the suspension has dissolved. A dark blue solution results. Since the reaction produces chlorine gas, it is necessary to employ a vapor-scrubber to remove the gas generated by scrubbing with an NaOH-solution in the packed scrubber (which forms NaClO + NaCl).

3. Solution of NH_4OH and H_2O_2

A solution of these two "titrants" is required at a ratio of 3.0:1.0. The total volume is 520 liters or 138.7 gallons.

This requires:

> NH_4OH (Concentrated reagent) = 390 liters or 104.1 gallons
> H_2O_2 (30%) = 130 liters or 34.7 gallons

These volumes are measured and placed in a 150 gallon glass-lined tank just before the precipitation process is to take place.

4. Precipitation of YVO $_4$:Eu

The following volumes of liquids are pumped into a 300 gallon glass-lined tank, while stirring continues in the tank:

     i. 130 liters, or 34.67 gallons of VO^{2+} solution, followed by 130 liters of deionized water.

     ii. 130 liters or 34.67 gallons, of Y^{3+} - Eu^{3+} solution is added to the vanadyl solution already in the tank. Then 130 liters of deionized water follows.

     iii. The total volume is now 520 liters or 137.38 gallons.

This solution is heated to 70 °C, but less than 80 °C. while stirring continues. The NH_4OH + H_2O_2 solution is then added at a steady rate of about 3.0 liters/minute, as feasible. The pH will attempt to decrease towards < 1.00 as the precipitation proceeds due to release of H^+ from the precipitate being formed. The change in pH should steadily increase with a minimum of oscillation, until a pH = 8.00 is obtained. Once this pH is reached, and it remains steady, the precipitate is stirred for about 45 minutes. The precipitate is now ready for processing.

5. Processing of Precipitate

In order to determine if the titration is complete, and that the precipitate is free of all polymeric forms, a small sample is taken from the tank, and the precipitate is filtered. The wet cake is then placed in a small silica crucible and fired for one hour in air at 1200 °C. The crucible is allowed to cool and is inspected. If any trace of brown (due to the presence of V_2O_5) is seen, the titration did not go to completion. In most cases, the "body-color" will be white and the precipitate can be further processed. (If it is not, the suspension must be reheated to 80 °C and more NH_4OH + H_2O_2 is added while stirring).

The contents of the tank are pumped into a filter box where the mother liquor is separated from the solids. The solids are then washed several times while still on the filter to remove residual traces of the caustic.

The wet cake is then dried for 8-12 hours at 140 °C. Following this, the powder is allowed to cool, and then is fired.

6. Firing

The dried powder will now be fluorescent, but needs to be fired to develop its full luminosity. The powder is loaded into silica trays and is fired:

6 hours at 1200 °C.

At the end of the firing, the trays are cooled, the phosphor is inspected and is then processed through a 325 mesh screen to remove all particles larger than about 44 μ.

The yield expected is about 96-98% of the lanthanides added, calculated as the vanadates. This method produces a phosphor whose brightness and luminosity in HPMV lamps is improved by as much as 125% over the phosphor made by the solid state process.

In order to compare this process to that of the solid state process, consider the following diagram which shows the actual manufacturing steps required. Note that only 6 steps are required in contrast to the several given in 4.2.73.

4.2.81.-

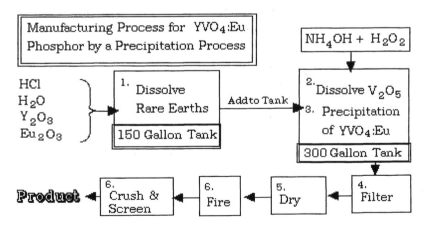

In actuality, either solution can be added to the other before precipitation. That is, the vanadyl solution can be added to the rare earth solution, or vice-versa.

But, as shown in the following table, a difference in assay of the precipitate occurs, probably due to the amount of NH_4Cl that is residual in the precipitate before firing. The yield in both cases was 97.6% of the added lanthanides. It is clear that the precipitation process produces not only a brighter phosphor, but also is easier to accomplish as well.

Table 4-9

Effect of Process Variation on Properties of the YVO_4:Eu Phosphor Produced

Variations in the Precipitation Process	Filtered	Fired	% Output*
Method - A	Hot	Covered	100
$[Y^{3+} \Rightarrow VO^{2+}] + [NH_4OH + H_2O_2] \Rightarrow YVO_4 \Downarrow$		Uncovered	130
(Assay of Precipitate = 84.6 %)	Cold	Covered	97
		Uncovered	108
Method - B	Hot	Covered	91
$[VO^{2+} \Rightarrow Y^{3+}] + [NH_4OH + H_2O_2] \Rightarrow YVO_4 \Downarrow$		Uncovered	100
(Assay of Precipitate = 73.3 %)	Cold	Covered	94
		Uncovered	105

* Compared to the Standard Solid State Product

Note that it is important to add the lanthanide to the vanadyl solution, not vice-versa before the shift in pH by titration is started. Also, filtering hot and then firing in uncovered trays produces the best phosphor. The reason for this is probably that the NH_4Cl normally present as a product in the precipitation is minimized in the filtered product since it remains more soluble in the hot solution as it is being separated from the solids by filtration. Open trays allow the excess NH_4Cl to escape early in the firing process. Although one might wonder why the wet cake might not be washed to totally remove the NH_4Cl, tests have shown that some NH_4Cl needs to be present during the firing process in order to produce the best phosphor. The optimum firing time can be obtained from the following equation:

4.2.82.- $\qquad T = -25 t + 1350 \,°C.$

which was derived from the following diagram:
4.2.83.-

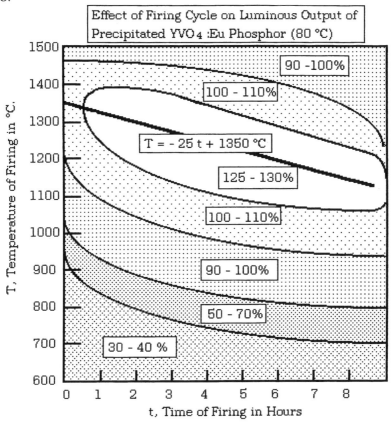

Effect of Firing Cycle on Luminous Output of Precipitated YVO$_4$:Eu Phosphor (80 °C)

iii. - MANUFACTURE OF YVO$_4$:Dy PHOSPHOR

Another phosphor that can be manufactured by this process is YVO$_4$:Dy^{3+}. The YVO$_4$:Dy phosphor is very interesting in that it possesses a constant-energy absorption band, like the sodium salicylate generally used for calibration of optical equipment for quantum efficiency measurements. The process is the same as that given above except that the optimum Dy^{3+} content is 0.003 mol Dy^{3+} / mol of YVO$_4$. This requires that 0.9970 mols of Y$_2$O$_3$ (225.15 gm./liter) and 0.003 mols of Dy$_2$O$_3$ (1.1190 gm./liter) be used. The working volume is 520 liters or 138.70 gallons.

Therefore, one weighs out:

$$Y_2O_3 = 117.08 \text{ kg.} = 258.00 \text{ lbs.} (520 \times 225.70 \text{ gm/mol})$$
$$Dy_2O_3 = 96.980 \text{ gm.} = 1.2822 \text{ lbs.} (520 \times 0.1865)$$

These are dissolved as given above and a vanadyl solution is prepared. The solution of rare earths is then added to the vanadyl solution and the titration with $NH_4OH + H_2O_2$ proceeds as detailed above. The spectra of this phosphor is shown in the following diagram:

4.2.84.-

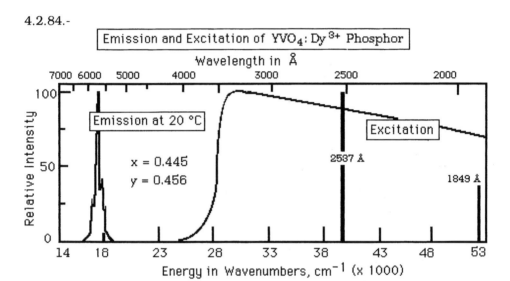

The major emission peak occurs at 5750 Å, the QE is 0.88, A= 0.96, the half-width of the emission band is 60 Å, and the average particle size is 9-15 µ. The quantum efficiency of this phosphor is 1.76 times that of the sodium salicylate phosphor which is usually quoted as 50%. However, the latter absorbs from about 4000 Å to 2000 Å whereas the YVO₄:Dy phosphor absorbs from about 3600-3700 Å to 2000 Å. Nevertheless, it is a constant quantum efficiency phosphor just like sodium salicylate (14).

iv.- MANUFACTURE OF Y(P,V)O₄:Eu PHOSPHOR

Still another phosphor used in the trade is the Y(P,V)O₄:Eu composition which is used mainly in HPMV lamps for color correction. This phosphor can be made by one of two methods. One has the choice of first preparing the (Y,Eu)PO₄ • 5/3 H₂ O precipitate and then adding precipitated (Y,Eu)VO₄ before calcining them together, or one can co-precipitate both compounds at the same time in the same solution.

The reasons for precipitating either $YPO_4 \bullet 5/3\ H_2O$ and then mixing it with precipitated YVO_4:Eu or directly precipitating $Y(V,P)O_4$:Eu stem from the fact that if a solid state reaction was used, the product was found to deviate from the expected mol ratio. That is, when the ingredients were added to formulate a specific mol ratio of VO_4 to PO_4, it was found that the fired product did not mirror the original mol ratio. However, if the precipitate, $YPO_4 \bullet 5/3\ H_2O$, is used, then the expected ratio results. This is shown in the following diagram, given as 4.2.45. on the next page, which shows the mol ratios added and the mol ratios that resulted in the fired product(14).

As shown in these diagrams, the reactions to form $Y(V,P)O_4$: Eu were:

A. Solid State Reaction = $Y_2O_3 + 2\ (NH_4)_2\ HPO_4 + YVO_4$ (ppt.)

B. Precipitated $YPO_4 \bullet 5/3\ H_2O + YVO_4$ (ppt.)

C. Precipitated $YPO_4 \bullet 5/3\ H_2O + Y_2O_3 + V_2O_5$

D. Solid State Reaction = $2\ Y_2O_3 + V_2O_5 + 2\ (NH_4)_2\ HPO_4$

The reason found for this behavior was that the phosphate acted as a template for the particle habit formed, the same type of mechanism presented for the $CaHPO_4$ particles used to prepare halophosphate phosphors, as described above.

Thus, to manufacture the $Y(P,V)O_4$:Eu phosphor, it is necessary to precipitate either the $YPO_4 \bullet 5/3\ H_2O$ precipitate and then mix other oxides to form the final composition or to co-precipitate both compounds together. Needless to say, the fired compositions from the co-precipitated compounds were determined to be exactly stoichiometric.

It should be clear that the deviation from stoichiometry observed in the solid state reactions is due to side-reactions that occur. It is only when the method of co-precipitating vanadates and phosphates together is used that a stoichiometric product obtains. Therefore, it is imperative to use one of the

470

4.2.85.-

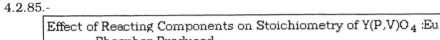

Effect of Reacting Components on Stoichiometry of Y(P,V)O$_4$:Eu
Phosphor Produced

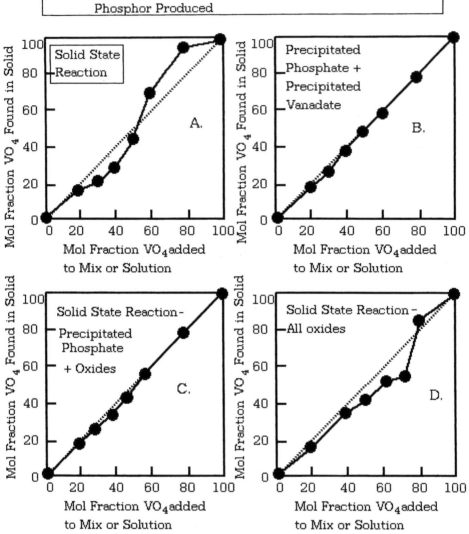

precipitation methods to produce a phosphor that is both stable and of
highest luminous efficiency in lamps.

The reason for using the phosphate-vanadate phosphor instead of the YVO$_4$:Eu
phosphor lies in the fact that the mixed phosphor has a better performance in
HPMV lamps.

1. Precipitation Method for YPO$_4$ • 5/3 H$_2$O

The precipitation of YPO$_4$ • 5/3 H$_2$O is accomplished as follows:

a. Rare Earth Solution at a 2.00 Molar Concentration

This requires 0.93023 mols of Y$_2$O$_3$ (210.06 gm./liter) and 0.06977 mols of Eu$_2$O$_3$ (24.560 gm./liter). The working volume is 520 liters or 138.70 gallons.

Weigh out the following amounts of rare earths for dissolution:

$$Y_2O_3 = 109.23 \text{ kg.} = 240.6 \text{ lbs. } (520 \times 210.86 \text{ gm/mol})$$

$$Eu_2O_3 = 12.77 \text{ kg.} = 28.10 \text{ lbs. } (520 \times 24.560)$$

Mix these oxides together thoroughly before use.

The following totals are the volumes required for dissolution of the oxides:

286 liters, or 75.6 gallons, of concentrated HCl

234 liters, or 61.8 gallons, of deionized water

The procedure is exactly the same as that given above in that about 1/4 of the oxides are first added slowly to about 45 gallons of the concentrated acid. then 25-30 gallons of concentrated acid are added to this solution and more oxides are added. This continues until all of the acid and oxides have been added. The solution will heat up as oxides are added to the acid. Do not let the solution go to boiling. When a total of 234 liters of concentrated HCl have been added, the volume will be close to 260 liters. If it is not, add sufficient water to adjust the volume to a total of 520 liters. The concentration of Ln^{3+} will now be exactly 1.00 molar. Heat the solution to about 60 °C while stirring the solution in the tank.

b. Preparation of Phosphoric Acid Solution

A 2.20 molar volume of H_3PO_4 is required. For a 300 gallon tank, a working volume of 520 liters (137.4 gallons) is required. 149.66 ml. of concentrated acid (85%) per liter is required. To 300 liters of deionized water (79.3 gallons), add 44.90 liters (11.97 gallons) of concentrated H_3PO_4 slowly with stirring. Dilute to a total working volume of 520 liters. Begin heating the acid solution to 80 °C.

c. Precipitation of Rare Earth Phosphates.

Heat the rare earth solution to 60 °C. while stirring. Set up the flow-rate control to deliver 1750 ml./minute of the rare earth solution to the acid-solution tank. Begin pumping the rare earth solution, i.e.-

4.2.86.-   1.00 molar Ln^{3+} ⇒ 2.20 molar H_3PO_4 solution at 80 °C.

At the beginning, the solution remains clear. Gradually, a milky condition is noted and then a coarse white precipitate obtains. (Note that the solubility of the $LnPO_4$ precipitates is inversely proportional to temperature. That is, they are less soluble at high temperature in solution than at room temperature. It is for this reason that the precipitation temperature must be in excess of 80 °C. in order to obtain a good yield).

At the end of the addition period, let the precipitate remain suspended by stirring and monitor the solution pH. When it becomes stable, continue stirring so as to promote crystal growth.

After about 45 minutes, obtain concentrated NH_4OH and add slowly to tank while the suspension is still being stirred. Stop the addition when the pH approaches 1.40. The final pH should be 1.40 ± 0.02 and up to 100 liters of concentrated ammonia may be required. Allow the precipitate to stir for 15 more minutes, still maintaining the temperature at 80 °C.

d. Final Processing of Precipitate

After the pH adjustment, stop stirring and allow the precipitate to settle. Remove the mother liquor by decantation and apply two cold water (~ 15-20 °C) washes by resuspension and settling procedures. Filter and dry for 16 hours at 105-110 °C.

The expected yield is:

	10 Gallon Batch	300 Gallon Batch
Precipitate:	3.237 Kg.	247.2 lb.
Assay:	85.04% (as $LnPO_4$)	---------
Yield:	97% of rare earths	94 - 98% of rare earths

It has been found that the usual precipitate is: $(Y,Eu)PO_4 \cdot 5/3\ H_2O$.

e. Notes on Procedure Used

   i. Optimized Ratios of Acid to Ln^{3+}

The reasons why the above procedure is specified comes from unpublished experimental data obtained by the author. If one attempts to precipitate a rare earth phosphate using H_3PO_4, the usual product is a gelatinous mess. It was determined that little precipitation occurred at room temperature, (if at all) and that a precipitate did not appear until solution temperatures rose above about 45 °C. Additionally, the use of 1:1 molar ratios when $Y^{3+} + Eu^{3+}$ was added to PO_4^{3-} also resulted in gelatinous materials which could not be filtered or otherwise processed even at elevated temperatures. By using an excess of H_3PO_4 over that actually needed to form the compound, it was determined that a crystalline material could be obtained. An equilibrium exists in solution between:

$$4.2.87.- \qquad Ln^{3+} + PO_4^{3-} \quad \Leftrightarrow \quad LnPO_4 \downarrow + 3\ H^+$$

Temperature insolubility drives the reaction to the right while the excess acid causes the reaction to go to the left. The result is that the particles grow in size due to decreased surface area of the larger particles which slows down the equilibrium reaction.

The amount of excess acid required was found to be determined by the original concentration of the rare earth solution itself. This is shown in the following diagram:

4.2.88.-

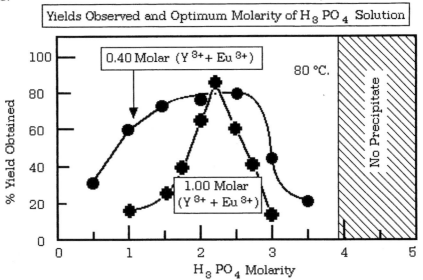

When a 0.40 molar solution of Y^{3+} + Eu^{3+} ions is used, the molarity of H_3PO_4 can be quite wide, i.e.- the mol ratio is not very critical. But when the concentration of Y^{3+} + Eu^{3+} ions is 1.00 molar, then the mol ratio becomes much more critical. If a ratio such as 0.4 molar Ln^{3+}/3.2 molar H_3PO_4 solution ratio is used, the particle size will be much larger than that produced from a 1.00 M Y^{3+} + Eu^{3+} / 2.40 M H_3PO_4 ratio.

ii. Yields from Solution

Since we wish to manufacture as much material as possible at one time, it is obvious that we should use the higher concentration. Note also that the yield is never more than about 80% of the added rare earths. It was determined that yields approaching 100% could be obtained if the final pH was adjusted. The following data shows results obtained when NH_4OH was added after the solution had been stirred for 10 minutes.

Adjusted pH	% Yield Obtained
Unadjusted = ~ 0.21	52.0
0.80	73.0
1.00	78.1
1.25	84.2
1.50	101.0

These data fit a straight line equation:

4.2.89.- $\%\ Yield = 39.76\ pH + 44.45 \quad (r^2 = 0.955)$

For 100% yield, an adjustment of the final pH to pH = 1.14 is required.

iii. Type of Hydrate Obtained

The ratio of acid to rare earth concentrations also determines the nature of the hydrate that is precipitated as shown in the diagram given as 4.2.90. on the next page.

Note that both a monohydrate and a dihydrate obtains at specific mol ratios. However, it is not known why the mixed hydrates are the preferred composition over much of the range when the precipitate is formed.

2. Co-Precipitation of Phosphate-Vanadate

To co-precipitate the phosphate-vanadate, one uses the same procedure as given above, except that H_3PO_4 is added to the solution before the titration with NH_4OH and H_2O_2 begins.

The procedure is:

4.2.90.-

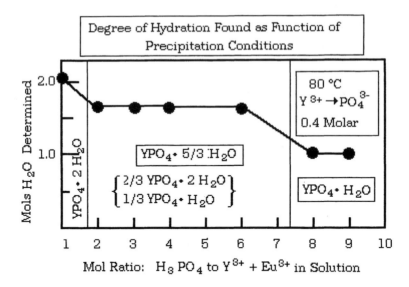

To co-precipitate the vanadate-phosphate, one follows these steps:

1. Rare Earth Solution

A 2.00 molar solution of Ln^{3+} is employed.. This requires 0.93023 mols of Y_2O_3 (210.06 gm./liter) and 0.06977 mols of Eu_2O_3 (24.560 gm./liter). The working volume is 520 liters or 138.70 gallons.

Weigh out the following amounts of rare earths for dissolution:

Y_2O_3 = 109.23 kg. = 240.6 lbs. (520 x 210.86 gm/mol)
Eu_2O_3 = 12.77 kg. = 28.10 lbs. (520 x 24.560)

The following totals are the volumes required for dissolution of the oxides:

286 liters, or 75.6 gallons, of concentrated HCl
234 liters, or 61.8 gallons, of deionized water

To a 150 gallon glass-lined tank, add 168.0 liters, or 45 gallons, of concentrated HCl to the tank containing about 10 gallons of deionized water. The resulting diluted solution will heat up. Maintain stirring and add small increments of the mixed oxide while stirring until about 1/4 of the total has been added. The solution will heat up proportionally to the amount of oxides added and how fast they are added. Do not allow the solution to boil.

When the rate of dissolution has slowed down and some of the rare earth oxides remain in suspension, add approximately 20-25 liters (5-6 gallons) of concentrated HCl and stir until dissolved. Add more oxide and continue the process until all of the mixed oxides are completely dissolved and a total of 286 liters of concentrated acid plus 234 liters of deionized water has been added. If the Ln_2O_3 dissolves before all of the acid has been used, dilute volume to final volume using deionized water.

2. Vanadyl (VO^{2+}) Solution

A 2.10 molar solution of vanadyl ion is required. Thus, 1.05 mols of V_2O_5 (191.0 gm/liter) must be dissolved in 1.00 liter of concentrated HCl. Since the working volume is 520 liters or 138.7 gallons, we will require 99.32 kg. or 218.7 lbs. of V_2O_5 (520 x 191.00 @ 95.2% assay).

The total volume of 138.7 gallons of concentrated HCl is placed in a 150 gallon glass-lined tank having an enclosed cover equipped with a vent structure and scrubber. The total weight of 218.7 lbs. of V_2O_5 is added to the tank with stirring. A suspension will first form. Heat the solution in the tank while continuing to stir until all of the suspension has dissolved. A dark blue solution results. Since the reaction produces chlorine gas, it is necessary to employ a vapor-scrubber to remove the gas generated by scrubbing with an NaOH-solution in the packed scrubber (which forms NaClO + NaCl).

3. Preparation of Phosphoric Acid Solution

A 2.10 molar volume of H_3PO_4 is required. For a 300 gallon tank, a working volume of 520 liters (137.4 gallons) is used. 142.86 ml. of concentrated acid (85%) per liter is required. To 300 liters of deionized water (79.3 gallons), add 42.86 liters (11.32 gallons) of concentrated H_3PO_4 slowly with stirring. Dilute to a total working volume of 520 liters.

4. Solution of Titrants

A solution of these two "titrants" is required at a ratio of 3.0:1.0. The total volume is 520 liters or 138.7 gallons. This requires:

NH_4OH (Concentrated reagent) = 390 liters or 104.1 gallons
H_2O_2 (30%) = 130 liters or 34.7 gallons

These volumes are measured and placed in a 150 gallon glass-lined tank just before the precipitation process is to take place.

Alternately, make up a NaOH- $NaBO_4$ solution (at a 10:1 ratio) as follows:

Weigh out 200 gm. of NaOH/liter of solution or 20.80 kg. (45.84 lb.) which is added to 520 liters or 138.7 gallons of deionized water with stirring. To this solution is added 48.90 gm/liter or 25.43 kg. (56.03 lb.) of $NaBO_4$. (NaOH = 5.00 molar and $NaBO_4$ = 0.50 molar in solution).

This volume is used in place of the NH_4OH + H_2O_2 (30%) solution for titrating the mixed phosphate-vanadyl solution. A solution of 11.0 molar NH_4OH (concentrated reagent) and 0.5 molar $NaBO_4$ can also be employed.

4. Precipitation of YVO_4:Eu

The following volumes of liquids are pumped into a 300 gallon glass-lined tank, while stirring continues in the tank:

i. 104 liters, or 27.48 gallons of VO^{2+} solution, followed by 156 liters of phosphoric acid solution. This gives a 60 mol% PO_4 - 40 mol% VO_4 precipitate.
ii. 130 liters or 34.67 gallons, of Y^{3+} - Eu^{3+} solution is added to the vanadyl solution already in the tank. Then 130 liters of deionized water follows.

iii. The total volume is now 520 liters or 137.38 gallons.

This solution is heated to 80 °C, but less than 85 °C. while stirring continues. The NaOH + NaBO$_4$ solution is then added at a steady rate of about 3.0 liters/minute, as feasible. The pH will attempt to decrease towards < 1.00 as the precipitation proceeds due to release of H$^+$ from the precipitate being formed. The change in pH should steadily increase with a minimum of oscillation, until a pH = 8.00 is obtained. Once this pH is reached, and it remains steady, the precipitate is stirred for about 45 minutes. Then, it is filtered hot and washed upon the filter. The wet cake is then dried for 16 hours at 125 °C. and then fired as before, using open silica trays and firing in air.

It has been determined that this method produces a superior phosphor to both solid state and precipitated YVO$_4$: Eu phosphor compositions, as shown by the following data:

Phosphor Composition

In Solution		In Phosphor			
PO$_4$	VO$_4$	PO$_4$	VO$_4$	% Output*	Yield
80	20	78	22	95	89%
60	40	60	40	130	96
50	50	47	53	115	91
40	60	42	58	111	94
20	80	18	82	101	92

* compared to the solid-state product

A comparison of the three methods given above has been made, with the following results:

Comparison of the Three Methods of Preparation

	QE*	A	x	y	Emission*
Solid State	0.82	0.92	0.672	0.327	6210 Å
Precipitated YVO$_4$	0.86	0.93	0.667	0.333	6320
Precipitated 40 mol% YVO$_4$ - 60 mol% YPO$_4$	0.92	0.94	0.665	0.335	6370

* Using HPMV excitation

It should be clear that the optimum phosphor for use in HPMV lamps is the mixed PO_4 - VO_4 phosphor, prepared by a precipitation process. Whether it is prepared by precipitation or by a solid state process has been a matter of preference by each individual manufacturer.

The type of equipment needed to precipitate these vanadate phosphors is similar to that already given in 3.4.3. of Chapter 3. However there are some important exceptions, as shown in the following diagram:

4.2.91.- Equipment Used for Precipitation of Y(P,V)O$_4$:Eu Phosphor

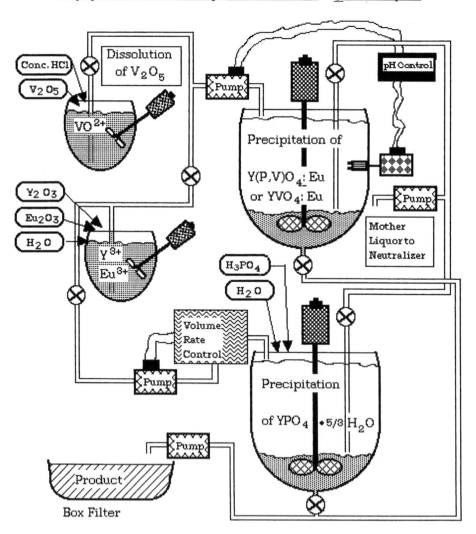

These exceptions include the need for a pH control and a rate control for adding rare earth solution to the reacting mixture in suspension. The equipment for automatic control of the pH is used to add base to change the pH which needs to be in constant flux during the reactions which occur to form the final composition, i.e.- (Y, Eu)VO$_4$.

Also, the volume rate control is needed to control the rate of addition of lanthanides to the phosphoric acid solution so as to allow the suspension of (Y,Eu)PO$_4$•5/3 H$_2$O to grow to a suitable particle size.

The other equipment needed such as an oven for drying the precipitate and a ribbon blender for final mixing to obtain a uniform product are not shown.

16. Manufacture of "TriColor" Component Phosphors

A high brightness lamp was introduced to the marketplace in the 1980's. This fluorescent lamp was usually "U-shaped" or had one of the shapes presented in a previous chapter concerning the so-called "miniature" fluorescent lamps. These lamps were of the "high-output" variety in which three line (or narrow-band) emitting phosphors were used to produce various "white" lamp-colors. Luminous output exceeded 3600 lumens/watt in some cases.The blue phosphor was usually Sr$_5$Cl(PO$_4$)$_3$:Eu^{2+} or (Ba,Mg)$_2$Al$_{16}$O$_{27}$:Eu^{2+} and the green phosphor was MgAl$_{11}$O$_{19}$:Ce^{3+}:Tb^{3+}, while the red one was Y$_2$O$_3$:Eu^{3+}.

a. BLUE-EMITTING PHOSPHORS

i. MANUFACTURE OF Sr$_5$Cl(PO$_4$)$_3$:Eu^{2+} PHOSPHOR

The formulation used for the manufacture of this phosphor is given as follows:.

4.2.92.- Formulation for Sr$_5$Cl(PO$_4$)$_3$:Eu^{2+} Phosphor

Compound	Mol Ratio	Assay	Grams	Lb/Lb
SrHPO$_4$	3.00	98.2%	560.9	0.655
SrCO$_3$	1.46	99.2	217.3	0.254
SrCl$_2$	0.45	99.8	71.48	0.083
Eu$_2$O$_3$	0.02	99.9	7.0384	0.0082

In this case, the cation to anion ratio used is: 4.95 : 3.00. Alternately, divalent europium carbonate, i.e.- $EuCO_3$ can be substituted for the oxide , Eu_2O_3 , at a mol ration of 0.04 mol $EuCO_3$/ mol of phosphor. This formulation is blended and hammermilled twice with blending in between the two hammering steps. The powder is then loaded into 2000 ml silica crucibles (with Covers) and then fired:

$$1100 \text{ °C. for 4 hours in 90\% } N_2 \text{ gas} + 10\% \text{ } H_2 \text{ gas}$$

The gas volume must be adjusted to the size of the furnace being used. A typical gas flow is: 900 ml of N_2 - 100 ml of H_2 gas per minute. At the end of the firing period, the crucible is removed from the furnace, the phosphor is cooled, inspected and then washed as given above for processing the halophosphates, except that 0.50 molar ethylenediamine tetraacetic acid (EDTA) in 2.0 molar NH_4OH is substituted in the process for the 1% HCl solution (see pp. 373 & 381). The final step is to crush the phosphor and screen it through a 325 mesh screen to remove particles larger than 44 μ.

Another formulation used to produce this phosphor substitutes $CaCl_2$ for $SrCl_2$, viz:

4.2.93- Formulation for (Sr,Ca) $_5Cl(PO_4)_3$:Eu^{2+} Phosphor

Compound	Mol Ratio	Assay	Grams	Lb/Lb
$SrHPO_4$	3.00	98.2%	560.9	0.672
$SrCO_3$	1.46	99.2	217.3	0.260
$CaCl_2$	0.45	99.8	50.05	0.060
Eu_2O_3	0.02	99.9	7.0384	0.0084

This mix is processed as before, except that the firing schedule is:

$$3.5 \text{ hours at 1150 °C in a 90\% } N_2 \text{ - 10\% } H_2 \text{ flowing gas mixture.}$$

After firing, the cake is inspected and crushed. The powder is then washed as given above and then dried 16 hours at 125 °C. Finally, the powder is screened through a 325 mesh screen. The spectral properties are:

4.2.94.-

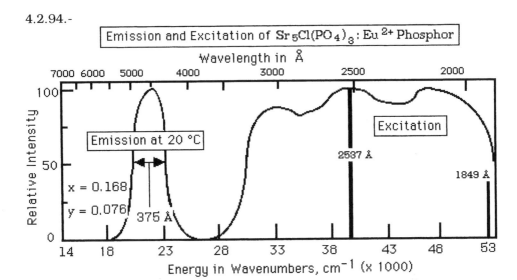

The QE is 0.91, A = 0.97, the luminosity is 0.14, the peak wavelength is 4450 Å and the usual particle size produced is 12-15μ.

This phosphor is one of the blue-components manufactured for use in high-output miniature fluorescent lamps. It is also used by one manufacturer as a blue component in a Lite White blend for use as an illuminating source in the display of vegetables in Supermarkets.

ii. MANUFACTURE OF $BaMg_2Al_{16}O_{27}:Eu^{2+}$ PHOSPHOR

Another phosphor that has been used as the blue-emitting component in these tri-color lamps is based upon another aluminate matrix. The formulation to be used is:

4.2.95.- <u>Formulation for $BaMg_2Al_{16}O_{27}:Eu^{2+}$ Phosphor</u>

Compound	Mol Ratio	Assay	Grams	Lb/Lb
$3MgCO_3 \bullet Mg(OH)_2 \bullet 3H_2O$	1.98 MgO	44.6% MgO	178.93	0.145
$BaCO_3$	1.06	0.992	209.2	0.169
Al_2O_3	8.00	0.975	836.6	0.678
$EuCO_3$	0.04	0.995	8.521	0.00691

This formulation is blended and hammermilled twice with blending in between the two hammering steps. The powder is then loaded into 2000 ml alumina crucibles and then fired:

    a. 1450 °C. for 6 hours in air

    b. Wet Milled for 2 hours.

    c. Refired 4 hours at 1150 °C in 90% N_2 - 10% H_2
    atmosphere in open alumina trays.

Between firings, the cakes are inspected and then crushed before screening through a 200 mesh screen prior to refiring. After wet milling at 3lbs./gallon of water, the slurry is filtered and the wet powder dried before refiring.

The excitation-emission spectra are given in the following diagram:

4.2.96.-

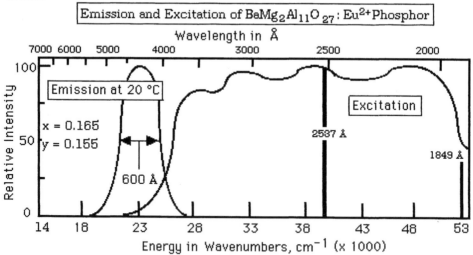

The QE is 0.88, A = 0.97, the luminosity is 0.21, the peak wavelength is 4540 Å and the usual particle size produced is 12-15μ. The cation/anion ratio is 3.04/8.00. This phosphor is also used as a blue component in the tri-phosphor high output lamps.

Another formulation that has been used by another manufacturer is:

4.2.97.- Formulation for $BaMg_2Al_{16}O_{27}:Eu^{2+}$ Phosphor

Compound	Mol Ratio	Assay	Grams	Lb/Lb
$3MgCO_3 \cdot Mg(OH)_2 \cdot 3H_2O$	1.90 MgO	44.6% MgO	171.70	0.141
MgF_2	0.10	0.985	6.327	0.0052
$BaCO_3$	0.86	0.992	171.08	0.141
Al_2O_3	8.00	0.975	836.59	0.688
$EuCO_3$	0.14	0.995	29.825	0.0245

This formulation is blended, hammermilled twice with blending in between the hammering steps. The powder is then loaded into 2000 ml alumina crucibles and then fired:

$$3 \text{ hours. at } 1200 \text{ °C in } 95\% \text{ } N_2 + 5\% \text{ } H_2$$

It is then refired:

Ball-mill for 2 hours. in a water slurry- Filter and dry powder

Refire 3 hours. at 1250 °C. in 95% N_2 + 5% H_2 + water-vapor(steam)

The water vapor removes any excess fluoride that may be present. After firing, the powder is wet milled and dried before refiring. The QE of this phosphor is said to be 100%, A = 0.90, the peak emission occurs at 4500 Å, the average particle size is about 12 μ and the activator content is 14 mol%.

iii.- MANUFACTURE OF $SrMg_2Al_{18}O_{39}:Eu^{2+}$ PHOSPHOR

Another blue-emitting phosphor is based on the above magnetoplumbite structure, but is considered to be a defect structure, perhaps containing $SrAl_2O_4$ (15). A discussion of these structures is given below.

This phosphor is manufactured using the following formulation:

4.2.98.- Formulation for $SrMg_2Al_{18}O_{39}:Eu^{2+}$ Phosphor

Compound	Mol Ratio	Assay	Grams	Lb/Lb
$3MgCO_3 \cdot Mg(OH)_2 \cdot 3H_2O$	2.00 MgO	44.6% MgO	180.73	0.129
$SrCO_3$	1.67	0.992	248.53	0.178
Al_2O_3	9.00	0.975	937.85	0.672
Eu_2O_3	0.083	0.995	29.341	0.0210

This formulation is blended, hammermilled twice with blending in between the hammering steps. The powder is then loaded into 2000 ml alumina crucibles and then fired:

> a. 4 hours at 1500 °C in air
> b. Wet Milled for 2 hours.
> c. Refired 4 hours at 1150 °C in 90% N_2 - 10% H_2 atmosphere in open alumina trays.

At the end of the refiring period, the powder is cooled in nitrogen gas atmosphere.

The excitation-emission of this phosphor is quite similar to that given for the $BaMg_2Al_{16}O_{27}:Eu^{2+}$ phosphor except that the emission band occurs at 4650 Å. The QE is said to be 100%, A = 0.90, the activator content is 20 mol%, and the average particle size is usually about 15 μ.

b. GREEN-EMITTING PHOSPHORS

The tri-color lamp is based on the use of red, green and blue-emitting phosphors which are used to simulate response by the human eye for the three "prime" colors.

i. MANUFACTURE OF $Mg_3Al_{11}O_{19}:Ce^{3+}:Tb^{3+}$ PHOSPHOR

This phosphor is one of the family of magnesium aluminates used in the tri-phosphor blend for high output miniature fluorescent lamps. Its preparation is accomplished as follows:

4.2.99.- Formulation for $Mg_3Al_{11}O_{19}:Ce^{3+}:Tb^{3+}$ Phosphor

Compound	Mol Ratio	Assay	Grams	Lb/Lb
$3MgCO_3 \cdot Mg(OH)_2 \cdot 3H_2O$	2.00 MgO	44.6% MgO	180.74	0.120
Al_2O_3	11.00	0.975	1150.32	0.762
Tb_4O_7	0.0825	0.995	61.995	0.0411
CeO_2	0.670	0.985	117.08	0.0775

Note that a major part of the cations present are also activators. Thus, the more correct phosphor formula would be: $(Ce,Tb)Mg_2Al_{11}O_{19}$. This formula is blended, hammermilled twice with blending in between hammermillings. The mix is then fired:

a. 1450 °C. for 6 hours in air in alumina crucibles

b. Refired 4 hours at 1150 °C in 90% N_2 - 10% H_2 atmosphere in open alumina trays.

The cake is inspected between firings and screened through a 200 mesh screen. Sometimes, it is hammermilled without a screen to effect a thorough mixing before the second firing. the final step is to screen the powder through a 325 mesh screen to remove large particles.

The emission - excitation spectra of this phosphor is given as follows:
4.2.100.-

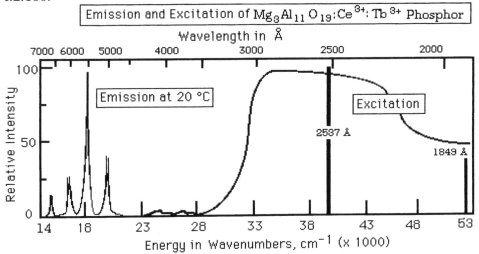

The QE of this phosphor is 0.92, A= 0.97, the luminosity is 0.70, and the average particle size produced is 12-15 μ. The emission, consisting of a set of four lines with the strongest peaking at 5520 Å. , is typical of Tb^{3+}. Note that Ce^{3+} is used as a sensitizer for the emission of Tb^{3+}. (Actually, Ce^{3+} is a good emitter and is the activator in yttrium aluminum garnet (YAG) as we will describe below).

A comparison of the relative radiant power per constant wavelength interval of this phosphor compared to that of the standard NBS green Zn_2SiO_4:Mn phosphor is given in the following diagram:

4.2.101.-

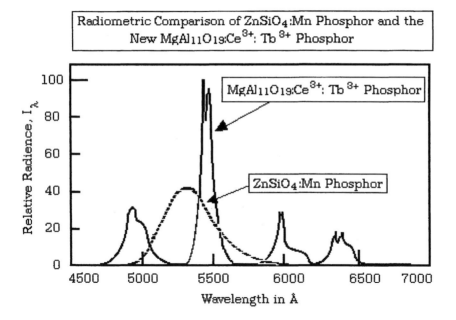

It is easily seen that this phosphor (QE = 80%) has considerably more luminous output than the green silicate (QE = 82%) phosphor that has been the workhorse for green emission in fluorescent lamps. However, the emission is discontinuous which may be a factor in its usage as a general replacement for green emission in fluorescent lamps, as we will discuss below. We will also show how the line emission of a blend of red, green, and blue emitters compares to that of the "normal" spectrum used in current fluorescent lamps.

ii.- MANUFACTURE OF $BaMg_2Al_{16}O_{27}{:}Eu^{2+}{:}Mn^{2+}$ PHOSPHOR

A variation of the phosphor given above as $BaMg_2Al_{16}O_{27}{:}Eu^{2+}$ is achieved by the addition of Mn^{2+} to the formulation. Since the structure is related to the cubic spinels, a green peak is seen.

The phosphor is prepared by mixing the following ingredients as shown in the following:

4.2.102.- Formulation for $BaMg_2Al_{16}O_{27}{:}Eu^{2+}{:}Mn^{2+}$ Phosphor

Compound	Mol Ratio	Assay	Grams	Lb/Lb
$3MgCO_3 \bullet Mg(OH)_2 \bullet 3H_2O$	1.90 MgO	44.6% MgO	171.70	0.140
MgF_2	0.10	0.986	6.319	0.00516
$BaCO_3$	0.86	0.992	171.08	0.140
Al_2O_3	8.00	0.975	836.59	0.684
$EuCO_3$	0.12	0.995	25.564	0.02091
$MnCO_3$	0.10	0.995	11.553	0.00945

This formula is blended, hammermilled twice with blending in between hammermillings.

The powder is then loaded into 2000 ml alumina crucibles and then fired:

3 hours. at 1200 °C in 95% N_2 + 5% H_2

Ball-mill for 2 hours. in a water slurry-Filter and dry powder

Refire 3 hours. at 1250 °C. in 95% N_2 + 5% H_2 + water-vapor(steam)

After firing, the powder is wet milled and dried before refiring. After the final firing, the phosphor is cooled in nitrogen and then crushed and screened through a 325 mesh screen to remove particles larger than about 44 μ.

The spectra for this phosphor is shown in the following diagram as given in 4.2.103. on the next page.

490

4.2.103.-

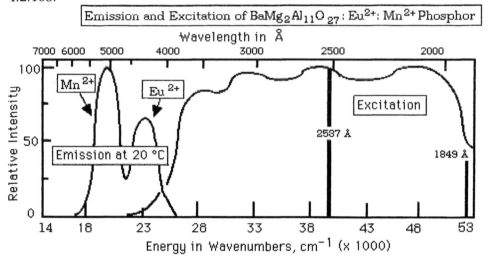

The QE of this phosphor is said to be 100% depending upon the Mn^{2+} content, A = 0.90, the peak emission occurs at 4500 Å, the average particle size is about 12 μ and the activator content is 14 mol%.

Actually, it is surprising that Eu^{2+} acts as a sensitizer for the Mn^{2+} emission, but as shown in the emission spectra, two peaks are seen whose relative intensity depends upon the Mn^{2+} content, just as in the Sb^{3+} - Mn^{2+} system in the alkaline earth apatites.

A comparison of the emission of several phosphors with varying Eu^{2+} and Mn^{2+} contents is given in the following:

Table 4-10

Relative Emission Peak Heights for the $BaMg_2Al_{16}O_{27}:Eu^{2+}: Mn^{2+}$ Phosphor

Activator	Peak Height	Activator	Peak Height	Quantum Eff.
Eu^{2+} = 0.10	70%	Mn^{2+} = 0.00	0 %	88 %
Eu^{2+} = 0.20	5 %	Mn^{2+} = 0.60	100%	100 %
Eu^{2+} = 0.10	20%	Mn^{2+} = 0.23	85%	92 %
Eu^{2+} = 0.10	45%	Mn^{2+} = 0.10	50%	80 %

Note, however, that the emission peaks are blue and green. Thus, the addition of a red-emitting phosphor will be required to produce a satisfactory "white" lamp.

These aluminate phosphors are based upon magnetoplumbites having a hexagonal structure with space group P6₃/mmc. They consist of spinel-like blocks, separated by intermediate layers of deviating structure, containing the large cation including the trivalent rare earths.

The following shows some of these types of compounds and their approximate formulas:

TABLE 4-11

Compositions of Spinels and Magnetoplumbites

Spinels	Magnetoplumbites	
$MgAl_2O_4$	$(Ba,Sr)Al_{12}O_{19}$	$(La,Ce)MgAl_{12}O_{19}$
$MgAl_2O_4$	$(Ba,Sr)Ga_{12}O_{19}$	$LaMgGa_{11}O_{19}$
$MgAl_2O_4$	$(Ba,Sr)Fe_{12}O_{19}$	$LaMgFe_{11}O_{19}$

Most of the phosphor formulations given above produce one of these structures. However, some may also contain spinels in solid solution with the magnetophumbate matrix, as well.

c. LAMP BLENDS MADE USING NARROW BAND EMITTERS

A number of Manufacturers have made lamp tests using the narrow band emitters just described above. The distribution of emission wavelengths of such a lamp shows gaps between the peaks of the three emitting colors, in contrast to the continuous spectrum usually seen in lamps currently being marketed.

What we are speaking of is illustrated in the following diagram, which shows the emission spectrum composed of three narrow band or line emitters, as shown in 4.2.104. on the next page.

According to Koedam and Opstelten (16), a "Deluxe" lamp spectrum with a high color-rendering index (CRI) involving three line emitters can be chosen. But, since the wavelengths of each of the three bands are interdependent,

492

4.2.104.-

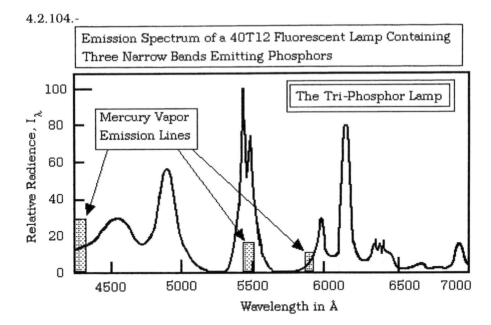

Emission Spectrum of a 40T12 Fluorescent Lamp Containing
Three Narrow Bands Emitting Phosphors

The Tri-Phosphor Lamp

Mercury Vapor
Emission Lines

Relative Radience, I_λ

careful choice of the phosphors must be made to produce such a lamp. This
spectrum can be compared to that of 4.2.6. wherein the spectrum is that of
Cool White Halophosphate.

The following Table, given on the next page, summarizes the spectral
properties of all of the line-emitting phosphors (whose manufacture was
described above). The lumen values given are those for the lamp containing
each phosphor, as are the color coordinates, x and y.

Note that the phosphors of choice are:

$$Blue = BaMg_2Al_{16}O_{27}:Eu^{2+}$$

$$Green = (Ce_{0.67}Tb_{0.33})MgAl_{11}O_{19}$$

$$Red = (Y_{0.9} Eu_{0.11})_2O_3$$

These are the phosphors which perform best in 40T12 fluorescent lamps.
However, we have not yet evaluated the CRI of such lamps.

TABLE 4-12

Properties of Various Line emitting Phosphors in 40T12 Fluorescent Lamps

Color	Phosphor Composition	QE	Reflect % A	Color x/ y	Lamp Output Lumens	Maint.
Red	$(Y_{0.9} Eu_{0.11})_2O_3$	88	0.10	0.597/0.331	1430	88%
Green	$(Ce_{0.67}Tb_{0.33})MgAl_{11}O_{19}$	80	0.08	0.597/0.331	2354	95%
	$(Zn_{0.90}Mn_{0.10})SiO_4$	75	0.15	0.246/0.622	2068	73%
	$BaMg_2Al_{16}O_{27}:Eu^{2+}: Mn^{2+}$	80	0.07	0.288/0.576	2310	65%
Blue	$Sr_5Cl(PO_4)_3:Eu^{2+}$	85	0.03	0.159/0.038	264	79%
	$BaMg_2Al_{16}O_{27}:Eu^{2+}$	100	0.10	0.151/0.066	528	92%
	$SrMg_2Al_{18}O_{39}:Eu^{2+}$	100	0.08	0.149/0.181	990	67%
	$BaMg_2Al_{16}O_{27}:Eu^{2+}: Mn^{2+}$	95	0.16	0.148/0.225	1100	86%

As shown in the following table, a Cool White blend can be made from various choices of blue, green and red phosphors. Note that a lamp having at least 3200 lumens in total luminous output has been achieved with a CRI of at least 85%. This is equal or superior to the 40T12 lamp in which the halophosphate phosphor system is being used.

TABLE 4-13

Spectral Output and Color of Cool White Phosphor Blends with Tri-Phosphors

PHOSPHOR BLENDS

BLUE	GREEN	RED	Lumen	CRI
$Sr_5Cl(PO_4)_3:Eu^{2+}$	$(Zn_{0.90}Mn_{0.10})SiO_4$	$Y_2O_3:Eu$	2916	79%
$BaMg_2Al_{16}O_{27}:Eu^{2+}$	$(Zn_{0.90}Mn_{0.10})SiO_4$	$Y_2O_3:Eu$	2934	80
$BaMg_2Al_{16}O_{27}:Eu^{2+}$	$BaMg_2Al_{16}O_{27}:Eu^{2+}: Mn^{2+}$	$Y_2O_3:Eu$	2835	82
$BaMg_2Al_{16}O_{27}:Eu^{2+}$	$(Ce_{0.67}Tb_{0.33})MgAl_{11}O_{19}$	$Y_2O_3:Eu$	3280	85
$BaMg_2Al_{16}O_{27}:Eu^{2+}: Mn^{2+}$	$(Ce_{0.67}Tb_{0.33})MgAl_{11}O_{19}$	$Y_2O_3:Eu$	3200	87
$SrMg_2Al_{18}O_{39}:Eu^{2+}$	$(Ce_{0.67}Tb_{0.33})MgAl_{11}O_{19}$	$Y_2O_3:Eu$	3240	91

Additionally, it is possible to manufacture all of the other Deluxe Color blends for the four major lamp colors being marketed. This is an important aspect if these phosphors are to find universal usage, and are to replace the Halophosphate phosphors currently in use in fluorescent lamps.

These are given in the following, shown on the next page, viz-

These are:

Lamp Color	x	y	Lumens	CRI
Warm White	0.480	0.405	3321	83 %
White	0.447	0.401	3360	84
Cool White	0.385	0.378	3281	85
Daylight	0.320	0.341	3240	82

It should be clear that these new phosphor blends are the wave of the future in the Lamp Industry. Nevertheless, the luminous output of these lamps consists of a series of lines spanning the visible spectrum. In contrast, the current lamps have a continuous spectrum. Since the human animal has matured in Sunlight, it is not certain how the human organism will respond when subjected to discontinuous line spectra for long periods of time (particularly office personnel where the only lighting is that within a confined space). It is for this reason that a complete substitution of the older lighting system has not been made to date. Indeed, some manufacturers have marketed lamps for indoor lighting based on an extended spectrum, in contrast to those lamps based on the tri-phosphor system. Only time will tell how well such lamps are accepted by the marketplace. However, for miniature lamps which are intended to replace incandescent lamp bulbs (because of reduced electrical costs), the tri-phosphor lamp has no peer.

D. MANUFACTURE OF SPECIALTY PHOSPHORS

By specialty phosphors, we mean those not used in fluorescent lamps for lighting purposes, but those made for use in fluorescent lamps for special purposes. A good example is the lamps made for plant-growth or the "black-light" type of fluorescent lamp. A number of these phosphors have evolved as the need for a special lamp was demanded by the Marketplace.

1. Manufacture of $BaSi_2O_5$:Pb Phosphor

This phosphor is a so-called "black-light" phosphor since it has an emission band close to 3550 Å. It is Pb^{2+} - activated and the formula usually employed is given in the following, shown as 4.2.105. on the next page.

4.2.105.- Formulation Used for Preparation of BaSiO $_5$:Pb Phosphor

Compound	Mol Ratio	Assay	Grams	Lb/Lb
BaCO$_3$	1.00	0.994	198.53	0.583
SiO$_2$	2.05	0.882	139.64	0.410
PbF$_2$	0.010	0.973	2.52	0.007$_4$

This mixture is processed as given above in 4.2.19. except that a covered silica crucible, as shown above on p. 393 is usually employed for firing. The solid state reaction is:

4.2.106.- $BaCO_3 + 2\ SiO_2 \Rightarrow BaSi_2O_5 + CO_2 \Uparrow$

The firing schedule used is:

4.0 hours in air at 1100 °C.

After firing, the cake is inspected, crushed, and sieved through a 325 mesh screen before usage thereof.

Another formulation used by a different manufacturer is shown as follows:

4.2.107.-

Compound	Mol Ratio	Assay	Grams	Lb/Lb
BaCO$_3$	0.88	0.994	174.71	0.528
SiO$_2$	2.00	0.882	136.24	0.411
PbF$_2$	0.08	0.973	20.16	0.061

This translates to a 1.08 : 2.27 cation to anion mol ratio, or 1.00 : 2.10, compared to the 1.00 : 1.03 ratio for the first formula given. In this case, the firing schedule is:

6.0 hours at 1085 °C. in air in a covered crucible

The higher activator content used apparently compensates for the longer firing time employed (since a small amount of volatile lead undoubtedly escapes from the covered crucible during firing). The fired cake is processed as given for the first case.

The excitation-emission curves for this phosphor are shown in the following diagram:

4.2.108.-

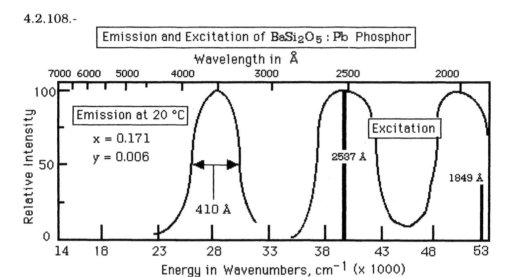

Emission and Excitation of $BaSi_2O_5$: Pb Phosphor

The QE is 0.76, absorbance is 0.94, average particle size produced is 10-15 μ the emission band peaks at 3510 Å, and peak excitation is at 2590 Å.

For many years, this phosphor was used as the standard to produce a lamp for medical, photocopy and "blacklight" usages. However, it has been superseded in some cases by other phosphors.

2. Manufacture of $(Ba,Sr)_2$ $(Mg,Zn)Si_2O_7$: Pb Phosphor

This phosphor emits in the ultraviolet range and has been used as a "blacklight" phosphor similar to the barium disilicate phosphor described above.

The solid state reactions involved are:

4.2.109.- $Ba(NO_3)_2 + ZnO + MgCO_3 + 2\ SiO_2 \Rightarrow (Ba,Mg,Zn)_3 Si_2 O_7 + 2NO_2 \uparrow$
$+ CO_2 \uparrow$

This produces a disilicate having the Ackermanite structure, i.e.- a tetragonal crystal in which there are two cationic sites. One site has tetragonal symmetry and the other has octahedral symmetry. The Si_2O_7 grouping consists of two silica tetrahedrons attached at one end of each tetrahedron.

The formula to be used is:

4.2.110.- Formulation for $(Ba,Sr)_2(Mg,Zn)Si_2O_7$: Pb Phosphor

Compound	Mol Ratio	Assay	Grams	Lb/Lb
$Ba(NO_3)_2$	1.15	0.999	300.59	0.455
BaF_2	0.15	0.999	26.30	0.040
$SrCO_3$	0.51	0.994	75.30	0.114
$3MgCO_3 \bullet Mg(OH)_2 \bullet 3H_2O$	1.00 MgO	44.6% MgO	94.83	0.143
ZnO	0.20	0.996	16.31	0.025
$CaCO_3$	0.05	0.992	5.00	0.007_6
SiO_2	2.06	0.882	139.64	0.211
PbO	0.015	0.973	3.35	0.005_1

This mix is blended first, and then hammermilled twice, blending in between. It is then fired in two stages:

a. Prefire in air, using a 2 liter crucible for 4 hours at 875 °C.

b. Dry mill for 4 hours, adding 6% $BaCl_2$ by weight of the powder as a flux.

c. Refire in an air-steam mixture for 4 hours at 985 °C.

Alternatively, the firing can be accomplished in a silica tray which is pushed through a tube-furnace so as to control the steam atmosphere more easily. At the end of the firing, the cooled cake is inspected, crushed and sieved through a 325 mesh screen.

Another formulation that has been used by another manufacturer to produce essentially the same phosphor is shown on the next page as 4.2.111.

Note that only minor differences may be discerned in these two formulas.

4.2.111.- <u>Formulation for (Ba,Sr) $_2$ (Mg,Zn)Si $_2O_7$: Pb Phosphor</u>

Compound	Mol Ratio	Assay	Grams	Lb/Lb
$BaCO_3$	1.10	0.999	217.29	0.397
BaF_2	0.15	0.999	26.30	0.048
$SrCO_3$	0.36	0.994	53.47	0.098
$3MgCO_3 \cdot Mg(OH)_2 \cdot 3H_2O$	1.00 MgO	44.6% MgO	94.83	0.173
$CaCO_3$	0.12	0.992	12.11	0.022
SiO_2	2.06	0.882	139.64	0.255
PbO	0.015	0.973	3.35	0.006_1

These components are blended and hammermilled as before, hammermilling twice, blending before and after each hammermilling.

The firing schedule to be used is:

> 6 hours in air at 800 °C.
> Mill crushed dry cake for 2 hours
> Refire in air in a closed 2 liter crucible for 4 hours
> at 985 °C.

An ultraviolet emitting phosphor is obtained as shown in the following diagram:
4.2.112.-

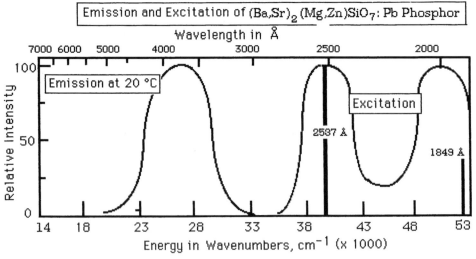

Alternatively, the firing can be accomplished in a silica tray which is pushed through a tube-furnace so as to control the steam atmosphere more easily. At the end of the firing, the cooled cake is inspected, crushed and sieved through a 325 mesh screen.

The formulation as given produces a phosphor having the Ackermanite structure. This crystal is composed of disilicate tetrahedra, i.e. $Si_2O_7^{6-}$, having two types of cationic sites, one with tetragonal symmetry, and the other with octahedral symmetry. As a result, the structure will accommodate two different types of cations, each occupying one or the other symmetry.

Actually, such crystal compositions are a part of the melilite group of alkaline earth silicates which form a series of tetragonal structures containing the two cationic sites. The space group of these structures is D_{2d}^3 The silica tetrahedra form points of tetrahedral coordination for two general types of cations, "Hardystonite" - exemplified by $(Ba,Sr)_2 ZnSi_2O_7$, and "Ackermanite" - exemplified by $(Ba,Sr)_2 MgSi_2O_7$. In the former, the Zn^{2+} cation occupies the tetragonal cationic site, while in the latter it is the Mg^{2+} cation. Both form solid solutions of infinite proportion. Therefore, it is possible to substitute **all** of the alkaline earth cations into these structures, plus those of Zn^{2+} and Cd^{2+}. Because of this factor, the optimal phosphor is not easy to determine because of the multitude of possible formulations. However, by studying the phase diagrams available (8), one can sort out the possible phosphor compositions, viz-

4.2.113.- Types of Silicates Present in the Melilite Series of Structures

a. Hardystonites
$(Ba,Sr)_2 ZnSi_2O_7$: Pb
$(Ba,Sr)_2 CdSi_2O_7$: Pb
$Ca_2 ZnSi_2O_7$: Pb
$Sr_2 ZnSi_2O_7$: Pb
$Ba_2 ZnSi_2O_7$: Pb
$Ba_2 CdSi_2O_7$: Pb
$Ba_2 CdSi_2O_7$: Pb
$Ba_2 CdSi_2O_7$: Pb

b. Ackermanites
$(Ba,Sr)_2 MgSi_2O_7$: Pb
$Ca_2 MgSi_2O_7$: Pb
$Sr_2 MgSi_2O_7$: Pb
$Ba_2 MgSi_2O_7$: Pb

500

These formulations represent nearly all of the phosphors which can be obtained, including those of intermediate composition, formed as solid-solutions between the two structures. In the following diagram is shown the effect of variation of the various cations and the range of optimal composition, as determined for this silicate phosphor system, viz-

4.2.114.-

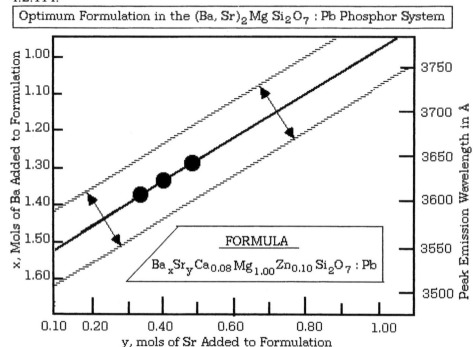

Note that in this diagram, a "band" of compositions exist which are more efficient than those lying "outside" of the "band". It is these which have the highest quantum efficiencies as well as the best performance in fluorescent lamps. Note also that the peak emission wavelength shifts somewhat with composition. This is due to the fact that the Pb^{2+} emission comes from this cation which is affected by how large the neighboring cations are in size.

Another way to shift the emission band is shown in the following diagram, given on the next page as 4.2.115.

4.2.115.-

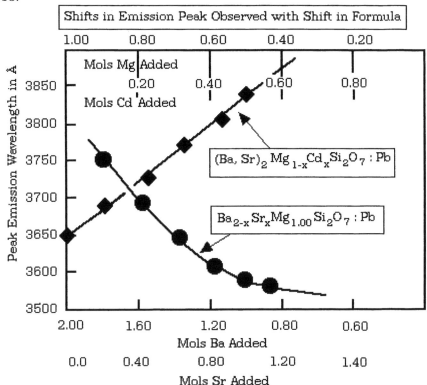

In this case, it is easy to see that one formulation is based on a pure Akermanite structure while the other is a mixed Akermanite-Hardystonite composition. It should be clear, then, that peak emission of these phosphors can be adjusted in the spectral region between about 3500- 4000 Å . The brightest phosphors in fluorescent lamps is that containing about 2 mol% of Pb^{2+} per mol of phosphor.

This, of course, depends upon the initial amount of PbO added and the firing history of the phosphor. Nevertheless, this phosphor system is very versatile since it is one of the few systems, other than halophosphates, whose emission bands can be changed by change in composition to suit the specific application.

3. Manufacture of $Y_3Al_5O_{11}$: Ce Phosphor

This phosphor is used in HPMV lamps as well as in high output miniature lamps because its emission does not degrade at high intensities. The same composition, but with Nd^{3+} as an activator, is grown as a single crystal from the melt to form the familiar YAG crystal used extensively in the Laser Industry for a multitude of applications and is regarded as the "work-horse" of the Industry.

The formulation to be used is:

4.2.116.-   Formulation for $(Y)_3 Al_5O_{11}$: Ce^{3+} Phosphor

Compound	Mol Ratio	Assay	Grams	Lb/Lb
$Y_2(C_2O_4)_3 \cdot 3\ H_2O$	3.00	0.445 as Y_2O_3	1522.2	0.742
Al_2O_3	5.02	0.975	525.0	0.256
$CeCO_3$	0.02	0.985	4.0635	0.0198

This formulation is blended, hammermilled twice with blending in between. The powder is then loaded into 2000 ml alumina crucibles and then fired:

1450 °C. for 6 hours in air

At the end of the firing period, the crucible cakes are inspected and crushed. The powder is then sieved through a 325 mesh screen. Alternately, it can be hammermilled, without the use of a screen in the hammermill to produce a free-flowing powder before it is sieved.

Some manufacturers refire the powder in a reducing atmosphere, viz-

Refire 4 hours at 1150 °C in 95% N_2 - 5% H_2 atmosphere in open alumina trays.

However, most regard this step as not necessary.

The excitation-emission spectra for this phosphor is given on the next page, viz-

4.2.117.-

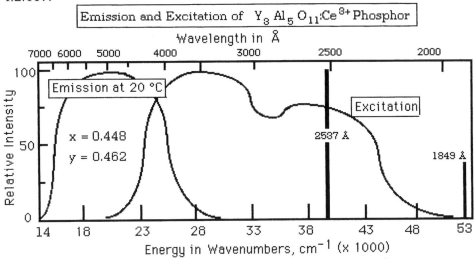

Emission and Excitation of $Y_3 Al_5 O_{11}:Ce^{3+}$ Phosphor

Note that the yellow emission band peaks at about 5000 Å and is unusually broad. It is so wide that it overlaps the excitation band to a significant degree. What this means is that the phosphor is "Daylight-Excited". That is- under normal illumination by sunlight, emission results. Some manufacturers have described this phosphor as having a "yellow-body" color. But, it does not. What is seen is the normal emission of the phosphor caused by sunlight illumination.

The QE of this phosphor is low, being about 0.45 at 2537Å excitation. But, its QE at 3650 Å is about 0.82, A = 0.85, and the normal particle size produced is about 12 μ. Thus, it is suitable for use as a HPMV phosphor. However, its contribution to improved color rendition of the uncorrected HPMV lamp is not high, and consequently this phosphor has not found extensive usage in the Marketplace.

Some manufacturers prefer to precipitate the Y^{3+} and Ce^{3+} together as oxalates, and then add the alumina just before firing. The procedure for doing this is:

   1. Rare Earth Solution: A HNO_3 solution is prepared by adding 2.35 gallons of concentrated HNO_3 to 25.0 gallons of deionized water in a 50.0 gallon glass-lined tank. This forms a 1.200 molar solution. To this

is added 5.13 kilograms (11 lb. 5 oz.) of Y_2O_3 and 30.78 gm. of $CeCO_3$ with stirring. The materials should be mixed beforehand and should be added slowly since the dissolution reaction is exothermic and the solution tends to boil over if the oxides are added too fast. The solution is allowed to cool while the oxalic acid solution is being prepared.

2. Oxalic Acid Solution: To 25.0 gallons of deionized water in a 30.0 gallon tank is added 16.771 kg. (36 lbs. 15 oz.) of oxalic acid with stirring. The oxalic acid solution needs to be about 40-50 °C in order to dissolve all of the oxalic acid. When the dissolution is complete, 6.5 gallons of ethyl alcohol are added to this volume. The solution is then heated to 80 °C. just before precipitation.

3. Precipitation: The rare earth solution is then heated to about 80 °C. and the hot (80 °C.) oxalic acid solution is then pumped into the Y^{3+} and Ce^{3+} solution at a rate of about 0.5 gallons per minute while stirring. There is an induction period before any precipitate will be seen. The total time of precipitation will be about 1.0 hour. The precipitate is allowed to settle and the mother liquor is removed by decantation. One water wash is given to the precipitate by resuspension. The suspension is then filtered on a drum-filter using a water wash wherein the cake is sprayed on the filter. The wet precipitate is then dried for 16 hours at 125 °C.

The weight of oxalate is then recorded, the oxalate is assayed and the particle size distribution is measured. The yield is then calculated and recorded. The precipitate is then substituted as given above for the yttrium oxalate in the formulation.

4. Manufacture of $Sr_2P_2O_7$:Eu Phosphor

This phosphor is used in lamps intended for photocopy devices such as Xerox™ machines. Its emission is close to the near-ultraviolet and as such is used to reflect printed characters from a written page onto a sensitized rotating drum (usually having a silicon layered film integral thereon). The drum-layer becomes charged and picks up black carbon particles so as to redeposit them upon a paper page, thereby reproductively duplicating the original written page. Sometimes, the sensitized layer on the drum is CdS. The phosphor formulation to be used for manufacture is:

4.2.118.- Formulation for $Sr_2P_2O_7$:Eu^{2+} Phosphor

Compound	Mol Ratio	Assay	Grams	Lb/Lb
$SrHPO_4$	2.00	94.6%	388.16	0.983
$EuCO_3$	0.02	99.5%	4.261	0.0108
$(NH_4)_2HPO_4$	0.04	99.2%	2.654	0.00672

This formulation is blended, followed by hammermilled twice with blending in between each hammermilling.

The powder is then loaded into 2000 ml silica crucibles and fired:

1250 °C. for 4 hours in air

Refired 4 hours at 1150 °C in 90% N_2 - 10% H_2 atmosphere in open alumina trays.

After firing, the cakes are inspected and then crushed before screening through a 200 mesh screen prior to refiring. Some manufacturers prefer to fire in a N_2 atmosphere on the first firing, particularly if $EuCO_3$ is being used as the activator addition. If Eu_2O_3 is used, then it is necessary to refire, as given above.

The excitation-emission properties of this phosphor are given in the following diagram, given as 4.2.119. on the next page.

The QE of this phosphor is 0.88, A = 0.94 and the peak emission occurs at 4200 Å. The ratio of cation to anion is: 2.00 : 2.00. Note that both the ß - $Sr_2P_2O_7$ and α - $Sr_2P_2O_7$ structures are activated by Eu^{2+} , but it is the latter which is the more efficient by a factor of five. Therefore, it is important to fire this phosphor at temperatures high enough to ensure the formation of the high-temperature form. This phosphor finds extensive usage as a lamp for photocopy purposes.

4.2.119.-

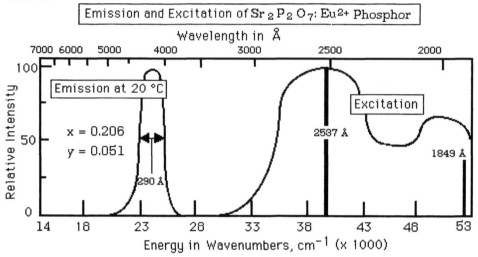

Alternately, a phosphor having an emission shifted from 4200 Å to 3920 Å can be prepared by substituting Mg^{2+} for part of the Sr^{2+} present without effect upon the QE of the phosphor so-produced. Whereas substitution of Sr^{2+} by either Ba^{2+} or Ca^{2+} results in phosphors considerably lower in efficiency, the use of Mg^{2+} substitution results in a phosphor having a slight shift in peak emission without affecting the QE. In some photocopy applications, the shift in peak wavelength is advantageous since the optical response of the transfer drum of the xerox™ machine is more efficient at this wavelength. Such a phosphor is produced using the following formulation:

4.2.120.- Formulation for $(Sr,Mg)_2P_2O_7$:Eu^{2+} Phosphor

Compound	Mol Ratio	Assay	Grams	Lb/Lb
$SrHPO_4$	1.80	94.6%	349.34	0.927
$3MgCO_3 \bullet Mg(OH)_2 \bullet 3H_2O$	0.20 MgO	44.6% MgO	18.073	0.048
$EuCO_3$	0.02	99.5%	4.274	0.0113
$(NH_4)_2HPO_4$	0.04	99.2%	5.324	0.0141

This formulation is blended, hammermilled twice with blending in between each hammermilling, taking care that complete mixing occurs.

The powder is then loaded into 2000 ml silica crucibles and then fired:

1250 °C. for 4 hours in air

Refired 4 hours at 1150 °C in 90% N_2 - 10% H_2 atmosphere in open alumina trays.

After firing, the cakes are inspected and then crushed before screening through a 200 mesh screen prior to refiring. Some manufacturers prefer to fire in a N_2 atmosphere on the first firing, particularly if $EuCO_3$ is being used as the activator addition. If Eu_2O_3 is used, then it is necessary to refire, as given above.

5. Manufacture of $YMg_2Al_{11}O_{19}$:Ce^{3+} Phosphor

This phosphor is an example of one whose emission occurs in the near ultraviolet. The formulation to be used is given as follows:

4.2.121..- Formulation for $YMg_2Al_{11}O_{19}$:Ce^{3+} Phosphor

Compound	Mol Ratio	Assay	Grams	Lb/Lb
$3MgCO_3 \bullet Mg(OH)_2 \bullet 3H_2O$	2.00 MgO	44.6% MgO	180.74	0.203
Y_2O_3	0.45	0.992	102.43	0.116
Al_2O_3	5.50	0.975	575.16	0.649
$CeCO_3$	0.14	0.995	28.159	0.0318

This formulation is blended, hammermilled twice with blending in between the hammering steps. The powder is then loaded into 2000 ml alumina crucibles and then fired. The first firing is:

4 hours at 1450 °C in air

At the end of the refiring period, the powder is cooled in nitrogen gas atmosphere. The cakes are then inspected and crushed. The powder is then wet-milled for 2.0 hours, using a slurry mixture of about 2.0 lbs. of powder/gallon of water. At the end of this time, the slurry is filtered and the wet powder is dried for 16 hours at 125 °C.

It is then refired:

> Refire for 4 hours at 1150 °C in 90% N_2 - 10% H_2 atmosphere in open alumina trays.

The cooled phosphor is then screened through a 325 mesh screen to remove the largest particles.

The spectral properties of this phosphor can be seen in the following:

4.2.122.-

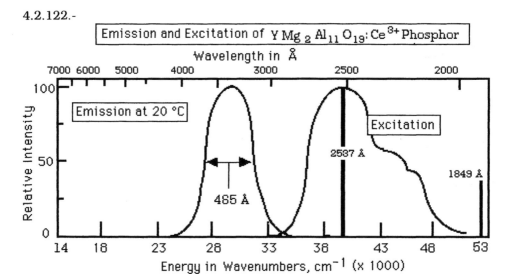

The QE is 70%, A = 0.10, the peak emission occurs at 3400Å, and the cation/anion ratio is 3.0:11.0 .

This phosphor may be compared to that given above in which the yttrium aluminum garnet composition (see above- 4.2.117. in particular) was used to produce a "yellow" emission. The addition of Mg^{2+} to the same formulation apparently changes the structure to one of the other magnetoplumbite series. This produces an ultraviolet peak because of the weaker internal crystal field present in this crystal. Note that in YPO_4 :Ce (having a much weaker crystal field), the emission occurs at about 3250 Å.

The $YMg_2Al_{11}O_{19}$:Ce^{3+} phosphor is used to produce an erythemal lamp, i.e.- a "Suntan" lamp for medical use.

6. Manufacture of MgGa$_2$O$_4$:Mn^{2+} Phosphor

This phosphor is another example of the type where atoms of the matrix act as the sensitizer for the green emission of Mn^{2+} in an octahedral site after the matrix has been formed and then subjected to a reducing atmosphere. The solid state reaction is:

4.2.123.- $MgO + Ga_2O_3 \Rightarrow MgGa_2O_4$

and the formulation to be used is:

4.2.124.- Formulation for Manufacture of MgGa$_2$O$_4$:Mn   Phosphor

Compound	Mol Ratio	Assay	Grams	Lb/Lb
3MgCO$_3$ • Mg(OH)$_2$ • 3H$_2$O	1.20 as MgO	44.6% MgO	497.02	0.5520
Ga$_2$O$_3$	1.00	0.985%	401.71	0.4461
MnCO$_3$	0.015	0.995 %	1.7329	0.001924

The mix is prepared by hammermilling twice, with blending in between, and then is fired in silica crucibles:

   a, First Firing (in Air): 4.0 hours @ 1100 °C

   b. Second Firing (in 80% N$_2$ - 20% H$_2$): 2.0 hours @ 1100 °C

After the first firing, the cakes are inspected and crushed. Sometimes, the powder is washed in a dilute acid solution (~ 5% HCl) to remove excess MgO that may be present, and water-washed twice before filtration and drying of the wet cake. The dried powder is then refired. At the end of this firing, the cooled cake is inspected, crushed and sieved through a 325 mesh screen to remove all > 44μ particles.

It is important that the first firing be accomplished in air to form the matrix. If a reducing atmosphere is used during the first firing, a black or discolored nonfluorescent mass is obtained, due to reduction of the Ga$_2$O$_3$ to Ga-metal. The reducing atmosphere used on the second firing has two effects: 1) it serves to reduce the Manganese activator to its divalent state, and 2) it causes the reduction of part, or all, of the Gallium to its monovalent state, i.e.- Ga$^+$.

510

The optimum cation to anion ratio has been shown to be about 1.00:1.20. If the ratio of these two components is varied, one obtains:

4.2.-125.-

MgO	Ga_2O_3	Products (by X-ray Analysis)
0.50 mol	1.00 mol	$MgGa_2O_4 + Ga_2O_3$
1.00	1.00	$MgGa_2O_4 + Ga_2O_3$
1.20	1.00	$MgGa_2O_4$
1.50	1.00	$MgGa_2O_4 + MgO$

Because of the solid state reaction requires that an excess of one or the other components is needed to obtain a complete reaction, it is easier to use an excess of MgO which is not absorptive but reflects the UV at the excitation peak. any excess of Ga_2O_3 would cause absorption of the UV and a lower efficiency when the phosphor is used in a fluorescent lamp. and, the excess MgO is easily removed by acid-washing.

The excitation-emission properties are shown in the following:

4.2.126.-

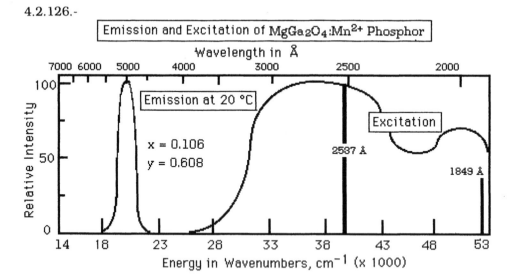

The QE of this phosphor is 0.82, A = 0.86, the peak emission occurs at 5040 Å, and the optimum cation to anion ratio is 1.20: 1.00. at an activator content of 1.5 mol%. This phosphor is the green phosphor generally used in fluorescent lamps in Xerox™ machines used for photocopy purposes.

Actually, SnO_2 may be substituted in part for the Ga_2O_3 present in the formulation, but the efficiency is lower. Additionally, the peak of the Mn^{2+} emission is affected slightly, that of $Mg_2SnO_4:Mn^{2+}$ being at 5000 Å and that of $MgGa_2O_4:Mn^{2+}$ being at 5040 Å.

7. Manufacture of $Sr_3(PO_4)_2:Eu^{2+}$ Phosphor

This phosphor is a blue emitting phosphor used in photocopy lamps. It is manufactured according to the following formulation:

4.2.127.- Formulation Used for $Sr_3(PO_4)_2:Eu^{2+}$ Phosphor

Compound	Mol Ratio	Assay	Grams	Lb/Lb
$SrHPO_4$	2.00	94.6%	388.16	0.775
$SrCO_3$	0.95	97.6%	75.63	0.151
$EuCO_3$	0.03		4.04	0.008

This formulation is weighed out, mixed together and then hammermilled twice, with blending in between hammermillings.

The mix is prepared and then fired:

 a, First Firing (in Air): 4.0 hours @ 1150 °C

 b. Second Firing (in 95% N_2 - 5% H_2): 2.0 hours @ 1175 °C

Let cool after the first firing, inspect cakes and crush into powder. After the second firing, the phosphor is inspected, crushed and sieved through a 325 mesh screen to remove all > 44µ particles.

The excitation-emission properties are shown in the following, given as 4.2.128. on the next page.

The QE is 0.90, A= 0.08, the peak emission occurs at 4200 Å, the cation to anion ratio is A similar phosphor can be prepared by substituting $BaHPO_4$ for the $SrHPO_4$ in the above formulation. However, its emission intensity is about 80% that of the strontium analogue.

512

4.2.128.-

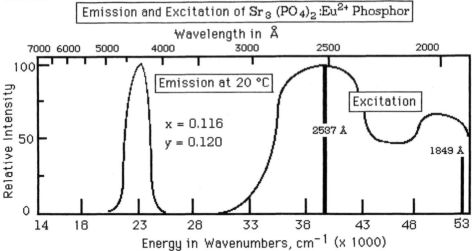

Emission and Excitation of $Sr_3(PO_4)_2:Eu^{2+}$ Phosphor

Wavelength in Å

Emission at 20 °C

x = 0.116
y = 0.120

Excitation

2587 Å

1849 Å

These phosphors can be compared to prior known phosphors, as follows:

4.2.129.- Radiometric Comparison of Phosphors

Phosphor	Peak in Å	Half-Width	Intensity	Peak Height Ratio
α - $Sr_2P_2O_7:Eu^{2+}$	4220	320	156	4.00
α - $Sr_3(PO_4)_2:Eu^{2+}$	4150	380 Å	104	2.67
α - $Ba_3(PO_4)_2:Eu^{2+}$	4150	340	84	2.15
$CaWO_4$: Pb	4460	1240	39	1.00
$MgWO_4$:W	4880	1420	45	1.15

It should thus be clear that these newer phosphors are more efficient than the older ones like $MgWO_4$:W. However, the older ones have continued in use in fluorescent lamps because the Eu^{2+} - activated phosphors have much narrower emission bands and do not blend well to give the desired color in certain phosphor blends where a broad band of emission is required. Also, they are not as stable during the lifetime of the fluorescent lamp as the older ones which have been used so long that the current lamp making processes have been optimized in their favor. Nevertheless, the fact that such phosphors exist which are more efficient (not necessarily more luminous) indicates that future fluorescent phosphor lamp blends will use these phosphors rather than the older ones.

At present, the newer Eu^{2+} - activated phosphors have been used primarily in specialty lamps.

REFERENCES CITED

1. RC Ropp, *Luminescence and the Solid State,* pp. 283-360, Elsevier Science Publ., NY & Amsterdam, (1991).

2. See for example: WB Fowler - Ed., "Physics of Color Centers", p. 53, Academic Press, NY and London (1968).

3. W. Kleeman, "Optical Transitions of Ag⁻ Centers in Alkali Halides", *Z. Phys.,***234** 362 (1970); See also: A Ranfagni, D Mugnai and M Bacci, "The Optical Properties of Tl⁺ - like Impurities in Alkali-Halide Crystals", *Adv. in Physocs,* **32** 823 (1993).

4. JC Platteeuw and G. Mayer, *Trans. Faraday soc.* **52** 1066 (1956).

5. H. Remy *Treatise on Inorganic Chemistry,* **Vol I.** 608, Elsevier Sci. Publ., NY (1956).

6. Koelmans and Cox, *Philips Res. Rpt.* **10** 145 (1954).

7. RC Ropp and RF Quirk, USP 3,348,961 (Oct. 1967)

8. WL Wanmaker and ML Verheyke, *Philips Res. Rpt.,* **11** 1-18 (1956).

9. G. Kemeny and CH Haake, *J.Chem. Phys.* **33** 783 (1960); See also: AH McKeag, "Some aspects of the Activation of Phosphors by Tetravelent Manganese", *Acta Physica Hungarica* **14** 14 (1962).

10. "Method for Preparing Rare Earth Oxide Phosphors", U.S.P. 3,449,258 by R.C. Ropp (1969).

11. "Method for Preparing Rare Earth Activated Rare Earth Metal Vanadate Phosphors", U.S.P. 3,511,785 by R.C. Ropp (1970).

514

12. "Method of Preparing Rare Earth Metal Orthovanadate Phosphors by Precipitation", U.S.P. 3,580,861 by R.C. Ropp (1971).

13. "Precipitation of Rare Earth Vanadates from Aqueous Solution", by R.C. Ropp and B. Carroll, J. Inorg. Nucl. Chem., **39**, 1303 (1977).

14. "Yttrium Phosphate-Yttrium Vanadate Solid Solutions and Vegard's Law", by R. C. Ropp and B. Carroll, *Inorg. Chem.*, **14**, 2199 (1975).

15. JMPJ Verstagen, *Paper Presented at Electrochemical Society Meeting, Abstract No.98-Spring Meeting*, **Vol. 74-1,** San Francisco (1974).

Chapter 5

MANUFACTURE OF CATHODE-RAY TUBE PHOSPHORS

This chapter will describe the complexities concerning the manufacture of phosphors used to form luminescent screens in cathode ray tubes. In general, most of these phosphors are based upon the zinc and cadmium sulfides, as mentioned in the last chapter. In that Chapter, we described the manufacture of fluorescent lamp phosphors, and also described some of the prevailing energy conversion mechanisms concerning both lamp and cathode ray tube (CRT) phosphors.

We will now present a more detailed description of the energy conversion mechanisms involving CRT phosphors, including why the energy efficiency of these phosphors does not exceed about 20-25%. In contrast, lamp phosphors may approach 100% in energy (quantum) efficiency. We will then describe how cathode-ray tube screens are prepared and the critical factors involved in their manufacture. Finally, we will delineate in detail the manufacture of CRT phosphors from the standpoint of the materials needed, the steps required, and the final properties sought for each individual phosphor in terms of their emission intensities and color, and persistence.

The original system of classifying CRT phosphors came from JEDEC, which is an acronym for the former Joint Electronic Device Engineering Council which is now the JT-31 Committee on Phosphor and Optical Characteristics within the Electronic Industries Association (EIA). JEDEC also established the initial code for such phosphors by establishing a JEDEC Register Code in terms of "P-numbers". Most manufacturers of cathode ray tubes refer to these numbers when specifying phosphor screen characteristics.

5.1. ENERGY CONVERSION EFFICIENCIES OF CRT PHOSPHORS

In general, the cathode ray tube consists of the following parts: 1. A glass tube or "envelope"; 2. An electron gun; 3. The phosphor screen; and 4. A high voltage anode. A diagram illustrating its construction is given below as 5.2.1. The cathode ray tube operates by generating a beam of electrons within the electron gun. This beam is then accelerated toward the phosphor screen by the high voltage anode, the electrons acquiring a kinetic energy proportional to the voltage drop across the tube. Upon encountering a phosphor particle, part of that energy is transferred to the lattice, and the electron continues until it is stopped by a number of succeeding interactions within the particle, or within other particles. In other words, it is a series of reflections of each individual electron that causes transfer of energy to the phosphor itself.

When a 20 KV electron strikes a CRT phosphor, part of its kinetic energy is absorbed by the lattice, resulting in band-gap excitation, i.e.- ~ 3.7 electron volts (e.v.) in energy for ZnS. This mechanism is sometimes described as an "exciton" wave, since a bound pair of positive and negative charges is involved. The "bound-charge" lattice energy then travels until it reaches an energy sink, usually the activator center. The overall energy efficiency for ZnS:Ag is about 20-22%. The next best cathode-ray tube phosphor is only about 6-8%, except for some of the newer rare earth phosphors, which can reach 10-15% in energy conversion efficiency. Thus, it is important that the lattice does not contain significant lattice defects since these can also function as non-radiative energy sinks.

The ultimate efficiency of a phosphor excited by an electron beam depends primarily upon the mean excitation energy needed to create, by collision, a free electron-hole pair. This in turn will depend upon the band gap energy, E_g , of the crystal. The transfer of energy from any given electron is due to scattering and transfer of energy to energetic carriers within the lattice. These carriers may then interact with phonons, defects etc. of the lattice before they reach the activator center. Each electron-hole pair delivers only the energy E_m to the center, as can be determined from the emission band. Hence, the energy efficiency, as given by Garlick (1), can be expressed as:

5.1.1.- $\eta = (1 - s) \, C \, E_m / \alpha \, E_g$

      where: C must be ≥ 1 if the recombination is to be radiative

                    $E_m =$ mean quantum energy of emission band

                    $\alpha \, E_g =$ minimum energy of a primary electron to create an electron-hole pair by collision.

                    $1\text{-s} =$ back-scattering probability

C is a factor denoting losses due to back-scattering of primary electrons and losses due to a finite probability of an electron-hole pair to recombine nonradiatively. For a good phosphor, C is usually between 0.90 and 1.00. The factor, (1-s), is related to back-reflection of the energetic electrons and can be estimated from Palluel's data (2). Since E_m is essentially constant for a given activator center, this leaves αE_g, i.e.- the energy required to form the electron-hole pair in the cascade following primary electron absorption, as the only quantity that affects efficiency.

The unknown in 5.1.1. is the constant, α , which depends upon the phosphor host material. It defines a multiple of the energy band gap, E_g , of the material. Values of α have been estimated by Shockley(3) for the case of pair production in silicon and his values can be applied also to other semiconductors such as ZnS. In the case of equal effective masses of free electrons, m_e, and free holes, m_h, α will equal 1.5. For the case where $m_e \neq m_h$, $\alpha > 1.5$. Since free holes are usually more massive than free electrons, α has been estimated to be about 3.0 in ZnS, which means on the average that:

5.1.2.- αE_g = 3.0 x 3.7 = 11.1 ev.

Thus, an energy of 11.1 ev. is required to create an electron-hole pair in the phosphor, ZnS:Zn⁻, by cascading primary electrons. It is for this reason that the accelerating voltage in a CRT needs to be several kilovolts. It is the effective mass of the "hole" created that is the most critical of all of the factors controlling the CRT efficiency. If we use the value of C =1.0, we can estimate values of α for several materials by substituting experimentally derived CRT efficiencies into the equation of 5.1.1. Some of these values are shown in the following Table, given on the next page.

Note that the efficiency values in parentheses shown in Table 5-1 represent those most often quoted (3). Using these values, we can obtain the following values for the constants in 5.1.1., as shown in Table 5-2, which is also given on the next page.

By close inspection of both tables, we can conclude that a multiple of the band-gap energy is required for the production of energy-carrier pairs which then produce cathodoluminescence. Thus, it should be clear that the rather low values of energy conversion experienced for CRT phosphors is caused by a severe limitation on the value of maximum energy conversion in CRT phosphors. This is due not to the band gap energy of the crystal but that

TABLE 5-1

Reported Efficiencies of Some CRT Phosphors

Phosphor	% Energy Efficiency
ZnS:Ag	17.8- 21.0 (19.6)
$(Zn,Cd)S_2$:Ag	12.0-21.0 (21.0)
ZnS:Cu	11.0-23.0 (17.5)
ZnS:Mn	(4.0)
ZnO:Zn	(7.0)
Zn_2SiO_4:Mn	4.7-8.5 (6.0)
Zn_3PO_4:Mn	6.0-8.0 (7.0)
$CaWO_4$:W	3.0-3.6 (3.3)
$MgWO_4$:W	2.0- 2.9 (2.5)
YVO_4:Eu	(7.1)
Y_2O_3:Eu	(6.4)
Y_2O_2S:Eu	(13.1)

TABLE 5-2

Values Calculated using the Garlick Equation: $\alpha = (1\text{-}s)\ C\ E_m/\ \eta\ E_g$

Values of:	ZnS:Cu	ZnO:Zn	Zn_2SiO_4:Mn	$CaWO_4$:W	YVO_4:Eu	CaS:Ce
s	0.26	0.23	0.23	0.30	0.19	0.22
C	1	1	1	1	1	1
η	0.23	0.07	0.085	0.036	0.071	0.22
E_m in ev.	2.3	2.4	2.4	2.9	2.03	2.3
E_g in ev.	3.7	3.25	4.5	5.0	3.5	4.8
α	2.1	8.1	4.8	11.3	6.6	1.7
$\alpha\ E_g$	7.88 ev	26.3 ev	21.6 ev	56.5 ev	23.1 ev	8.2 ev

energy required to produce the bound-pair in the crystal, as related to the "effective mass" of the hole. The other factor operating is that of defects present in the crystal lattice which function as non-radiative energy "sinks". It should be evident that the bound-pair energy is closely related to the type of bonding prevalent in the crystal. However, the concept of the effective mass of the hole requires some further explanation.

When a energetic electron encounters an inorganic crystal, what it sees is a mass of oxygen atoms whose lattice configuration depends upon what other metallic cations are present, as we have already described in a prior work (4). This is due to the fact that lattice structures, in general, depend upon the space-filling characteristics of the largest atom present, which is most often the oxygen (oxide) atom. The same will be true for ZnS, where the sulfide atom is the dominant feature. Thus, the type of bonding present in the crystal directly affects the dependence of α on the effective masses of free energy carriers, and has a direct consequence on the CRT response of these phosphors. Therefore, it is probable that the effective mass of the bound-pair is directly related to the electronegativity of the anion. This in turn has a significant effect upon the free mass of E_h, and consequently on the energy efficiency of the phosphor. Pertinent electronegativities of various anions are:

$$O^= = 3.5 \qquad Se^= = 2.4$$
$$S^= = 2.5 \qquad F^- = 4.0$$

Electronegativities of anions such as $SiO_3^=$, $SO_4^=$, PO_4^{3-} and VO_4^{3-} are even higher. High electronegativity means strong binding of a charge to an anion, i.e.- a high degree of charge localization in the anion sub-lattice and a correspondingly high effective mass of a free hole. Hence, a dependence of a on the hole mass, m_h, means in effect a dependence of α on the electronegativity of the anion present in the compound. It is for this reason that the most efficient CRT phosphors have been determined to be those based upon anions such as sulfides, selenides and fluorides, and **not** upon oxides and oxide-based anions. That this is true can be easily ascertained by comparing the two phosphors, ZnO:Zn⁻ and ZnS:Zn⁻, where the ratio of CRT efficiency is: 0.63 to 1.00. Note also that both are semi-conductors. Even lower CRT efficiencies are found for the phosphors with insulating properties.

Because of these factors, as outlined above, most CRT phosphors lose more than 80% of the energy of the electrons incident upon the phosphor screen via non-radiative pathways, including those involving heat losses and vibrational processes within the lattice.

Let us now proceed to a description of the manufacture of cathode ray tube (CRT) screens. These include both television and data display CRT screens.

5.2. MANUFACTURE OF CATHODE-RAY TUBES AND PHOSPHOR SCREENS

Two types of cathode ray tubes are manufactured, those used for information display (both black & white and color) and those suitable for use in black & white or color television receivers. In general, the cathode ray tube consists of the specific operative parts as shown in the following diagram. Herein, a general outline of the working parts of a typical cathode ray tube is presented, viz-

5.2.1.-

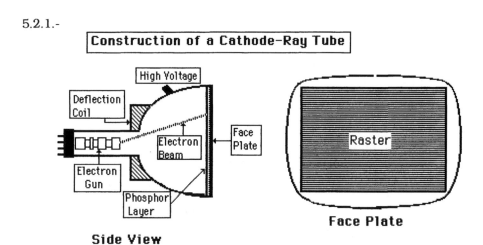

As shown in this diagram, a beam of electrons is generated by the electron gun and the electrons are accelerated toward the phosphor layer, i.e.- the "phosphor screen", by application of high voltage to the anode. It is customary to "aluminize" the back side of the phosphor screen by evaporating aluminum metal onto it as a coating and then connecting the aluminum film to the high voltage supply. This ensures that the accelerating voltage is uniformly distributed over the entire area of the "face-plate" of the tube. The purpose of the deflection coil is to cause the beam to sweep back and forth across the screen to form the "raster" (which is an interleaved series of lines "painted upon the phosphor screen" to cause it to emit visible radiation). By modulating the beam, one can "paint" a moving picture of various objects. The total number of lines within the raster is about 525, but the separate raster lines are interwoven so that they overlap to form the complete picture. The electron-gun can be modulated so that "bunches of electrons" reach the screen. If the screen is white-emitting, this gives a black & white picture (black = no excitation). The raster itself is formed in real time so that we see moving pictures (provided that the phosphor has a proper "decay time"). The human eye has a movement perception of about 1/20 of a second. The raster is reformed about every 1/30 of a second, giving the appearance of continuous motion. We shall be concerned with how the phosphor screen is prepared and how the phosphors are used in order to make such screens.

Two types of CRT phosphor screens have been used, monochrome and tricolor. The former is used for computer information display and for radar and other display screens. The latter is employed in both color television and some computer screens. Whereas the black & white screen is continuous, the tricolor screen is discontinuous. That is, the black & white screen is formed as a single continuous layer of phosphor particles. However, the tricolor screen is composed of discontinuous "dots" or "lines" of phosphor particles. It is because of these differences that the two methods of forming phosphor screens have evolved.

Color screens can be composed of one or the other of the following arrangement of the three primary colors. i.e.- red, green and blue emitting phosphors, shown in two patterns in the following diagram:

5.2.2.-

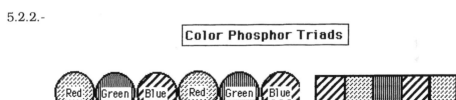

Color Phosphor Triads

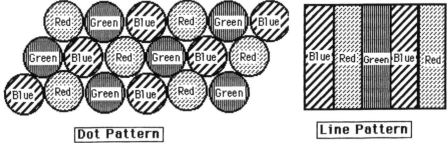

Dot Pattern   **Line Pattern**

In general, tricolor tubes are built so that a "shadow-mask" is interposed between the phosphor screen and the three electron guns (one for each color). In a tricolor tube containing a dot pattern, the phosphor dots are usually about 0.017 inch in diameter on the average, arranged in a triad pattern on the face-plate. The shadow mask contains holes numbering exactly 1/3 of the total number of dots actually present on the phosphor screen. Each shadow-mask hole is positioned exactly at the interstice of each three-dot triad. Then, each electron-gun can excite just one color of phosphor. This means that the shadow mask must be "indexed" at a precise position during the formation of each color screen. Thus, we have three color-rasters which are superimposed on each other, each raster being generated by its own electron-gun controlled by the color signal being transmitted to the color T.V. set.

The line-raster system is similar to the dot-raster system except that vertical lines are etched in the shadow-mask and a single electron-gun can be used. The line pattern exposure is generally more open than that of the hole pattern exposure so that more electrons get through the vertical line shadow mask, resulting in a brighter picture at the face-plate of the television tube. A general design of the shadow-mask tube is shown in the following diagram:

5.2.3.-

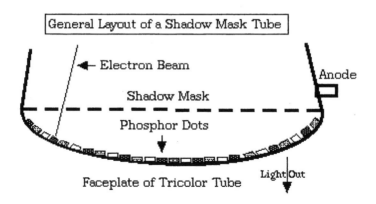

General Layout of a Shadow Mask Tube

The most recent color T.V. tubes have eliminated the shadow mask and use a single gun. The position of the electron beam, i.e.- the instantaneous color being generated, is controlled by a secondary phosphor layer which is on top of the emitting phosphor layer. The control phosphor is usually a U.V. phosphor with a nanosecond decay. Since the color phosphors decay in milliseconds, there is time for the feed-back signal to be detected and used to control the emission color being generated in real time, as each raster is being formed. The control-phosphor tube, since it does not have a shadow-mask, can be over five times higher in screen brightness than similar shadow-mask tubes.

We will not describe all of the various methods that have been used to improve brightness of the screen emission, spectral color, or contrast except to say that it has become common to surround the "dots" or "lines" of phosphor with a black "surround" to improve contrast (Contrast can be defined as the ability to distinguish exact outlines of displayed objects on the screen and is a function of phosphor particle size as well).

From this discussion, it should be obvious that the two most important properties of cathode-ray phosphors are the response to electron-beam excitation (brightness) and the decay time. We require a long-decay phosphor for radar applications and a short-decay phosphor for T.V. and display usage. Nearly **all** the cathode-ray phosphors are based on the zinc and cadmium sulfides because they exhibit the highest efficiency to cathode-ray excitation.

A. <u>Manufacture of Monochrome Screens for Display Purposes</u>

This method was developed for Black & White television tubes before color television came to the forefront. It consists of a "settling" method in which the screen is formed directly upon the inner side of the glass face-plate of the tube by allowing the phosphor particles to settle to form a layer which comprises the "screen". The physical setup is shown in the following diagram, given as 5.2.4. on the next page.

In this case, a potassium silicate is used as the binding agent, i.e.- K_2SiO_4 , which is combined with the phosphor and a gelling agent. The potassium silicate is actually one having a special formula, viz-

$$K_2SiO_{3.78} \quad \text{or} \quad K_2(SiO_{3.78})_n$$

where "n" may be several hundred. This arises because the SiO_4 tetrahedra are corner-bound to each other to form a rather long chain. The actual formulation is thus a polymeric soluble silicate which is called "Kasil[TM]" in the trade (Kasil[TM] is available from The Philadelphia Quartz Co.- i.e.- the PQ Co.). The specific gravity of Kasil[TM] is usually about 1.249 and it contains about 27.5% solids. When a phosphor suspension is made, the soluble silicate is

5.2.4.-

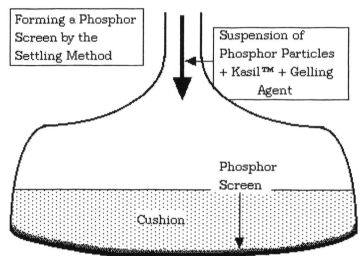

deliberately gelled so as to make it adhere to the phosphor surface. When the phosphor particles settle, the bound-silicate then binds to the glass, thereby forming an adherent screen.

One general method used to form a cathode ray screen consists of the following steps:

> 1. Wash glass tube thoroughly using an acid solution, i.e.- 5% HCl, followed by a water wash.
>
> 2. Preparation of Kasil™ binder solution
>
> 3. Preparation of water "cushion"
>
> 4. Addition of phosphor to Kasil™ solution to form suspension
>
> 5. Addition of "gelling agent" to cause Kasil™ to deposit silica on phosphor particles.
>
> 6. Addition of phosphor suspension to "cushion"
>
> 7. Let settle to form screen

8. Decant settling solution and dry at 50 °C.

9. Bake the screen to form final adherent phosphor layer

A dilution of pure water and Kasil™ is made, the volume depending upon the size of the screen to be prepared. A cushion solution is prepared by adding water to the cathode ray tube to be coated. A given weight of phosphor is measured so as to give a "powder weight" of about 2.5 to 6.5 mg/cm^2 of glass surface. Obviously, the total weight will depend upon the size of the screen to be deposited. The phosphor powder is added to Kasil™ solution, dispersing it by application of ultrasonic energy. The gelling agent solution is then added to the Kasil™ suspension, and the total volume is added to the cushion layer in the tube, filtering it through a 325 mesh screen before it dilutes the cushion volume (This removes any free gel particles that may have formed). Allow to settle for 30 minutes before decanting the solution, leaving an adherent layer behind.The "green" screen is then dried for 1.0 hour at 50 °C. Finally, the screen is baked at 150-350 °C for an hour to form the final adherent screen.

As a specific example, a 6" x 6" glass plate is to be coated at 6.5 mg/cm^2 as a "coating weight", in a 2000 ml. beaker having a bottom-area of 410 cm^2 . A phosphor weight of 2.6678 grams is required. In the following recipe, strontium acetate is used as the gelling agent. The procedure to be used is:

1. Dilute Kasil™ by half to 14% solids content by adding an equal amount of water and Kasil™ to form a stock solution. This solution cannot be kept for more than 24 hours before it begins to deteriorate. (However, the concentrated solution is stable).

2. Make up a 0.50% Sr(OAc)$_2$ stock solution by adding 5.0 gm to 1000 ml. of water. (Ba(OAc)$_2$ is also sometimes used, depending upon the Manufacturer).

3. Add powder to 80 ml. of Kasil solution, using ultrasonics to thoroughly disperse the powder to form the suspension.

4. Add 102 ml. of the Sr(OAc)$_2$ solution to the Kasil™ suspension and then add 818 ml. of water to form a total volume of 1000 ml.

This gives a final concentration of 0.51 gm/liter of Sr(OAc)$_2$ in the suspension. This volume is calculated from the following:

Total Vol./(Conc./liter) x gm/liter = 1000 (.51)/5.0 = 102 ml.

The Sr^{2+} sensitizer solution is added first, followed by the water.

5. The total volume is then poured through a 325 mesh screen into the 2000 ml. beaker containing the glass plate and the suspension of particles is allowed to settle for 30 minutes.

6. At the end of this time, the liquid is decanted carefully, so as not to disturb the phosphor layer formed on the bottom of the beaker and the glass plate.

7. The glass plate is carefully removed and placed in an oven for 1.0 hour at 50 -80 °C. to dry. It is then baked for 1.0 hour at 350 °C.

The purpose of the gelling agent is to cause the silicate to be neutralized, thereby forming a colloidal silicate film on the phosphor particles. The drying and baking steps then causes the silica coating to bind to the glass.

It has been shown that mono-valent ions like Na$^+$ or K$^+$ as nitrates or chlorides are not very effective in causing the formation of the colloidal silicate on the surface of the phosphor particles. And trivalent ions cause the reaction to occur too fast. Only divalent ions like Sr^{2+} and Ba^{2+} are effective in forming the screen at a slow pace. This produces a screen that can be handled during subsequent operations to finish the manufacture of the cathode ray tube. Thus, the most important factor concerning the screen is the "dry adherence" of the screen. The factors most affecting dry adherence are shown in the following diagram, given as 5.2.5. on the next page.

In this case, a barium nitrate solution was used as the gelling agent. The "dry adhesion" was determined by use of a jet of air from a vertical nozzle whose pressure could be varied from 0 - 65 psi. The pressure was raised in steps until phosphor particles were dislodged from the glass.

5.2.5.-

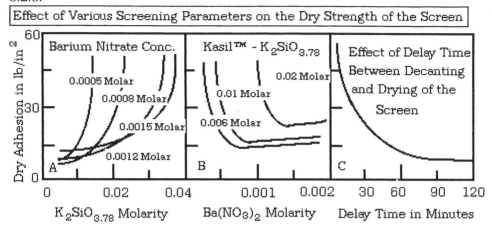

This was repeated and the average pressure value observed was recorded as the "dry adhesion". Reproducibility was found to be ±10%. A dry adhesion above 45 psi. was found to be adequate for most subsequent screening operations.

In the 5.2.5. diagram, several combinations of gelling agent, i.e.- $Ba(NO_3)_2$ and adhesive (Kasil™), can be found in "A" that produce adequate dry strength. However, as the ratio of gelling agent to adhesive in "B" becomes too high, the adherence drops precipitously. This behavior results because the gelation occurs too rapidly and the "wet" strength becomes too low for decantation of the settling solution to take place. We conclude that it is essential to choose a combination wherein the gelling process takes at least 20 minutes to occur. It is for this reason that the ratio of Kasil™ to $Sr(C_2H_3O_2)_2$ in the above example is 27:1 (0.067 molar to 0.0025 molar).

In general, a minimum ratio of 20:1 is required to make satisfactory phosphor screens by the settling method. Note also that if the wet screen is not dried immediately, a loss in screen adherence occurs. The reason for this is that when the screen has settled out, it still contains pores between the particles. When the liquid is decanted, some free silicate remains behind within these pores and, if the screen is dried immediately, it serves to complete the bonding process of the screen to the glass surface. However, if the screen is not dried immediately, a loss in screen strength and adherence is noted.

528

Thus, if the screen is allowed to remain wet over a period of time, the slow reaction between gelling agent and silicate continues to form gel particles, the screen adherence drops rapidly with time, and the "wet" strength is affected deleteriously. Even after drying and baking, the dry strength will be low.

The factors critical to the process are:

1. The ratio of Kasil™ to gelling agent must be kept above 20 to 1 so that the gelling process is slow. The silicate then has time to adhere to the particles, which then adhere to the glass.

2. Screen drying needs to be done immediately to stop the gelling action from proceeding further.

3. A water-cushion must be used in order to disperse the particles so as to form a uniform screen across all of the face-plate of the tube.

4. The sequence of addition to the tube must be: a) add the water cushion (sometimes this also contains a small amount of Kasil™ added); b) Make up a phosphor particle dispersion by adding powder to a dilute solution of Kasil™; c) Make up a gelling solution by adding the chosen salt to water to form a solution in the range of 0.00025 to 0.0025 molar; d) The amount of gelling agent must be in the range of 1/30 to 1/10 (1/20 is preferred) of that of the binder concentration which can range between 0.01 to 0.04 molar; e) The sequence of addition must be Kasil + phosphor, gelling agent, and then water before applying the total volume to the tube.

Once the screen has been made, the sequence of events to finish the cathode ray tube is:

1. The screen is "re-wet" with water and an organic lacquer consisting of a polyvinyl alcohol solution is applied to the screen by a spraying technique and then dried at 80 °C. The purpose of the lacquer film is to provide a smooth surface

for deposition of the aluminum metal film and a smooth reflective surface when the tube manufacture is completed.

2. The tube is "dagged", that is - a colloidal graphite suspension is used to paint the entire interior of the glass tube to form a conductive layer inside. Care must be taken not to get the black coating on the screen. The "paint" used is called "AquaDag™". The tube is then dried.

3. A tungsten coil containing Al wire is placed within the tube and a vacuum is then applied to the tube. The coil is then heated to evaporate the Al to form a continuous metal film on top of the lacquer film. The tube is then baked at 400 °C to decompose the organic film so as to leave an adherent aluminum metal layer behind. One purpose of the Al metal-film is to reflect the luminescence of the phosphor screen in a forward direction to enhance the screen brightness.

4. Finally, the electron gun assembly is placed within the neck of the tube, which is then sealed off at a pressure less than 1.0 microns of mercury, i.e.- about 6×10^{-7} mm.

5. A base is applied to the electron gun to complete the construction of the cathode ray tube.

Each CRT phosphor has an optimum "screen-weight" which produces the highest "screen-brightness". This, of course, depends upon the particle size of the phosphor, its crystal density, and how the phosphor particles pack together to form the screen. One CRT manufacturer has determined certain values to be "optimum" for these commercially available CR phosphors, as shown in the following Table, given on the next page.

Values are given in this Table are results of actual CRT screens made to establish optimum screen weights by one manufacturer. They are given in terms of both "P-numbers" (which are phosphor characteristics as established by JEDEC) and their chemical composition.

TABLE 5-3

Optimum Screen Weights for Certain Commercially Available CRT Phosphors

JEDEC #	Composition	Use	Screen Weight	* % Trans.
P-1	$ZnSiO_4$:Mn	Display	2.0 mg/cm^2	44.0
P-2	ZnS:Cu	Radar	6.8	40.0
P-4	ZnS:Ag + $ZnCdS_2$:Ag	B&W Display	4.0	44.0
P-11	ZnS:Ag:Ni	Photographic	3.0	48.0
P-12	$ZnMgF_2$:Mn	Radar	4.0	42.0
P-16	$Ca_2MgSi_2O_7$:Ce	Flying spot scanner	2.0	47.0
P-19	$KMgF_3$:Mn	Radar	6.2	43.0
P-20	$ZnCdS_2$:Ag	Display Tubes	4.2	42.0
P-22	ZnS:Ag+$ZnCdS_2$:Ag+Y_2O_3:Eu	Color Television	4.0	44.0
P-31	ZnS:Cu:Ni	Displays	5.0	44.0
P-33	(K,Na)MgF_3:Mn	Radar	2.0	44.0
P-36	$ZnCdS_2$:Ag:Ni	Display	2.1	44.0
P-37	ZnS:Ag:Ni	Display	1.4	44.0
P-38	$MgZnF_2$:Mn	Radar	6.0	50.0
P-39	$ZnSiO_4$:Mn:As	Radar	2.2	44.0
P-40	$ZnCdS_2$:Ag + $ZnCdS_2$:Cu		3.8	45.0

* "Trans" is the % transmittance of the dry screen before it is aluminized.

B. Manufacture of Tricolor Screens for TV and Display Purposes

In this case, a completely different approach has been developed. Basically, it consists of the use of a water-based phosphor slurry using soluble polyvinyl alcohol as the binding agent. The slurry containing suspended phosphor particles is spread on the face-plate of the tube in a thin layer. The shadow mask is then placed into its indexed position, and the polyvinyl alcohol (PVA) slurry is exposed to an ultraviolet-emitting point source through the holes in the shadow mask. Where the UV light falls on the screen, the PVA is polymerized and becomes insoluble in water. The soluble part of the screen is then removed by washing with water, leaving a series of dots behind on the face-plate. This procedure is repeated for a total of three times, one for each color of phosphor. The face-plate is then baked for 1.0 hour at 400 °C. to form the final tricolor screen. It is then sealed to the back part of the tube to form the the final construction, using a lead-frit to seal the glass parts together.

The exact methods used to manufacture tricolor screens have varied according to each Manufacturer, who has developed his own proprietary protocols. One such method involves:

1. The components used to form the screening slurry are used in a specific ratio, viz-

Ratio of Weights

	Ratio of Weights
1.00 gm. of phosphor powder	1.00 gm.
1.59 ml. of PVA solution (@ 8.5% by weight)	0.135 gm.
0.25 ml. of $(NH_4)_2Cr_2O_7$	0.0162 gm.
Emulsifier- Tween 20™	0.00135 gm.
Defoamer- Triton CF-54™	0.00068 gm.

The ammonium dichromate is used as the polymerization initiator for the PVA. These solutions are made up as follows:

a. PVA = 8.50% solution by weight, dissolved in water to give a viscosity of 55.0 centipoise. Dissolve 85.0 gm/liter of solution.

b. Use 10 ml. of water/ 4.0 gm. of phosphor

c. $(NH_4)_2Cr_2O_7$ = 64.8 gm in 1000 ml of water

d. Emulsifier- Tween 20™ = 20.0 gm. per liter of water

e. Defoamer- Triton CF-54™ = 10.0 gm. per liter of water

2. The following quantities are added to prepare a 1-quart mill batch, which is milled for 1.0 hour:

200 gm. phosphor
100 ml. of pure water
318 ml. PVA solution
50 ml of $(NH_4)_2Cr_2O_7$ solution
19.5 ml of Tween 20™ solution
19.5 ml. of Triton CF-54™ solution

The procedure used to prepare the phosphor screen is:

1. The face-plate is first washed with detergent, then an HF wash (~5% solution), followed by another water wash and then drying.

2. The phosphor slurry is spread in a thin layer on the face-plate and the layer is then exposed to the UV source. About 20-25 gm. of slurry is required for a 25 inch face-plate. The time of exposure is critical since the size of the phosphors dots being formed is a direct function of the exposure time. Usually, about 7-8 minutes is sufficient to form 0.017 inch diameter dots, i.e.- 17.0 mils.

3. The face-plate is then spun to remove excess unexposed slurry which is then reclaimed for further usage. The face-plate is then washed to remove traces that remain of the coating slurry.

4. This procedure is repeated twice more, one for each color of phosphor.

It has been determined that each color field and color phosphor has special problems associated with its application to the glass panel. The first (green) is sensitive to dirt-particle contamination and must be handled somewhat differently during the process. The second color applied is most susceptible to "spoke patterns" that arise from channeling of the flow through the dot pattern of the first applied color. The last color applied is likely to cause any cross-contamination of colors that might occur. It is for this reason that some Manufacturers prefer to apply red as the second color, while others prefer to apply blue as the second color. At any rate, it is clear that the required procedure must be modified for each color.

1. Green Phosphor - This is applied first since it is less sensitive to contamination by the other colors. 50.0 lbs of phosphor are milled in 15.0 gallons of water added to 9.53 gallons of PVA solution, along with 2.200 liters of Triton CF-54™ and 2.200 liters of Tween 20™ solution, for 30-60 minutes. The viscosity should now be about 24 centipoise. This slurry is then sieved through a 325 mesh screen. Just before use, 0.11 gm. of $(NH_4)_2 Cr_2O_7$ per gram of PVA is added to the slurry with stirring. This is 1.50 gallons of the $(NH_4)_2 Cr_2O_7$ solution. About 22 gm.

of the slurry is required to coat a 27 inch diagonal face-plate. The applied thin layer is then processed as before by exposure to UV light through the shadow mask. It takes about 7.0 minutes to develop a 17 mil dot size.

2. Red Phosphor - 50 lbs. of phosphor is milled in water alone, i.e.- 15.0 gallons, for 30 minutes. This suspension is then added to 9.5 gallon of PVA solution containing 2.200 liters of Triton CF-54™ and 2.200 liters of Tween 20™ solution. This mixture is then milled for 15 more minutes. The viscosity should now be about 42 centipoise. Just before use, 0.12 gm. of $(NH_4)_2 Cr_2O_7$ per gram of PVA is added to the slurry with stirring. This is 1.64 gallons of the $(NH_4)_2 Cr_2O_7$ solution. The slurry is then sieved as before and coated upon the green phosphor dots already present on the face-plate. Exposure then takes place, but a longer exposure time is usually required, about 8-9 minutes to develop a 17 mil dot size. The face-plate is then spun to retrieve excess slurry for reuse, and the face-plate is then washed and dried before applying the final phosphor layer.

3. Blue Phosphor - The same proportions are used as in the green phosphor formula. Final viscosity should be about 27 centipoise before use. Exposure is then monitored so as to obtain a dot size to entirely cover the screen. This is usually about 17-19 mils for the blue dot diameter and takes about 6-7 minutes to complete. The face-plate is spun to retrieve excess slurry for reuse and then washed with water.

4. Finally, the screen is washed with a dilute hydrogen peroxide solution (a 2% solution is used) to decompose any remaining dichromate to CrO_3^{2-}. The latter is more soluble and not colored orange. A final water wash serves to remove any remaining chromate that might be present. The screen is then dried at 50-60 °C and then baked at 400 °C. for 30 minutes.

5. A method for determining the amount of residual chromate present in the phosphor after the tricolor tube has been completely processed involves the following:

Dissolve 0.10 to 0.15 gm. of phosphor in 12.0 ml. of 1:1 HNO3. Add 1.0 ml. of concentrated H2SO4 and evaporate to fumes of SO3. Dilute residue to 75 ml. with water and add 1.0 ml. of 0.10 N AgNO3 solution and 1.50 gm. of K2S2O7 with stirring. Boil for 15 minutes. Cool, transfer to a 100 ml. volumetric flask and dilute to the mark with water. Place a 10.0 ml. aliquot in a 25 ml. volumetric flask and add 1.0 ml. of H3PO4. Add 1.0 ml. of).25% Diphenylcarbazide in acetone and dilute to the mark with water. Mix and measure absorbance at 5400 Å in 2 cm. cells immediately.

The following diagram shows a standard absorbance to be used to relate Cr^{3+} concentration to that present in the phosphor. Such an analysis gives the manufacturer a method to check on how efficient the washing process is in removal of dichromate from each screen during processing to form the final cathode ray tube tricolor screen. One color may retain chromate more readily than another.

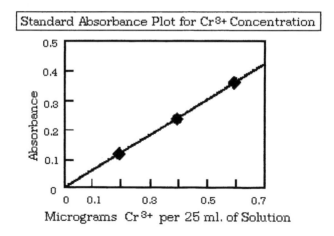

The shadow mask TV tube has undergone many changes since it first appeared in 1953. As shown in the following table, given as Table 5-4 on the next page, the tube brightness has steadily increased over the years that it has been manufactured. The improvements made in color TV tubes have included:

1. Change in hole size in the shadow mask wherein those in the center are about 40% larger than those on the periphery.

TABLE 5-4

Year	Phosphor Screen Blue - Green - Red	Efficiency in Lumens per Watt
1953	Sulfide-Silicate-Phosphate	0.60
1955	Sulfide-Silicate-Phosphate	1.00
1959	Sulfide-Silicate-Phosphate	1.40
1961	All Sulfide	1.60
1964	Sulfide-Sulfide-Rare Earth	6.90
1975	Sulfide-Rare Earth-Rare Earth	8.20
1991	Sulfide-Rare Earth-Rare Earth	8.60

2. Use of a black surround to enhance screen brightness

3. Improvements in the slurrying process.

4. Development of new methods of screen deposition including "dusting" of the phosphor, i.e.- a "dry" deposition method.

In the last method, the screen is formed by one of two methods:

1). The PVA solution, sans phosphor, is applied as a thin film to the face-plate. Then, either the phosphor particles are "dusted" over the PVA layer and the layer is exposed through the shadow mask to form the dots, or:

2) The layer is first exposed and the excess PVA is washed off, leaving "tacky-dots" behind. The phosphor is then air-dusted over these dots, and the screen is washed to remove excess phosphor particles which are not stuck on the dots.

The screen containing the phosphor-dots is then dried and then processed as before, including application of a lacquer, formation of the aluminized metal layer, and baking to remove the organic binder. The advantage to this process of "air-dusting" is that a brighter and more luminous screen results.

It has become common to manufacture tricolor tubes on an automated conveyor which carries the face-plates to each step required to complete the manufacture of the face-plate. The sequence of events that occur and the operations applied to each phosphor screen are shown in the following diagram. Here is shown the individual steps that occur as each glass face-plate proceeds through the conveyer and each series of phosphor dots are applied to form the final finished screen, viz-

5.2.6.-

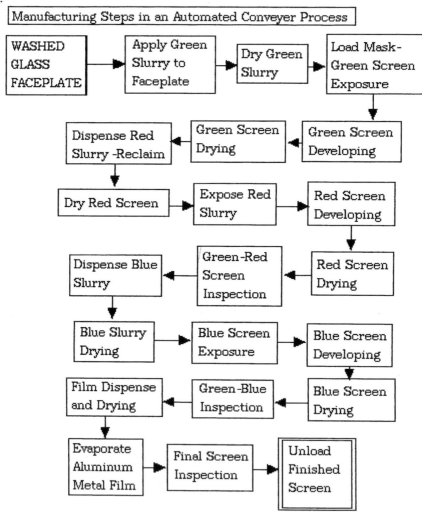

| Manufacturing Steps in an Automated Conveyer Process |

WASHED GLASS FACEPLATE → Apply Green Slurry to Faceplate → Dry Green Slurry → Load Mask- Green Screen Exposure

Dispense Red Slurry -Reclaim ← Green Screen Drying ← Green Screen Developing

Dry Red Screen → Expose Red Slurry → Red Screen Developing

Dispense Blue Slurry ← Green-Red Screen Inspection ← Red Screen Drying

Blue Slurry Drying → Blue Screen Exposure → Blue Screen Developing

Film Dispense and Drying ← Green-Blue Inspection ← Blue Screen Drying

Evaporate Aluminum Metal Film → Final Screen Inspection → Unload Finished Screen

Millions of color TV tubes have been manufactured by this, or a similar, method.

The "optimum" screen weight for most tri-color tubes lies between:

$$2.0 \text{ mg./cm}^2 < \text{ screen weight } < 4.5 \text{ mg./cm}^2$$

At this point, the finished screen transmission will be 42-44%, before the aluminizing step.

C. Methods of Coating Cathode Ray Phosphors

It has been determined that any milling of the phosphors used to manufacture cathode ray screens causes a loss of screen brightness. It is for this reason that most Manufacturers prefer phosphors that can be directly stirred or dispersed into the screen coating vehicle. One way to produce such a phosphor is to coat its surface so that it is free-flowing. At least two different methods have been used, the most common being that of "silica-coating" by applying a colloidal silica to the surface of the phosphor.

1. Process for Silica Coating Cathode Ray Phosphors

The phosphor to be coated is slurried in water, a solution of colloidal silica is added (called Ludox™ in the Trade - available from DuPont) and then an electrolyte is added to cause the colloidal solution to deposit on the surface of the phosphor particles. Ludox™ is usually manufactured from a sodium silicate solution by electrolysis of the cation, Na^+, using a mercury anode, to form stable silica particles of colloidal dimensions in suspension. The addition of any electrolyte causes the destabilization of the colloid which then deposits upon whatever surface is present. Since most phosphor surfaces are charged (or become charged) in an aqueous suspension, the silica sol deposits preferentially upon that surface. One procedure that has been used is:

a. Suspend 400 gm. of phosphor in 1000 ml. of pure water

b. Dilute the Ludox™ by a 10:1 volume in water and add 10 ml. of the diluted volume to the suspension. This volume is not stable and must be used within 24 hours.

c. Stir for 15 minutes. then add 20.0 ml. of a 1.00 molar $MgSO_4 \cdot 7 H_2O$ solution. Stir for 15 more minutes.

d. Let phosphor settle to the bottom of the container, decant liquid, and then filter phosphor, adding wash water to remove all of the powder.

e. Dry for 12 hours at 110 °C. The powder so=produced will now be free-flowing and will disperse into a suspension readily.

The same procedure may be used for larger batches of phosphor:

For 200 lbs. of phosphor:

Use 70 gallons of water,
Add 2.30 liters of the 10:1 Ludox dilution,
Add 4.5 liters of $MgSO_4 \bullet 7 H_2O$ solution.

2. Alumina Coating Procedure for Cathode Ray Phosphors

This method uses a colloidal alumina to coat the surfaces of the phosphor particles. The material used is called Baymal™ and is available from DuPont. It comes as a powder and must put into suspension by stirring. To do so, one weighs out 5.00 gram and suspends it in 95 ml. of water (to which 1.00 gram of $MgCl_2 \bullet 6H_2O$ has been added) using a high speed stirrer (Caution, the powder is extremely fine, about 0.02 μ in dimension, and must be handled in a fume-hood to prevent inhalation of the dust). A suspension of phosphor is made by adding 1000 grams to 5 liters of water. To this is added the 95 ml. of Baymal™ suspension. This mixture is then stirred for 15 minutes before allowing the phosphor to settle. the liquid is decanted, and the powder is filtered and dried for 10 hours at 110 °C. The powder will now be free-flowing.

5.3.- MANUFACTURE OF CATHODE RAY PHOSPHORS

Phosphors intended for use in cathode ray screens have been classified according to their decay time, emission color and their relative brightness. Decay time is measured as the time for the initial luminous response of the phosphor to decay to 10% of its initial value after excitation has ceased (Note that 1/e is not used, i.e.- time to 37% of the initial steady-state brightness, like in lamp phosphors).

The qualitative counterpart for decay time as defined by JEDEC is given in the following:

5.3.1.- <u>JEDC Definitions for Decay Characteristics of CRT Phosphors</u>

		Very long	= over 1.0 second
100 Milliseconds	<	**Long**	< 1.0 second
1.0 millisecond	<	**Medium**	< 100 milliseconds
10 microseconds	<	**Medium-Short**	< 1.0 millisecond
1.0 microseconds	<	**Short**	< 10 microseconds
		Very Short	< 1.0 microsecond

The first phosphor to be registered was "P-1", which was a medium decay persistence phosphor having the composition, $Zn_2SiO_4:Mn$. About 35-40 phosphors of differing chemical composition have been registered but the registration was always in terms of phosphor screen composition, and may consist of two or more phosphors. Although several of the P-screens are now obsolete and no longer used, we will describe their manufacture for the historical perspective. Only a few will not be described, but merely catalogued as being no longer used. About 1975, JEDEC changed to the JT-31 committee within the EIC organization. Since then, the cataloging system was changed to one wherein the screen color, persistence, and the Manufacturer was identified. Since we are interested in the manufacture of phosphors, we will adhere to the JEDEC system of P-numbers, even though the current number of screens registered by JT-31 exceeds the total P-numbers by a wide margin. We do this because the original P-number system evolved in an effort to describe the phosphors used to prepare the screen. But when the registrations of combinations of both new and old phosphors literally inundated the JT-31 committee, they decided to register screen properties, and not that of individual phosphors used to make the screens.

We will describe the preparation of cathode ray (CR) tube phosphors in terms of:

    a. Emission Color and Emission Band(s)
    b. Persistence of CR Screen
    c. CIE Color Coordinates, i.e. - $\overline{x}$ and $\overline{y}$
    d. Peak Emission Wavelength
    e. Fluxes Used (if any)

f. Ratio of ZnS to CdS (if any)

g. Activator Concentration

In addition we will specify the steps required to manufacture each phosphor, viz-

5.3.2.- Steps Required in the Preparation of Phosphors

a. Selection and Assay of Materials

b. Preparation of Phosphor Mix

c. Blending Steps

d. Firing Steps

e. Washing Steps (if any)

f. Post-washing Steps, including Coatings Applied

All of these steps will be delineated as they apply to manufacture of each specific phosphor. Most of these steps have already been described in detail in the last Chapter and the reader is referred to pages 388-393 there. In many cases, we will present the relative brightness of each phosphor in terms of a selected standard, P-20, which is a $ZnCdS_2$:Ag phosphor whose emission band peaks exactly at the 5500 Å wavelength of the photonic response of the human eye, i.e.- the standard CIE luminosity curve peaking at 5500 Å.

The equipment required for the manufacture of CRT phosphors, using a continuous process is shown in the following diagram, presented on the next page as 5.3.3. Note that it is a repeat of that given in the last Chapter as 4.2.10. We have already described the manufacture of both ZnS and CdS in Chapter 3 (see pages 345- 349). If a continuous firing process is to be used, the equipment and procedures to be used is nearly identical to that used for lamp phosphor manufacture. The only difference is that the raw materials used are ZnS and CdS precipitates plus the fluxes added before firing.

However, if production proceeds via a "batch" schedule, the steps that apply to lamp phosphors manufactured as blend components also applies here. This is given as 5.3.4. , as presented also on the following page.

5.3.3.- Equipment Required for Continuous CRT Phosphor Manufacturing

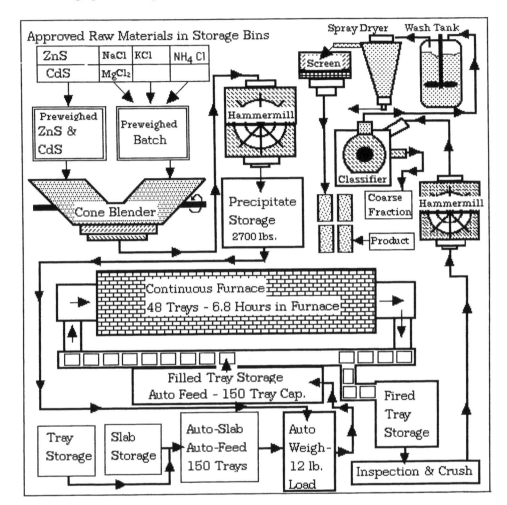

5.3.4.- Procedure for Manufacture of CRT Phosphors via a Batch Process

    a. Assay all materials

    b. Weigh out components

    c. Mix by blending and then by hammermilling all components together.

    d. Load silica trays or crucibles with mixed powder.

e. Fire according to a stipulated firing schedule, maintaining whatever firing atmosphere is considered essential for formation of the phosphor within the furnace being used.

f. Cool fired trays and phosphor in specified atmosphere until they reach room temperature. Many times, this will be air.

g. Inspect fired phosphor "cake" under ultraviolet radiation and scrape off all inert material. Crush cake.

h. Sometimes, the phosphor must be refired. Place in trays and re-fire as specified in the procedure. Cool as before.

i. Reinspect and crush phosphor cake. Sift through 325 mesh stainless steel screen.

j. Wash powder by suspension in an aqueous solution if so specified. Filter powder to separate wash solution and to form wet cake.

k. Dry overnight in oven.

l. Measure relative luminous efficiency, weigh amount produced and record.

m. Place in plastic-lined drums for future use.

Let us now proceed to a description of the manufacture of each individual cathode ray phosphor.

A. Manufacture of P-1 Phosphor

This phosphor has a $Zn_2 SiO_4$:Mn:Pb composition with spectral properties quite similar to those described for fluorescent lamp applications. Its formulation differs somewhat since its intended CR usage requires a specific persistence when used as a cathode ray tube (CRT) screen. Some Manufacturers add a small amount of Pb^{2+} to the formulation, others do not.

The formulation to be used is given as 5.3.5. on the next page:

5.3.5.- Formulation for a P-1 Phosphor

	Mols/ Mol Phosphor	Weight in	
		Grams	Lbs./Lb.
ZnO	2.00	162.78	0.637
SiO_2	1.13 (as SiO_2)	67.90	0.266
$MnCO_3$	0.20	22.99	0.0899
PbF_2	0.00816	2.00	0.0078

Some Manufacturers use silicic acid, i.e.- H_3SiO_3 instead of SiO_2. This mix is blended thoroughly following all the steps given above, including two hammermilling steps, with blending in between. The assay of SiO_2 will vary according to the nature of the silica being used. In general, it must be a pure grade with no more than about 5 ppm total transition metal elements present. A large 2-liter silica crucible is generally used for firing:

3 hours @ 1235 °C. in air in a covered silica tray.

After firing, the phosphor cake is inspected under 2527 Å light, crushed and then sieved through a 325 mesh screen. It has been found that the addition of Pb^{2+} increases the luminous efficiency of the phosphor, as does the addition of As^{3+}. When the latter was added, the phosphor screen was registered as P-39. It is likely that As^{3+} introduces a large increase in the decay of the phosphor, probably by introduction of vacancies which mandatorily form upon the introduction of a trivalent cation into a divalent cationic lattice. The emission is green and the persistence is medium as shown in the following diagram, presented as 5.3.6. on the next page.

Also presented on the next page is a table which gives a summary of the P-1 phosphor properties, including those spectral and physical properties considered essential to the proper formation and operation of the P-1 screen.

Note that the screen brightness is only 56% that of P-20, a $ZnCdS_2$:Ag composition which is the brightest phosphor in the P-series. This arises because the Zn/Cd ratio in P-20 has been adjusted so that its emission coincides very closely with that of the y-coordinate of the CIE standard (which is close to the peak of the average human eye response curve to green light).

5.3.6.-

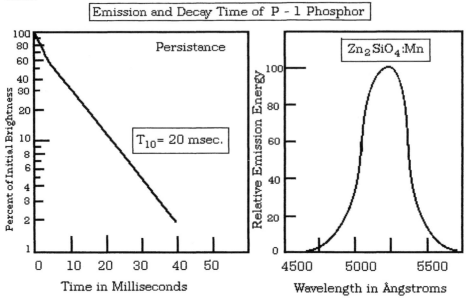

Emission and Decay Time of P - 1 Phosphor

Spectral and Decay Properties of P-1 Phosphor	
Emission Color	Green
Emission Peak	5250 Å
CIE Coordinates	x=0.228 ; y = 0.710
Persistence - $T_{10\%}$	20 milliseconds
Decay Class	Medium
Activator Concentration	20 mol%
Zn/Cd Ratio	NA
Usual Particle Size	9.5 μ
Body Color	White
CRT Brightness vs: P-20	56%
Optimum Screen Weight	4.0 mg/cm^2

B. Manufacture of P-2 Phosphor

The composition of this phosphor is essentially that of copper-activated zinc sulfide. But, some manufacturers have modified the composition to:

$$ZnS:Cu:Ag \quad or \quad ZnCdS_2 : Cu$$

This phosphor is made either by using a co-precipitated $(Zn,Cu)S_2$ material in which the ratio of Cu to Zn is : 0.015 : 0.985, or by using a mixture of ZnS, $(Zn_{0.88}Cu_{0.12})S_2$, plus fluxes. We will give all of these formulas to show the variations possible. As shown in the following, a "concentrate" has been used in two of these formulas, and a total of three different formulas are given.

5.3.7.- Formulations for a P-2 Phosphor

Formula	Compound	Mol Ratio	Assay	Grams	Lb/Lb
a.	ZnS	0.985	0.995	96.46	0.938
	CuS	0.015	0.995	1.441	0.0140
	$MgCl_2$	3.0 % by Weight of sulfides		2.938	0.0286
	NaCl	2.0 % by Weight of sulfides		1.958	0.0190
b.	ZnS	0.900	0.995	88.14	0.857
	$(Zn_{0.88}Cu_{0.12})S_2$	0.100	0.995	9.771	0.0950
	$MgCl_2$	3.0 % by Weight		2.937	0.0286
	NaCl	2.0 % by Weight		1.958	0.0190
c.	ZnS	0.938	0.995	91.86	0.881
	CdS	0.050	0.989	7.304	0.0700
	CuS	0.012	0.995	1.147	0.0110
	$MgCl_2$	2.0 % by Weight		2.006	0.0192
	NaCl	2.0 % by Weight		2.006	0.0192
d.	ZnS	0.880	0.995	86.18	0.854
	$(Zn_{0.88}Cu_{0.12})S_2$	0.100	0.995	9.771	0.0969
	$(Zn_{0.88}Ag_{0.12})S_2$	0.0100	0.995	1.0305	0.0102
	$MgCl_2$	2.0 % by Weight		1.9396	0.0192
	NaCl	2.0 % by Weight		1.9396	0.0192

The purpose of the fluxes is to promote crystal growth during firing. This procedure is quite common, and each Manufacturer has his own proprietary preferences. Although the ones that we have shown are to be regarded as typical, one should keep in mind that other flux mixtures can also be used.

If a flux is **not** used during firing, the particle size (which is of the order of 1.0-3.0 μ from the precipitation process) increases very slowly with firing time. It has been found that particle sizes necessary to produce good screens by settling are difficult to achieve without the use of fluxes during firing. Also, the phosphor efficiency can be lower by a factor of two or more. This has led some researchers to speculate that Cl^- plays a role in the formation of the activator center, but this hypothesis has never been rigorously established. Our conviction is that the activator center is composed of a Ag^- or Cu^- ion which resides upon a sulfide site, viz-

$$(Zn,Zn^+{}_I) \; S_{1-x} : Ag^-{}_S=$$
$$\text{or} \quad (Zn,Zn^+{}_I) \; S_{1-x} : Cu^-{}_S=$$

and that the fluxes only serve to promote crystal growth by serving as a melt-phase during firing. It is for this reason that various chloride mixtures have been used, since the eutectic melting of the mixture can be varied by changing the ratio and total content of the chloride components.

The mixes given above in 5.3.7. are first blended, then hammermilled twice, with blending in between hammermillings. They are then fired for:

4 hours at 910 °C in air

using a covered 2-Liter silica crucible. After firing, the cakes are inspected, crushed and then sieved through a 325 mesh screen. It is common to wash such phosphors after firing to remove the excess chloride flux. This is monitored, using $AgNO_3$, to determine when all of the chloride had been separated from the crystals. Once the washing step is completed, it is common to "silica-coat" the phosphor, using the procedure described above.

The spectral properties of P-2 phosphors are shown in the following diagram, given on the next page as 5.3.8.

Note that three different decay times are shown as well as three differing emission bands. These are due to the differences in formulation as given above. One is classified as "medium" while the other two are classified as "long". Depending upon the end-use of the CRT screen, one or the other of these formulations are to be used.

5.3.8.-

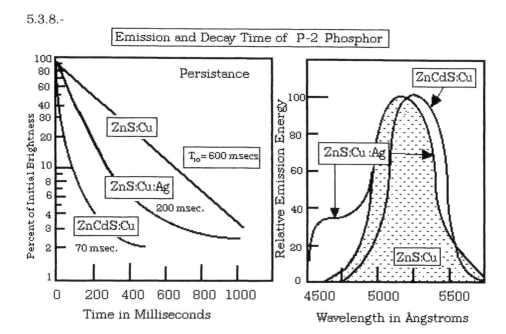

We can also summarize these properties as shown in the following table:

TABLE 5-5

Decay and Spectral Properties of the Various P-2 Formulations

Composition	Decay Time	Decay Type	x & y	Emission Peak
ZnS:Cu	600 msec.	Long	0.279, 0.534	5200 Å
ZnS:Cu:Ag	200 msec.	Long	0.205, 0.445	4550 & 5250 Å
$ZnCdS_2$:Cu	70 msec.	Medium	0.279. 0.544	5350 Å

The following table, presented on the next page, summarizes the properties of one of these P-2 phosphors. Note that the "body-color", that is, the reflected light color is slightly "yellowish". This is due to the fact that solid solutions of ZnS and CdS are colored according to the amount of CdS actually present (CdS itself has a strong orange reflective color). The compound, $AlCl_3$, is sometimes used as an additive to change the decay properties of P-2 phosphors. As stated above, the introduction of a trivalent cation into a divalent cationic lattice causes the formation of vacancies in the lattice, viz-

$$(Zn^{2+}, Al^{3+}, V^-_S{}^=)S:Cu^-$$

Spectral and Decay Properties of P-2 Phosphors	
Emission Color	Green
Emission Peak	5350 Å
CIE Coordinates	x=0.279 ; y = 0.544
Persistence - $T_{10\%}$	200 milliseconds
Decay Class	Long
Activator Concentration	1.2 mol%
Zn/Cd Ratio	0.94/0.05
Usual Particle Size	9.5 μ
Body Color	Yellowish
CRT Brightness vs: P-20	59 %
Optimum Screen Weight	6.5 mg/cm^2

Although some regard the Al^{3+} cation to be an "activator", it is not because it has no energy levels capable of doing so, i.e.- $Al^{3+} = 1s^2 2s^2 2p^6$ (This cation has a closed electron shell). However, its addition does cause a change in the decay properties of this phosphor because of the introduction of a lattice vacancy whose formation is mandatory to preserve the lattice charge compensation prevalent in this type of solid.

C. Manufacture of P-3 Phosphor

This phosphor is now obsolete, due primarily to its composition which incorporates beryllium in a zinc silicate, like the lamp phosphors used before Halophosphates were introduced, i.e.- $Zn_2 BeSiO_4 :Mn$.

The formulation to be used in its manufacture is as follows:

5.3.9.- Formulation for $Zn_2 BeSiO_4 :Mn$ Phosphor

	Mols/ Mol Phosphor	Weight in	
		Grams	Lbs./Lb.
ZnO	1.80	146.49	0.655
BeO	0.20	5.002	0.0224
SiO_2	1.08 (as SiO_2)	64.88	0.290
$MnCO_3$	0.06	6.90	0.0309
PbF_2	0.00053	0.130	0.000582

This mix is blended thoroughly following all the steps given above. The assay of SiO_2 will vary according to the nature of the silica being used. In general, it must be pure with no more than about 5 ppm total transition metal elements present. Some Manufacturers use silicic acid, i.e.- H_3SiO_3 instead of SiO_2.

The mix is then hammermilled as before and then fired in air. A large 2-liter silica crucible is generally used for firing:

Fire 6.0 hours in air @ 1100 °C using a covered crucible.

The fired cake is then processed as given above in 5.3.3. Finally, the washed powder is coated to form a free-flowing material.

The cathode-ray emission and decay properties are shown in the following diagram:
5.3.10.-

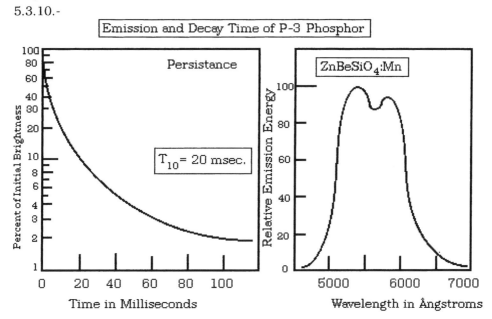

It has been found that Be^{2+} functions in this formulation to introduce a redder peak into the normally "green" emission of the Mn^{2+} activator. However, instead of merely shifting the Mn^{2+} peak, a new peak is observed which is due to the Mn^{2+} activator as modified by the presence of the Be^{2+} ion as a nearest neighbor in the crystal structure. Thus, one peak is close to 5400 Å while the position of the second peak is dependent upon the amount of Be^{2+} actually present in the crystal structure.

For this reason, the color coordinates of P-3 can vary, depending upon the Manufacturer and the amount of Be^{2+} actually present, i.e.- x can vary from 0.469 - 0.476 & y from 0.514 to 0.523 . As an alternative, the phosphor, $Cd_5Cl(PO_4)_3$:Mn can be mixed with P-1, i.e.- $ZnSiO_4$:Mn, to serve as a substitute for the P-3 phosphor. The spectral and decay properties are nearly identical for the mixture. The preparation of the $Cd_5Cl(PO_4)_3$:Mn phosphor was given in the last chapter.

The decay is classified as medium. Note that two emission peaks are present, one at 5400 Å and the other at 5800 Å. A summary of the physical and spectral properties of the P-3 phosphor is given as follows, viz-

Spectral and Decay Properties of P-3 Phosphors	
Emission Color	Greenish Yellow
Emission Peak	5400 Å & 5800 Å
CIE Coordinates	x=0.279 ; y = 0.544
Persistence - $T_{10\%}$	20 milliseconds
Decay Class	Medium
Activator Concentration	6.0 mol%
Zn/Cd Ratio	NA
Usual Particle Size	12.5 μ
Body Color	White
CRT Brightness vs: P-20	46 %
Optimum Screen Weight	4.5 mg/cm^2

As stated, the P-3 phosphor - $ZnBeSiO_4$:Mn, is no longer in use, but it was used earlier in CRT's before many of the more recently used phosphors were discovered.

D. Manufacture of P-4 Phosphor

The P-4 screen is the standard black & white screen, still in use today. It consists of both blue-emitting and yellow-emitting phosphors, mixed together to form a "white" emitting screen. In general, the emission color of the Black

& White screen is 9300 °K, which actually is a bluish-white. However, some Manufacturers prefer a P-4 screen having a color temperature of about 11,000 °K.

At least three different types of P-4 screens have been made, all-sulfide, sulfide + silicate and a silicate-silicate. The blue-emitting component is usually a ZnS:Ag phosphor, and the yellow-emitting component is a $ZnCdS_2$:Ag phosphor. A blue-emitting $CaMg(SiO_3)_2$:Ti phosphor was also used at one time. The manufacture of these phosphors was accomplished by use of specific sulfide precipitates, using "concentrates" as described in the P-2 procedure. The formulations to be used for preparing such phosphors are given in the following:

5.3.11.- Formulations for Preparation of P-4 Phosphors

1. Blue-Emitting Components

a.

Compound	Mol Ratio	Assay	Grams	Lb/Lb
ZnS	0.90	0.995	88.14	0.827
$(Zn_{0.834}, Ag_{0.166})S$	0.10	0.995	10.50	0.0986
$MgCl_2$	6% by weight of sulfides		5.919	0.0556
$LiSO_4$	2% by weight of sulfides		1.973	0.0185

b.

Compound	Mol Ratio	Assay	Grams	Lb/Lb
$CaCO_3$	1.05	0.986	106.59	0.295
$3MgCO_3 \bullet Mg(OH)_2 \bullet 3H_2O$	0.95 MgO	44.6% MgO	85.85	0.238
H_2SiO_3	1.83	75.05% SiO_2	146.50	0.405
TiO_2	0.035	0.985	2.839	0.0078
NH_4Cl	0.168	----------	8.987	0.0249
$(NH_4)_2SO_4$	0.081	----------	10.703	0.0296

2. Yellow Emitting Component

Compound	Mol Ratio	Assay	Grams	Lb/Lb
ZnS	0.455	0.995	66.06	0..437
$(Zn_{0.68}Ag_{0.32})S$	0.080	0.995	8.9276	0.0590
CdS	0.465	0.995	67.52	0.446
$MgCl_2$	6% by weight of sulfides		6.603	0.0436
$LiSO_4$	2% by weight of sulfides		2.201	0.0145

These components are mixed by blending, then hammermilled twice, with blending in between hammermillings.

The blue-emitting sulfide is fired in air in a 2 liter silica crucible for:

3.5 hours at 1085 °C

The yellow-emitting sulfide phosphor is fired in air in a 2 liter silica crucible for:

3.0 hours at 915 °C.

The blue-emitting silicate is fired in air in a 2 liter silica crucible for:

1.50 hours at 1180 °C.

When the phosphor cakes are cool, they are inspected and crushed. The blue-emitting silicate is then screened through a 325 mesh screen, and then is suspended in water. A silica coating is applied, as described above.

The sulfides are first screened through 325 mesh screen, and then are washed free of excess flux, using the same procedure described above for the P-2 phosphor. When dry, the body-color of the blue-emitting phosphor will be white while that of the yellow-emitting phosphor will have a slight yellowish cast. This is due in part to the florescence caused by exposure to daylight, as previously described for the $Y_3Al_5O_{11}$: Ce phosphor in the last chapter. The final step is the application of a silica-coating, as described above.

The following diagram, given as 5.3.12. on the next page, shows the emission of these phosphors and their decay times.

In this diagram, we have shown the properties of the all-sulfide P-4 CRT screen. Obviously, a non-sulfide will have some properties that differ.

In the following table, also given on the next page, is shown a summary of the all-sulfide P-4 phosphor properties, including those physical properties necessary for proper formation of the screen, viz-

5.3.12.-

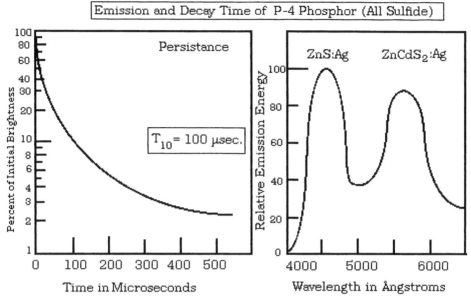

Emission and Decay Time of P-4 Phosphor (All Sulfide)

Spectral and Decay Properties of P-4 Phosphors	
Emission Color	White
Emission Peak	4500 Å & 5800 Å
CIE Coordinates	x=0.285 ; y = 0.285
Persistence - $T_{10\%}$	120 microseconds
Decay Class	Medium Short
Activator Concentration	2.0 mol%
Zn/Cd Ratio	53 Zn/47 Cd (yellow)
Usual Particle Size	8.5 - 12.0 µ
Body Color	White(very slight yellow)
CRT Brightness vs: P-20	62 %
Optimum Screen Weight	4.0 mg./cm²

As we stated, at least three different P-4 screens have been made in the past.
These are summarized in 5.3.13. presented on the next page. Obviously, the
spectral properties of each of these screens will vary somewhat, depending
upon the components used to form the screen.

5.3.13.- P-4 Compositions Used in the Past for CRT-Screens

$$\text{All sulfide} = ZnS{:}Ag + ZnCdS_2{:}Ag$$
$$\text{Sulfide Blue + Yellow Silicate} = ZnS{:}Ag + Zn_2BeSiO_4{:}Mn$$
$$\text{Blue Silicate + Yellow Silicate} = CaMg(SiO_3)_2{:}Ti + Zn_2BeSiO_4{:}Mn$$

Note that only the all-sulfide screen has survived, due primarily to the abandonment of the Be-containing formulation and the fact that the blue-emitting silicate is not as efficient as the sulfide composition. However, the blue-emitting silicate is still used in some radar screens because of its superior stability and resistance to "burn" under cathode-ray excitation.

E. Manufacture of P-5 Phosphor

This phosphor is essentially a $CaWO_4{:}W$ composition and is manufactured very like that of its lamp counterpart. However, it is fired at a temperature slightly higher than that of the lamp phosphor in order to ensure the formation of the high temperature polymorph.

5.3.14.- Formulation for CaWO4:W Phosphor

Compound	Mol Ratio	Assay	Grams	Lb/Lb
$CaCO_3$	1.005	98.6%	102.02	0.2830
WO_3	1.000	97.4%	238.04	0.6604
NaCl	6% by weight	-------	20.40	0.0566
$PbCO_3$	0.00002	98.6%	0.00534	0.00002

The mix is hammermilled twice as given before and fired in 2 liter silica crucibles for:

5 hours at 1085 °C. in a covered silica tray

After firing, the cake is processed as given above, including a washing step to remove the excess chloride ion that may be present. While the cake is still wet, it is "silicized", using the procedure given above. A free-flowing powder results. The decay is medium short and the emission is a deep-blue.

The emission spectra and decay are given on the next page as 5.3.15. The P-5 screen was originally developed for use with photographic film. It is now obsolete and rarely used. A summary of P-5 properties is given in the following table, also shown on this page:

5.3.15.-

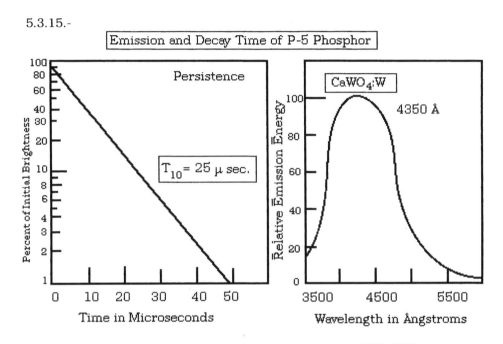

Emission and Decay Time of P-5 Phosphor

Spectral and Decay Properties of P-5 Phosphors	
Emission Color	Deep Blue
Emission Peak	4350 Å
CIE Coordinates	x=0.169 ; y = 0.132
Persistence - $T_{10\%}$	25 microseconds
Decay Class	Medium Short
Activator Concentration	Not Known
Zn/Cd Ratio	NA
Usual Particle Size	12.5 μ
Body Color	White
CRT Brightness vs: P-20	15 %
Optimum Screen Weight	4.5 mg/cm^2

Note that the spectral properties of the P-5 phosphor are quite like those of its lamp phosphor counterpart. The only difference is that the persistence, i.e.- decay time, is specified because of the criticality of decay to proper display in a cathode ray tube. Note also that the sensitizer concentration added, i.e. - the Pb^{2+} content, is much lower than that used for the lamp phosphor. In general, the activator concentrations used for CRT phosphors are lower by a factor of ten or more than corresponding lamp phosphor compositions. Some Manufacturers prefer not to add Pb^{2+} to the mix, since this results in a change of the decay properties.

F. Manufacture of P-6 Phosphor

This phosphor preceded the black & white P-4 screen, being registered in 1946, compared to the later registration of P-4 sulfides in 1952 (The reason for this is not known to the author). The P-6 screen was composed of two sulfide phosphors. In comparison to P-4, the emission of the blue phosphor was shifted from 4350 Å to 4600 Å, and that of the yellow one from 5600 Å to 5750 Å., accomplished by changes in composition. This gave a screen color of about 6000 °K, compared to the 9300 °K screen of P-4 (and the 11,000 °K screen being manufactured today). The following describes a formulation used for manufacture of these phosphors:

5.3.16.- Formulations Used for Preparation of P-6 Phosphors

a. Blue-Emitting Phosphor

Compound	Mol Ratio	Assay	Grams	Lb/Lb
ZnS	0.786	0.995	76.59	0.652
$(Zn_{0.88}Ag_{0.12})S_2$	0.100	0.995	9.771	0.0831
CdS	0.114	0.995	16.47	0.1401
$MgCl_2$	6% by weight of sulfides		6.003	0.0511
$LiSO_4$	2% by weight of sulfides		2.001	0.0170

b. Yellow-Emitting Phosphor

Compound	Mol Ratio	Assay	Grams	Lb/Lb
ZnS	0.490	0.995	47.75	0.378
$(Zn_{0.85}Ag_{0.15})S_2$	0.100	0.995	10.254	0.0811
CdS	0.410	0.995	59.23	0.468
$MgCl_2$	4% by weight of sulfides		6.880	0.0544
NaCl	2% by weight of sulfides		2.291	0.0181

Each mix is hammermilled twice, with blending in between, as stated above. It is important that each component be handled separately so that no cross-contamination can take place. The phosphors are fired in 2 liter silica crucibles with lids at:

Blue phosphor = 3.5 hours @ 955 °C

Yellow phosphor = 3.0 hours @ 900 °C.

After firing, each component is processed separately. the fired cake is crushed and then washed free of excess chloride. While the powder is still suspended in the final wash water, it is silica-coated using the procedure described above.

The specific spectral and decay properties of this screen are shown in the following diagram:

5.3.17.-

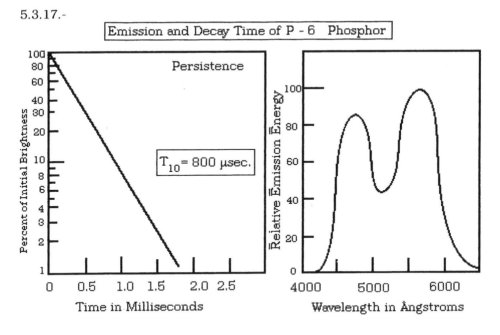

The P-6 screen was the forerunner of the Black & White screens now in use today. The emission color of about 6000 °K was considerably more yellowish than the 9,300 or 11,000 °K screens that are manufactured today. Note that

558

the yellow peak in the P-6 screen is much higher than that of the P-4 screen. At the time that the P-6 screen was registered, it was felt that a slightly yellowish Television screen would be more acceptable to the Public. Such has not proven to be the case. It is for this reason that the P-6 screen has become obsolete, while the P-4 screen has survived.

The properties of the P-6 phosphor are shown in the following table:

Spectral and Decay Properties of P-6 Phosphor	
Emission Color	5500 °K White
Emission Peaks	4650 & 5700 Å
CIE Coordinates	x=0.338 ; y = 0.374
Persistence - $T_{10\%}$	800 microseconds
Decay Class	Medium Short
Activator Concentration	1.2 mol%
Zn/Cd Ratio	Blue = 0.90 Zn/0.084 Cd Yellow=0.635 Zn/0.35 Cd
Usual Particle Size	9.5 µ
Body Color	Yellowish White
CRT Brightness vs: P-20	36%
Optimum Screen Weight	4.0 mg/cm^2

It should be noted that the same spectral properties could be achieved by use of a blue-emitting cubic ZnS:Ag phosphor whose emission band is at 4550 Å. For some reason, some manufacturers preferred to use the $ZnCdS_2$:Ag formulation shown above. Whether this was due to superior decay properties or total luminous output is not known.

G. Manufacture of P-7 Phosphor

This phosphor screen is used mainly in radar tubes and consists of two differing phosphors, one with a long decay and the other with a shorter decay. The purpose of this was to produce a screen which could be used to discriminate between transient and stable artifacts displayed on the screen. The P-7 phosphor screen was the first in a series of screens, and phosphors, designed to do just that.

These phosphors are manufactured using the following formulations:

5.3.18.- Formulations for Preparation of P-7 Phosphors

a. Blue Emitting Phosphor

Compound	Mol Ratio	Assay	Grams	Lb/Lb
ZnS	0.900	0.995	88.14	0.829
$(Zn_{0.834}Ag_{0.166})S_2$	0.100	0.995	10.502	0.0973
$MgCl_2$	6% by weight of sulfides		5.909	0.0553
$LiSO_4$	2% by weight of sulfides		1.970	0.0185

b. Yellow Emitting Phosphor

Compound	Mol Ratio	Assay	Grams	Lb/Lb
ZnS	0.755	0.995	73.94	0.654
$(Zn_{0.923}Cu_{0.077})S_2$	0.100	0.995	9.730	0.0860
CdS	0.145	0.995	21.053	0.1861
$MgCl_2$	3% by weight of sulfides		3.142	0.0278
NaCl	2% by weight of sulfides		2.094	0.0185
$BaCl_2$	3% by weight of sulfides		3.142	0.0278

These components are first mixed by blending, then hammermilled twice, with blending in between hammermillings. Care must be taken not to expose one formulation to the other so that cross-contamination does not occur. The blue-emitting phosphor formulation is particularly susceptible to this aspect, and any source of copper, however small, will cause a change in the emission characteristics.

The blue-emitting sulfide is fired in air in a covered 2 liter silica crucible for:

3.5 hours at 1075 °C

The yellow-emitting phosphor is fired in air in a 2 liter silica crucible for:

5.0 hours at 1150 °C.

using a cover on the crucible. When the phosphor cakes are cool, they are inspected and crushed. The blue-emitting phosphor is then screened through a 325 mesh screen, and then is suspended in water. A silica coating is applied as described above.

The other sulfide is first screened through 325 mesh screen, and then is washed free of excess flux, using the same procedure described above. When dry, the body-color of the blue-emitting phosphor will be white while that of the yellow-emitting phosphor will have a slight yellowish cast.

The two can be mixed together just prior to forming the screen in a 50/50 ratio. However, it was discovered that a superior P-7 screen can be made by first depositing the yellow component, followed by the blue component as a separate "top" layer. This forms a "cascade" screen which has superior properties as a detector, compared to the screen made by first mixing the two components, and then settling them together to form the screen.

The spectral properties of the P-7 screen are given in the following diagram:

5.3.19.-

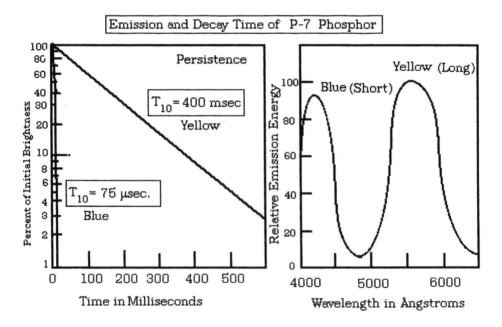

A radar tube is manufactured so that an electron beam rotates, forming a revolving raster on the screen. When the electron beam strikes the screen, both components are excited and emit their characteristic color, giving a white appearance to the raster screen. However, the blue-emitting component soon decays, leaving the yellow component still phosphorescing. Note that in the following diagram, the object at the top of the screen has decayed and lost part of its outline on the screen. Upon refreshment of the object, it again becomes distinct.

5.3.20.-

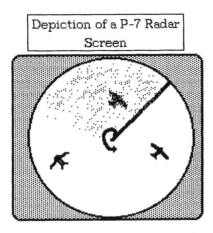

Depiction of a P-7 Radar Screen

Thus the screen first appears as "white", but leaves behind a trace of any objects defined by the electron beam excitation as "yellow". In radar, this has the advantage that slow moving objects can be tracked by the trace left behind on the screen as the radar dish rotates. The next rotational excitation caused by the dish (which is synchronized to the electron beam) then shows if, and how, the object has moved in relation to its former position.

The properties of the P-7 screen are summarized in the following Table, given on the next page.

The screens, **P-8 , P-9, & P-10** are obsolete and will not be described, except as follows:

P-8 : No data available- this was replaced by the P-7 screen.

P-9: Reservation of this number was withdrawn

P-10: This screen was a "dark-trace" screen, made from KCl.

Spectral and Decay Properties of P-7 Phosphors	
Emission Color	White
Emission Peaks	4350 Å & 5550 Å
CIE Coordinates	Blue: x = 0.285; y = 0.285
	Yellow: x = 0.357; y = 0.537
Persistence - $T_{10\%}$	Blue: 75 microseconds.
	Yellow: 400 milliseconds.
Decay Class	Blue: Medium Short
	Yellow: Long
Activator Concentration	Blue = 0.0166 Ag^- /mol ZnS
	Yellow = 0.007 Cu^- /mol $ZnCdS_2$
Zn/Cd Ratio	Blue: NA
	Yellow: 77.5 Zn/14.5 Cd
Usual Particle Size	Blue = 31 μ; Yellow =36 μ
Body Color	White(very slight yellow)
CRT Brightness vs: P-20	62 %
Optimum Screen Weight	4.0 mg./cm^2

H. Manufacture of P-11 Phosphor

This phosphor is one of the "workhorses" of the monochrome screens used, and was originally developed for cases where a photographic record was desired. It was used in Oscilloscope screens because its emission more closely matched the response curve of the then available photographic films, thus giving high exposure speeds to capture transient phenomena (Nowadays, all high-speed oscilloscopes feature "refresh rates" so that high speed events can be kept on the raster until a record is made of them).

This phosphor is essentially a blue-emitting ZnS:Ag composition, "quenched" by the addition of Ni^{2+}. That is, the normal decay rate of Ag^- is decreased by the addition of nickel to the overall composition.

The formulation to be used for its manufacture is given on the next page as follows:

5.3.21.- <u>Formulation for Preparation of P-11 Phosphor</u>

Compound	Mol Ratio	Assay	Grams	Lb/Lb
ZnS	0.90	0.995	88.14	0.831
$(Zn_{0.987}Ag_{0.100}, Ni_{0.003})S_2$	0.100	0.995	10.043	0.0947
$MgCl_2$	6% by weight of sulfides		5.919	0.0556
$LiSO_4$	2% by weight of sulfides		1.973	0.0185

This mix is first blended, then hammermilled twice, with blending in between hammermillings. It is then fired in a 2 liter covered silica crucible for:

4 hours at 850 °C in air

using a covered 2-Liter silica crucible. After firing, when the phosphor cakes are cool, they are inspected and crushed. This blue-emitting phosphor is then screened through a 325 mesh screen, and is washed free of excess flux, using the same procedure described above and a silica coating is applied. The slurry is then filtered and the wet cake is dried for 24 hours at 125 °C.

The P-11 phosphor has the following spectral characteristics, shown in the following diagram:

5.3.22.-

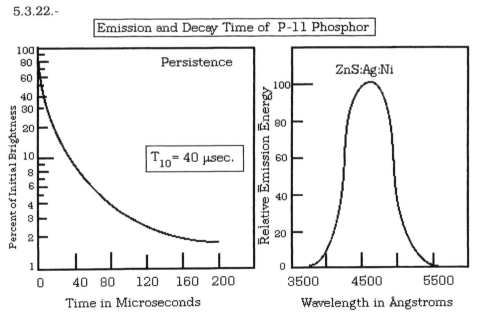

Data concerning both the spectral and physical properties are also shown in the following table. The emission is blue and the decay is classified as short.

Spectral and Decay Properties of P-11 Phosphors	
Emission Color	Blue
Emission Peaks	4550 Å
CIE Coordinates	Blue: x = 0.144; y = 0.121
Persistence - $T_{10\%}$	Blue: 40 microseconds.
Decay Class	Short
Activator Concentration	Blue = 0.0100 Ag⁻ /mol ZnS + 0.00032 Ni^{2+} /mol ZnS
Zn/Cd Ratio	NA
Usual Particle Size	6-13 μ
Body Color	White
CRT Brightness vs: P-20	38 %
Optimum Screen Weight	3.0 mg./cm^2

The P-11 phosphor screen has been used extensively with the P-1 phosphor to produce a screen used for measurement purposes where the measurement graticule is composed of one phosphor (green) superimposed upon a blue screen (P-11).

The preparation of such a tube is described by the following procedure:

First, the P-11 screen is prepared by a settling technique, as given above. Once the P-11 screen has been finished, it is wet with PVA and then dried. The P-1 phosphor is then "dusted" upon the dried PVA + P-11 combination, and then exposed, using ultraviolet light as described above. Alternately, the screen can be made by two selective settling techniques, forming the P-11 screen first, forming the graticule using PVA plus exposure, and then settling the P-1 screen upon the P-11 gradicule. The specifics of this process are:

5.3.23. Process for Manufacture of a Phosphor Graticule

a. Preparation of Solutions

PVA - Make up a 3.5% solution (21.0 grams of PVA in 579 ml. of water).

Sensitizer : Make up fresh for each use: 3.0 gm. $NH_4 Cr_2O_7$ in 25.0 ml of water.

Stock PVA solution is mixed with the sensitizer at the ratio of: 1.0 ml. of dichromate per 6.50 ml. of PVA. Work the PVA through a filter mesh just before use to remove any lumps that may be present. 15.0 ml. is adequate to wet down a 5.0 inch round screen.

b. Specific Steps in the Procedure

1. Wet down the prepared P-11 screen with the sensitized PVA solution and allow to soak for 5.0 minutes. Place on tilt table to pour off excess liquid. Dry partially at 110 °C for about 1/2 hour.

2. Apply the prepared mask, indexing it into the proper position on the inner part of the face-plate, next to the P-11 screen. Expose the dried PVA to an ultraviolet arc light for about 70.0 seconds.

3. Soak exposed screen with water, washing out the dichromate color and wipe the wet unexposed phosphor + PVA to help remove it. thoroughly wash tube, and then thoroughly dry the face-plate on the dryer.

4. The face-plate is then sealed to the neck of the glass tube, and a P-1 screen is then settled on top of the P-11 gradicule. This cascade screen is then dried as described above. Since the P-1 particles will settle on top of the P-11 gradicule. it will be necessary to brush the back of the cascade screen to remove the P-1 particles. A 2537 Å light can be used to locate these particles to aid in their removal.

5. The CR tube is then finished in the usual manner by aluminizing, sealing in the electron gun, and then evacuating the tube to about 10^{-7} mm. of internal mercury pressure.

When finished, a phosphor graticule face-plate has the following appearance when it is activated by the electron beam to form the raster, as shown on the next page. Note that only those phosphors which are not cross-contaminated by one another can be used in the above described manner to manufacture a phosphor graticule screen.

5.3.24.-

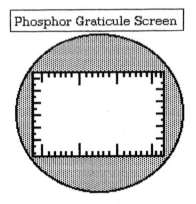

Phosphor Graticule Screen

I. Manufacture of P-12 Phosphor

This phosphor was originally developed for use in radar tubes where a moderate decay time was desirable. Its composition consists essentially of ZnF_2, modified by the addition of a minor amount of MgF_2, and is activated by Mn^{2+}.

It is manufactured using the following formulation:

5.3.25.- Formulation for P-12 Phosphor

Compound	Mol Ratio	Assay	Grams	Lb/Lb
$ZnF_2 \bullet 4H_2O$	0.845	0.579	256.04	0.9591
MgF_2	0.120	0.985	7.590	0.0284
MnF_2	0.035	0.975	3.336	0.0125

Sometimes a flux such as 6% NaCl is added to these components, depending upon the particle size desired. The whole is mixed together by blending, then hammermilled twice, with blending in between hammermillings.

The resulting mix is then placed in a 2 liter silica crucible and fired in air using a covered crucible for:

3.5 hours at 785 °C in air

After the cakes are cool, they are inspected and discolored surface areas are removed. Since this phosphor does not respond well to ultraviolet light excitation, ordinary light is used during the inspection procedure. The cakes are then crushed, sieved through a 325 mesh screen and then washed (if flux is used). Finally, the powder is suspended in water and then silica coated as described above. Some Manufacturers prefer to use an alumina crucible instead of the silica crucible. This is due to the fact that fluorides attack silica to form fluosilicic acid. However, after the first or second firing, the surface of the silica crucible becomes coated with the phosphor, and further attack then ceases during individual firings.

The spectral properties are shown in the following diagram. The emission is orange and the decay is classified as long.

5.3.26.-

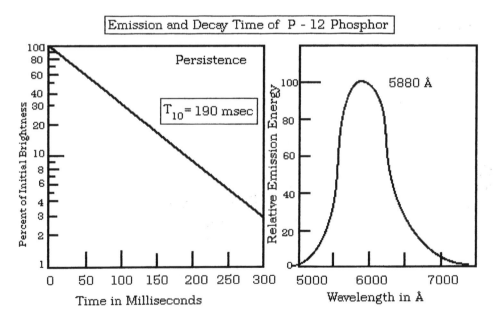

A summary of the spectral and physical properties of the P-12 phosphor is presented in the following table, as given on the next page.

Spectral and Decay Properties of P-12 Phosphors	
Emission Color	Orange
Emission Peak	5880 Å
CIE Coordinates	x = 0.547; y = 0.451
Persistence - $T_{10\%}$	Orange: 190 milliseconds.
Decay Class	Long
Activator Concentration	0.035 mol Mn^{2+}/ mol
Zn/Cd Ratio	NA
Usual Particle Size	11-19 μ
Body Color	White
CRT Brightness vs: P-20	20 %
Optimum Screen Weight	4.0 mg./cm^2

This phosphor was one of the first long decay phosphors to be developed, based on Mn^{2+} emission in a fluoride lattice. As we shall see, other compositions were also developed to achieve the same purpose.

J. Manufacture of P-13 Phosphor

This phosphor, a magnesium silicate activated by Mn^{2+}, has a reddish orange emission and a medium decay time. It is manufactured using the following formulation:

5.3.27.- Formulation for P-13 Phosphor

Compound	Mol Ratio	Assay	Grams	Lb/Lb
$3MgCO_3 \cdot Mg(OH)_2 \cdot 3H_2O$	1.00 MgO	44.6% MgO	90.37	0.568
SiO_2	1.05	0.925	68.20	0.428
$MnCO_3$	0.0050	0.950	0.6050	0.0038

The mix is first blended and then hammermilled twice as given before and fired in air in 2 liter silica crucibles for:

4 hours at 1205 °C. in a covered crucible

After firing, the cake is processed as given above, including inspection, crushing and sieving through a 325 mesh screen. The powder is then suspended in water and is "silicized", using the procedure given above. A free-flowing powder results. The decay is medium and the emission is a deep-red. The spectral properties are shown in the following diagram:

5.3.28.-

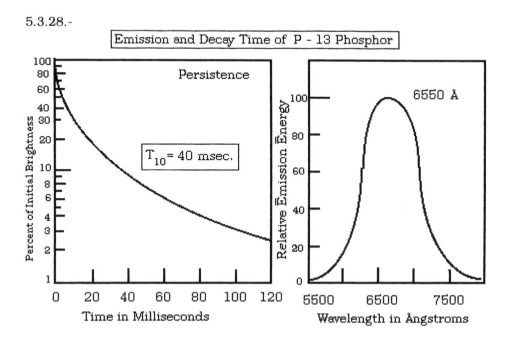

The emission is deep red and the decay is medium. A summary of the P-13 phosphor properties is given in the following table:

Spectral and Decay Properties of P-13 Phosphors	
Emission Color	Deep Red
Emission Peak	6550 Å
CIE Coordinates	x = 0.670; y = 0.329
Persistence - $T_{10\%}$	40 milliseconds.
Decay Class	Medium
Activator Concentration	0.005 mol Mn^{2+}/ mol
Zn/Cd Ratio	NA
Usual Particle Size	5-9 µ
Body Color	White
CRT Brightness vs: P-20	3 %
Optimum Screen Weight	4.2 mg./cm^2

This phosphor was used primarily in military radar screens, but is now nearly obsolete, except for its use in military radar tubes.

K. Manufacture of P-14 Phosphor

This phosphor screen is another example of the "cascaded" screen. The initial screen color is a bluish purple, composed of a blue emitting short-decay phosphor and an orange emitting long-decay phosphor. Once the blue emission has decayed, the phosphorescence is orange. The P-14 phosphor is manufactured using the following formulations:

5.3.29.- Formulation for P-14 Phosphors

a. Blue-Emitting Component

Compound	Mol Ratio	Assay	Grams	Lb/Lb
ZnS	0.90	0.995	88.14	0.827
$(Zn_{0.834}, Ag_{0.166})S$	0.10	0.995	10.50	0.0986
$MgCl_2$	6% by weight of sulfides		5.919	0.0556
$LiSO_4$	2% by weight of sulfides		1.973	0.0185

b. Orange Emitting Component

Compound	Mol Ratio	Assay	Grams	Lb/Lb
ZnS	0.397	0.995	38.68	0.307
$(Zn_{0.923}Cu_{0.077})S_2$	0.100	0.995	9.779	0.0776
CdS	0.503	0.995	72.67	0.577
$MgCl_2$	3.0 % by Weight		2.937	0.0223
NaCl	2.0 % by Weight		1.958	0.0155

c. Orange Emitting Component (Formulated by Another Manufacturer)

Compound	Mol Ratio	Assay	Grams	Lb/Lb
ZnS	0.702	0.995	68.40	0.592
$(Zn_{0.933}Cu_{0.067})S_2$	0.100	0.995	9.675	0.0847
CdS	0.198	0.995	28.75	0.249
$MgCl_2$	3.0 % by Weight		3.207	0.0278
NaCl	2.0 % by Weight		2.139	0.0185
$BaCl_2$	3.0 % by Weight		3.207	0.0278

These components are mixed by blending, then hammermilled twice, with blending in between hammermillings. Care must be taken not to expose one formulation to the other so that cross-contamination does not occur. The blue-emitting phosphor formulation is particularly susceptible to this aspect. Either of the orange-emitting phosphor formulations can be used, depending upon the preference of the Manufacturer.

The blue-emitting sulfide is fired in air in a covered 2 liter silica crucible for:

3.5 hours at 1075 °C

The orange-emitting phosphors are fired in air in a 2 liter silica crucible for:

Formula b. -   3.0 hours at 1150 °C.

Formula c. -   5.0 hours at 1150 °C.

using a cover on the crucible. When the phosphor cakes are cool, they are inspected and crushed. The blue-emitting phosphor is then screened through a 325 mesh screen, and then is suspended in water. A silica coating is applied as described above.

The orange-emitting sulfide is first screened through 325 mesh screen, and then is washed free of excess flux, using the same procedure described above. When dry, the body-color of the blue-emitting phosphor will be white while that of the orange-emitting phosphor will have a slight yellowish cast.

The spectral properties of this screen are shown in the following diagram, given as 5.3.30. on the next page. A summary of its spectral and physical properties is also given on the next page.

The P-14 phosphor was one of those used originally in radar screens. Even though it had high brightness when used in a radar screen, it was found to have several deficiencies such as a rather short time of viewing before the screen brightness decayed to a point where the objects being tracked by the radar installation could not be observed clearly, even in a closed and dark room.

5.3.30.-

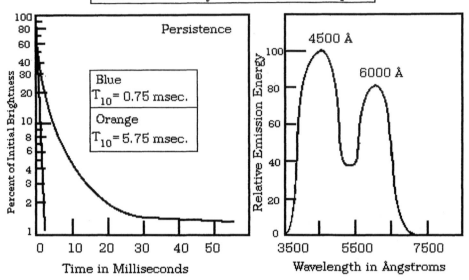

Emission and Decay Time of P-14 Phosphor

Spectral and Decay Properties of P-14 Phosphors	
Emission Color	Bluish Purple (Both excited)
Emission Peaks	4600 Å & 6000 Å
CIE Coordinates	Blue: x = 0.148; y = 0.071
	Orange: x = 0.511 ; y = 0.466
Persistence - $T_{10\%}$	Blue: 75 µsec.
	Orange: 5.5 milliseconds.
Decay Class	Blue: Medium Short
	Orange: Medium
Activator Concentration	Blue = 0.0166 mol Ag⁻/mol
	Orange = 0.0077 Cu⁻/mol
Zn/Cd Ratio	NA
Usual Particle Size	5-9 µ
Body Color	Blue = White
	Orange = Yellowish
CRT Brightness vs: P-20	Blue = 13 %
	Orange = 17 %
Optimum Screen Weight	4.2 mg./cm^2

573

It was for this reason that some of the newer phosphors like P-19, P- 26 and P-38 were developed. These phosphors then allowed viewing times of up to several minutes before the screen needed to be refreshed.

L. Manufacture of P-15 Phosphor

This phosphor is the $ZnO:Zn^0$ phosphor that we have described in more detail in Chapter 3. as having the $ZnO_{1-x}:xZn^0{}_0^=$ composition, where the activator is the zinc atom in the metallic state, situated on an oxide site in the lattice made vacant by a reduction process (see page 370). The P-15 phosphor is easily prepared by firing ZnO in a reducing atmosphere. In general, one takes a quantity of pure ZnO and fires it in a silica tray for:

3.0 hours at 855 °C

using a flowing atmosphere composed of 95% N_2 gas and 5% H_2 gas. At the end of the firing period, the crucible is allowed to cool in the flowing atmosphere, before being removed for further processing. The cake is then inspected, crushed and then sieved through a 325 mesh screen. Sometimes, it is then silicized before use to form the CRT screen. The following diagram shows the spectral properties of this phosphor:
5.3.31.-

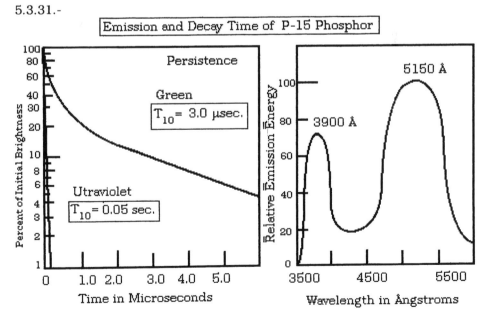

The spectral and physical properties are given in the following table:

Spectral and Decay Properties of P-15 Phosphors	
Emission Color	Green
Emission Peak	5000 Å
CIE Coordinates	x = 0.245; y = 0.408
Persistence - $T_{10\%}$	3.5 microseconds
Decay Class	Short
Activator Concentration	"Self-activated"
Zn/Cd Ratio	NA
Usual Particle Size	5-11 μ
Body Color	White
CRT Brightness vs: P-20	17 %
Optimum Screen Weight	4.5 mg./cm^2

The P-15 phosphor was originally used in a flying-spot scanner and is applicable as a screen to any CRT requiring a highly visible screen and a fast decay. It is easily prepared and has also found use in the preparation of solid state resistors since it is a semi-conductor as well as a phosphor.

Actually, a reduced ZnO has found many uses in Industry, including that of a filler in synthetic rubber used in automobile tires. It is also luminescent but is used for its conductive properties rather than for any spectral properties.

M. Manufacture of P-16 Phosphor

This phosphor is an ultraviolet-emitting silicate activated by Ce^{3+}. Since its decay is very fast, it has been used in flying spot scanners and other devices where a very fast decay is desired. It is manufactured using the following formulation:

5.3.32.- Formulation for P-16 Phosphor

Compound	Mol Ratio	Assay	Grams	Lb/Lb
$CaCO_3$	2.00	0.982	203.85	0.453
$3MgCO_3 \cdot Mg(OH)_2 \cdot 3H_2O$	1.00 MgO	44.6% MgO	90.37	0.201
SiO_2	2.05	0.925	133.15	0.296
$Ce(NO_3)_3 \cdot 6 H_2O$	0.050	0.950	22.854	0.0508

The mix is hammermilled twice as given before, with blending in between hammermillings, and then fired in 2 liter silica crucibles for:

4 hours at 1225 °C. in a covered crucible

After firing, the cake is processed as given above, including inspection and removal of dark spots (since the emission occurs in the near-ultraviolet, a 2537 Å lamp cannot be used for inspection), crushing and sieving through a 325 mesh screen. The powder is then suspended in water and is "silicized", using the procedure given above. A free-flowing powder results.

The spectral properties of the P-16 screen are given in the following diagram:

5.3.33.-

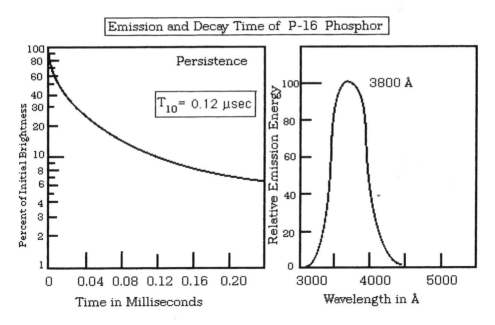

A description of its physical and spectral properties is given in the table shown on the next page.

More recently, another type of phosphor was developed having an even faster rate of decay. This phosphor is based upon $YPO_4 : Ce^{3+}$ and/or $LaPO_4 : Ce^{3+}$.

Spectral and Decay Properties of P-16 Phosphors	
Emission Color	NA
Emission Peak	3800 Å
CIE Coordinates	NA
Persistence - $T_{10\%}$	0.14 microseconds.
Decay Class	Very short
Activator Concentration	0.05 mol Ce^{3+}/ mol
Zn/Cd Ratio	NA
Usual Particle Size	15-19 µ
Body Color	White
CRT Brightness vs: P-20	NA
Optimum Screen Weight	4.2 mg./cm^2

Such phosphors are prepared by the coprecipitation process already described in Chapter 4 - page 466. Details given there involved production sized batches. Since this procedure has already been presented, we will address only the barest outline to be followed to prepare small batches. If a production batch is desired, one must follow those given in Chapter 4.

5.3.34. Process for Preparing YPO_4:Ce and $LaPO_4$:Ce Phosphors

a. Prepare a 1.00 molar solution of Y^{3+} and Ce^{3+}, by dissolving 0.95 mols of Y_2O_3 in HCl plus water. 0.05 mols of $Ce(NO_3)_3 • 6 H_2O$ are also added. This involves 214.52 grams of the Y_2O_3 and 21.71 grams of the cerium salt.

b. A 2.20 molar H_3PO_4 solution is also required. Add 150.5 ml. of 85% H_3PO_4 to 1000 ml. of water with stirring.

c. The Ln^{3+} solution is pumped into the phosphoric acid solution, i.e.- 1.00 molar Ln^{3+} ⇒ 2.20 molar H_3PO_4 solution at 80 °C. The precipitate is then filtered and dried at 125 °C. overnight.

The powder is next milled in water for about 45 minutes, using 100 gm. of powder to 300 ml. of water in a 2 quart mill containing 1/2 inch flint stones and then dried again overnight. It is then placed in a 2 liter covered silica crucible and fired for:

3.0 hours in air at 1175 °C.

After firing, the cooled cakes are inspected, crushed and then sieved through a 325 mesh screen. The powder is then silicized according to the procedure already given above.

The peak emission occurs in the ultraviolet region near to 3500 Å, with a secondary peak near 3300 Å. This emission is not visible to the human eye.

A comparison of the decay rates and emission bands of both $YPO_4 : Ce^{3+}$ and P-16 phosphors is given in the following diagram:

5.3.35.-

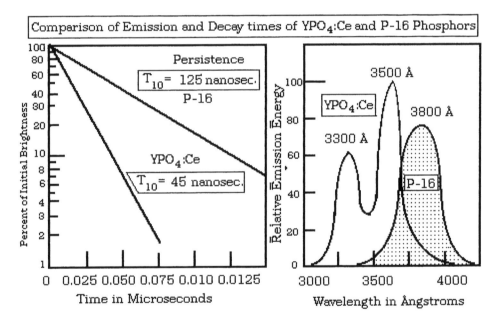

Note that the $YPO_4:Ce^{3+}$ phosphor is about 3 times faster in decay time as the P-16 phosphor and has nearly 150% of the emission intensity of that phosphor. The $YPO_4:Ce^{3+}$ phosphor was used as the control phosphor for the instantaneous color being generated with the single-gun non-shadow mask tube mentioned above.

N. Manufacture of P-17 Phosphor and Screen

The P-17 screen was made with the P-7 yellow phosphor, with a P-15 $(ZnO:Zn^0)$ phosphor as a cascade screen. A bluish white screen color resulted with a yellow phosphorescence. The preparation of both of these phosphors has already been given. This screen is no longer used to any great extent, having been replaced by newer screens like P- 32 (see below)

O. Manufacture of P-18 Phosphor

This screen is outdated, but was prepared as the all-silicate P-4 screen, where the components were limited to silicates (or phosphates) for preparation of a black & white television screen. It was made using either the P-3 yellow or $Cd_5Cl(PO_4)_3:Mn$ phosphors and the blue-emitting $CaMg(SiO_3)_2:Ti$ phosphor. The preparation of these phosphors has already been given above.

P. Manufacture of P-19 Phosphor

This phosphor has a $KMgF_3:Mn$ composition and a yellow emission under CRT excitation. The formulation used in its preparation is:

5.3.36.- Formulation for P-19 Phosphor

Compound	Mol Ratio	Assay	Grams	Lb/Lb
KF	1.00	0.895	64.91	0.475
MgF_2	1.00	0.875	71.20	0.521
MnF_2	0.005	0.945	0.4917	0.00360

These materials must be thoroughly dry before they are weighed and blended. They are then hammermilled twice, with blending between hammermillings.

The resulting mix is then fired in a 2 liter covered silica crucible in air for:

3.0 hours at 835 °C.

After firing, when the phosphor cakes are cool, they are inspected and crushed. A 3650 Å lamp is used even though this phosphor is not very responsive to ultraviolet excitation. The powder is then sifted through a 325 mesh screen and then is suspended in water, followed by silicizing, filtering the wet cake and drying it to obtain a free-flowing powder.

The spectral properties are shown in the following diagram:

5.3.37.-

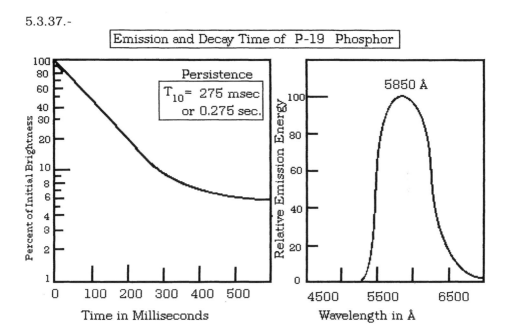

Emission and Decay Time of P-19 Phosphor

A summary of the properties of this screen are given in the following table:

Spectral and Decay Properties of P-19 Phosphors	
Emission Color	Orange
Emission Peak	5850 Å
CIE Coordinates	x = 0.592 ; y = 0.408
Persistence - $T_{10\%}$	275 milliseconds.
Decay Class	Long
Activator Concentration	0.005 mol Mn^{2+}/ mol
Zn/Cd Ratio	NA
Usual Particle Size	8-11 µ
Body Color	White
CRT Brightness vs: P-20	17 %
Optimum Screen Weight	6.2 mg./cm^2

This phosphor has an orange emission and the decay is long. It has been used as a screen for radar CR tubes for military applications. When a long decay phosphor screen is desired, this phosphor is generally used. However, as noted above, the P-19 phosphor has been outmoded by some of the newer phosphors, notably P-26 and P-38, which are improved versions of the P-19 phosphor. These compositions have decay times that can be observed for several minutes in the dark, and their T_{10} decay rate is as long as 3.0 seconds.

Q. Manufacture of P-20 Phosphor

This phosphor is very similar to the green emitting $ZnCdS_2$:Ag phosphor used in color television tubes. It is manufactured by using the following formulation:

5.3.38.- Formulation for P-20 Phosphor

Compound	Mol Ratio	Assay	Grams	Lb/Lb
ZnS	0.540	0.995	52.87	0.428
$(Zn_{0.715}Ag_{0.285}) S_2$	0.100	0.995	9.771	0.0791
CdS	0.360	0.995	52.01	0.421
$MgCl_2$	4% by weight of sulfides		6.603	0.0535
NaCl	2% by weight of sulfides		2.201	0.0178

These components are first mixed, then hammermilled twice, with blending in between. The resulting mix is then fired in air in a 2 liter silica crucible for:

3.0 hours @ 935 °C.

After firing, the cooled cakes are inspected under 3650 Å light, crushed and sieved through a 325 mesh screen before washing to remove excess flux. In this case, complete removal of flux is essential to obtain the highest brightness screen in a cathode ray tube. The wet cake is then resuspended and silicized as described above.

The spectral properties are given in the following diagram, given on the next page as 5.3.39. When a high brightness screen is desired, it is usually the P-20 screen that is specified.

5.3.39.-

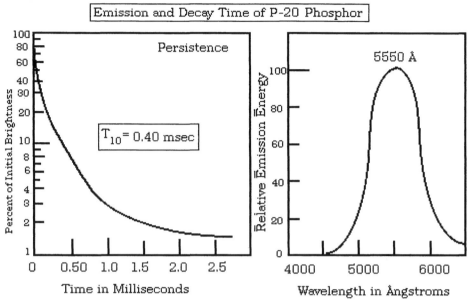

The following is a summary of the physical and spectral properties of the P-20 phosphor.

Spectral and Decay Properties of P-20 Phosphors	
Emission Color	Greenish Yellow
Emission Peak	5550 Å
CIE Coordinates	x = 0.381 ; y = 0.575
Persistence - $T_{10\%}$	400 microseconds.
Decay Class	Medium Short
Activator Concentration	0.0285 mol Ag^-/ mol
Zn/Cd Ratio	0.855 Zn/0.145 Cd
Usual Particle Size	7-10 μ
Body Color	Yellowish
CRT Brightness vs: P-20	100 %
Optimum Screen Weight	4.2 mg./cm^2

Note that this formulation for P-20 has been adjusted so that the peak of the emission band coincides exactly with that of the y-coefficient of the CIE diagram. This is desirable since this peak wavelength conforms to the eye response curve of the human eye and produces the highest possible brightness response in human vision. It is for this reason that the P-20 phosphor is used as the standard of comparison. Because of differences in both ZnS and CdS precipitates, it may be necessary to adjust the Zn/Cd ratio to obtain the required emission peak. This is done by experimental means, measuring the emission peak as a function of small adjustments in ratio.

R. Manufacture of P-21 Phosphor

The screen made from the P-21 phosphor has a long decay and an orange emission. It has been used primarily in radar screens for military applications. The phosphor itself is manufactured by the use of the following formulation:

5.3.40.- Formulation for the P-21 Phosphor

Compound	Mol Ratio	Assay	Grams	Lb/Lb
MgF_2	0.9455	0.985	59.80	0.971
LiF	0.0500	0.984	1.318	0.02141
MnF_2	0.0045	0.945	0.4425	0.00719

These components are blended, and then hammermilled twice, with blending in between hammermillings. The mix is then placed in a covered silica crucible and fired for:

3.5 hours at 815 °C in air

At the end of the firing period, the cakes are cooled in air and then inspected as described for the preparation of P-19 phosphor. The powder is the suspended in water and then silicized as described above.

The emission of P-21 phosphor is an orange red, like the P-19 phosphor, but the decay is longer, as shown in the following diagram, given as 5.3.41. on the next page.

The emission is an orange red in color and the decay time is about 800 milliseconds, i.e.- 0.8 seconds to 10% of its original intensity under steady-state CRT excitation.

5.3.41-

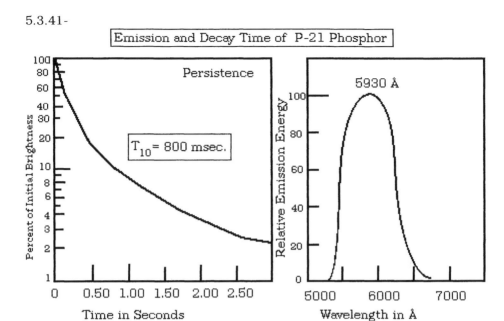

Emission and Decay Time of P-21 Phosphor

A summary of its physical and spectral properties is given in the following table, viz-

Spectral and Decay Properties of P-21 Phosphors	
Emission Color	Orange Red
Emission Peak	5930 Å
CIE Coordinates	x = 568 ; y = 0.432
Persistence - $T_{10\%}$	800 microseconds.
Decay Class	Long
Activator Concentration	0.0285 mol Mn^{2+}/ mol
Zn/Cd Ratio	NA
Usual Particle Size	12-18 μ
Body Color	White
CRT Brightness vs: P-20	17 %
Optimum Screen Weight	4.5 mg./cm^2

The P-21 screen was developed in response to the request for a longer decay phosphor for use in radar tubes, and the like. As such, it has replaced the P-19 phosphor in several applications. This phosphor is susceptible to damage by water, unless it is fired to a highly crystalline form before coating, which forms a protective layer on the outer surface of the particles. Some effort has been made to do the coating procedure in a non-aqueous media, but this has not been totally successful.

S. Manufacture of P-22 Phosphor

The P-22 phosphor screen was registered for use as the color television screen. As such, it has undergone many changes since the original registration, caused mainly by the introduction of the Y_2O_3 :Eu^{3+} phosphor by the author in 1959-1960. We will first describe the preparation of the all-sulfide screen phosphors and then the modifications that have been used.

The original red-green-blue mixture of sulfides used to manufacture early color television screens were obtained by changing the ratio of ZnS/CdS as activated by Ag^-. The green and the blue phosphors are still being used (although they have been improved so as to more closely conform with the demands of current color CRT manufacturers, i.e.- color saturation, brightness, particle size and coatings). We will describe a general preparation so as to show how the sulfide compositions can be manipulated to control emission color.

The manufacture of the original sulfide phosphors used for color television uses the following formulations:

5.3.42.- Compositions to be Used for Tri-Color Emitting Sulfide Phosphors

a. Blue-Emitting Phosphor

Compound	Mol Ratio	Assay	Grams	Lb/Lb
ZnS	0.736	0.995	72.08	0.606
$(Zn_{0.88}Ag_{0.12})S_2$	0.100	0.995	9.771	0.0969
$MgCl_2$	6% by weight of sulfides		6.603	0.0556
$LiSO_4$	2% by weight of sulfides		2.201	0.0185

Sometimes, $AlCl_3$ is added to the above mix as an additional flux. The quantity to be added is: 2% by weight = 2.201 gram = 0.0185 lb./lb.

b. Green-Emitting Phosphor

Compound	Mol Ratio	Assay	Grams	Lb/Lb
ZnS	0.540	0.995	52.87	0.428
$(Zn_{0.715}Ag_{0.285})S_2$	0.100	0.995	9.771	0.0791
CdS	0.360	0.995	52.01	0.421
$MgCl_2$	4% by weight of sulfides		6.603	0.0535
NaCl	2% by weight of sulfides		2.201	0.0178

a. Red-Emitting Phosphor

Compound	Mol Ratio	Assay	Grams	Lb/Lb
ZnS	0.280	0.995	27.28	0.221
$(Zn_{0.88}Ag_{0.12})S_2$	0.100	0.995	10.254	0.0831
CdS	0.720	0.995	104.54	0.847
$MgCl_2$	4% by weight of sulfides		6.603	0.0535
NaCl	2% by weight of sulfides		2.201	0.0178

As we stated before, the choice of flux components is arbitrary. The ones that we have shown are those used by one Manufacturer to produce these phosphors. It is important to prevent cross-contamination of each color from the others. Therefore, each phosphor is treated as a separate entity and the equipment used for one is **not used** for the other.

These phosphors are fired in air in a covered 2 liter silica crucible as follows:

Blue phosphor- 2.5 hours @ 1075 °C.

Green phosphor- 3.0 hours @ 915 °C.

Red phosphor- 3.0 hours @ 800 °C.

Note that as the Cd/Zn ratio goes up, the optimum firing temperature goes down. The fired cakes are then processed as given above in 5.3.3. Finally, the washed powder is silica-coated to form a free-flowing material.

The solid state chemistry of the zinc and cadmium sulfides is fairly complex. ZnS occurs as one of two dimorphic compounds, depending upon the firing temperature:

ZnS - cubic: Firing Temperature   ~ 850 - 950 °C.

ZnS - hexagonal: Firing Temperature ~ 1050 - 1200 °C.

However, if CdS is fired with ZnS, the product always has a hexagonal crystal structure. What we are speaking of is illustrated by the following, viz-

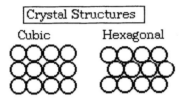

| Crystal Structures |

Cubic   Hexagonal

Note that the same layers of atoms are involved, except that the second layer has slipped into the interstices of the bottom layer. Since the Cd^{2+} atom is slightly larger than the Zn^{2+} atom, the hexagonal arrangement has a lower entropy, and a cubic structure is not possible. The ZnS:Ag⁻ phosphor has its primary emission band in the blue region of the spectrum, but its peak emission depends upon which crystal form prevails. For ZnS:Ag⁻, these peaks are:

Cubic = 4490 Å = 850 °C Firing temperature

Hexagonal = 4350 Å = 1100 °C Firing Temperature

However, when CdS is introduced, the emission band shifts, the amount depending upon the mol ratio of CdS to ZnS present. The amount of shift is almost linear, as shown in the following diagram, given as 5.3.44. on the next page.

The linearity in this diagram can be expressed as:

5.3.43.-Mol % CdS Required = 0.352 Peak Wavelength(in millimicrons) -153.6

But, this equation does not produce the exact quantity of CdS that is required

5.3.44.-

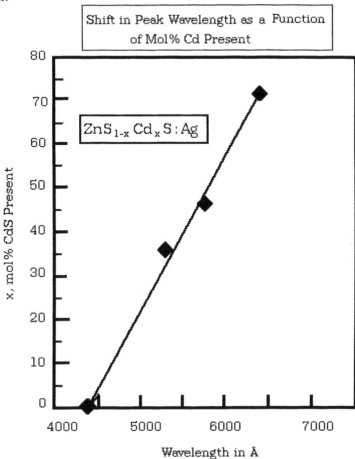

Shift in Peak Wavelength as a Function
of Mol% Cd Present

$ZnS_{1-x}Cd_xS:Ag$

x, mol% CdS Present

Wavelength in Å

to shift the emission to the selected wavelength. However, it can be used to establish the range of CdS required to obtain the desired peak emission of the phosphor. Because of differences in physical properties of the ZnS and CdS manufactured, it is usually necessary to prepare several test-batches of phosphor before actual phosphor production can begin.

Many cathode ray phosphor manufacturers specify weight % instead of mol%. Because of the differences in molecular weight of the two components, it is easy to become confused.

The following table gives a comparison of these two factors, viz-

Peak Wavelength	Amount of CdS Present	
	Weight %	Mol %
4350 Å	0.00	0.00
5300	42.0	36.0
5800	44.6	46.5
6400	79.8	72.0

Note that the differences in emission color observed lie in the ratio of ZnS to CdS in the final phosphor composition. Thus, a "green" phosphor has about 35-40 mol % of CdS present, a "yellow" phosphor has about 45-50 mol% in the formulation, while a "red" emitting phosphor has about 60 mol% CdS/40 mol% ZnS. Note that the peak wavelengths given above are compositions that are generally used to prepare specific CRT screen-colors. i.e.-

Blue = 4350 Å   Green = 5300 Å

Yellow = 5800Å   Red = 6400 Å

The spectral properties of the P-22 phosphors are given in the following diagram:

5.3.45.-

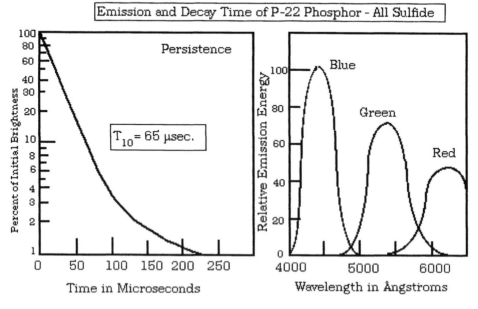

These characteristics are perhaps better seen in the   following diagram, where the CIE color coordinates produced as a function of CdS in the ratio, i.e.-CdS/ZnS , are given:

5.3.46.-

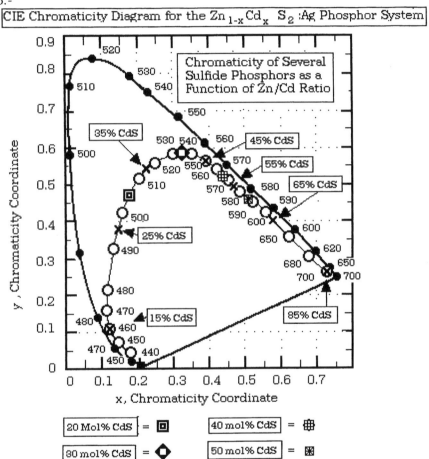

In this diagram, the CdS content is given in both weight-% and mol%. These values are the result of an experimental investigation and can be regarded as absolute values. Note that the peak emission can be shifted from 4350 Å to nearly 7500 Å, depending upon the weight-% CdS present. However, it should also be noted that the actual ZnS:CdS ratio required is a function of the type of Process used to manufacture these sulfides. That used for the sulfides in the above diagram was the Sulfate Process. Slight differences in ZnS/CdS ratios required to produce a specific peak emission may be noted for sulfides produced by the Chloride Process.

The emission of these sulfide phosphors decays in about 0.20 milliseconds to 1.0% of the original brightness, as shown above in 5.3.45. Actually, each component of the all-sulfide tri-color screen decays somewhat unequally, depending upon the Cd/Zn ratio of the phosphor. This is shown in the following table concerning these phosphors. Also given are the CIE color coordinates and the usual particle size produced.

5.3.47.-

Spectral and Decay Properties of P-22 Phosphors	
Overall Emission Color	White
Emission Peaks	4450 Å , 5300 & 6400 Å
CIE Coordinates	BLUE: x=0.146; y = 0.078
	GREEN: x=0.303 ; y = 0.587
	RED: x=0.665 ; y = 0.335
Persistence - $T_{10\%}$	BLUE: T_e = 67 μsec.
	GREEN: T_e = 50 μsec.
	RED: T_e = 42 μsec.
Decay Class	Medium Short
Activator Concentration	BLUE: 0.012 mol Ag$^+$ / mol
	GREEN: 0.012 mol Ag$^+$ / mol
	RED: 0.012 mol Ag$^+$ / mol
Zn/Cd Ratio (in mols/mol)	BLUE: 1.00 Zn/0 Cd
	GREEN: 0.55 Zn/0.45 Cd
	RED: 0.20 Zn/0.80 Cd
Usual Particle Size	BLUE: 4- 8 μ GREEN: 7 - 12 μ
	RED: 10 - 21 μ
Body Color	BLUE: White
	GREEN: White(very slight yellow)
	RED: Yellowish
CRT Brightness vs: P-20	BLUE: 83 % GREEN: 94 %
	RED: 62 %
Optimum Screen Weight	4.0 gm./cm^2

In 1964, an other red emitting phosphor was introduced, YVO_4 :Eu^{3+} , as a substitute for the red $ZnCdS_2$:Ag phosphor. Actually, this was in response to the development of the Y_2O_3 :Eu^{3+} phosphor which was 5.8 times brighter than the sulfide red-emitting phosphor. While the YVO_4 :Eu^{3+} phosphor was only about 2.9 times brighter than the red-sulfide, it also cost about 1/2 that of the oxide phosphor. Thus, it was used first in color television screens. As shown in the following diagram, the emission consists of several lines centered around the 6250 Å region of the spectrum. Also shown is the emission for Y_2O_2S : Eu^{3+} and the YVO_4 :Eu^{3+} phosphor.

5.3.48.-

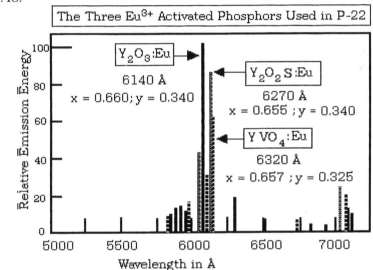

The Y_2O_2S : Eu^{3+} phosphor is a relative of the oxide phosphor, and has its main emission line shifted slightly redder compared to that of the main oxide line (see 5.3.48. for details). However, if one prepares the oxide phosphor, i.e.- Y_2O_3:Eu^{3+}, using the H_2O_2 technique described in the last chapter, one obtains a phosphor whose overall emission is just as red as the oxysulfide phosphor.

To manufacture the oxysulfide phosphor, one follows this procedure:

a. A Y_2O_3 : $Eu_{0.065}$ phosphor is first prepared, using techniques already described. These steps include precipitation of the oxalate, followed by firing in air to form the phosphor. It is then mixed with sulfur, according to the ratio given in the following formulation:

5.3.49.- Formulation for the Red Emitting Oxysulfide Phosphor

Compound	Mol Ratio	Assay	Grams	Lb/Lb
$Y_2O_3:Eu_{0.065}$	0.90	------	344.05	0.991
S	0.10	0.985	3.2548	0.00937

This mix is then hammermilled together, using the same techniques already described before.

The mix is then fired in air in a covered silica crucible for:

3.0 hours at 975 °C

After firing, the crucibles and their contents are cooled in air and the phosphor cakes are then crushed, sieved through a 325 mesh screen, suspended in water, and silicized as described above.

The oxysulfide phosphor usually produced has the solid-solution composition:

0.907 mol% Y_2O_3 + 0.033 mol Eu_2O_3 + 0.060 mol% Y_2S_3

Actually, these components could be used to form the phosphor, but only if the solid state reaction was done in a neutral atmosphere. The Y_2S_3 compound is not stable in air and will decompose at the firing temperature. The sulfide forms an infinitely solid solution with the oxide, and the specific stoichiometric formulation of the optimized phosphor is:

$(Y_{0.967}, Eu_{0.033})_2(O_{0.94}, S_{0.06})_3$

Although 10 mol% S is added, part is lost due to oxidation, and only 6.0 mol% ends up in the phosphor composition. In some cases, depending upon the firing conditions, it may be necessary to determine the actual level of sulfide incorporated into the phosphor. This is done by a thermogravimetric method, as shown in the following diagram, given on the next page, viz-

5.3.50.-

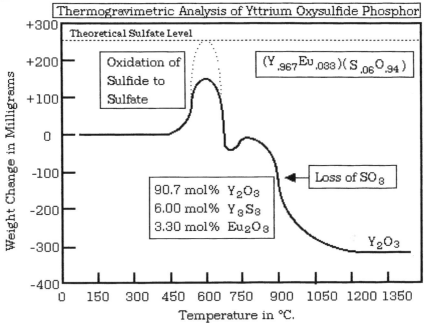

Thermogravimetric Analysis of Yttrium Oxysulfide Phosphor

Theoretical Sulfate Level

+300
+200 — Oxidation of
+100 — Sulfide to
Sulfate
$(Y_{.967}Eu_{.033})(S_{.06}O_{.94})$

0

-100 — Loss of SO$_3$

90.7 mol% Y$_2$O$_3$
6.00 mol% Y$_3$S$_3$
3.30 mol% Eu$_2$O$_3$

Y$_2$O$_3$

Weight Change in Milligrams

-200
-300
-400

0   150   300   450   600   750   900   1050 1200 1350

Temperature in °C.

Note that there is an exothermic reaction when the sulfide present in the crystal oxidizes at 475 °C to sulfate, i.e. - $S^= + 2O_2 \Rightarrow SO_4^{2-}$. The sulfate then decomposes at about 600 °C, leaving the pure oxide behind at 1200 °C.

Alternately, the oxysulfide phosphor could be produced by firing together:

$$0.94 \text{ mol of } (Y_{0.934}, Eu_{0.066})_2O_3 + 0.06 \text{ mol of } (Y_{0.934}, Eu_{0.066})_2S_3$$

However, it is easier to add sulfur to the oxide, and then fire this mixture than to prepare the sulfides by precipitation of the soluble lanthanides with H$_2$S. (One will find that precipitated lanthanide sulfides are extremely difficult to handle because of their physical properties). In some cases, depending upon firing conditions, it may be necessary to increase, or decrease, the amount of sulfur present in the original formulation before firing.

However, the "optimum" sulfide content still needs to be 6.00 mol% in the final composition, i.e.- $(Y_{0.967}, Eu_{0.033})_2(O_{0.94}, S_{0.06})_3$.

T. Manufacture of P-23 Phosphor

This phosphor screen was a low temperature "white" screen having color coordinates of:

$$x = 0.342 \qquad y = 0.369$$

This is approximately 5000 °K in color and was made by varying the ratio of the blue/yellow components of the P-4 screen to obtain the specified color temperature (with some red-emitting sulfide also added). This screen is also called "Sepia Tone" and is used in television monitors.

The following diagram shows the spectral properties of this screen:
5.3.51.-

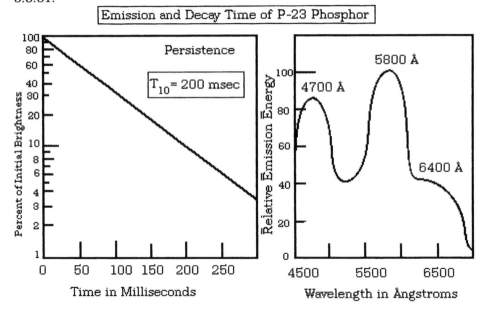

Note that three phosphors have been used to prepare this screen, a blue-emitter, a green-yellow emitter and a red emitter. These are all sulfide compositions, with the Zn/Cd ratio adjusted to give the desired emission peak, and the % of each component adjusted to give the proper emission characteristics. The red phosphor is the P-22 component given above, the yellow component is the P-4 yellow, both of whose manufacture have already been described. Also used as the red component is P-27 (see below). Actually, the P-27 phosphor is a much deeper red and is more pleasing to the human eye than the red P-22 sulfide phosphor. The blue-green emitting component of the P-23 screen is manufactured as follows:

5.3.52.- Formulation for P-23 Phosphor - Blue-Green Component

Compound	Mol Ratio	Assay	Grams	Lb/Lb
ZnS	0.879	0.995	86.34	0.705
$(Zn_{0.88}Ag_{0.12})S_2$	0.100	0.995	9.771	0.0798
CdS	0.121	0.995	17.57	0.143
$MgCl_2$	4% by weight of sulfides		6.603	0.0539
NaCl	2% by weight of sulfides		2.201	0.0180

This mix is hammermilled twice, with blending in between, as described above. It is then fired in a 2 liter silica crucible in air, using a lid, for:

3.0 hours @ 915 °C.

After firing, the cakes are inspected, crushed and sieved through a 325 mesh screen, before being silicized as described above. The physical and spectral properties are given in the following table, viz-

Spectral and Decay Properties of P-23 Phosphor	
Emission Color	"Sepia-Tone"
Emission Peaks	Blue Green =4700 Å Yellow Green = 5800 Å Red = 6400 Å
CIE Coordinates	Blue: x=0.155 ; y = 0.055 Green: x=0342.; y = 0.369 Red: x=0.665 ; y = 0.335
Persistence - $T_{10\%}$	200 milliseconds
Decay Class	Medium Short
Activator Concentration	1.2 -2.5 mol%
Zn/Cd Ratio	Blue-Green: 88 Zn/12 Cd
Usual Particle Size	10.5 - 14.0 µ
Body Color	White(very slight yellow)
CRT Brightness vs: P-20	83 %
Optimum Screen Weight	4.5 mg./cm^2

U. Manufacture of P-24 Phosphor

This screen is essentially a P-15 screen, but the $ZnO:Zn^0$ phosphor had been modified slightly to eliminate the ultraviolet peak present in the P-15 phosphor (see p.563 for the spectrum of P-15).

This is accomplished by use of the following formula:

5.3.53.- Formulation for the P-24 Phosphor

Compound	Mol Ratio	Assay	Grams	Lb/Lb
ZnO	1.00	0.985	82.54	0.9605
$LiCO_3$	0.05	0.986	3.395	0.0395

Note that $LiCO_3$ is added to eliminate the UV-peak. Other components such as Sb_2O_3, Bi_2O_3 and PbO have also been used but the best seems to be LiO.

The mix is hammermilled twice as given before, with blending in between hammermillings, and then fired in in a silica tray for:

3.0 hours at 855 °C

using a flowing atmosphere composed of 95% N_2 gas and 5% H_2 gas.

At the end of the firing period, the crucible is allowed to cool in the flowing atmosphere, before being removed for further processing. The cake is then inspected, crushed and then sieved through a 325 mesh screen. Sometimes, it is then silicized before use to form the CRT screen. The spectral and physical properties are identical to those of the P-15 phosphor, except that the ultraviolet peak is missing. These values may be found on page 579.

V. Manufacture of P-25 Phosphor

This phosphor has a $CaSiO_3:Pb:Mn$ composition and is similar to the lamp phosphor except that the activator contents are much lower.

It is manufactured via the following formulation:

5.3.54.- <u>Formulation for P-25 Phosphor</u>

Compound	Mol Ratio	Assay	Grams	Lb/Lb
$CaCO_3$	1.00 CaO	0.569	56.95	0.452
SiO_2	1.05	0.925	68.20	0.542
$MnCO_3$	0.0050	0.950	0.6050	0.00481
PbF_2	0.0006	0.973	0.1512	0.00120

This mix is first blended, then hammermilled twice as described above. with blending in between. It then is fired in air in a 2 liter silica crucible for:

4.0 hours at 1200 °C.

At the end of the firing period, the cakes are cooled, inspected and crushed. The powder is then wet milled for 4 hours in a ball mill, using about 400 grams of powder per liter of deionized water. The powder is then filtered and then resuspended in water. A silica coating is then applied as described above. Filter and dry powder at 110 °C. for 12 hours.

The spectral properties of the P-25 phosphor are shown in the following diagram:

5.3.55.-

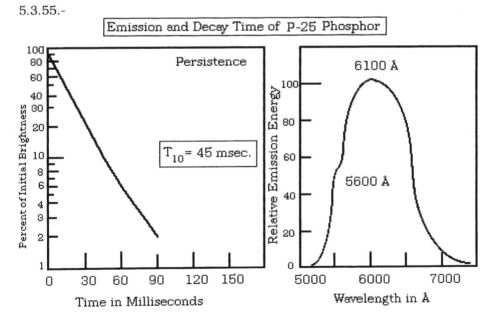

Emission and Decay Time of P-25 Phosphor

Note that a small peak appears at 5600 Å. This is probably due to the small amount of Pb^{2+} added to enhance the main Mn^{2+} peak at 6100 Å.

The following table summarizes the spectral and physical properties of this phosphor.

Spectral and Decay Properties of P-25 Phosphor	
Emission Color	Orange
Emission Peaks	6100 Å & 5600 Å
CIE Coordinates	x=0.560; y = 0.540
Persistence - $T_{10\%}$	45 milliseconds
Decay Class	Medium
Activator Concentration	0.5 mol%
Zn/Cd Ratio	NA
Usual Particle Size	6.5 - 14.5μ
Body Color	White
CRT Brightness vs: P-20	9 %
Optimum Screen Weight	4.5 mg./cm^2

The emission is orange and consists of two peaks. The decay is medium with 45 milliseconds required to reach 10% of its original intensity after excitation has ceased. Note that this phosphor has a brightness that is only 9% that of P-20. It has been used in military screens for display where it is necessary to view scenes for 10 seconds to two minutes after excitation is removed.

W. Manufacture of P-26 Phosphor

This phosphor has a composition of $KMgF_3$:Mn which is an improved P-19 phosphor composition. Its emission is orange and the decay is long. The formulation used in its preparation is:

5.3.56.- Formulation to be Used for P-26 Phosphor

Compound	Mol Ratio	Assay	Grams	Lb/Lb
KF	1.00	0.895	64.91	0.459
MgF_2	1.00	0.875	71.20	0.503
MnF_2	0.055	0.945	5.4089	0.0382

These materials must be thoroughly dry before they are weighed and blended. They are then hammermilled twice, with blending between hammermillings. The resulting mix is then fired in a 2 liter covered silica crucible for:

3.5 hours at 835 °C. in air

After firing, when the phosphor cakes are cool, they are inspected and crushed. A 3650 Å lamp is used even though this phosphor is not very responsive to ultraviolet excitation. The powder is then sifted through a 325 mesh screen and then is suspended in water, followed by silicizing, filtering the wet cake and drying it for 16 hours at 125 °C to obtain a free-flowing powder.

The spectral properties are shown in the following diagram, viz-

5.3.57.-

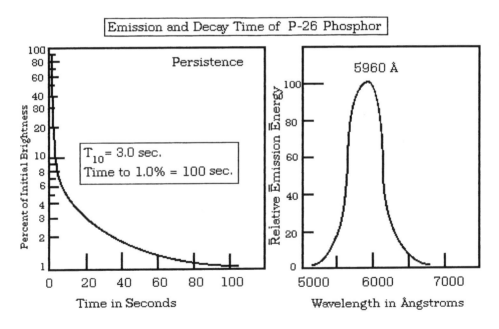

The emission is deep orange and the decay is very long. This phosphor also has a "long-tail" in its decay like P-25, but the "tail" is much longer. It is used in radar displays where viewing of the image for several minutes is desired, after the CRT excitation has ceased.

A summary of the physical and spectral properties of the P-26 phosphor is shown in the following table:

Spectral and Decay Properties of P-26 Phosphor	
Emission Color	Orange
Emission Peaks	5960 Å
CIE Coordinates	x=0.550; y = 0.425
Persistence - $T_{10\%}$	3.0 seconds
Decay Class	Very long
Activator Concentration	5.5 mol%
Zn/Cd Ratio	NA
Usual Particle Size	6.5 - 14.5μ
Body Color	White
CRT Brightness vs: P-20	14 %
Optimum Screen Weight	6.1 mg./cm²

P-26 is used extensively in Europe for radar tubes and the like, whereas P-19 is used more in the United States. P-26 is also used in CRT's where a low refresh-rate is desired.

X. Manufacture of P-27 Phosphor

This phosphor has a $Zn_3(PO_4)_2 : Mn^{2+}$ composition and has been used mostly in display tubes. It was used as the red emitting component in a P-6 screen but is not used for this purpose at this time. Lately, P-27 phosphor has been used in some specialty screen blends, such as P-23. It is manufactured according to the following formulation:

5.3.58.- Formulation for P-27 Phosphor

Compound	Mol Ratio	Assay	Grams	Lb/Lb
$ZnNH_4PO_4$	2.085	0.955	389.47	0.808
ZnO	1.000	0.985	82.62	0.171
$MnCO_3$	0.085	0.985	9.919	0.0206

These components are blended, and then hammermilled twice, with blending in between hammermillings. The mix is then placed in a covered silica crucible and fired for:

5 hours at 925 °C.

After firing, the cooled cakes are inspected and crushed. The powder is then wet milled for 2 hours in a ball mill, using about 400 grams of powder per liter of deionized water. The powder is then filtered and then resuspended in water. A silica coating is then applied as described above. Filter and dry powder at 125 °C. for 12 hours.

The spectral properties of the P-27 phosphor are shown in the following diagram:

5.3.59.-

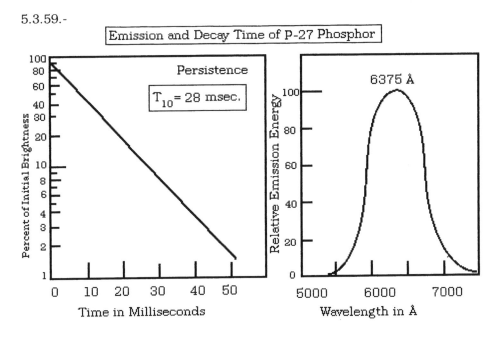

The P-27 phosphor is also used extensively in color television monitors not only because of its pleasing color but also because it is more stable than either of the red P-22 phosphors and does not "burn" to the same extent during long hours of operation, i.e.- the red screen brightness is maintained over a much longer period during continuous operation of the color television monitor.

A summary of the physical and spectral properties of the P-27 phosphor is shown in the following table:

Spectral and Decay Properties of P-27 Phosphor	
Emission Color	Deep Red
Emission Peaks	6375 Å
CIE Coordinates	x=0.674; y = 0.326
Persistence - $T_{10\%}$	28 milliseconds
Decay Class	Medium
Activator Concentration	8.5 mol%
Zn/Cd Ratio	NA
Usual Particle Size	7.4 - 13.2 µ
Body Color	White
CRT Brightness vs: P-20	12 %
Optimum Screen Weight	6.5 mg./cm^2

Y. Manufacture of P-28 Phosphor

The P-28 phosphor has a $ZnCdS_2$:Cu:Ag composition. Its decay is long and the Cd/Zn ratio has been adjusted so as to produce the main emission peak at 5500 Å. The formulation to be used for manufacture of the P-28 phosphor is:

5.3.60.- Formulation for a P-28 Phosphor

Compound	Mol Ratio	Assay	Grams	Lb/Lb
ZnS	0.500	0.995	48.96	0.395
$(Zn_{0.85}Cu_{0.15})S_2$	0.100	0.995	9.765	0.0788
$(Zn_{0.85}Ag_{0.15})S_2$	0.100	0.995	10.433	0.0841
CdS	0.360	0.995	43.56	0.351
$MgCl_2$	6.0 % by Weight		6.763	0.0545
NaCl	4.0 % by Weight		4.508	0.0364

The components are first blended, and hammermilled twice, with blending in between each hammermilling.

It is then fired in a 2 liter covered silica crucible for:

4 hours at 850 °C in air

using a covered 2-Liter silica crucible. After firing, when the phosphor cakes are cool, they are inspected and crushed. The phosphor is then screened through a 325 mesh screen, and is washed free of excess flux, using the same procedure described above and a silica coating is applied.

The P-28 phosphor has the following spectral characteristics, as shown in the following diagram:

5.3.61.-

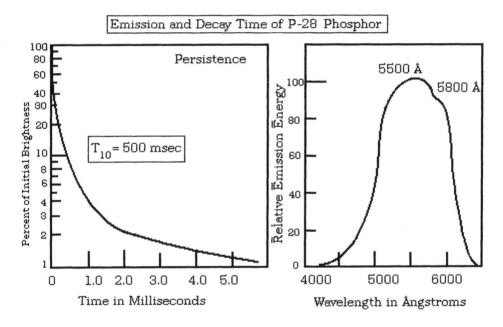

The physical and spectral properties are given in the following table, as shown on the next page.

The P-28 phosphor has been used mostly in radar displays. Note that the initial brightness and color is due mostly to the Ag⁻ activator, but its luminance soon decays, leaving the Cu⁻ activator emission as a long "tail" which remains clearly visible for several seconds.

Spectral and Decay Properties of P-28 Phosphor	
Emission Color	Yellowish green
Emission Peaks	5500 Å & 5800 Å
CIE Coordinates	x=0.370; y = 0.540
Persistence - $T_{10\%}$	500 milliseconds
Decay Class	Long
Activator Concentration	0.015 mol% Ag⁻ 0.015 mol% Cu⁻
Zn/Cd Ratio	0.36 Cd/0.64 Zn
Usual Particle Size	6.5 - 14.5µ
Body Color	Yellowish
CRT Brightness vs: P-20	63 %
Optimum Screen Weight	5.5 mg./cm^2

Z. Manufacture of P-29 Phosphor

The P-29 screen is no longer used. It consisted of two colors derived from a linear array of alternate strips or lines of P-2 and P-25 phosphors. CRT's incorporating this phosphor screen were used as indicators in aircraft as target indicators, navigational beacons, terrain clearance, elevation indicators, and collision course indicators.

AA. Manufacture of P-30 Phosphor

This screen was reserved and then the registration request was withdrawn.

BB. Manufacture of P-31 Phosphor

This phosphor is essentially a ZnS:Cu⁻ phosphor "quenched" by the addition of Ni^{2+} to lower the decay time. This phosphor has been used extensively as a display phosphor in screens used in black and white cathode ray tubes.

The formulation to be used in its manufacture is shown on the next page as 5.2.63. The Ni^{2+} added to diminish the normal long decay of the Cu⁻ activator. The green emission band is close to the maximum response of the human eye and therefore presents a high screen brightness.

5.3.62.- Formulation for a P-31 Phosphor

Compound	Mol Ratio	Assay	Grams	Lb/Lb
ZnS	0.800	0.995	78.34	0.741
$(Zn_{0.88}Cu_{0.12})S_2$	0.100	0.995	9.771	0.0925
$(Zn_{0.933}Ni_{0.067})S_2$	0.100	0.995	9.748	0.0922
$MgCl_2$	2.0 % by Weight		1.957	0.0185
NaCl	3.0 % by Weight		2.936	0.0278
$BaCl_2$	3.0 % by Weight		2.936	0.0278

These components are blended together, and then hammermilled twice, with blending in between the hammermilling steps. The mix is then placed in a covered silica crucible and fired for:

3 hours @ 1135 °C. in air

After firing, when the phosphor cakes are cool, they are inspected under ultraviolet light and crushed. This green-emitting phosphor is then sifted through a 325 mesh screen, and is washed free of excess flux, using the same procedure described above. Finally, a silica coating is applied, using the process described above.

Some Manufacturers add a small amount of CdS to the above formulation to shift the emission as near as possible to the CIE y-coefficent peak of the curve, i.e.- a 5550 Å peak. In that case, the formulation to be used is:

5.3.63.- Formulation for a High-Brightness P-31 Phosphor

Compound	Mol Ratio	Assay	Grams	Lb/Lb
ZnS	0.750	0.995	78.34	0.703
CdS	0.050	0.995	5.724	0.0514
$(Zn_{0.88}Cu_{0.12})S_2$	0.100	0.995	9.771	0.0877
$(Zn_{0.933}Ni_{0.067})S_2$	0.100	0.995	9.748	0.0875
$MgCl_2$	2.0 % by Weight		1.957	0.0178
NaCl	3.0 % by Weight		2.936	0.0264
$BaCl_2$	3.0 % by Weight		2.936	0.0264

This mix is processed in a manner identical to that given above. The P-31 phosphor has the following spectral characteristics:

5.3.64.-

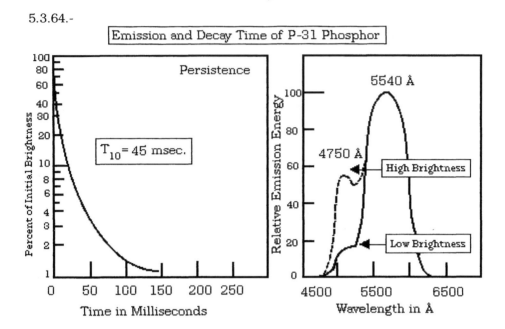

Emission and Decay Time of P-31 Phosphor

Note that another peak is evident in 5.3.64., in addition to the main peak at 5540 Å. This may be due to the Cu⁻ activator in a cubic environment in the crystal, since the firing temperature is intermediate to that temperature which causes formation of the hexagonal phase of ZnS.

The P-31 phosphor is a good example of the conventional display phosphor. It is very efficient and has good visibility and clarity to the human eye. It has therefore been used in computer screens as well as other display screens.

However, many manufacturers prefer to use the P-1 phosphor for a display screen since it tends not to burn to the same degree as does the P-31 phosphor under the same operating conditions.

The physical and spectral properties of the P-31 phosphor are summarized in the following table:

Spectral and Decay Properties of P-31 Phosphor	
Emission Color	Green
Emission Peaks	4750 Å & 5540 Å
CIE Coordinates	x=0.248; y = 0.583
Persistence - $T_{10\%}$	45 milliseconds
Decay Class	Medium
Activator Concentration	1.2 mol%
Zn/Cd Ratio	NA (may be: 0.05 CdS/0.95 ZnS)
Usual Particle Size	6.5 - 14.5µ
Body Color	White
CRT Brightness vs: P-20	114 %
Optimum Screen Weight	5.0 mg./cm^2

Note that we have given the usual formulation and method of firing generally used for the P-31 phosphor. However, each manufacturer has his own methods, developed over the period of several years to manufacture this phosphor.

CC. Manufacture of P-32 Phosphor

The P-32 screen is another cascade screen used primarily in military radar installations. It is composed of the blue-emitting $CaMg(SiO_3)_2$:Ti phosphor used in P-4 screens and a $ZnCdS_2$:Cu phosphor having a very long decay. The preparation of these phosphors uses the following formulations:

5.3.65.- Manufacture of P-32 Phosphor Components

1. Blue-Emitting Component

Compound	Mol Ratio	Assay	Grams	Lb/Lb
$CaCO_3$	1.05	0.986	106.59	0.293
$3MgCO_3 \bullet Mg(OH)_2 \bullet 3H_2O$	0.95 MgO	44.6% MgO	85.85	0.236
TiO_2	0.058	0,985	4.705	0.0129
H_2SiO_3	1.83	75.05% SiO_2	146.50	0.403
NH_4Cl	0.168	----------	8.987	0.0247
$(NH_4)_2SO_4$	0.081	----------	10.703	0.0295

2. Yellow Emitting Component

Compound	Mol Ratio	Assay	Grams	Lb/Lb
ZnS	0.621	0.995	60.81	0.548
$(Zn_{0.68}Cu_{0.32})S$	0.275	0.995	26.768	0.2414
CdS	0.104	0.995	15.100	0.1362
$MgCl_2$	6% by weight of sulfides		6.161	0.0556
$LiSO_4$	2% by weight of sulfides		2.054	0.0185

The yellow-emitting sulfide phosphor is fired in air in a 2 liter silica crucible for:

3.0 hours at 915 °C.

The blue-emitting silicate is fired in air in a 2 liter silica crucible for:

2.50 hours at 1180 °C.

When the phosphor cakes are cool, they are inspected and crushed, taking care to keep the two phosphors separate from one another.

The blue-emitting silicate is then screened through a 325 mesh screen, and is then milled in water, using a 400 gram load in a one-quart mill along with 150 ml. of water. The slurry is filtered and then is suspended in water. A silica coating is applied, as described above.

The yellow-emitting sulfide is first screened through a 325 mesh screen, and then is washed free of excess flux, using the same procedure described above for the P-2 phosphor. The final step is the application of a silica-coating, as described above. When dry, the body-color of the blue-emitting phosphor will be white while that of the yellow-emitting phosphor will have a slight yellowish cast. This is due in part to the florescence caused by exposure to daylight, as previously described for the $Y_3Al_5O_{11}$: Ce phosphor in the last chapter. The screen is prepared by first settling the yellow sulfide, followed by the blue-emitting silicate phosphor.

The spectral characteristics of the P-32 screen are shown in the following diagram, given on the next page as 5.3.66.

5.3.66.-

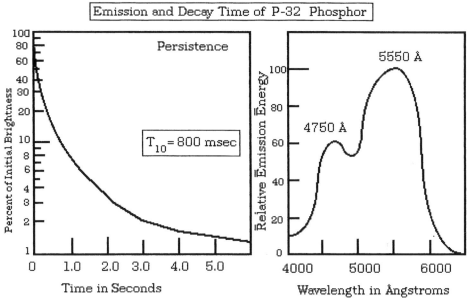

Note that a long "tail" is evident in the decay of the screen. Note also that the decay color is a yellowish green and that the phosphor emission has been adjusted to coincide with the maximum wavelength response of the Human eye, i.e.- the maximum of the CIE y-coordinate curve. Thus, this screen is actually an improved P-7 type of screen.

The screen is prepared by first settling the yellow emitting phosphor, and then settling the blue screen on top of that. Then, the whole is dried, lacquered and aluminized as described previously before the electron gun is mounted, the tube sealed off and the end-cap put into place to complete the manufacture of the radar tube.

A summary of the spectral and physical properties of the P-32 phosphor is given in the following table, viz-

Spectral and Decay Properties of P-32 Phosphors	
Emission Color	Purplish Blue (Phosphorescence is a greenish yellow)
Emission Peaks	4750 Å & 5550 Å
CIE Coordinates	Blue: x = 0.170; y = 0.124 Yellow: x = 0.340; y = 0.515
Persistence - $T_{10\%}$	Blue: 75 microseconds. Yellow: 800 milliseconds.
Decay Class	Blue: Short Yellow: Very Long
Activator Concentration	Blue = 0.058 Ti^{3+}/mol Silicate Yellow =8.8 mol% Cu^- per mol of $ZnCdS_2$
Zn/Cd Ratio	Blue: NA Yellow: 90.6 Zn/10.4 Cd
Usual Particle Size	Blue = 31 μ; Yellow = 36 μ
Body Color	White(very slight yellow)
CRT Brightness vs: P-20	74 %
Optimum Screen Weight	5.5 mg./cm^2

DD. Manufacture of P-33 Phosphor

This screen is an improved version of the P-26 phosphor. Its emission is orange and the decay is long. The formulation used in its preparation is given in the following formulation:

5.3.67.- Formulation Used for P-33 Phosphor

Compound	Mol Ratio	Assay	Grams	Lb/Lb
KF	0.90	0.895	58.42	0.418
MgF_2	1.035	0.875	71.20	0.510
MnF_2	0.065	0.945	5.4089	0.0387
NaF	0.100	0.896	4.686	0.0336

These materials must be as dry as possible before they are weighed and blended.
They are then hammermilled twice, with blending between hammermillings.

The resulting mix is then fired in a 2 liter covered silica crucible for:

3.5 hours at 885 °C. in air

After firing, when the phosphor cakes are cool, they are inspected and crushed. A 3650 Å lamp is used even though this phosphor is not very responsive to ultraviolet excitation. The powder is then sifted through a 325 mesh screen and then is suspended in water, followed by silicizing, filtering the wet cake and drying it for 16 hours at 125 °C to obtain a free-flowing powder.

The spectral properties are shown in the following diagram:

5.3.68.-

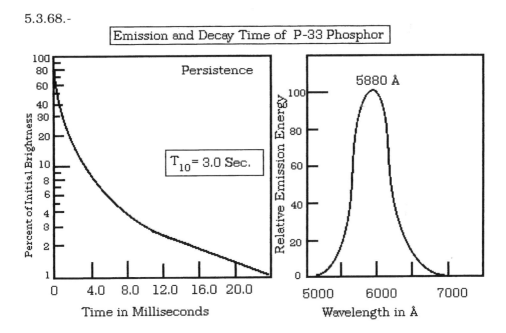

This phosphor is not as stable as the P-26 phosphor and tends to "burn" as the screen is being used. The emission is a strong orange and the decay is very long. The P-33 phosphor also has a "long-tail" in its decay like P-26, but the "tail" is much longer. Thus, its main use has been in radar displays where viewing of the image for several minutes is desired, after the CRT excitation has ceased.

The following table shows the physical and spectral properties of the P-33 phosphor.

Spectral and Decay Properties of P-33 Phosphor	
Emission Color	Orange
Emission Peaks	5860 Å
CIE Coordinates	x=0.559; y = 0.440
Persistence - $T_{10\%}$	3.0 seconds
Decay Class	Very long
Activator Concentration	6.5 mol%
Zn/Cd Ratio	NA
Usual Particle Size	8.5 - 19.5μ
Body Color	White
CRT Brightness vs: P-20	14 %
Optimum Screen Weight	2.0 mg./cm^2

EE. Manufacture of P-34 Phosphor

This phosphor has a composition, $(Zn^{2+}, Pb^{2+})S:Cu^-$, which was originally developed to produce a screen in which the decay is slow, but more importantly in which the image could be maintained and/or restored by application of infra-red radiation after the excitation beam was stopped.

The P-34 phosphor can be manufactured using the following formulation:

5.3.69.- Formulation for a P-34 Phosphor

Compound	Mol Ratio	Assay	Grams	Lb/Lb
ZnS	0.620	0.995	60.72	0.526
$(Zn_{0.75}Pb_{0.25})S_2$	0.200	0.995	26.712	0.2312
$(Zn_{0.85}Cu_{0.15})S_2$	0.180	0.995	17.578	0.1522
$MgCl_2$	6.0 % by Weight		6.301	0.0545
NaCl	4.0 % by Weight		4.200	0.0364

This mix is first blended together, then hammermilled twice, with blending in between hammermillings. It is fired in a separate furnace in air, using a 2 liter silica crucible with a lid, for:

3.5 hours at 915 °C

After firing, the cakes are inspected under "Blacklight" and the cakes are crushed. The powder is then sifted through a 325 mesh screen. The phosphor is then suspended in water and washed to remove excess flux. It is then filtered and dried at 125 °C overnight.

Alternately, the powder is first milled in water, using the proportions of 400 gram of phosphor in a half-gallon mill (equipped with about 8 lbs. of 1/2 inch flint stones, i.e.- the mill is filled to about 50% of its total volume) with 150 ml. of water added. The mill-time is 1.5 hours at about 20-25 rpm. At the end of the milling, the phosphor slurry is decanted and the mill is washed with water to remove all of the powder. The phosphor is then washed to remove all excess flux. The slurry is then decanted and filtered. The wet cake is then dried at 125 °C for 16 hours. The spectral properties of this screen are shown in the following diagram:

5.3.70.-

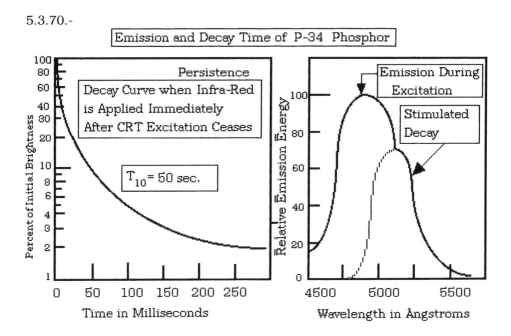

The spectral and physical properties of the P-34 phosphor are given in the table on the following page, viz-

Spectral and Decay Properties of P-34 Phosphor	
Emission Color	Greenish blue to Green
Emission Peaks	4750 Å and 5250 Å
CIE Coordinates	x=0.559; y = 0.440
Persistence - $T_{10\%}$	3.0 seconds
Decay Class	Very long
Activator Concentration	Pb^{2+} = 5.0 mol%
	Cu^- = 2.7 mol%
Zn/Cd Ratio	NA
Usual Particle Size	8.5 - 19.5μ
Body Color	White
CRT Brightness vs: P-20	88 %
Optimum Screen Weight	4.5 mg./cm^2

The emission is greenish blue with the stimulated emission by infra-red having a green color. The decay is very long with a T_{10} of 3.0 seconds. This phosphor has found use mostly in radar screens where the objects need to viewed over a time period of several seconds. As stated, this is accomplished by application of infra-red radiation to the screen after it has been excited in the CRT.

FF. Manufacture of P-35 Phosphor

The P-35 screen was composed of a $Zn(S,Se):Ag^-$ composition. It is prepared using the following formulation:

5.3.71.- Formulation for a P-35 Phosphor

Compound	Mol Ratio	Assay	Grams	Lb/Lb
ZnS	0.820	0.995	80.30	0.1845
$(Zn_{0.85}Ag_{0.15})S_2$	0.180	0.995	17.578	0.0404
ZnO	1.00	0.986	82.53	0.1896
Se	3.00	-------	236.88	0.5441
$MgCl_2$	6.0 % by Weight		10.824	0.0249
NaCl	4.0 % by Weight		7.216	0.0165

Note that this formulation contains a mixture of ZnO and Se. These react to form ZnSe according to the following reaction:

5.3.72.- $2 ZnO + 3 Se \Rightarrow 2 ZnSe + SeO_2 \Uparrow$

The volatile SeO_2 is toxic and the firing must be carried out in a well-ventilated hood, equipped with an air-scrubber using a caustic wash. Even the soluble Na_2SeO_3 is toxic and must be disposed of carefully, by precipitation as an alkaline earth selenate. Although one could prepare the ZnSe directly by precipitation using H_2Se gas, the latter is so toxic to the human body that this procedure is almost never tried. It is the solid state method that is more amenable to atmospheric control of the firing products. Se itself is volatile and condenses within the furnace as a coating in the cooler parts of the furnace. Usually, one employs a tube furnace for this purpose so that the firing is done within a confined space, even though the firing is being carried out in air atmosphere.

The above mix is first blended together, then hammermilled twice, with blending in between hammermillings. This procedure needs to be carried out in an enclosed space to control the dust which contains Se-particles.

The mix is fired in a separate furnace in air, using a 2 liter silica crucible with a lid, for:

3.5 hours at 935 °C

After firing, the cakes are inspected under "Blacklight" and the cakes are crushed. The powder is then sifted through a 325 mesh screen. The phosphor is then suspended in water and washed to remove excess flux. It is then filtered and dried at 125 °C overnight. Sometimes, a silica coating is applied before this phosphor is processed to form the CRT screen.

The following diagram, given on the next page as 5.3.73., shows the spectral properties of the P-35 phosphor. Note that two emission bands are present. the 4660 Å peak is likely to be due to the Ag^- activator with sulfide nearest neighbors, while that at 5200 Å is probably due to the Ag^- activator in a site with selenide nearest neighbors.

616

5.3.73.-

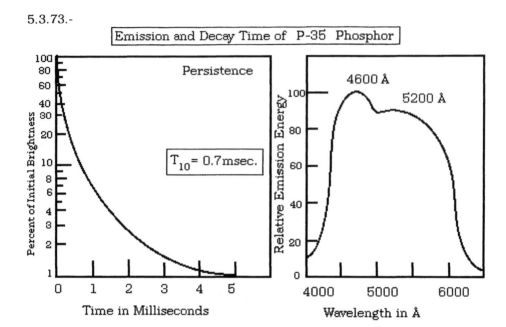

The spectral and physical properties of this phosphor is summarized in the following table, viz-

Spectral and Decay Properties of P-35 Phosphors	
Emission Color	Yellow Green
Emission Peak	4660 Å
CIE Coordinates	x = 0.200 ; y = 0.245
Persistence - $T_{10\%}$	0.70 milliseconds.
Decay Class	Medium Short
Activator Concentration	0.027 mol Ag⁻/ mol
Zn/Cd Ratio	NA
Usual Particle Size	7-10 µ
Body Color	Almost White
CRT Brightness vs: P-20	48 %
Optimum Screen Weight	2.8 mg./cm²

This is the only sulfo-selenide phosphor registered with JEDEC, due perhaps to the dangers of preparation encountered. Needless to say, the P-35 phosphor is not used in the United States anymore, and is rarely used in Europe.

GG. Manufacture of P-36 Phosphor

When a need for a fast decay screen with a green emission arose, the P-36 phosphor was developed. It is used mostly in display screens and has a $ZnCdS_2$:Ag⁻:Ni composition and an emission band at 5320 Å. This phosphor is manufactured using the following formulation:

5.3.74.- Formulation for a P-36 Phosphor

Compound	Mol Ratio	Assay	Grams	Lb/Lb
ZnS	0.440	0.995	43.09	0.347
CdS	0.360	0.995	52.27	0.421
$(Zn_{0.715}Ag_{0.285})S_2$	0.100	0.995	9.771	0.0787
$(Zn_{0.933}Ni_{0.067})S_2$	0.100	0.995	9.748	0.0786
$MgCl_2$	2.0 % by Weight		2.298	0.0185
NaCl	3.0 % by Weight		3.446	0.0278
$BaCl_2$	3.0 % by Weight		3.446	0.0278

These components are blended together, and then hammermilled twice, with blending in between the hammermilling steps. The mix is then placed in a covered silica crucible and fired for:

3 hours @ 1035 °C. in air

After firing, when the phosphor cakes are cool, they are inspected under ultraviolet light and crushed. This green-emitting phosphor is then sifted through a 325 mesh screen and then is washed to remove all of the excess flux. Finally, it is silicized, using the procedure outlined above.

The spectral properties are shown in the following diagram, given as 5.3.75. on the next page. The emission is green and the decay time is very fast.

A summary of the spectral and physical properties of the P-36 phosphor is given in the following table, which is also on the next page.

618

5.3.75.-

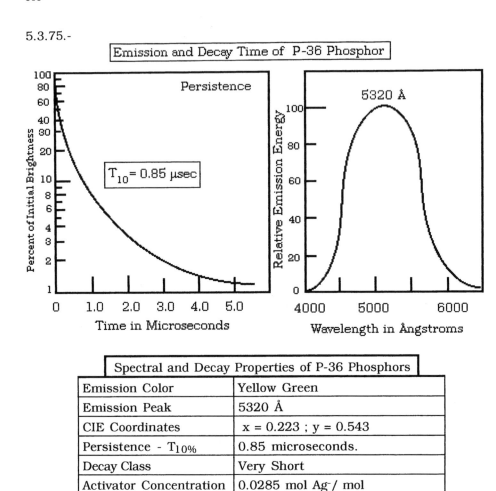

Emission and Decay Time of P-36 Phosphor

Persistence

T_{10} = 0.85 µsec

Percent of Initial Brightness

Time in Microseconds

Relative Emission Energy

5320 Å

Wavelength in Ångstroms

Spectral and Decay Properties of P-36 Phosphors	
Emission Color	Yellow Green
Emission Peak	5320 Å
CIE Coordinates	x = 0.223 ; y = 0.543
Persistence - $T_{10\%}$	0.85 microseconds.
Decay Class	Very Short
Activator Concentration	0.0285 mol Ag$^-$/ mol 0.0067 mol Ni^{3-}/ mol
Zn/Cd Ratio	0.64 Zn/0.36Cd
Usual Particle Size	7-10 µ
Body Color	Yellowish
CRT Brightness vs: P-20	104 %
Optimum Screen Weight	2.1 mg./cm^2

HH. Manufacture of P-37 Phosphor

This phosphor has a ZnS:Ag:Ni composition. It is used as a fast decay blue-emitting phosphor for display purposes. It is manufactured using the following formulation:

5.3.76.- Formulation for a P-37 Phosphor

Compound	Mol Ratio	Assay	Grams	Lb/Lb
ZnS	0.750	0.995	73.47	0.679
CdS	0.050	0.995	7.256	0.0670
$(Zn_{0.715}Ag_{0.285})S_2$	0.100	0.995	9.771	0.0903
$(Zn_{0.933}Ni_{0.067})S_2$	0.100	0.995	9.748	0.0900
$MgCl_2$	2.0 % by Weight		2.005	0.0185
NaCl	3.0 % by Weight		3.007	0.0278
$BaCl_2$	3.0 % by Weight		3.007	0.0278

Some Manufacturers do not use CdS in their formulation. Then, the total amount of ZnS used is 0.80 mol. These components are blended together, and then hammermilled twice, with blending in between the hammermilling steps. The mix is then placed in a covered silica crucible and fired for:

4 hours @ 1135 °C. in air

After firing, when the phosphor cakes are cool, they are inspected under ultraviolet light and crushed. This blue-emitting phosphor is then sifted through a 325 mesh screen and then is washed to remove all of the excess flux. Finally, it is silicized, using the procedure outlined above.

The spectral properties of the P-37 screen are shown in the following diagram, given as 5.3.77. on the next page. Note that the emission peak occurs at 4660 Å and the color is a greenish-blue. The following table, also presented on the next page, summarizes the properties of the P-37 phosphor.

This phosphor is actually an improved version of the P-11 phosphor. It has been found that the addition of Al^{3+} to the ZnS composition has a definitive effect upon its decay properties under CRT excitation. This effect was discovered in the early 1960's. Since then, it has become quite common to add Al^{3+} to a given sulfide phosphor formulation, especially since the spectral characteristics remain unchanged. Just the decay properties are affected. The reason for this lies in the fact that a trivalent cation is being added to the divalent ZnS matrix which becomes an interstitial and prevents the formation of a cation vacancy.

620

5.3.77.-

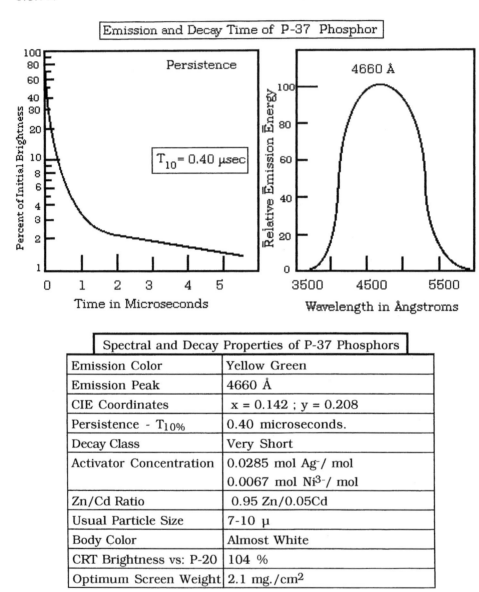

Emission and Decay Time of P-37 Phosphor

Spectral and Decay Properties of P-37 Phosphors	
Emission Color	Yellow Green
Emission Peak	4660 Å
CIE Coordinates	x = 0.142 ; y = 0.208
Persistence - $T_{10\%}$	0.40 microseconds.
Decay Class	Very Short
Activator Concentration	0.0285 mol Ag$^-$/ mol 0.0067 mol Ni^{3-}/ mol
Zn/Cd Ratio	0.95 Zn/0.05Cd
Usual Particle Size	7-10 µ
Body Color	Almost White
CRT Brightness vs: P-20	104 %
Optimum Screen Weight	2.1 mg./cm^2

Since the activator usually ends up as Ag$^-$ or Cu$^-$ on a sulfide site in the ZnS lattice, this addition has the effect of charge-compensating the activator, e.g.-

$$(Zn, Zn^+{}_I) S : Ag^-{}_S= \quad or: \quad (Zn, Al^+{}_I) S : Ag^-{}_S=$$

where the Ag^- activator occupies a sulfide lattice site. The decay properties are thus affected as:

$$(Zn, V_{Zn})S: Ag^-_S= \qquad T_{10} = 27 \text{ milliseconds}$$
$$(Zn, Al)S: Ag^-_S= \qquad T_{10} = 2.0 \text{ milliseconds}$$
$$(Zn, Ni)S: Ag^-_S= \qquad T_{10} = 0.18 \text{ milliseconds}$$

It is the vacancy which has the effect of prolonging the decay time of the phosphor. It is much more difficult to add Ni^{2+} to control the decay time than to add Al^{3+} to the formulation because of the extreme criticality of the Ni^{2+} concentration in the ZnS composition. If $AlCl_3$ is used, it works both as a flux and an decay-control agent in the composition. The P-37 formulation to be used in this case is:

5.3.78.- Formulation for a P-37 Phosphor

Compound	Mol Ratio	Assay	Grams	Lb/Lb
ZnS	0.800	0.995	78.34	0.7412
$(Zn_{0.715}Ag_{0.285})S_2$	0.100	0.995	9.771	0.0925
$(Zn_{0.933}Ni_{0.067})S_2$	0.100	0.995	9.748	0.0922
$MgCl_2$	2.0 % by Weight		1.957	0.0185
NaCl	3.0 % by Weight		2.936	0.0278
$AlCl_3$	3.0 % by Weight		2,936	0.0278

This mix is handled as given above and is fired in an identical manner. As stated, the spectral properties are identical, but the decay time is lower by a factor of 10 or more. Note that the addition of Al^{3+} will have the same effect on the decay properties of the Cu^- activator in both ZnS and $ZnCdS_2$ hosts.

II. Manufacture of P-38 Phosphor

This phosphor was developed as an improved P-21 phosphor which in turn had been developed as an improved P-12 phosphor. The P-38 phosphor has a composition composed mostly of MgF_2 with some ZnF_2 present (Note that this is almost the exact opposite of the P-19 phosphor).

The P-38 phosphor is manufactured using the following formulation:

5.3.79.- Formulation for P-38 Phosphor

Compound	Mol Ratio	Assay	Grams	Lb/Lb
MgF_2	0.815	0.985	7.590	0.0284
$ZnF_2 \cdot 4H_2O$	0.150	0.579	256.04	0.9591
MnF_2	0.035	0.975	3.336	0.0125

These components need to be as dry as possible. Even then, they are slightly hygroscopic and need to be assayed just before use. The components are mixed together by blending and then hammermilled twice, with blending in between. The mix is then placed in a 2 liter silica crucible and fired in air using a covered crucible for:

3.5 hours at 885 °C

After the cakes are cool, they are inspected and discolored surface areas are removed. Since this phosphor does not respond to ultraviolet light excitation, ordinary light is used during the inspection procedure. The cakes are then crushed, sieved through a 325 mesh screen and then washed (if flux is used). Finally, the powder is suspended in water and then silica coated.

The spectral properties are shown in the following diagram:
5.3.80.-

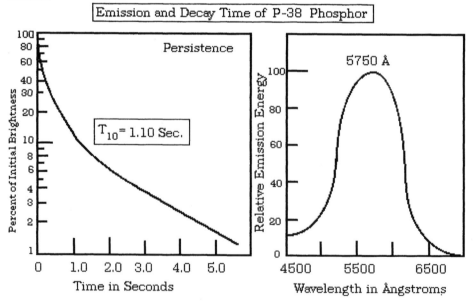

The spectral and physical properties are summarized in the following table:

Spectral and Decay Properties of P-38 Phosphor	
Emission Color	Orange
Emission Peak	5750 Å
CIE Coordinates	x = 0.570 ; y = 0.420
Persistence - $T_{10\%}$	1.10 seconds.
Decay Class	Very Long
Activator Concentration	0.035 mol Mn^{2+}/ mol
Zn/Cd Ratio	NA
Usual Particle Size	11-19 μ
Body Color	White
CRT Brightness vs: P-20	28 %
Optimum Screen Weight	6.0 mg./cm^2

If a longer decay is desired, the following formulation can be used:

5.3.81.- Formulation for P-38 Phosphor with a Longer Decay

Compound	Mol Ratio	Assay	Grams	Lb/Lb
MgF_2	0.815	0.985	7.590	0.0281
$ZnF_2 \cdot 4H_2O$	0.150	0.579	256.04	0.9466
MnF_2	0.072	0.975	6.8629	0.0253

This composition is processed as given above. The spectral properties are shown in the following diagram, given on the next page as 5.3.82.

The P-38 phosphor was developed to replace the P-21 phosphor where a long decay phosphor was needed. These phosphors are based upon various alkali and alkaline earth fluorides. A comparison of their compositions is made in the following table, given as 7.3.83. It should be noted that the P-38 phosphor was one of the last radar phosphors to be registered.

Note also that two types of compositions have been employed for this type of screen, namely $KMgF_3$ and $ZnMgF_4$.

5.3.82-

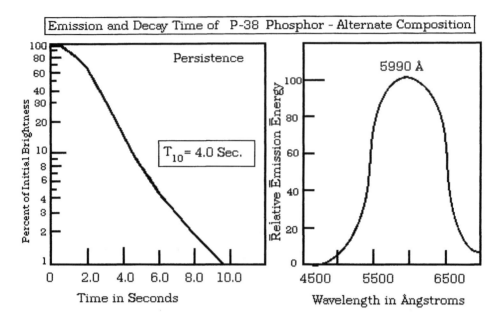

Emission and Decay Time of P-38 Phosphor - Alternate Composition

Persistence

T_{10} = 4.0 Sec.

5990 Å

7.3.83.- Comparison of Various Fluoride Based Phosphors

P-Number	Composition	Emission Peak	Decay to T_{10}
P-12	$Zn_{0.845}Mg_{0.120}F_4:Mn_{0.035}$	5880 Å	190 msec.
P-19	$K_{1.0}Mg_{1.0}F_3:Mn_{0.005}$	5850 Å	275 msec.
P-21	$Mg_{0.9455}Li_{0.050}F_3:Mn_{0.0045}$	5930 Å	800 msec.
P-26	$K_{1.0}Mg_{1.0}F_3:Mn_{0.055}$	5960 Å	3.0 sec.
P-33	$K_{0.90}Mg_{1.035}F_3:Mn_{0.0325}$	5880 Å	3.0 sec.
P-38	$Mg_{0.815}Zn_{0.150}F_4:Mn_{0.035}$	5750 Å	1.1 sec.

Compositions used to prepare such phosphors with suitable decay properties have included:

1) alkaline earth fluorides activated by Mn^{2+}

2) zinc and cadmium sulfides activated by Cu^-.

In both cases, these compositions were deliberately "overfired" to promote the formation of lattice-vacancies which serve as "traps" and therefore promote the very long decay times desired for this application.

JJ. Manufacture of P-39 Phosphor

This phosphor is essentially a P-1 phosphor to which As^{3+} has been added to promote a longer decay. The following formulation is used in its manufacture:

5.3.84.- Formulation for a P-39   Phosphor

	Mols/ Mol Phosphor	Weight in Grams	Lbs./Lb.
ZnO	2.00	162.68	0.633
SiO_2	1.13 (as SiO_2)	67.90	0.264
$MnCO_3$	0.18	20.69	0.0805
PbF_2	0.00816	2.000	0.00778
As_2O_3	0.0188	3.720	0.01448

This mix is blended as before, followed by hammermilling twice, with blending done in between hammermillings. This mix is then placed in a large 2-liter silica crucible and is fired for:

3 hours @ 1235 °C. in air

After firing, the phosphor cake is inspected under 2527 Å light, crushed and then sieved through a 325 mesh screen. It has been found that the addition of Pb^{2+} increases the luminous efficiency of the phosphor, as does the addition of As^{3+}. It is the As^{3+} that introduces a large increase in the decay of the phosphor, probably by introduction of vacancies which mandatorily form upon the introduction of a trivalent cation into a divalent cationic lattice.

The emission is green and the persistence is long as shown in the following diagram, given as 5.3.85. on the next page. Also shown on the next page is a table which summarizes the spectral and physical properties of the P-39 phosphor.

5.3.85.-

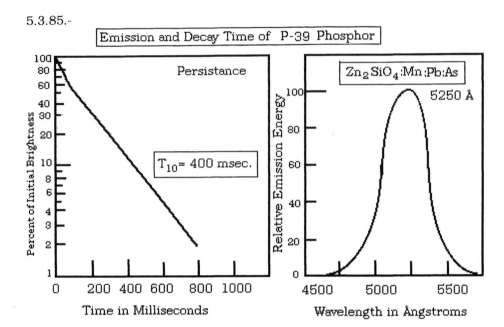

Emission and Decay Time of P-39 Phosphor

Spectral and Decay Properties of P-39 Phosphor	
Emission Color	Green
Emission Peak	5250 Å
CIE Coordinates	x=0.194 ; y = 0.710
Persistence - $T_{10\%}$	400 milliseconds
Decay Class	Medium
Activator Concentration	20 mol%
Zn/Cd Ratio	NA
Usual Particle Size	9.5 μ
Body Color	White
CRT Brightness vs: P-20	68 %
Optimum Screen Weight	4.0 mg/cm^2

Note that the emission is exactly the same as the P-1 phosphor, and that the only difference in the spectral properties of the P-39 phosphor, compared to that of the P-1 phosphor, is the persistence, or decay time. The decay time of the P-39 phosphor is increased by a factor of 20 times over that of the P-1 phosphor.

KK. Manufacture of P-40 Phosphor

This screen is composed of $ZnCdS_2:Ag$ and $ZnCdS_2:Cu$ and is actually an improved P-14 screen where the yellow component has been modified so that its peak emission occurs at 5500 Å while the blue component peaks at 4450 Å. This screen has also been used as a "penetration-phosphor" screen. For this type of screen, the voltage of the electron beam is varied so that the top layer is preferentially excited, but not the bottom layer. By increasing the instantaneous voltage, one can excite both layers simultaneously, thereby changing the emission color of the screen. This has certain advantages when this type of screen is used for display purposes in that specific data displayed can be highlighted.

The P-40 phosphors are manufactured using the following formulations.

5.3.86.- Formulations for the P-40 Phosphor

a. Blue-Emitting Component

Compound	Mol Ratio	Assay	Grams	Lb/Lb
ZnS	0.800	0.995	78.34	0.702
CdS	0.100	0.995	14.52	0.1301
$(Zn_{0.834}, Ag_{0.166})S$	0.100	0.995	10.50	0.0941
$MgCl_2$	6% by weight of sulfides		6.202	0.0556
$LiSO_4$	2% by weight of sulfides		2.067	0.0185

b. Yellow-green Emitting Component

Compound	Mol Ratio	Assay	Grams	Lb/Lb
ZnS	0.640	0.995	62.67	0.4786
$(Zn_{0.923}Cu_{0.077})S_2$	0.100	0.995	9.779	0.0747
CdS	0.360	0.995	52.27	0.3991
$MgCl_2$	3.0 % by Weight		3.741	0.0286
NaCl	2.0 % by Weight		2.494	0.0190

These components are mixed by blending, then hammermilled twice, with blending in between hammermillings. Care must be taken not to expose one formulation to the other so that cross-contamination does not occur. The blue-emitting phosphor formulation is particularly susceptible to this aspect. The blue-emitting sulfide is fired in air in a covered 2 liter silica crucible for:

3.5 hours at 1075 °C

The yellow-green emitting phosphor is fired in air in a 2 liter silica crucible for:

5.0 hours at 1150 °C.

using a cover on the crucible. When the phosphor cakes are cool, they are inspected and crushed. The blue-emitting phosphor is then screened through a 325 mesh screen, and then is suspended in water. A silica coating is applied as described above.

The other sulfide is first screened through 325 mesh screen, and then is washed free of excess flux, using the same procedure described above. When dry, the body-color of the blue-emitting phosphor will be white while that of the orange-emitting phosphor will have a slight yellowish cast.

The spectral properties of this screen are shown in the following diagram: 5.3.87.-

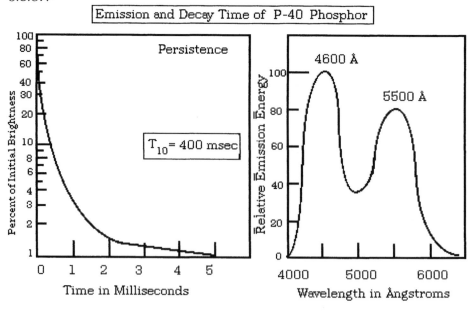

The spectral and physical properties of the P-40 screen are summarized in the following table, viz-

Spectral and Decay Properties of P-40 Phosphor	
Emission Color	Greenish blue to Green
Emission Peaks	4600 Å and 5500 Å
CIE Coordinates	x=0.276 ; y = 0.312
Persistence - $T_{10\%}$	3.0 seconds
Decay Class	Very long
Activator Concentration	Ag^- = 1.6 mol% Cu^- = 0.77 mol%
Zn/Cd Ratio	NA
Usual Particle Size	8.5 - 19.5µ
Body Color	White
CRT Brightness vs: P-20	88 %
Optimum Screen Weight	4.5 mg./cm^2

This screen was registered in early 1967 as an alternative to the P-34 screen.

LL. Manufacture of P-41 and P-42 Screens

The P-41 screen was composed of:

$$ZnMgF_2{:}Mn \text{ and } CaMg(SiO_3)_2{:}Ce$$

The latter is manufactured using a formulation very similar to that of the P-4 blue emitting phosphor except that 0.045 mol Ce^{3+} is substituted for the 0.035 mol of TiO_2. The resulting screen is an orange-emitting one having an ultraviolet component as well. It is used for triggering a light pen as part of a specialized display.

The P-42 screen was composed of:

$$ZnS{:}Cu \text{ and } ZnSiO_4{:}Mn^{2+}{:}As^{3+}$$

The manufacture of both of these phosphors was described above. The P-42 screen has been used for anti-flicker and low frame rate displays.

About this time, the JEDEC Registration Committee began to be inundated with requests for screen registration. From the period of 1967 to 1974, at least 20-25 more screens were actually registered, with many more requests still waiting. It was during this time that the following were registered:

$$P-43 = Gd_2O_2S:Tb^{3+}$$
$$P-44 = La_2O_2S:Tb^{3+}$$
$$P-45 = Y_2O_2S:Tb^{3+}$$
$$P-46 = Y_3Al_5O_{12}:Ce^{3+}$$
$$P-47 = Y_2SiO_5:Ce$$
$$P-48 = Y_3Al_5O_{12}:Ce^{3+} + Y_2SiO_5:Ce$$

This situation arose following the announcement of the $Y_2O_3:Eu^{3+}$ and $YVO_4:Eu^{3+}$ phosphors that revolutionized the P-22 screens, and color television in 1960-1964 by increasing the screen brightness by a factor of four to five. This led to a major effort in research concerning new phosphors, based mostly on Ln^{3+} activation of various hosts.

Thus, a great number of new types of CRT phosphors were developed during the period of 1964 to 1975. It was also during this period that many of the domestic CRT manufacturers in the United States ceased manufacture of color television tubes, due in part to the business climate that prevailed in Japan where labor costs of manufacture were lower and quality was high. Essentially, the U.S. manufacturers were forced out of the business because they could not meet either the prices of the Japanese manufacturers or their quality. However, a small number remained in the CRT manufacturing business supplying military tubes used in airplane cockpits as well as specialty CRT's for various display purposes.

5.4- OTHER CATHODE RAY PHOSPHORS BEING MANUFACTURED

About 1985, a change was brought about in the Registration format. A newer Committee was formed (JT-31 referred to above), and a new classification of CRT phosphors was initiated, based on both screen color and usage. Part of these applications arose because of new types of data display applications had arisen such as computer screens and the like. Many of the new registrations involved usage of prior known phosphors in new combinations. Rather than list all of these screens, we will further describe the new phosphor compositions that have been developed and will relate how such phosphors are manufactured.

A. Manufacture of $Ln_2O_2S:Ln^{3+}$ Phosphors

The host, Ln_2O_2S, has many interesting properties. First of all, the introduction of a small amount of sulfur as sulfide into the cubic defect-fluorite lattice causes the structure to change to that of hexagonal packing. This sulfide introduction into the Ln_2O_3 lattice also has a profound effect on the spectral properties of the phosphors produced. Although Ln_2O_2S, where Ln may be Y^{3+}, Gd^{3+} or La^{3+}, is not fluorescent itself, i.e.- it is not "self-activated", the introduction of various Ln^{3+} cations produces phosphors having very high efficiencies under cathode ray excitation. This is due to the presence of an additional charge-transfer (CT) band that extends from about 3400 Å to 2000 Å, as shown in the following diagram:

5.4.1.-

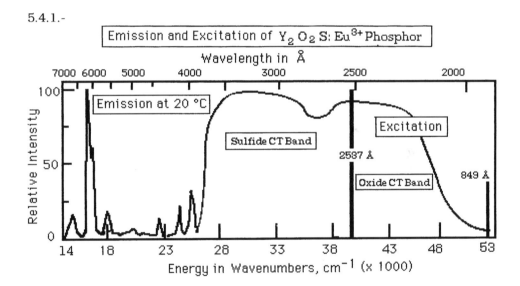

Note that two dominant bands are present at 3400 Å and 2400 Å. These are the sulfide and the oxide CT bands, respectively. The overall CT-band is due to the sulfide-oxide composition, and is always dominant in the spectrum regardless of the activator present. The Y_2O_3 host also supports efficient emission from a few activators, but its CT band is found nearer to 2100 Å. In contrast, the $LnVO_4$ composition has a CT-band extending from 3400 to 2000

Å, but supports only a limited number of such activators. Only a few have correspondingly good CRT response. It is well known (4) that Ln^{3+} activators respond remarkably well to excitation energy in a host which possesses a wide phonon energy spectrum. This occurs because of the many energy levels available to such activating cations are strongly coupled to the phonon spectrum via vibronic coupling at the activator site in the crystal. That this is true is supported by the large number of activators observed in the Ln_2O_2S composition having efficient CRT response. Note that we have already discussed the role of the host in such diverse phosphors as $YVO_4:Dy^{3+}$ wherein it was shown that this phosphor exhibits constant-quantum efficiency due to a CT type of activation energy which is transferred via vibronic coupling from the vanadate to the activator after the host lattice has been excited by the electron beam via phonon absorption processes.

The general method of preparing oxysulfide phosphors has already been given. In general, one prepares the mixed oxalates of the lanthanides and then fires them to obtain the corresponding oxides in solid solution. One then mixes these oxides with sulfur and refires this mixture. The following presents a representative formulation for one of these new phosphors, viz-

5.4.2.- Formulation for Gd_2O_2S: Tb^{3+} Phosphor

Compound	Mol Ratio	Assay	Grams	Lb/Lb
$Gd_2O_3:Tb_{0.075}$	0.90	------	344.05	0.991
S	0.10	0.985	3.2548	0.00937

This mix is then hammermilled together, using the same techniques already described before. The mix is then fired in air in a covered silica crucible for:

3.0 hours at 975 °C

After firing, the crucibles and their contents are cooled in air, the phosphor cakes are then crushed, sieved through a 325 mesh screen, suspended in water, and silicized as described above.

We have already described the manufacture of the oxalates, using yttrium as the example. In the following table is shown the relative amounts of oxides required to form the oxalate precipitates required for each formulation:

5.4.3.- Relative Quantities to be Used for Oxalate Precipitations

Composition	Weight	Activator	Mol Ratio	Weight
Y_2O_3	225.81 gm.	Eu_2O_3	0.065 Eu^{3+}	11.437 gm.
La_2O_3	325.81	Tb_4O_7	0.055 Tb^{3+}	10.281
Gd_2O_3	362.50	Pr_6O_{11}	0.0155 Pr^{3+}	2.639
		Sm_2O_3	0.025 Sm^{3+}	4.360

To make a 250 gallon production batch as described in Chapter 4, page 458, one multiplies the above weights by 227.2 and the exact same procedures are followed, including the firing temperatures used to form the mixed oxides.

The general procedure to be used is:

1. Rare Earth Solution: A HNO_3 solution is prepared by adding 23.5 gallons of concentrated HNO_3 to 250 gallons of deionized water in a 500 gallon glass-lined tank. This forms a 1.200 molar solution. To this is added the requisite rare earth oxides to form a soluble solution.

2. Oxalic Acid Solution: To 250 gallons of deionized water in a 300 gallon tank is added 167.71 kg. (369 lbs. 9 oz.) of oxalic acid with stirring. The solution needs to be about 40-50 °C in order to dissolve all of the oxalic acid. When the dissolution is complete, 65 gallons of ethyl alcohol are added to this volume. The solution is then heated to 80 °C. just before precipitation.

3. The hot (80 °C.) oxalic acid solution is then pumped into the Y^{3+} and Eu^{3+} solution at a rate of about 5.0 gallons per minute while stirring. The precipitate is allowed to settle and the mother liquor is removed by decantation. One water wash is given to the precipitate by resuspension. The suspension is then filtered on a drum-filter using a water wash wherein the cake is sprayed on the filter. The wet precipitate is then dried for 16 hours at 125 °C.

4. The dried oxalates are then fired 6 hours at 1275 °C in an open silica tray. At the end of the firing period, the trays are allowed to cool and the cakes are inspected, crushed and then sieved through a 325 mesh

screen to remove particles larger than about 44 μ. This product is then used to prepare the oxysulfide.

The spectral properties of the Tb^{3+} activated phosphors are given in the following diagram:

5.4.4.-

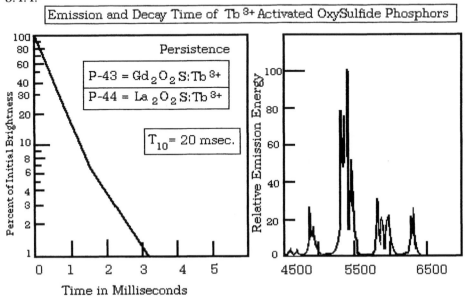

Emission and Decay Time of Tb $^{3+}$ Activated OxySulfide Phosphors

Persistence

P-43 = $Gd_2O_2S{:}Tb^{3+}$

P-44 = $La_2O_2S{:}Tb^{3+}$

T_{10} = 20 msec.

Percent of Initial Brightness

Relative Emission Energy

Time in Milliseconds

Note that these phosphors have the same emission lines in all cases, as well as the same decay curve. The decay time at T_{10} is 1.2 milliseconds. The green emission seen is typical for Tb^{3+} which shows four sets of lines centered at about 4950 Å, 5495 Å , 5890 Å and 6230 Å. The only difference to be seen is the overall brightness, the gadolinium-based phosphor being about 135% brighter than the lanthanum-based oxysulfide.

These phosphors were registered as :

	Composition	Decay Class	Emission Peak	Color
P-43	$Gd_2O_2S{:}Tb^{3+}$	Medium Short	5445 Å	x = 0.333; y = 0.556
P-44	$La_2O_2S{:}Tb^{3+}$	Medium Short	5475	x = 0.300; y = 0.596
HDA	$Y_2O_2S{:}Tb^{3+}$	Medium Short	5445	x = 0.341; y = 0.586

A major use of these phosphors has been in color television. At very high current densities, i.e.- > 0.10 amp./cm^2, the P-43 and P-44 phosphors surpass the zinc and cadmium sulfide phosphors normally used. Thus, $Y_2O_2S:Tb^{3+}$ and $Gd_2O_2S:Tb^{3+}$ can be used for projection color television sets where the CRT is run at very high current densities and high screen brightness. Also falling into this category is the phosphor $YPO_4:Tb^{3+}$ which has identical spectral and luminous output under excessive beam currents. However, the optimum Tb^{3+} concentration is 0.26 mol Tb^{3+}/ mol YPO_4 , which makes it too expensive as compared to the oxysulfide phosphors.

The phosphor, $Gd_2O_2S:Tb^{3+}$, has also found extensive use as an x-ray intensifying screen for use either in a fluoroscope or with x-ray film. Because of the strong x-ray absorption of the Gd^{3+} cation, a layered film of this phosphor is put in close contact with the unexposed x-ray film. The x-ray dosage required for complete exposure of the x-ray film is then cut by a factor of up to ten. Neutron radiography is also made possible by $Gd_2O_2S:Tb^{3+}$ layers incorporated within a plastic film which is put between the neutron source and the part to be analyzed.

An analogous phosphor can be prepared using a lower molar ratio of Tb^{3+} in the phosphor, i.e.- $Y_2O_2S: Tb_{0.0050}$. As shown in the following diagram, given on the next page as 5.4.5., the emission spectrum looks quite different from that given above in 5.4.4.

Note that several sets of lines are present in this spectrum. The composition, as indicated is $Y_2O_2S: Tb_{0.0050}$ where the Tb^{3+} content is 1/10 that shown in 5.4.3. above. The prominent emission lines observed are centered at: 3850 Å , 4200 Å, 4600-4800 Å, 5450 Å and 5800 Å. Not shown is a set at 6250 Å having a relative intensity of 20%. The screen has a resulting whitish appearance with color coordinates of:

$$x = 0.253 ; \ y = 0.312$$

The usual Tb^{3+} spectrum consists of sets of lines at :

4950 Å	5450 Å
5890 Å	6230 Å.

5.4.5.-

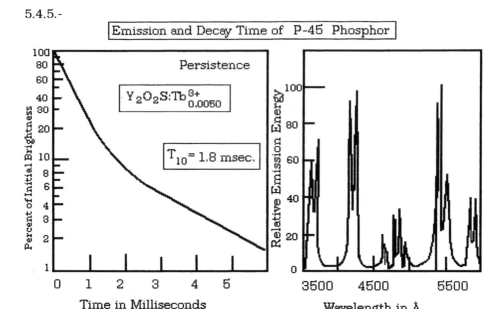

To explain this difference, consider the following: Tb^{3+} has two possible emitting states, $^5D_0$ and $^5D_1$. The transition from excited to ground state at high Tb^{3+} concentrations is $^5D_0 \Rightarrow {}^7F_J$. However, at low Tb^{3+} concentrations, emission comes from both the $^5D_0$ and $^5D_1$ excited states, resulting in the spectrum shown above.

Still another oxysulfide phosphor that has been used is $Y_2O_2S:Pr^{3+}$. The spectral properties of the composition indicated in 5.4.3. are given in the following diagram, given as 5.4.6. on the next page.

This screen is a light green with a medium short decay. It is used for green traces on a CRT screen where both fast decay and line emission is needed. Its use as a component in a two or three-color screen with a liquid-crystal color shutter has allowed generation of multi-color displays. The phosphors are chosen based on their decay characteristics so that the shutter can discriminate among them, and allow specific colors to be observed upon the instantaneous viewing screen. Although the $Y_2O_2S:Pr^{3+}$ phosphor does not have the equivalent efficiency and brightness of its relative, the $Y_2O_2S:Tb^{3+}$ phosphor, its decay characteristics are superior for this application.

5.4.6.-

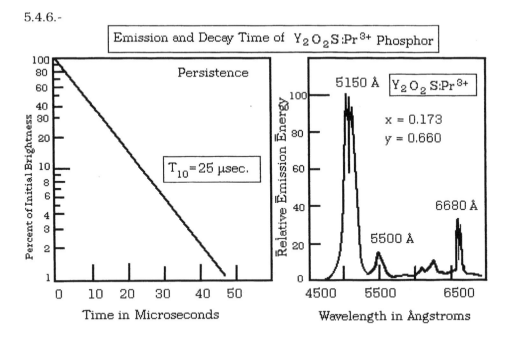

Emission and Decay Time of $Y_2O_2S:Pr^{3+}$ Phosphor

Persistence

$T_{10} = 25 \mu sec.$

5150 Å $Y_2O_2S:Pr^{3+}$

x = 0.173
y = 0.660

6680 Å

5500 Å

Time in Microseconds

Wavelength in Angstroms

Still another oxysulfide phosphor used as a CRT screen is the $Y_2O_2S:Tb^{3+}:Sm^{3+}$ composition. It is manufactured using the following formulation:

5.4.7.- Formulation for the $Y_2O_2S:Tb^{3+}:Sm^{3+}$ Phosphor

Compound	Mol Ratio	Assay	Grams	Lb/Lb
$Y_2O_3:Tb_{0.075}$	0.45	------	106.34	0.3156
$Y_2O_3:Sm_{0.025}$	0.45	------	227.34	0.6747
S	0.10	0.985	3.2548	0.0097

This mix is handled as described above, including the firing temperature and time.

The spectrum of the $Y_2O_2S:Tb^{3+}:Sm^{3+}$ phosphor is shown in the following diagram, given as 5.4.8. on the next page.

5.4.8.-

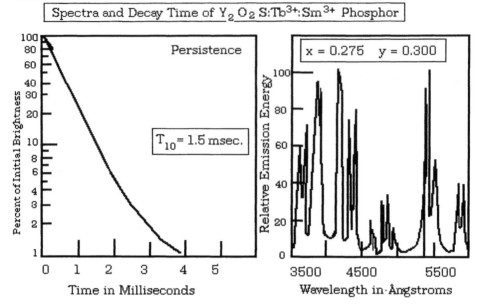

Spectra and Decay Time of $Y_2 O_2 S:Tb^{3+}:Sm^{3+}$ Phosphor

This screen has a near white color with a medium short decay. It has been used as a display screen capable of withstanding high current densities without exhibiting "screen- burn". However, its screen brightness is low compared to more conventional screens.

B. Tb^{3+}, Ce^{3+} and Ti^{3+} (P-52) Activated Silicates

Several other lanthanide activated phosphors have found use as CRT screens. One such is the $Y_2SiO_5:Tb^{3+}$ phosphor. It is manufactured using the following formulation:

5.4.9.- Formulation for the $Y_2SiO_5 : Tb^{3+}$ Phosphor

Compound	Mol Ratio	Assay	Grams	Lb/Lb
$Y_2O_3:Tb_{0.075}$	1.00	------	106.34	0.4932
H_2SiO_3	1.05	75.05% SiO_2	109.27	0.5068

These components are blended, and then hammermilled twice, with blending in between hammermillings. The mix is then placed in a covered silica crucible and fired for:

4.5 hours at 1225 °C.

After firing, the cooled cakes are inspected and crushed. The powder is then wet milled for 2 hours in a ball mill, using about 400 grams of powder per liter of deionized water. The powder is then filtered and then resuspended in water. A silica coating is then applied as described above. Filter and dry powder at 125 °C. for 12 hours. Alternately, the precipitated oxalates can be mixed with the requisite amount of silica, as shown in the above formulation and the mix is fired as shown above to form the phosphor composition.

The spectrum of this phosphor is very like that already given for the oxysulfide phosphors. Sets of lines appear at:

4920 Å	5450 Å
5850 Å	6200 Å

The color coordinates are: x = 0.334 and y = 0.581. The T_{10} decay rate is 5.8 microseconds. Thus, the major difference seen for this phosphor as compared to others such as the oxysulfides is the slightly longer decay found in the $Y_2SiO_5:Tb^{3+}$ phosphor. This screen has been used for high brightness, high loading rasters where the phosphor brightness maintenance is critical. Such applications include projection color television tubes.

Another composition that has been registered is the composition:

$$Y_2(Si, Ge)O_5:Tb^{3+}$$

where up to 50% of the SiO_2 can be substituted by GeO_2. If this is done, the firing schedule to be used is:

3.5 hours at 1135 °C.

The spectral properties obtained are nearly identical to the pure silicate phosphor except that its color coordinates are:

x = 0.337   y = 0.575

making the silico-germanate phosphor slightly more saturated in its greenish emission than the silicate phosphor given above. For certain display purposes, this superior saturation is more useful and is not as tiring to the human eye.

Still another similar phosphor is the composition: Zn_2SiO_4:Ti. This phosphor was registered as P-52 in 1977.

5.4.10.- Formulation for a P-52 Phosphor

Compound	Mol Ratio	Assay	Grams	Lb/Lb
ZnO	2.00	0.965	168.66	0.6714
SiO_2	1.13 (as SiO_2)	0.966	70.280	0.2798
TiO_2	0.12	0.985	9.7340	0.0387
NaF	0.06	0.988	2.5489	0.0101

This mix is blended thoroughly following all the steps given above, including two hammermilling steps, with blending in between. The assay of SiO_2 will vary according to the nature of the silica being used. In general, it must be a pure grade with no more than about 5 ppm total transition metal elements present.

A large 2-liter silica crucible is generally used for firing:

3 hours @ 1235 °C. in air in a covered silica tray.

After firing, the phosphor cake is inspected under 2527 Å light, crushed and then sieved through a 325 mesh screen. Sometimes, this composition is "silicized', using the procedure described above.

The emission is a purplish blue and the persistence is medium as shown in the following diagram, given as 5.4.11. on the next page.

The decay is classified as medium with a deep purplish emission color. The P-52 screen is used in some military applications.

5.4.11.-

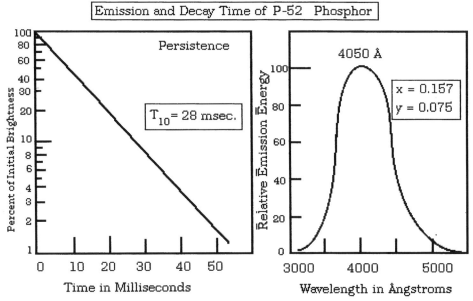

Emission and Decay Time of P-52 Phosphor

Persistence

T_{10} = 28 msec.

Percent of Initial Brightness

Time in Milliseconds

4050 Å

x = 0.157
y = 0.075

Relative Emission Energy

Wavelength in Ångstroms

Still another phosphor used is the composition: Y_2SiO_5 :Ce which also has a deep blue emission. It is manufactured using the following formulation, viz-

5.4.12.- Formulation for the Y_2SiO_5 : Ce^{3+} -Phosphor

Compound	Mol Ratio	Assay	Grams	Lb/Lb
Y_2O_3 :$Ce_{0.055}$	1.00	------	106.34	0.4932
H_2SiO_3	1.05	75.05% SiO_2	109.27	0.5068

These components are blended, and then hammermilled twice, with blending in between hammermillings. The mix is then placed in a covered silica crucible and fired for:

4.5 hours at 1205 °C.

After firing, the cooled cakes are inspected and crushed. The powder is then wet milled for 2 hours in a ball mill, using about 400 grams of powder per liter of deionized water. The powder is then filtered and then resuspended in water. A silica coating is then applied as described above. Filter and dry powder at 125 °C. for 12 hours. Alternately, the precipitated oxalates can be mixed with the requisite amount of silica, as shown in the above formulation and the mix is fired as shown above to form the phosphor composition. The spectral properties of the Y_2SiO_5:Ce phosphor are shown in the following diagram:

5.4.13.-

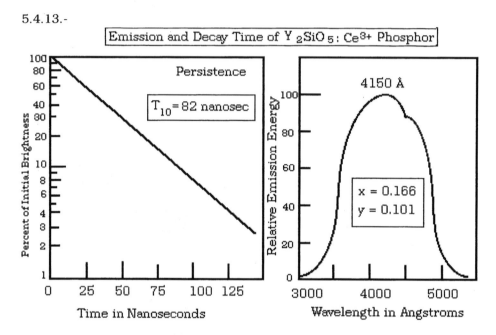

The emission is a purplish blue and the decay is very fast. This screen is used wherever a very fast decay is needed such as flying spot scanners. This phosphor has the advantage that it can be used as a visible screen for capturing transient electrical phenomena when employed within an oscilloscope, for example. Actually, it is an improved version of the P-16 and YPO_4:Ce^{3+} phosphors referred to above. The following table gives a comparison of these three phosphors:

P-16= $CaMgSiO_3$:Ce^{3+}	3800 Å	140 nanosec.
YPO_4:Ce^{3+}	3300 & 3500 Å	45 nanosec.
Y_2SiO_5:Ce	4150 Å	82 nanosec.

The fastest decay is that of the phosphate but its emission is not visible like that of the ytrrium silicate composition.

C. Manufacture of Rare Earth Activated Aluminates

Several CRT phosphors are based on aluminate hosts. Such a phosphor is the $Y_3Al_5O_{12}:Ce^{3+}$ composition, i.e.- the "YAG" phosphor which we have already described on page 510 of the last chapter. It is prepared using the formulation:

5.4.14.-   Formulation for $(Y)_3 Al_5 O_{11}: Ce^{3+}$ Phosphor

Compound	Mol Ratio	Assay	Grams	Lb/Lb
$(Y_{0.988}, Ce_{0.012})_2(C_2O_4)_3 \cdot 3\ H_2O$	3.00	0.445 as Y_2O_3	1522.2	
Al_2O_3	5.02	0.975	525.0	

This formulation is blended, hammermilled twice with blending in between. The powder is then loaded into 2000 ml alumina crucibles and then fired:

1360 °C. for 6 hours in air

At the end of the firing period, the crucible cakes are inspected and crushed. The powder is then sieved through a 325 mesh screen. Alternately, it can be hammermilled, without the use of a screen in the hammermill to produce a free-flowing powder before it is sieved. Finally, it can be coated using the "Baymal™" procedure given on page 547.

Some manufacturers refire the powder in a reducing atmosphere, viz-

Refire 4 hours at 1150 °C in 95% N_2 - 5% H_2 atmosphere in open alumina trays.

However, this usually is not necessary. The spectral characteristics are given in the following diagram, given as 5.4.15. on the next page.

Note that the emission band is very wide. This is essentially the same phosphor as the lamp phosphor except that the "optimum" activator concentration is lower for the CRT phosphor. The decay time is in nanoseconds and the emission color is a yellow green.

644

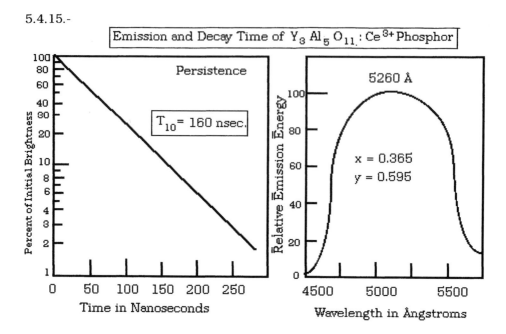

A combination of the $Y_3Al_5O_{12}:Ce^{3+}$ and the $Y_2SiO_5:Ce^{3+}$ phosphors was registered as a P-48 screen.

If Tb^{3+} is either substituted for the Ce^{3+} activator or added to the composition, a green emitting phosphor is obtained. There is some controversy as to whether the $Y_3Al_5O_{12}:Tb^{3+}$ phosphor or the $Y_3Al_5O_{12}:Ce^{3+}:Tb^{3+}$ is superior. However, the resulting emission is typical of the Tb^{3+} cation and the Ce^{3+} band emission is not seen, as shown in the following diagram, given as 5.4.16. on the next page.

Note that the emission lines are "bunched together" around 5480 Å. and that a continuum may be present. Whether this is due to the Ce^{3+} activator is not exzactly certain. However, it is not present in the phosphor solely acticvated by Tb^{3+}. Thus, it is likely that the low level band is due to Ce^{3+} which acts as a "sensitizer" for the Tb^{3+} emission.

The emission is a saturated green and the decay is classified as medium.

5.4.16.-

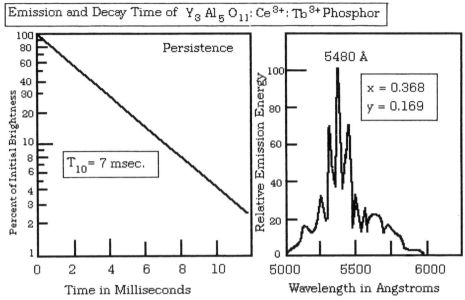

Emission and Decay Time of $Y_3 Al_5 O_{11}$: Ce^{3+}: Tb^{3+} Phosphor

Persistence

$T_{10} = 7$ msec.

Time in Milliseconds

5480 Å

x = 0.368
y = 0.169

Wavelength in Angstroms

D. Manufacture of Rare Earth Activated Borates

Several new phosphors have made the CRT screen scene, based on an orthoborate composition. Such compositions can be manufactured by either a solid state method or by a precipitation method, with the latter being a superior one.

One such borate has the composition: $InBO_3$:Tb^{3+} while another has the composition, $InBO_3$:Eu^{3+}.

To maufacture these phosphors using a solid state method, one combines the following ingredients:

5.4.17.- Formulation for a $InBO_3$:Tb^{3+} Phosphor

Compound	Mol Ratio	Assay	Grams	Lb/Lb
In_2O_3	1.00	0.988	230.40	0.7232
H_3BO_3	1.10	0.875	77.733	0.2440
Tb_4O_7	0.01375	0.985	10.437	0.0328

This mix is first blended, then hammermilled twice, with blending in between hammermillings. It is then loaded into a 2-liter silica crucible and is fired in air for:

3.5 hours at 1175 °C.

After firing, the cooled cakes are inspected and crushed. The powder is then wet milled for 2 hours in a ball mill, using about 400 grams of powder per liter of deionized water. The powder is then filtered and then resuspended in water. A silica coating is then applied as described above. Filter and dry powder at 125 °C. for 12 hours. Alternately, the co-precipitated oxalates, $In_{0.945}$, $Tb_{0.055}$)$_2$ Ox_3 can be mixed with the requisite amount of H_3BO_3, as shown in the above formulation and the mix is fired as shown above to form the phosphor composition. The spectrum of the $InBO_3:Tb^{3+}$ phosphor is shown in the following diagram:

5.4.18.-

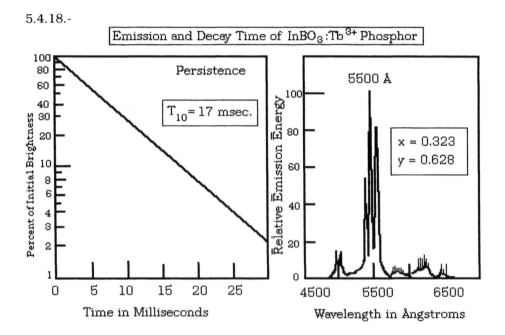

Note that the main emission peak occurs exactly at 5500 Å, the peak of the eye response curve of the human eye. Note also that the other emission peaks, characteristic of the Tb^{3+} activator, are suppressed in this spectrum, compared to the "normal" one expected. This has the effect of producing a very "green" phosphor and the screen made from it has very high brightness, especially under very high current conditions.

To manufacture the $InBO_3:Eu^{3+}$ phosphor, one uses the following formulation:

5.4.19.- Formulation for a $InBO_3:Eu^{3+}$ Phosphor

Compound	Mol Ratio	Assay	Grams	Lb/Lb
In_2O_3	1.00	0.988	230.40	0.6879
H_3BO_3	1.10	0.875	77.733	0.2321
Eu_2O_3	0.075	0.985	26.796	0.0800

This mix is handled as given above in 5.4.17. and fired in a similar manner, except the firing cycle is:

5.5 hours at 1175 °C.

After firing, the cooled phosphor cakes are handled as given above. A silica coating is then applied as described above. Filter and dry powder at 125 °C. for 12 hours.

The spectrum of the $InBO_3:Eu^{3+}$ phosphor is shown in the following diagram:
5.4.20.-

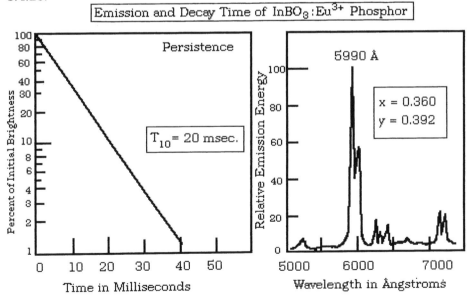

Emission and Decay Time of $InBO_3:Eu^{3+}$ Phosphor

Note that the main emission peak of the Eu^{3+} activator is considerably more orange than other Eu^{3+} activated phosphors such as the oxysulfide shown above. Because of this, the $InBO_3:Eu^{3+}$ phosphor is not used as a red component in any color television screens but the combination:

$$InBO_3:Eu^{3+} \quad + \quad InBO_3:Tb^{3+}$$

having main peaks at 5500 Å and 5990 Å has been registered as a useful screen for certain display purposes.

Alternately, the co-precipitated oxalates, $(In_{0.945}, Eu_{0.075})_2 Ox_3$ can be mixed with the requisite amount of H_3BO_3, as shown in the above formulation and the mix fired as shown above to form the phosphor composition.

If one wishes to prepare a production batch of these phosphors, using a co-precipitation method, one does so using the procedure set forth as follows:

5.4.21.- Procedure for Precipitating (In,Ln)BO $_3$ Compositions

For a 250 gallon batch:

a. Rare Earth Solution at a 2.00 Molar Concentration

This requires 0.9450 mols of In_2O_3 (262.368 gm./liter) and 0.0275 mols of Tb_4O_7 (20.562 gm./liter) per liter of solution. The working volume is 520 liters or 138.70 gallons.

Weigh out the following amounts of oxides for dissolution:

    In_2O_3 = 136.32 kg. = 300.780 lbs. (520 x 262.368 gm/liter)
    Tb_4O_7 = 10.692 kg. = 23.562 lbs. (520 x 24.560), or:

    Eu_2O_3 = 13.725 kg. = 30.244 lbs. (520 x 26.394)

(NB: Use one or the other of these activator concentrations. Do not use them together in the same formulation!)

Mix two of these oxides together thoroughly before use.

The following totals are the volumes required for initial dissolution of these oxides:

286 liters, or 75.6 gallons, of concentrated HCl

234 liters, or 61.8 gallons, of deionized water

One first adds about 1/4 of the oxides slowly to this acidic solution. Then 25-30 gallons of concentrated acid are added to this solution and more oxides are added. This continues until all of the acid and oxides have been added. The solution will heat up as oxides are added to the acid. Do not let the solution go to boiling. When a total of 234 liters of concentrated HCl has been added, the volume will be close to 260 liters. If it is not, add sufficient water to adjust the volume to a total of 520 liters. The concentration of Ln^{3+} will now be exactly 1.00 molar. Heat the solution to about 60 °C while stirring the solution in the tank.

b. Preparation of Boric Acid Solution

A 2.20 molar volume of H_3BO_3 is required. For a 300 gallon tank, a working volume of 520 liters (137.4 gallons) is required. Add 520 liters of deionized water to the tank. Begin heating the water while weighing out the boric acid. The weight of H_3BO_3 required is:

2.20 mols/ liter x 61.83311 x 520 = 70.737 Kg. = 155.88 lbs.

When the water temperature reches 55 °C, begin adding the acid slowly with stirring. Continue adding the acid and heat the solution to 80 °C. When the addition is complete, hold this temperature preparatory to the precipitation of the borates.

c. Precipitation of Indium Borates.

Heat the In^{3+} + Tb^{3+} solution to 60 °C. while stirring (At 40 °C., the solubility limit of H_3BO_3 is 1.518 molar, at 60 °C it is 2.500 molar, while at 80 °C, it is 3.482 molar). Set up the flow-rate control to deliver 3500 ml./minute of the rare earth solution to the acid-solution tank. Begin pumping the rare earth solution, i.e.-

5.4.22.-   2.00 molar In^{3+} ⇒ 2.20 molar H$_3$BO$_3$ solution at 80 °C.

At the beginning, the solution remains clear. Gradually, a coarse white precipitate obtains. (Note that the solubility of the InBO$_3$ precipitates is inversely proportional to temperature. That is, they are less soluble at high temperature in solution than at room temperature. It is for this reason that the precipitation temperature must be in excess of 80 °C. in order to obtain a good yield).

At the end of the addition period, let the precipitate remain suspended by stirring and monitor the solution pH. When it becomes stable, continue stirring so as to promote crystal growth.

d. Final Processing of Precipitate

After stirring to allow crystal growth in solution, stop stirring and allow the precipitate to settle. Remove the mother liquor by decantation and apply two cold water(~ 15-20 °C) washes by resuspension and settling procedures. Filter and dry for 16 hours at 105-110 °C.

The expected yield is:

	10 Gallon Batch	300 Gallon Batch
Precipitate:	3.237 Kg.	80.14 Kg. or 176.6 lbs.
Assay:	81.82% (as InBO$_3$)	---------
Yield:	95% of In^{3+}	96 - 98% of In^{3+}

It has been found that the usual precipitate is: (In,Ln)BO$_3$ • 2H$_2$O. The precipitate is then filtered on a rotary filter press while being washed with hot water at 80 °C. (This removes any excess borate that may be present). It is then put into an oven at 125 °C. and dried overnight. A fluffy white powder results. This powder is then placed in a 2-liter silica crucible and fired for:

3.5 hours at 980 °C.

The same phosphor results as shown above, but it is found that an increase in CRT efficiency is obtained by using the precipitation · method to produce these phosphors.

E. Manufacture of Alkaline Earth Sulfide Phosphors

Although the alkaline earth sulfides have been known to provide excellent hosts as CRT phosphors, their main drawback has been the deleterious hydrolysis behavior experienced when one trys to prepare a CRT screen. That is, they are chemically unstable in water and other solutions generally used in the manufacture of CRT's. Thus, none of these compositions received much attention until the recent past. Of all of these sulfides, only CaS is stable enough to be used, and only if it is prepared in a suitable manner.

1. Preparation of Luminescent Grade CaS

Two methods can be used to prepare CaS:

1) the "Nitrate" method

2) the "Carbonate" method.

The following is a description of the "Nitrate" method, viz-

a. The Nitrate Method

Starting materials are:

$Ca(NO_3)_2 \cdot 4 H_2O$
Mg metal ribbon
H_2SO_4 - AR grade

One starts by dissoling 454 gram of $Ca(NO_3)_2 \cdot 4 H_2O$ in about 600 ml. of deionized water. The solution is then heated to near boiling. A few pieces of Mg metal ribbon are added to the solution with stirring. The Mg does not dissolve entirely, but causes heavy metal impurities left in the solution to precipitate preferentially on the metal or to precipitate as hydroxides due to the $Mg(OH)_2$ present in solution. This method is especially successful in removing any Mn^{2+} that may be present which would cause a deviation in the expected spectral properties of the CaS phosphors.

The solution is then filtered and 110 ml. of H_2SO_4, diluted with 600 ml. of deionized water, is added to precipitate $CaSO_4 \bullet 2\ H_2O$. The white precipitate is then allowed to settle, is washed with pure water several times, and the supernatent liquid is decanted. Washing continues until the excess sulfate is no longer detected. The precipitate is then filtered, washed in alcohol, and then dried in a vacuum oven at about 100 - 125 °C. overnight.

Dry $CaSO_4 \bullet 2\ H_2O$ is fired in an open silica boat in a tube furnace in an atmosphere of H_2-gas (at about 50 cc. per minute gas flow for a 4 inch tube) for 0.5 to 1.0 hour at 1000 °C. At the end of this period, the gas flow is changed to that of H_2S (about 100 cc. per minute for a 4 inch tube) and the firing continues for another 45 minutes at 1000 °C. At the end of tis period, the boat is pushed into the cool zone and allowed to cool in the H_2S atmosphere. The result is a white, very fluffy CaS powder with no traces of oxide, sulfate, etc. (Note that the use of the H_2S atmosphere is very dangerous and adequate air-scrubbing equipment must be used. It is also advisable that the above procedure be accomplished by technically trained personnel).

b. The Carbonate Method

$CaCO_3$ is placed in a silica boat and is fired in a flowing argon atmosphere in a tube furnace (about 200 cc of gas per minute) for about 1.0 hour at 1100 °C. At the end of this period, the argon gas is replaced by H_2S gas, using a flow of about 100 cc. per minute through a 4 inch tube, and the firing is continued for at least 1.0 more hour. At the end of a cool zone and allowed to cool. (The same safety precautions discussed above apply).

The reactions involved are:

$$CaCO_3 \Rightarrow CaO + CO_2 \Uparrow$$
$$CaO + H_2S \Rightarrow CaS + H_2O \Uparrow$$

The result is again a fluffy white powder devoid of oxide, etc. However, it may still contain some Mn^{2+} which imparts a weak yellow fluorescence under either 2537 Å or 3650 Å excitation (This property may be used to detect its presence).

It is important to store the CaS powder under dry conditions before it is used to prepare the phosphor compositions. CaS itself is somewhat hygroscopic before use and needs to be protected from the atmosphere.

2. Preparation of the CaS:Ce^{3+} Phosphor

The nominal composition of this phosphor is: $CaS:Ce^{3+}_{0.050}:Cl$. If chloride (or other halide) is not used in the formulation, the resulting composition will be non-luminescent.

5.4.23.- Formulation for the CaS:Ce^{3+} Phosphor

Compound	Mol Ratio	Assay	Grams	Lb/Lb
CaS	1.00	0.995	72.50	0.9581
CeO$_2$	0.0050	0.995	0.8649	0.0114
N H$_4$Cl	0.0055	0.988	0.2942	0.0389
S	0.0624	0.995	2.0106	0.0266

This mix is blended as before and then is hammermilled twice, with blending in between the hammermilling steps.

The mix is then placed in a covered silica crucible and fired in a tube furnace with a flowing H$_2$S atmosphere for:

2.0 hours at 1200 °C.

The H$_2$S gas flow is sized for the furnace being used. About 5% HCl gas may also be introduced into the firing atmosphere. After firing, when the phosphor cakes are cool, they are inspected under ultraviolet light and crushed. This green-emitting phosphor is then sifted through a 325 mesh screen and then is washed in alcohol and then is put into a 1 quart mill , filled to 1/2 with 1/2 inch flint pebbles. About 400 gram of phosphor is used to 200 ml. of alcohol. The mill is run at 120 rpm. for 30 minutes. The slurry is then removed from the mill, which is washed out with alcohol to remoce all of the phosphor. It is then filtered and dried at 125 °C. in air for 12 hours.

The spectral properties of this phosphor is given in the following diagram:

5.4.24.-

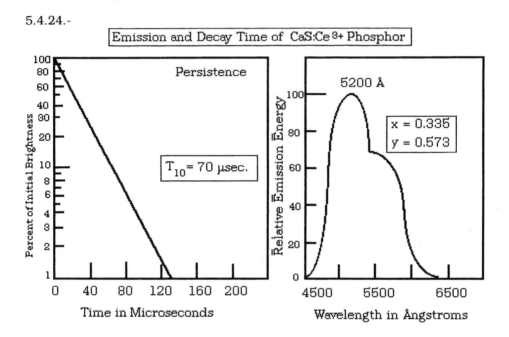

The emission is green and the decay is classified as medium short.

Although several other phosphor compositions can be made with the CaS composition, none of them have been registered as a CRT screen. Only CaS:Ce is presently being used as a CRT screen.

A good red phosphor can be made using the following ingredients:

$$1.00 \text{ mol of CaS}$$
$$0.10 \text{ mol of Eu}_2\text{O}_3$$
$$0.01 \text{ mol of CeO}_2$$
$$0.10 \text{ mol of NH}_4\text{Cl}$$
$$0.20 \text{ mol of S}$$

The same manufacturing procedure is used and the resulting phosphor has its emission peak near to 6500 Å. In this case, the broad emission band is due to the Eu^{2+} activator in the CaS host. Although it had a deep red saturation in color and a high screen brightness due its high cathode ray efficiency, it has not been used as a CRT screen to any great extent..

In conclusion, we have presented most of the CRT phosphors that have been, or are being, manufactured for use as screens for the various applications cited above. Although many critics have sounded the death knell for the CRT, it is clear that in 1993 the CRT will remain on the display scene for many years. The newest displays being developed have included:

> Electroluminescent panel displays
> Active matrix liquid crystal displays
> Vaccum fluorescent displays
> Field emitter displays
> Plasma panel displays

Each of the developers involved believe that their technology is superior to any of the others. Of those currently being developed, only the liquid-crystal displays are close to being commercial and are likely to be marketed as a flat-panel color display capable of producing color television pictures. Recently, an efficient field emission flat panel device which has the screen brightness of a similar CRT was announced (5). Its only advantage is that the total device thichness is less than two inches. Nevertheless, the CRT will remain on the display screen because of its versatility and because its technology has been refined to the point where any competing display finds it difficult to rival the screen brightnes of the CRT.

REFERENCES CITED

1. GFJ Garlick *Luminescence in Inorganic Crystals* , p. 699, P. Goldberg- Ed., Academic Press, NY (1966).

2. P. Palluel, *Compt. Rend.* **224** 1492 (1947).

3. See for example: A. Bril and HA Klasens, *Philips Res. Repts.* **7** 401 (1952). ; A Bril in "Luminescence of Organic and Inorganic Compunds", Edited by HP Kollman and GM Spruch, p. 479, Wiley & sons, NY (1962).; G. Gergeley, I Hamgos, K Toth, J Adam, & G Pozsgay, *Zeit. fur Phys.* **210** 11 (1959); Ibid, **211** 274 (1959); CW Ludwig and JD Kingsley, *J. Electrochem. Soc.* **117** pp. 348 & 353 (1970).

4. RC Ropp in "Luminescence and the Solid State", pp. 4-6, Elsevier Sci. Publ., Amsterdam & New York (1991).

5. "The Field-Emitter Display" by HF Gray in *Information Display* **3** 9 (1993).

SUBJECT INDEX

SUBJECT INDEX

SUBJECT INDEX

SUBJECT INDEX

SUBJECT INDEX

SUBJECT INDEX

SUBJECT INDEX

SUBJECT INDEX